Tsunamiites

Cover credit

Front main: Numerous coral boulders deposited by the 2004 Indian Ocean tsunami at Pakarang Cape, Thailand. Photo taken by K. Goto.

Front right: Miocene tsunami-induced upper bathyal conglomerates, cropped out at a seaside of Cita Peninsular, Central Japan. Photo taken by T. Shiki.

Front left: A ship carried inland by the 2011 Tohoku oki earthquake tsunami. Kesennuma, Northeast Japan. After Kawabe et al. 2013.

Front middle: Buildings turned over and pulled a little toward inland by the 2011 tsunami. Onngawa, northern Japan After Suzuki et al. 2013.

Back right: Tsunami deposits showing continuous facies change on a mega-trench wall in Nemuro low-land, eastern Hokkaido, Japan. Photo taken by F. Nanayama.

Back left: A thick peat layer of about 2.2 m associated with six Holocene ash and twelve tsunami sand beds in the Gakkarahama Coast, Hokkaido, Japan. Photo taken by F. Nanayama.

Tsunamiites
Features and Implications

Second Edition

Edited by

Tsunemasa Shiki
Uji, Kyoto, Japan

Yoshinobu Tsuji
Earthquake and Tsunami Disaster-Prevention Strategy Institute,
Ryugasaki, Ibaraki, Japan

Teiji Yamazaki
Toyonaka, Osaka, Japan

Futoshi Nanayama
Geological Survey of Japan, National Institute of Advanced Industrial
Science and Technology (AIST), Tsukuba, Ibaraki, Japan

ELSEVIER

Elsevier
Radarweg 29, PO Box 211, 1000 AE Amsterdam, Netherlands
The Boulevard, Langford Lane, Kidlington, Oxford OX5 1GB, United Kingdom
50 Hampshire Street, 5th Floor, Cambridge, MA 02139, United States

Copyright © 2021 Elsevier B.V. All rights reserved.

No part of this publication may be reproduced or transmitted in any form or by any means, electronic or mechanical, including photocopying, recording, or any information storage and retrieval system, without permission in writing from the publisher. Details on how to seek permission, further information about the Publisher's permissions policies and our arrangements with organizations such as the Copyright Clearance Center and the Copyright Licensing Agency, can be found at our website: www.elsevier.com/permissions.

This book and the individual contributions contained in it are protected under copyright by the Publisher (other than as may be noted herein).

Notices

Knowledge and best practice in this field are constantly changing. As new research and experience broaden our understanding, changes in research methods, professional practices, or medical treatment may become necessary.

Practitioners and researchers must always rely on their own experience and knowledge in evaluating and using any information, methods, compounds, or experiments described herein. In using such information or methods they should be mindful of their own safety and the safety of others, including parties for whom they have a professional responsibility.

To the fullest extent of the law, neither the Publisher nor the authors, contributors, or editors, assume any liability for any injury and/or damage to persons or property as a matter of products liability, negligence or otherwise, or from any use or operation of any methods, products, instructions, or ideas contained in the material herein.

Library of Congress Cataloging-in-Publication Data
A catalog record for this book is available from the Library of Congress

British Library Cataloguing-in-Publication Data
A catalogue record for this book is available from the British Library

ISBN: 978-0-12-823939-1

For information on all Elsevier Science publications visit our website at https://www.elsevier.com/books-and-journals

Publisher: Charlotte Kent
Acquisitions Editor: Peter Llewellyn
Editorial Project Manager: Naomi Robertson
Production Project Manager: Sruthi Satheesh
Cover Designer: Matthew Limbert

Typeset by TNQ Technologies

Contents

Contributors .... *xv*

Preface to the second edition .... *xvii*

Chapter 1: Introduction: why a Book on Tsunamiites **1**
    T. Shiki, K. Minoura, Y. Tsuji and T. Yamazaki

  References 4

Chapter 2: The term "Tsunamiite" **5**
    T. Shiki and T. Yamazaki

  References 7

Chapter 3: Tsunamis and tsunami sedimentology **9**
    D. Sugawara, K. Minoura and F. Imamura

  1. Introduction 9
  2. Generation, propagation, and quantification 10
    2.1 Generation of tsunamis 11
    2.2 Propagation of tsunamis 15
    2.3 Quantification of tsunamis 18
  3. Tsunami sedimentology 20
    3.1 The mechanics of sediment transport 22
    3.2 Characteristics of tsunami deposits 26
    3.3 A review of onshore tsunami sedimentation 32
    3.4 The occurrences of tsunamis and tsunamiites 40
  4. Concluding remarks 47
  References 47

Chapter 4: Bedforms and sedimentary structures characterizing tsunami
          deposits **53**
    O. Fujiwara

  1. Introduction 53
  2. Differences of waveforms between tsunami- and storm-induced waves 54
  3. Bedforms and sedimentary structures reflecting the tsunami waveform 55
  4. Single-bed deposits 55

Contents

5. Multiple-bed deposits..56
  5.1 Succession of sand sheets capped by mud drapes.........................57
  5.2 Repeated reversal of current directions...................................57
  5.3 Fining and thinning upwards series of sand sheets.......................59
6. Depositional model in shallow water...60
7. Conclusions 61
References...62

Chapter 5: Tsunami depositional processes reflecting the waveform in a small bay: interpretation from the grain-size distribution and sedimentary structures... 65

  O. Fujiwara and T. Kamataki

1. Introduction 65
2. Regional setting...66
  2.1 The Paleo-Tomoe Bay and its Holocene deposits.........................66
  2.2 Storm waves and tides around the Southern Boso Peninsula...............67
3. Sedimentary facies of the tsunami deposits...................................67
  3.1 Unit Tna...67
  3.2 Unit Tnb...70
  3.3 Unit Tnc...71
  3.4 Unit Tnd...71
4. Grain-size distribution of the tsunami deposits 71
  4.1 Sampling and methodology...71
  4.2 The T3 tsunami deposit at location 58 72
  4.3 The T3 tsunami deposit at location 45 74
5. Discussion...74
  5.1 Tsunami waveform..74
  5.2 Relationship between grain-size distribution and tsunami waveform.......75
  5.3 Discriminating tsunami deposits from storm deposits 78
  5.4 Tsunami deposits with a saw-toothed grain-size distribution 79
6. Conclusions 81
Acknowledgments...82
References...82

Chapter 6: Deposits of the 1992 Nicaragua tsunami 85

  B. Higman and J. Bourgeois

1. Introduction 85
2. 1992 tsunami deposits along the Nicaragua coast.............................87
  2.1 Site-by-site observations..89
3. Tsunami deposits near Playa de Popoyo 92
4. Grading of the tsunami deposits...96
  4.1 Landward grading...96
  4.2 Vertical grading...101

vi

5. Discussion... 105
Appendix A... 105
   Field and laboratory protocols 105
Acknowledgments.. 107
References.. 107

Chapter 7: Sedimentary characteristics and depositional processes of onshore tsunami deposits: an example of sedimentation associated with the July 12, 1993 Hokkaido-Nansei-Oki earthquake tsunami 109
F. Nanayama

1. Introduction 109
2. General setting 110
3. Methods 112
   3.1 Field survey... 113
   3.2 Sedimentary description... 114
4. Results 115
   4.1 General characteristics.. 115
   4.2 Sedimentary structures... 116
   4.3 Sedimentary units and facies.. 116
5. Discussion.. 120
   5.1 Sedimentary characteristics and facies of the 1993 onshore tsunami deposits... 120
   5.2 An ideal model of the 1993 tsunami sedimentation................... 122
6. Conclusions 124
Acknowledgments.. 124
References.. 124

Chapter 8: Distribution and significance of the 2004 Indian Ocean tsunami deposits: initial results from Thailand and Sri Lanka 127
K. Goto, F. Imamura, N. Keerthi, P. Kunthasap, T. Matsui, K. Minoura, A. Ruangrassamee, D. Sugawara and S. Supharatid

1. Introduction 127
2. Localities and methods of study.. 129
3. Distribution and significance of the tsunami deposits........................ 132
   3.1 Pakarang Cape, Thailand... 132
   3.2 Bang Sak beach, Thailand 135
   3.3 Garanduwa, Sri Lanka 139
4. Discussion.. 141
5. Conclusion... 142
Acknowledgments.. 143
References.. 143

Contents

Chapter 9: Thickness and grain-size distribution of Indian Ocean tsunami deposits at Khao Lak and Phra Thong Island, South-Western Thailand **145**

S. Fujino, H. Naruse, A. Suphawajruksakul, T. Jarupongsakul, M. Murayama and T. Ichihara

1. Introduction 145
2. Study areas 146
3. Impact of the tsunami 146
4. Thickness and grain-size distribution 146
5. Discussion 150
6. Conclusions 153
Acknowledgments 153
References 153

Chapter 10: Lessons from the 2011 Tohoku-oki tsunami: implications for Paleotsunami research **155**

D. Sugawara

1. Introduction 155
2. The 2011 Tohoku-oki tsunami and its precursors 156
3. Lessons learned from the Tohoku-oki tsunami 159
   3.1 Limitation of marine materials as evidence for tsunami inundation ... 159
   3.2 Larger extent and lower preservation potential of offshore tsunami deposits 162
   3.3 Possible false dating of Paleotsunami events due to tsunami-induced erosion 165
   3.4 Uncertainties in tsunami inundation distance based on deposit extent 168
   3.5 Spatial variability of deposit thickness and its relation to flow depth 170
   3.6 Challenges to estimating earthquake size and extent from tsunami deposits 172
4. Conclusions 175
Acknowledgments 176
References 176

Chapter 11: An overview on offshore tsunami deposits **183**

P.J.M. Costa, L. Feist, A.G. Dawson, I. Stewart, K. Reicherter and C. Andrade

1. Introduction 183
2. Offshore tsunami deposits — current knowledge 184
3. Offshore tsunami deposits — their features 185
   3.1 Internal architecture 186
   3.2 Textural and compositional aspects 187
   3.3 Geochemical inferences 187
   3.4 Palaeontological features 188
   3.5 Differentiation from other high-energy events 189

viii

Contents

4. Concluding remarks 189
Acknowledgments... 190
References... 190

Chapter 12: Combined investigation of tradition archives and sedimentary relics of tsunami hazards — with reference to the Great 1700 Cascadia tsunami and other examples 193

Y. Tsuji

1. Introduction 193
2. The tsunami traces of the 1700 Cascadia earthquake and corresponding archives in Japan 194
   2.1 An enormous tsunami told in stories down through tradition by native tribes of Canada 194
   2.2 Modern traces of sudden subsidence of Cascadia coast and an accompanying earthquake.. 195
   2.3 Sedimentary traces of tsunami flow and their formative influence 196
   2.4 Date identification of the 1700 tsunami from documents of towns along the Japanese coast 197
   2.5 Estimation of the occurrence time of the Great 1700 Cascadia earthquake... 199
   2.6 Estimation of magnitude of the 1700 Great Cascadia earthquake 200
   2.7 Discussions on the 1700 Great Cascade earthquake 200
3. Examples from the Japanese islands 200
   3.1 Documented Jogan tsunami of July 13, 869 CE and its sedimentary records 201
   3.2 Other examples in Japan and their lessons 202
4. Lack of sedimentary relic — A supplementary discussion............................. 202
5. Conclusive remarks 203
Acknowledgments... 204
References... 204

Chapter 13: Deep-sea homogenites: sedimentary expression of a prehistoric megatsunami in the Eastern Mediterranean 207

M.B. Cita

1. Introduction 207
2. Deep-sea homogenites 208
   2.1 Type A homogenites 209
   2.2 Type B homogenites 213
3. Discussion... 215
   3.1 Data 216
   3.2 Absence of tephra Z-2 in our data set 216
   3.3 Absence of homogenites in the Herodotus abyssal plain.............................. 216
   3.4 Comparison of type A and type B homogenites 219
4. Conclusions 220

ix

Contents

5. Post scriptum...221
Acknowledgments..222
References..222

Chapter 14: Tsunami-related sedimentary properties of mediterranean homogenites as an example of deep-sea tsunamiite 225
T. Shiki and M.B. Cita

1. Introduction 225
2. Setting, types, and distribution of homogenites....................................226
3. Sedimentary properties 227
   3.1 Structures and grain-size distribution..227
   3.2 Constituents and their relation to other features...........................230
4. Discussion of sedimentological problems...230
   4.1 Erosion of the deep-sea bottom...230
   4.2 Genesis of the sandy division of the type B homogenite 231
   4.3 High tsunami-induced suspension cloud..231
   4.4 Accumulation rate of the suspended load.......................................233
   4.5 Records of shuttle movement and backwash current of tsunamis...................233
5. Comparison with other deep-sea tsunamiites..234
6. Concluding remarks 235
Appendix: reflections on terminology 235
Acknowledgments..236
References..236
Additional comment to Chapter 14 238
Further reading 238

Chapter 15: Volcanism-induced tsunamis and tsunamiites 239
Y. Nishimura

1. Introduction 239
2. Volcanism-induced tsunamis...239
   2.1 The 1640 CE Hokkaido-Komagatake eruption and tsunami................240
   2.2 The 1741 CE Oshima-Ohshima eruption and tsunami......................241
   2.3 The 1792 CE Unzen eruption and tsunami 242
   2.4 The 1883 CE Krakatau eruption and tsunami 242
   2.5 The 1888 CE Ritter tsunami 243
   2.6 The 1994 CE Rabaul eruption and tsunamis..................................244
3. Volcanism-induced tsunamiites...245
   3.1 Managua tsunami deposits of 3000—6000 BP 246
   3.2 Santorini tsunami deposits of 3500 BP..247
   3.3 Aniakchak tsunami deposits of 3500 BP..249
   3.4 The 1640 CE Komagatake tsunami deposits 250
   3.5 The 1883 CE Krakatau tsunami deposits 251
   3.6 The 1994 CE Rabaul tsunami deposits...252
   3.7 The 1996 CE Karymsky tsunami deposits..255

4. Discussion and summary 256
Acknowledgments 282
References 258

Chapter 16: Tsunamiites—conceptual descriptions and a possible example at the Cretaceous—Paleogene boundary in the Pernambuco Basin, Northeastern Brazil 263

G.A. Albertão, P.P. Martins and F. Marini

1. Prologue 263
2. Introduction and previous studies 264
3. A theoretical approach toward the identification of tsunamiites 266
4. Methods and data analyses 272
5. Geological setting of the study area 273
6. Characteristics of the Poty quarry K/Pg boundary 274
7. The controversy about the position of the K/Pg boundary 282
8. Bed D, the possible tsunamiite and a tentative model 286
   8.1 Characteristics of bed D 286
   8.2 Semiquantitative modeling of the depositional process of bed D 291
   8.3 Discussion of the tsunami process 294
9. Updated information on the area (K/Pg boundary at Poty quarry) 295
   9.1 Discussion on alternative interpretations 295
   9.2 Present situation of the K/Pg boundary section at Poty quarry 297
10. Concluding remarks 297
Acknowledgments 299
References 299

Chapter 17: Deep-sea tsunami deposits in the proto-Caribbean sea at the Cretaceous/Tertiary boundary 305

K. Goto, R. Tada, E. Tajika and T. Matsui

1. Introduction 305
2. Paleogeography of the proto-Caribbean sea and geological setting of the study sites 307
3. Sedimentary processes of the Peñalver Formation 308
   3.1 Stratigraphic setting and studied localities 308
   3.2 Lithology and petrography at the type locality near Havana 310
   3.3 Lateral and vertical variations in lithology, composition and grain size 312
   3.4 Origin and sedimentary mechanism 314
4. Comparison of K/T-boundary deep-sea tsunami deposits in the proto-Caribbean sea 316
   4.1 Cacarajícara Formation 317
   4.2 Moncada Formation 318
   4.3 DSDP sites 536 and 540 320

Contents

4.4 Comparison of the thickness and sedimentary structures of the K/T-boundary deep-sea tsunami deposits in the proto-Caribbean sea 323

4.5 Compositional variations of the K/T-boundary deep-sea tsunami deposits in the proto-Caribbean sea 324

5. Implications for the genesis and number of tsunami currents at the K/T boundary 324

6. Conclusions 327

Acknowledgments... 327

References.. 327

Chapter 18: The genesis of oceanic impact craters and impact-generated tsunami deposits 331

K. Goto

1. Introduction 331
2. Morphology of oceanic impact craters... 332
3. The generation of tsunamis by oceanic impacts.................................. 333
   3.1 Crater-generated tsunamis 334
   3.2 Landslide-generated tsunamis... 336
4. Impact-generated tsunami deposits inside and outside the oceanic impact craters 336
   4.1 Deposits formed by water flowing into an oceanic impact crater ... 336
   4.2 Impact-generated deposits outside an oceanic impact crater 337
5. Distribution and significance of the K/T-boundary tsunami deposits around the Chicxulub crater 339
   5.1 The K/T-boundary impact event.. 339
   5.2 On the edge of the Yucatán platform... 340
   5.3 The Gulf of Mexico region 343
   5.4 The proto-Caribbean sea region (DSDP sites, Haiti and Cuba) 344
   5.5 The Atlantic Ocean... 346
   5.6 The Pacific Ocean 347
6. Significance and distribution of the K/T-boundary tsunami deposits 347
7. Summary 348

Acknowledgments... 348

References.. 349

Chapter 19: Tsunami boulder deposits — a strongly debated topic in paleo-tsunami research 353

A. Scheffers

1. Introduction 353
   1.1 A short glance on tsunami and paleotsunami research 353
   1.2 Costal boulders, tsunami boulders, and the role of their depositional environment on dislocation process and history 357

Contents

2. Examples of tsunami boulder deposition 359
   2.1 Documented case studies of recent tsunami events 360
   2.2 Boulders deposited by paleotsunami events 363
3. Conclusion 372
References 374

Chapter 20: Characteristic features of tsunamiites 383
   T. Shiki, T. Tachibana, O. Fujiwara, K. Goto,
   F. Nanayama and T. Yamazaki

1. Introduction 383
2. Characteristics of tsunamis and tsunami deposition 384
   2.1 Diastrophic nature of tsunamis and tsunamiite deposition 384
   2.2 Length of tsunami waves and sedimentary structures in tsunamiites 385
   2.3 Shuttle movement of tsunamis and its records 387
   2.4 Shuttle movement versus gravity flows and tsunamiite variety 388
   2.5 Tsunamiites as marker horizons 389
   2.6 Associated sediments 390
3. Sedimentary structures in tsunamiite beds 390
   3.1 Sedimentary structures in a sedimentary set 391
   3.2 Grading in a layer and fining upward through stacked layers 392
   3.3 Current structures 393
4. Constituents of tsunamiites 394
5. Tsunamiite features in various environments 395
   5.1 Coastal plain 395
   5.2 Coastal lacustrine basin 396
   5.3 Beach 397
   5.4 Nearshore 397
   5.5 Bottom of bays 397
   5.6 Shallow sea including continental shelf 398
   5.7 Deep-sea environments 399
6. Conclusive remarks 400
Acknowledgments 401
References 401

Chapter 21: Sedimentology of tsunamiites reflecting chaotic events in the geological record — significance and problems 405
   T. Shiki and T. Tachibana

1. Introduction 405
2. Tsunamiites as records of ancient events 406
   2.1 Studies of impact-induced tsunamiites 406
   2.2 Earthquake-induced tsunamiites and tectonics in geological time 407
   2.3 Volcanic-eruption-induced tsunamiites 408
   2.4 Submarine slides and tsunamiites 409

xiii

Contents

3. Patterns of tsunamiite occurrence in time..409
   3.1 Meteorite-impact frequency and tsunamiites 410
   3.2 Earthquake-induced tsunamiite occurrences and sea-level change...................411
   3.3 Slump-induced tsunamiites and sea-level change............................416
4. Preservation potential of tsunamiites 416
5. Conclusive remarks and future studies 418
Acknowledgments..418
References...418

Selected bibliography..423
Index 453

Contributors

G.A. Albertão   PETROBRAS (UN-BC/RES/GGER), Macaé, Rio de Janeiro, Brazil

C. Andrade   Instituto D. Luiz, Faculdade de Ciências, Universidade de Lisboa, Lisboa, Portugal

J. Bourgeois   Earth and Space Sciences, University of Washington, Seattle, WA, United States

M.B. Cita   Dipartimento di Scienze, della Terra 'Ardito Desio', Università di Milano, Milano, Italy

P.J.M. Costa   Department of Earth Sciences, Faculty of Science and Technology, University of Coimbra, Coimbra, Portugal;   Instituto D. Luiz, Faculdade de Ciências, Universidade de Lisboa, Lisboa, Portugal

A.G. Dawson   Geography and Environmental Science, School of Social Sciences, Dundee, United Kingdom

L. Feist   Neotectonics and Natural Hazards Group, RWTH Aachen University, Aachen, Germany

S. Fujino   Graduate School of Science, Kyoto University, Kyoto, Japan

O. Fujiwara   Geological Survey of Japan, National Institute of Advanced Industrial Science and Technology (AIST), Tsukuba, Ibaraki, Japan

K. Goto   Department of Earth and Planetary Science, Graduate School of Science, The University of Tokyo, Bunkyo-ku, Tokyo, Japan

B. Higman   Earth and Space Sciences, University of Washington, Seattle, WA, United States

T. Ichihara   Fukken Co., Ltd, Hiroshima, Japan

F. Imamura   International Research Institute of Disaster Science, Tohoku University, Sendai, Miyagi, Japan

T. Jarupongsakul   Faculty of Science, Chulalongkorn University, Bangkok, Thailand

T. Kamataki   Center for Regional Revitalization in Research and Education, Akita University, Akita, Japan

N. Keerthi   Deepwell Drilling Technologies (PVT) Ltd., Narahenpita, Colombo, Sri Lanka

P. Kunthasap   Department of Mineral Resources, Environmental Geology Division, Ratchtawi, Bangkok, Thailand

F. Marini   Ecole Nationale Polytechnique de Geologie—ENSG, CNRS/CRPG Centre de Recherches Petrographyques et Geochimiques, Vandoeuvre-les-Nancy, France - In memorian

P.P. Martins, Jr.   Universidade Federal de Ouro Preto, Departamento de Geologia, Campus do Morro do Cruzeiro, Ouro Preto, Minas Gerais, Brazil

T. Matsui Department of Complexity Science and Engineering, Graduate School of Frontier Science, The University of Tokyo, Bunkyo-ku, Tokyo, Japan

K. Minoura Professor Emeritus, Tohoku University, Sendai, Miyagi, Japan

M. Murayama Center for Advanced Marine Core Research, Kochi University, Kochi, Japan

F. Nanayama Geological Survey of Japan, National Institute of Advanced Industrial Science and Technology (AIST), Tsukuba, Ibaraki, Japan

H. Naruse Graduate School of Science, Kyoto University, Kyoto, Japan

Y. Nishimura Institute of Seismology and Volcanology, Hokkaido University, Sapporo, Japan

K. Reicherter Neotectonics and Natural Hazards Group, RWTH Aachen University, Aachen, Germany

A. Ruangrassamee Department of Civil Engineering, Chulalongkorn University, Patumwan, Bangkok, Thailand

A. Scheffers Southern Cross University, Southern Cross GeoScience, School of Environment, Science and Engineering, Lismore, Australia

T. Shiki Uji, Kyoto, Japan

I. Stewart School of Geography, Earth and Environmental Sciences, Faculty of Science and Engineering, University of Plymouth, United Kingdom

D. Sugawara Museum of Natural and Environmental History, Shizuoka, Japan; International Research Institute of Disaster Science, Tohoku University, Sendai, Miyagi, Japan

S. Supharatid Natural Disaster Research Center, Rangsit University, Lak-Hok, Pathumtani, Thailand

A. Suphawajruksakul Faculty of Science, Chulalongkorn University, Bangkok, Thailand

T. Tachibana Soil Engineering Corporation, Higashi-ku, Okayama, Japan

R. Tada Department of Earth and Planetary Science, Graduate School of Science, The University of Tokyo, Bunkyo-ku, Tokyo, Japan

E. Tajika Department of Earth and Planetary Science, Graduate School of Science, The University of Tokyo, Bunkyo-ku, Tokyo, Japan

Y. Tsuji Earthquake and Tsunami Disaster Prevention Strategy Institute, Ryugasaki, Ibaraki, Japan

T. Yamazaki Toyonaka, Osaka, Japan

Preface to the second edition

More than 10 years have passed since the first edition of our book "Tsunamiites—Features and Implications" was published by Elsevier. A few years ago (2017), Dr. Amy Shapiro, who is the acquisitions editor for science books at Elsevier Publication, informed us, the editors of the first edition, that the book had performed very well, both in sales and in downloads on ScienceDirect. Dr. Shapiro also pointed out that it would be great timing now to publish a new edition.

We, the editors, think this a wonderful suggestion. We are concerned, however, about the huge explosion of tsunamiite studies since the publication of the first edition of our book; consequently, including all relevant aspects dealt with in these new studies would be too great a challenge. In addition, there is another problem for us, editors. We were informed that one of our fellow editors, Professor K. Minoura, unfortunately does not feel well enough at present to undertake the task of such further editing.

We have considered all this thoroughly for several months, and we came to the conclusion that it would certainly be worth to put effort into a revision of the first edition. We would nevertheless like to remind any researcher in this field that our first edition still provides a wealth of fundamental knowledge and thoughts regarding tsunamiites and tsunami disaster prevention. We are convinced that both editions may tempt students and young researchers to participate in research into this interesting and socially important subject: Tsunamiite Sedimentology. With respect to the absence of K. Minoura, we are happy to inform the reader that we found F. Nanayama willing to take his place.

Most of the chapters of the first edition did not need any change, except concerning a matter of geochronologic terminology: the Tertiary is no longer a formal geochronologic unit. The original introduction of the first edition, published in 2008, is to be kept without any changes, that we, the editors, indicate in this preface, which is meant to illustrate the need for a new edition.

Among these chapters, Chapters 13 and 14, which are concerned with the Mediterranean homogenite, were also kept unchanged in the new edition, but some additional brief comments have been added, as it appeared that several readers wish to find more information about recent studies of other thick homogeneous layers developed in the Mediterranean Sea. Providing the information about this, however, is out of the subject of this book.

Some years have passed since the first edition of Tsunamiites was published. It is understandable that there is now a need for some new chapters. Undoubtedly, readers of the second edition will welcome the new information that we provide about the results of new tsunamiite studies that were published since the first edition.

Preface to the second edition

On March 11th, 2011, a large seismically induced tsunami invaded the Pacific coast of northeastern Japan and caused a widespread disaster. Research into the various tsunamiites deposited from this tsunami is an important addition to the second edition. This contribution was added by D. Sugagawa, who was also already among the contributors of the first edition.

It has been well recognized that historical documents often offer interesting information about tsunamis and tsunamiites in the past. Studies concerning this subject have not been discussed in the previous edition, except for a monograph and discussion about the Mediterranean homogenite. In the new edition, a new chapter on this subject is presented by Y. Tsuji, one of the original editors. This contribution demonstrates the remarkable role that historical documents can play in the study of tsunamis and their sediments.

In addition these new chapters, some other chapters have been significantly improved. In particular, the previous Chapter 10, which was written by A. Dawson and I. Stewart, has now been rewritten, as Chapter 11, by P.J.M. Costa and other several new contributors. G. Albertao and his coauthors have revised their original chapter. The previous Chapter 14, now Chapter 16, provides extensive additional descriptions and some interesting discussions. The previous Chapter 17 (Tsunami Boulder Deposits) has now been rewritten, with an additional subtitle, by the original author, A. Scheffers; it is now Chapter 19. All these new and additional texts bring this new edition up-to-date and relevant.

Each new chapter has, like the largely rewritten or revised ones, been reviewed by two experts as well as by the editors. Valuable improvements have resulted from their comments. All reviewers except for those who wished to remain anonymous are listed here in alphabetical order: B. Atwater, J. Clague, M. Engel, K. Ikehara D. King, K. Nakajima, Y. Ogawa, A. Polonia, T. Suzuki, and A.J. van Loon.

We, the editors, would like to thank these reviewers wholeheartedly. Many thanks also to Mr. D. Mcmillian for his kind linguistic advice and help. We are indebted to Mrs. H. Shiki for her long support and patience. We would also like to thank Drs. A. Shapiro and N. Robertson of Elsevier Co. for their stimulating advice and patience. Our editorial work was unexpectedly complex and more time- and energy-consuming than envisaged. Without the kind help of all these helpers, we could not have carried out this hard editorial task.

Tsunemasa Shiki
Kitabatake, Kohata, Uji, Kyoto, Japan

Yoshinobu Tsuji
Earthquake and Tsunami Disaster Prevention Strategy Institute,
Matsuba, Ryugasaki, Ibaragki, Japan

Teiji Yamazaki
2-choume, Kita-midorigaoka, Toyonaka, Osaka, Japan

Futoshi Nanayama
Geological Survey of Japan, National Institute of Advanced Industrial
Science and Technology (AIST), Tsukuba, Japan

CHAPTER 1

Introduction: why a Book on Tsunamiites*

T. Shiki[1], K. Minoura[2], Y. Tsuji[3], T. Yamazaki[4]

[1]Uji, Kyoto, Japan; [2]Professor Emeritus, Tohoku University, Sendai, Miyagi, Japan; [3]Earthquake and Tsunami Disaster Prevention Strategy Institute, Ryugasaki, Ibaraki, Japan; [4]Toyonaka, Osaka, Japan

The purpose of the present volume is to overview the state of the art in tsunamiite sedimentology and to point out any problems and subjects that need additional investigation, as well as to provide an insight into the direction and potential for future researches.

Tsunamiite sedimentation has found its way into the area of modern coastal tsunami research so as to provide information about tsunami run-up for use in disaster prevention. In addition, the study of recent tsunami deposits has induced research in tsunami-induced or -related sediments from the geological past. This volume deals with both topics and tries to place them in the framework of the entire field of geosciences.

How terrible the December 2004 Indian Ocean tsunami disaster was! Everybody now realizes the importance of the study of tsunamis and tsunami disasters. The need to study about tsunamiites and the need to publish reports of the recent progress in tsunamiite studies have also been recognized. Most pre-1990 researches on tsunami deposits were focused on the recognition of run-up height, inundation limit, and particularly finding recurrent interval times of coastal on-surge tsunamis. Now, however, tsunamiite sedimentology has developed a much wider interest as is demonstrated in this volume.

As stressed by Shiki et al. (2000), catastrophic events as well as various other patterns of time-series changes in nature, including gentle evolution and rhythmic change, are important subjects in the geosciences for understanding the whole truth of natural phenomena and history of the Globe (Earth). In fact, the analysis of tsunamis and their sediments, resulting from meteorite impact, has raised general interest in the past 10 years. Earthquakes and the resulting tsunamis are also remarkable examples of chaotic events that occurred in the geological past, as has been brought into light in the past several years.

---

* Written in 2008 for the publication of the 1st edition.

Tsunamiites. https://doi.org/10.1016/B978-0-12-823939-1.00001-X
Copyright © 2008 Elsevier B.V. All rights reserved.

It seems that the study of tsunamiites is more difficult than that of other event deposits. Many people want to be informed about criteria for recognizing tsunamiites from the remote geological past and from the prehistoric times. However, such criteria have not yet been established. This is the most important reason for publishing this book.

In this volume, we intend to provide and discuss some instances of new information about how to create a systematic procedure for reading the sedimentary records of tsunamis induced by earthquakes, submarine slide, volcanic eruption, and meteorite impact. For the purpose of this discussion, however, general characteristics of tsunamis must first be sufficiently understood. Some important characteristics of tsunamis that cause particular features in tsunami-induced and -reworked sediments are explained first in the chapter presented by Sugawara and others.

Next, several papers follow concerning sedimentary records of onshore and/or inundated inland tsunamis also discussing their significances. Among the contributions, a few report on the sediments generated by the 2004 Indian Ocean tsunami.

Studies of shallow sea tsunami deposits are represented by only a few papers written with the contribution of Fujiwara and Kamataki, but the importance of these studies is emphasized in the review chapters by Shiki and Tachibana and by Keating and others. All the aforementioned contributions are primarily concerned with seismic tsunamiites. The significance of studies of bedforms and sedimentary structures in tsunamiites is discussed by Fujiwara. Tsunamis and tsunamiites induced by submarine or coastal slumps are mentioned very briefly in a few contributions. A tsunamiite directly generated by volcanic eruption is dealt with in one contribution (by Nishimura) only. However, the Mediterranean homogenites that were induced by the collapse of continental-shelf sediments due to a gigantic volcanic eruption in the Minoan age are examined in two contributions. These sediments provide an interesting example of some special features of deep-sea tsunamiites. Deep-sea tsunami deposits are also dealt with in another contribution, which reports on the K/T boundary tsunamiite. In addition, there are two more contributions that discuss the K/T boundary meteorite impact-induced tsunamiites.

Several contributions, including a few of the aforementioned, discuss and present ideas concerning the characteristic sedimentary records of tsunamis. Dawson and Stwart point out the forgotten important role of offshore current deposition for tsunamiite deposition. Scheffers supplies new information on tsunami-derived boulder deposits. The significance of studies on tsunamiites in the geological past is illustrated in one chapter (by Shiki and Tachibana). Shiki and others describe and discuss the characteristic features of various tsunamiites that are helpful in establishing criteria for distinguishing between the various genetic types of tsunamiites. Finally, the serial papers, the "Tsunami Deposits Data Base" from Hawaii University are kindly presented.

In addition, a bibliography of books, articles, and reports on tsunamiites is presented, which can facilitate tsunamiite studies by both students and researchers.

We hope that the information and discussions provided in this volume will help those who wish to answer the question of "what are the characteristic sedimentological features of the many kinds of tsunamiites that need to be observed and described." We also hope that it will encourage new developments in the study of tsunamiites and other event deposits.

We are aware that much more research is needed to significantly increase our understanding of the processes that lead to the formation of various types of tsunamiites in different environments. While browsing through the bibliography on tsunamiites, the readers may get the impression that tsunamiites have neither yet been studied from every possible point of view nor they have been studied in the remote areas of our planet. This holds certainly if the tsunamiite studies are compared with those of turbidites, as described and analyzed by Bouma and many others (Bouma, 1962; Bouma and Brouwer, 1964). In addition, our interpretation and discussion may be challenged. The illustrations and discussions provided in this volume should, however, be considered as a stimulus and a starting point for future field and laboratory researches.

This volume was first planned in 2003. Most of the manuscripts were submitted first earlier than the end of 2005. The completion of this volume, however, has caused wear and tear to the editors and some contributors. Many of the manuscripts were revised several times and some were even more times. And it took more than 4 years to accomplish writing, compiling, and editing the manuscripts. Unexpected problems including the "university crisis" in Japan were found to be considerable obstacles to our work. In addition, the terrible December 2004 Indian Ocean tsunami kept editors and some contributors dreadfully busy for a long time.

On the other hand, tsunamiite sedimentology spread widely and made remarkable progress and development during these period. The Indian Ocean tsunami particularly provided a lot of research objects, subjects, and new information for tsunamiite sedimentology. Thus, the resulting delay in the publication of this volume should not be seen as a negative aspect only. On the contrary, we actually had more chance and time to obtain and include a lot of new insights, knowledge, and results of discussions in our book. Nevertheless, it is clear that some of the contributors who submitted their views and research results early were victims. We owe them an apology.

Each paper has been critically evaluated by at least two scientists as well as by the editors. Many valuable improvements resulted from their comments. The reviewers are listed as follows.

K. Chinzei, J. Clague, K. Giles, C. Goldfinger, D. S. Gorsline, C. B. Jaffe, G. Jones, D. King, S. Kiyokawa, F. Kumon, W. Maejima, F. Masuda, H. Okada, G. Prasetya, J.

4 Chapter 1

Schnyder, K. Suzuki, and R. Walker. Several other reviewers preferred to remain anonymous. In addition, some contributors of this volume were asked to review manuscripts for the volume. We, the editors, would like to thank all those reviewers of the manuscripts wholeheartedly.

We are thankful to Hatsuko Shiki, to our colleagues in tsunamiite sedimentology, F. Nanayama, O. Fujiwara, K. Goto, T. Tachibana, and to our friends R. Doba, J. Ross, D. Mcmillian for their help and cooperation. We, the volume editors, are greatly indebted to A. J. van Loon, the series editor of Developments in Sedimentology for his ever-abounding enthusiasm in critical reading of the manuscripts and his many helpful suggestions throughout the task. We would also thank F. Wallien, J. Bakker, L. Versteeg-Buschman, and others of Elsevier Publication Co., for their patience and support.

References

Bouma, A.H., 1962. Sedimentology of Some Flysch Deposits. A Graphic Approach to Facies Interpretation. Elsevier, Amsterdam, p. 164.
Bouma, A.H., Brouwer, A., 1964. Turbidites. Elsevier, Amsterdam, p. 264.
Shiki, T., Cita, M.B., Gosline, D.S., 2000. Sedimentary features of seismites, seismo-turbidites and tsunamiites—an introduction. Sediment. Geol. 135, vii–ix.

CHAPTER 2

The term "Tsunamiite"

T. Shiki[1], T. Yamazaki[2]
[1]Uji, Kyoto, Japan; [2]Toyonaka, Osaka, Japan

We are aware that the term "Tsunamiite" is new to many readers who have not worked on the sediments. We want to explain here the history of proposal and definition of the term "tsunamiite."

According to "Georeff," the oldest reference of "tsunamiite" is by a Chinese author Gong-Yiming in 1988 (Shanmugan, personal communication). We have also heard that someone used the word "tsunamiite," though without a clear definition at some earlier time.

The word "tsunamite" was used by Yamazaki et al. (1989) to describe the Miocene tsunami-worked boulder sediments in an upper bathyal environment in central Japan. Through the kind advice of Professor Ales Smith of England, the wrong spelling of the term was revised as "Tsunamiite" in Shiki and Yamazaki (1996). At that time, it was stated that the term "tsunamiite" should be used not only for the sediments transported by a tsunami itself but also for the sediments transported and deposited by a tsunami-induced current. It was also pointed out that the term is defined essentially based on a threshold mechanism just like the cases of "tempestites" and "seismites." By contrast, the term "turbidites" (deposits transported by turbidity currents) does not indicate anything about the trigger mechanism that caused the transport of these sediments. Moreover, Shiki and Yamazaki (1996) also proposed that the term "tsunamiite" could be used in a slightly wider sense than the word "tsunami deposit" (or "tsunami sediment"). "Tsunami-worked deposit" (including tsunami-reworked deposits) or "tsunami-worked sediment" (including tsunami-reworked sediment) has almost the same meaning as "tsunamiite." And because of the initiation mechanism and the following effect by tsunami during the flow downslope and the plume up of the fine-grained sediments, the so-called Mediterranean homogenite can be included in the term "tsunamiite." On the contrary, a kind of homogenite ["B-type(mega)-turbidite"] is a kind of "sediment gravity flow deposit" and can be called a "turbidite" in a wide sense by some researchers.

Some objections against the use of the term "tsunamiite" may arise from two different viewpoints.

Tsunamiites. https://doi.org/10.1016/B978-0-12-823939-1.00002-1
Copyright © 2008 Elsevier B.V. All rights reserved.

6   Chapter 2

1. Some researchers are of the opinion that any interpretive names for sediments are to be avoided because interpretations may change and different people may have different interpretations. They consider the history of the interpretive term "turbidite" as an example—even 50 years after the term had been invented, people are still arguing about certain examples. We believe, however, that the confusion about the interpretation is caused fundamentally by the difficulty of correct and reliable identification as a result of still insufficient investigation rather than by terminology.

There have, actually, been numerous interpretive terms used in geology just like in other natural sciences and in the social sciences, for example, tsunami deposits, seismite, storm deposits, tempestite, tidal deposit, debris-flow deposits, pyroclastic sediment, welded tuff, subduction slab, back-arc basin, and so on.

There are, to be more precise, different types of interpretative terms even if only sedimentology is considered. They are based on the following: (1) sort of external agencies, such as tsunamiites, tempestites, and current ripples; (2) state of transportation mechanism such as turbidites, debris-flow deposits, density-current deposits; and (3) the settings or environments such as onshore tsunamiites, bathyal sediments, trench deposits, and so on. Earth scientists who are against a terminology based on interpretation should answer the question: "Do you also deny the use of the terms 'storm deposits,' 'subduction slab,' etc. which are obviously based on interpretation?"

2. Another more reasonable objection could be that some people prefer the classical term "tsunami deposit" rather than "tsunamiite" and that they consider that the term "tsunamiite" is synonymous with "tsunami deposit." The present authors are, however, aware that the term "tsunami deposit" has been used commonly only for coastal uprush tsunami deposit. There are, however, various tsunami-related deposits in other environments. The upper bathyal tsunami-worked sediments in Japan, the Mediterranean homogenites mentioned earlier, and the Cretaceous/Tertiary boundary deep-sea tsunamiites are the examples that ought to be included in the term "tsunamiites."

We have one more reason to propose the use of the term "tsunamiite," though it is not essential. For example, the term "Middle Miocene upper bathyal tsunami-induced current conglomeratic deposit" can be used to indicate only one single deposit. Obviously, this term is far too long. The present authors think that even the contraction of only one word is more appropriate in this case.

There are many more terminological aspects that might deserve discussion here. However, further philosophical discussion is out of the scope of this contribution. We intend only to state that the term "tsunamiite" is appropriate and useful.

References

Shiki, T., Yamazaki, T., 1996. Tsunami-induced conglomerates in Miocene upper bathyal deposits, Chita Peninsula, central Japan. Sediment. Geol. 104, 175–188.

Yamazaki, T., Yamaoka, M., Shiki, T., 1989. Miocene offshore tractive current-worked conglomerates—Tsubutegaura, Chita Peninsula, central Japan. In: Taira, A., Masuda, M. (Eds.), Sedimentary Facies and the Active Plate Margin. Terra Publication, Tokyo, pp. 483–494.

CHAPTER 3

Tsunamis and tsunami sedimentology

D. Sugawara[1,2], K. Minoura[3], F. Imamura[2]

[1]Museum of Natural and Environmental History, Shizuoka, Japan; [2]International Research Institute of Disaster Science, Tohoku University, Sendai, Miyagi, Japan; [3]Professor Emeritus, Tohoku University, Sendai, Miyagi, Japan

1. Introduction

Tsunamis belong to the most catastrophic oceanic wave motions; they form due to the processes that result in shock waves, such as submarine earthquakes and slides, volcanic activity, and asteroid impacts. The mechanisms of the wave excitation give tsunamis the character of long waves, which is associated with the characteristic processes of tsunami propagation. In the case of a large-scale event, tsunamis become a giant surge of several tens of meters high near the coast. Tsunamis and related phenomena have attracted much social and scientific attention, since tsunamis frequently cause devastation of coastal areas, resulting in loss of human life and property and damage to ecosystems.

Geological investigations of tsunamis started in the 1960s, after the 1960 Chilean earthquake tsunami that attacked the coasts of Chile and Japan. Kon'no (1961) discovered that the tsunami deposited layers of marine sand and silt on the Pacific coast of north-east Japan. Wright and Mella (1963) found that a veneer of sand covered the alluvial plain of south-central Chile after the tsunami incursion. Both research groups reported that the coastal topography was dramatically altered due to the tsunami. Severe erosion of sandbars and sandpits was observed on both the Chilean and the Japanese coasts. Since then, tsunamis have been known as the agents of erosion and the deposition in coastal environments.

Early works on tsunami sedimentology have increased the understanding of the genesis of these deposits. For example, Minoura and Nakaya (1991) have modeled tsunami sedimentation based on the observation of the 1983 Japan Sea earthquake tsunami. They suggested that traces of a tsunami invasion can be preserved as tsunami deposits in the sedimentary succession of, for example, coastal flats, marshes, and lakes. They reconstructed a history of ancient tsunami events in north-east Japan based on the tsunami deposits identified within lacustrine successions. The risks of tsunamis for the coast of the

Tsunamiites. https://doi.org/10.1016/B978-0-12-823939-1.00003-3
Copyright © 2008 Elsevier B.V. All rights reserved.

North Sea have been discussed on the basis of prehistoric tsunami deposits along the coast of Scotland (Dawson et al., 1988; Long et al., 1990).

The major purpose of tsunami sedimentology is to identify historic and earlier tsunami events and to reconstruct these tsunami invasions. Ancient geological events that caused tsunamis, particularly volcanic eruptions and asteroid impacts, can be better understood by interpreting the resulting tsunami deposits. Estimation of the recurrence interval and magnitude of ancient tsunamis is an important subject, particularly from the standpoint of risk assessment. Geomorphological interest may focus on the clarification of the role of tsunamis in coastal zones, such as the development of coastal topography. In this context, many tsunami deposits have been studied from various aspects.

Descriptions of tsunami deposits show a wide diversity of tsunami deposits. This includes variations in the distribution pattern, grain-size composition, and sedimentary facies. For example, onshore tsunami deposits are typically composed of a sheet of well or poorly sorted sand (e.g., Dawson et al., 1988). On the other hand, an accumulation of conglomerates, ranging from pebbles to boulders, is a different type of tsunami deposit (e.g., Kato et al., 1988; Moore, 2000). The intricate behavior of tsunami run-up and backwash in various sedimentary environments may be responsible for the various variations. The diversity frequently causes a problem in the identification of tsunami deposits and even in their recognition. There are no obvious criteria available for identification, and there are no simple procedures for the interpretation either. Only an understanding of the nature of tsunamis may provide an appropriate interpretation of tsunami sedimentation.

This contribution describes the nature of tsunamis and the characteristics of tsunami deposits as an introduction to this volume. The nature of tsunamis is described from their hydrodynamic aspects. The characteristics of tsunami deposits described here are based on numerous works, in order to illustrate their diversity. Onshore tsunami sedimentation is discussed in particular.

2. Generation, propagation, and quantification

Tsunamis are a transient wave motion in oceans or enclosed basins generated by a variety of phenomena, of which atmospheric (meteorologic) phenomena are excluded. When a particular external force disturbs an oceanic or lacustrine area, a mass of water is forced to move temporarily. The displacement of the water mass propagates outward as a wave motion. Such a wave is characterized as a large wavelength that depends on the dimensions of the wave source, which is proportional to the magnitude of the external force. The large wavelength is responsible for the characteristic behavior of the wave. For example, the typical height of a wave in the deep sea is only several tens of centimeters.

However, once the wave arrives at the shallow sea, the height grows precipitously. In some cases, a wave reaching the coast becomes a giant surge of several tens of meters in height. The giant wave finally runs up on the land and frequently has a huge impact on the coastal environment. These processes of wave generation, propagation, run-up and backwash together define a "tsunami."

"Tsunami" means "harbour wave" in Japanese. This name originates from the observation that the tsunamis are usually not distinct in deep water, but surge deeply on the coast, especially bays and harbors, causing severe damage. "Seismic sea wave" is often used instead of "tsunami" because most tsunamis are associated with earthquakes. Tsunami that places emphasis on coastal behavior is a more appropriate name than seismic sea wave, however, because tsunamis can also result from nonseismic triggers.

In the following text, a summarized review is presented of the cause and effect of tsunamis, referring to several typical or famous examples. After that, we briefly discuss the hydrodynamics of tsunamis. Definitions of physical parameters that measure a tsunami are noted at the end of this chapter.

2.1 Generation of tsunamis

Subaqueous earthquakes and slides are major triggers; volcanic activity and asteroid impacts are occasional factors. Tsunamis can be classified into four types, according to their cause.

2.1.1 Earthquake-induced tsunamis

The occurrence of a submarine earthquake is explained by plate tectonics, the theory that is based on an earth model characterized by numerous lithospheric plates. Viscous underlayers on which lithospheric plates float are called the "asthenosphere." These plates that cover the entire surface of the earth and contain both the continents and sea floor move relative to each other at the rates of up to 10 cm per year. The region where two plates are in contact is called a "plate boundary," and the way in which one plate moves relative to another determines the type of boundary: divergent, where the two plates move apart from each other; convergent, where the two plates move toward each other; and transform, where the two plates slide horizontally past each other. Convergent boundaries commonly form subduction zones with deep ocean trenches in which one plate slides underneath the other. This type of plate boundary is especially important from the viewpoint of tsunami study. Submarine earthquakes triggered 9 of 10 tsunamis over the past 200 years (Imamura, 1996), and almost all of these earthquakes occurred in subduction zones along the plate boundaries. The Pacific Rim is a critical tsunami hazard region because of the many subduction zones in the region.

12   Chapter 3

Recently, the worst tsunami catastrophe in human history occurred. The Indian Ocean tsunami on December 26, 2004, triggered by the M9.3 Sumatra earthquake, struck the coastal areas of the countries around the Indian Ocean. The epicenter of the earthquake was located on the Sunda Trough, where the Indian—Australian plate slides beneath the Eurasian plate. The shallow focal depth of 10 km and the total fault length of 1500 km as a tsunami source may be responsible for the significant development of—and the extensive damage caused by—the tsunami. Field investigations of the tsunami reported an average tsunami height of 10 m along the surrounding coast of the Indian Ocean, and the maximum tsunami height exceeded 30 m in the north of Sumatra Island. The total number of victims and missing people due to the earthquake and the tsunami was estimated to be around 300,000. It is reported that the tsunami was observed even in Antarctica and the western coast of the United States, still having a height of several tens of centimeters. The tsunami induced characteristic sedimentation in the coastal areas of the Indian Ocean. Some preliminary results of geological and sedimentological investigations are presented in this book, and a few specific aspects are dealt with in this chapter.

Not all earthquakes generate tsunamis. To generate a tsunami, the fault plane must be near or just beneath the sea floor and must cause vertical displacement of the sea floor up to several meters over a large area, up to 100,000 km^2. Shallow-focus earthquakes, with a focal depth of less than 70 km, along subduction zones are responsible for most large-scale tsunamis.

2.1.2 Tsunamis induced by submarine sliding

Earthquakes and volcanic temblors often accompany landslides and submarine slides. Even a relatively small earthquake can induce slumping of a coastal landform and liquefaction of the sea bottom and the consequent massive movement of sediment. Submarine slides occur especially where huge piles of unstable sediment build up. The immediate trigger of a slide may be a small earthquake or new sediment supplied by a river. A sudden change in sea-bottom topography may also trigger a tsunami; therefore, submarine slides are considered to be one of the possible sources of several tsunami events. Recent catastrophic tsunami events on Flores Island in 1992, Skagway in 1994, Papua New Guinea in 1994 (Keating and McGuire, 2000; Tappin et al., 1999), and Turkey in 1999 have significantly increased the scientific interest in submarine slides and slide-induced tsunamis. The dimensions of the sea-bottom disturbance of submarine slides are usually much smaller than those of seismic sea-bottom displacement. Thus, the typical wavelength of the tsunamis generated by a submarine slide is also smaller. This may give the tsunami a particular feature, which is the peaked distribution of the tsunami height at a specific part of the coastline (Bardet et al., 2003).

The breakdown of gas hydrates is another possible cause of submarine slides. Gas hydrates are ice-like solids composed of water and gas. Hydrates can exist only under certain temperatures and pressures. For example, methane hydrates are found under the temperature and pressure conditions that correspond to near and just beneath the sea floor, with a water depth of 300–500 m, and the hydrates can exist up to depths of about 3000 m below the ocean floor. Many large gas-hydrate fields have been identified worldwide in recent marine sediments. Earthquakes and drops in sea level during glacials may trigger the decomposition of gas hydrates. As the hydrates fall apart into water and gas, they are likely to disturb an enormous volume of sea-bottom sediment (megaturbidite; e.g., Rothwell et al., 1998).

The Storegga slides on the continental slope off western Norway are a typical example. Three major slide events have been identified as responsible for the Storegga slide deposits (Bugge et al., 1987; Jansen et al., 1987). The total volume of sediment involved in the slides was 3880 km^3 for the first slide and 1700 km^3 for the second and third slides. The age of the first slide has been estimated to be more than 30,000 years, and the second and last slides occurred between 8000 and 5000 BP. Seismic activity in the neighboring area of the slide is considered to have been the trigger of the second Storegga slide. In addition, the decomposition of gas hydrates due to seismic activity might have resulted in the liquefaction of sediment and a submarine slide. Jansen et al. (1987) suggested that a large-scale "hydraulic effect" (in fact, this may mean a large-scale tsunami) could have been induced as a consequence of the massive movement of sediment. Geological evidence of a large-scale tsunami, corresponding in age to that of the second Storegga slide (~7000 BP), has been found in the coastal flats of Scotland (Dawson et al., 1988; Long et al., 1989, 1990) and the coastal lakes of western Norway (Bondevik et al., 1997).

2.1.3 Volcanism-induced tsunamis

Explosive eruptions of a volcanic island or submarine volcano involve the generation of tsunamis, although few historical examples exist. Human beings have experienced devastating tsunami disasters originating from such volcanic eruptions at least twice: after the eruptions of (1) Thera (Santorini) in the Aegean Sea during the late Bronze Age and (2) the Krakatau in the Sunda Sea in 1883. For example, the August 26–27, 1883 volcanic explosion of the Krakatau caused two-thirds of the island to disappear under the sea and triggered a tsunami of 37 m high. The waves raised large ships to 10 m above sea level and carried them as far as 3 km inland. The tsunami struck more than 250 villages in western Java and eastern Sumatra, and about 36,000 people lost their lives (Beget, 2000). Although the generation of this volcanic tsunami is complicated and still controversial, its origin has been associated with the formation of an extensive submarine caldera, the discharge of pyroclastic flows into the sea, and a submarine explosion due to the interaction of magma and sea water (e.g., Nomanbhoy and Satake, 1995; Self and Rampino, 1981; Sigurdsson et al., 1991).

14 Chapter 3

Several mechanisms of volcanic activity are involved in the generation of tsunamis, although precise physical models and direct observations of the interactions and processes are not yet available in full detail. An explosive eruption may lead to the loss of support of a volcanic cone by the magma chambers below the cone, resulting in a sudden collapse of the cone and the formation of a submarine caldera. Large masses of sea water surrounding the volcano rush into the caldera and the sea water finally overflows as a tsunami. Pyroclastic flows and debris avalanches also generate impulse waves as they enter a sea or lake by displacing huge volumes of water. Data from recent studies revealed that volcanic debris avalanches cause tsunamis (e.g., Maramai et al., 2005a; Nishimura and Miyaji, 1995). The impact of tsunamis from volcanic debris avalanches is quite significant in some cases; the origins of several historical tsunami disasters with large numbers of casualties are associated with the discharge of volcanic debris avalanches into the sea (Beget, 2000). In addition, it is suggested that tsunamis are also generated by submarine slides of vast edifices of lavas on the weak substrate of oceanic sediment around volcanic islands (e.g., Maramai et al., 2005a,b).

2.1.4 Tsunamis induced by asteroid impact

Since two-thirds of the earth is covered by oceans, asteroid impacts can be a cause of tsunamis. Although some examples are known of tsunamis generated by the impact of a meteorite, the magnitudes of these cases may surpass those of earthquake- or volcanism-induced tsunamis. It is commonly accepted that the impact of a meteorite generated the greatest tsunami in the earth's history, around the end of the Cretaceous and the beginning of the Tertiary (the K/T) boundary, about 65 million years ago. The diameter of the meteorite is estimated to have been about 10 km, and the collision velocity was probably about 25 km/s (Alvarez et al., 1980). The meteorite hit at Chicxulub on the Yucatan Peninsula, and a giant crater of 180 km in diameter was formed (Hildebrand et al., 1991). The mean sea level during the late Cretaceous was 200 m higher than that at present (Haq et al., 1981). The impact site is considered to have been a shallow-water platform at the time of the impact. The mechanism of tsunami generation by this meteorite impact was probably fairly similar to that of a volcanic tsunami. Sea water surrounding the impact site may have rushed into and flowed out of the 180-km-diameter submarine crater. Seismic waves due to the meteorite impact caused a collapse of the margin of the Yukatan platform and consequent submarine slides. Each process involved the generation of a gigantic tsunami. According to the numerical reconstruction of the K/T tsunami (Fujimoto and Imamura, 1997; Matsui et al., 2002), the tsunami height was calculated as reaching much as 200 m at the North American coast, and the run-up distance from the palaeocoastline was more than 300 km. The tsunami propagated eastward and westward from the point of impact and the two wave fronts encountered each other in the Indian Ocean 27 h later. Characteristic sedimentary successions at the K/T boundary, which suggest a strong current oscillation on the deep-sea floor and massive gravity flows of sediment from the continental slope, have been found in north-western Cuba (Kiyokawa et al., 2002; Tada et al., 2002; Takayama et al., 2000), around the Gulf of Mexico (Alvarez et al., 1992; Bourgeois et al., 1988; Smit et al., 1992, 1996), and in north-eastern Brazil (Albertao and Martins, 1996; Stinnesbeck and Keller, 1996).

2.2 Propagation of tsunamis

In the deep sea, tsunamis are typically around several tens of centimeters to a few meters in height and are several tens to hundreds of kilometers in wavelength ("long wave"). It is therefore difficult to identify the wave form of tsunamis in the open sea. Owing to the properties of long waves, the energy of tsunamis distributes from the bottom to the surface of the sea. As a tsunami approaches the shallow sea, the energy is compressed into a much shorter distance and a much more limited depth and therefore shows a significant increase of the tsunami height.

2.2.1 Tsunamis in the deep sea

Tsunamis have the characteristics of long waves. The definition of a long wave is that the wavelength is large and that the wave height is small compared with the water depth. (It is considered that the wavelength should be ~ 20 times greater than the water depth.) For the large-scale earthquake-induced tsunamis that have significant effects, the area of the sea-bottom displacement must range from tens to hundreds of kilometers, and the vertical displacement of the sea floor must be several meters, while the water depth of the wave source must be several thousands of meters. Thus, tsunamis can be treated as long waves. According to the long-wave approximation, the vertical acceleration of fluid is much slower than the gravity acceleration. Thus, the water pressure is similar to that of still water. Because the pressure does not vary with depth, the horizontal movement does not change with depth either (Fig. 3.1). The motion of a tsunami in the deep sea can be described by linear long-wave equations:

$$\frac{\partial \eta}{\partial t} = -h\frac{\partial u}{\partial x} \tag{3.1}$$

$$\frac{\partial u}{\partial t} = -g\frac{\partial \eta}{\partial x} \tag{3.2}$$

where η is the water elevation relative to the still-water level, h is the water depth, u is the current velocity, and g is the gravity acceleration. Here, u can be eliminated by combining the two equations.

$$\frac{\partial^2 \eta}{\partial t^2} = C^2\frac{\partial^2 \eta}{\partial x^2}, C = \sqrt{gh} \tag{3.3}$$

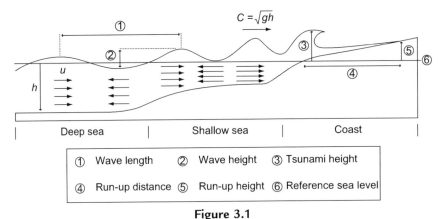

Figure 3.1
Schematic model of tsunami propagation and quantities defining a tsunami.

This equation is a classical one-dimensional wave equation and water elevation, η, travels in the $\pm x$ directions at propagation speed C. Thus, the propagation speed of a tsunami depends only on water depth. This is one of the most important characteristics of tsunamis. When a tsunami travels through a sea of 4000 m depth, the propagation speed is around 200 m/s (720 km/h), approximately the speed of a jet plane. Eq. (3.3) then becomes.

$$\eta_1 = f_1(x - Ct),\ \eta_2 = f_2(x + Ct) \tag{3.4}$$

where both f_1 and f_2 are arbitrary functions. Eq. (3.4) can be substituted into Eqs. (3.1) and (3.2):

$$u_1 = \frac{C}{h} f_1(x - Ct),\ u_2 = -\frac{C}{h} f_2(x + Ct) \tag{3.5}$$

Therefore, the current velocity, u, is described by the equation:

$$u = \pm \eta \sqrt{\frac{g}{h}} \tag{3.6}$$

For instance, when a tsunami with an amplitude of 1 m and a propagation speed of 200 m/s travels through water of 4000 m depth, the current velocity is only 5 cm/s. Although the surface oscillation propagates quickly, the water particles therefore move slowly. The current velocity is uniform throughout the water column, except near the sea bottom.

The direction of traveling of a tsunami is affected by the submarine topography because the propagation speed of a tsunami depends only on the water depth. The propagation speed, C, is proportional to the square root of the water depth, so the wave front travels faster in a deeper area than in a shallower area. Consequently, the wave front of a tsunami approaches the coast parallel to isobathymetric lines. For this reason, the energy of a tsunami is concentrated on projecting coastal features such as a cape.

Tsunamis are reflected by an undulating sea bottom or where the water depth changes sharply, but if the ratio of water depth to wavelength is much larger than the rate of change in the water depth, no reflection occurs. The following inequality represents this condition:

$$\frac{h}{\lambda} \gg \frac{\partial h}{\partial x}$$

(3.7)

where λ is the wavelength and h is the water depth.

Refraction, reflection, and diffraction of a tsunami often involve an unexpected excitation of wave height, such as around a circular island.

2.2.2 Tsunamis in shallow seas

When a tsunami proceeds from a deep sea to a shallow sea, the propagation speed decreases and the current velocity increases. The intervals between wave trains become short and the wave height increases (Fig. 3.1). When the water depth changes from h_0 to h, the amplification of the wave is

$$Hh^{1/4} = H_0 h_0^{1/4}$$

(3.8)

Where H_0 and H are the wave heights corresponding to h_0 and h, respectively. Eq. (3.8) suggests a constant energy transmission of tsunamis and is applicable under the assumption that the wave heights are small compared with the water depth. The energy of the tsunami is concentrated over a much shorter distance and a much shorter water column, and the tsunami will begin to interact significantly with the sea bottom. This phenomenon is referred to as "shallow-water deformation." In this phase, the linear long-wave approximation (Eqs. 3.1 and 3.2) is no longer applicable. Shallow-water long-wave theory is available for the evaluation of tsunami propagation in shallow seas:

$$\frac{\partial \eta}{\partial t} + \frac{\partial M}{\partial x} + \frac{\partial N}{\partial y} = 0$$

(3.9)

$$\frac{\partial M}{\partial t} + \frac{\partial}{\partial x}\left(\frac{M^2}{D}\right) + \frac{\partial}{\partial y}\left(\frac{MN}{D}\right) + gD\frac{\partial \eta}{\partial x} + \frac{gn^2 M}{D^{7/3}}\sqrt{M^2 + N^2} = 0$$

(3.10)

$$\frac{\partial N}{\partial t} + \frac{\partial}{\partial y}\left(\frac{N^2}{D}\right) + \frac{\partial}{\partial x}\left(\frac{MN}{D}\right) + gD\frac{\partial \eta}{\partial y} + \frac{gn^2 N}{D^{7/3}}\sqrt{M^2 + N^2} = 0$$

(3.11)

where η is the water elevation relative to the still-water level, D is the total depth ($= h + \eta$), g is the gravity acceleration, n is Manning's relative roughness, and M ($= uD$) and N ($=vD$) are the flux of the x and y directions, respectively. Eqs. (3.10) and (3.11) are a variation of the Navier–Stokes equation, which describes the motion of incompressible viscous fluid. Since they are nonlinear partial differential equations, a general solution to the wave height (η) and the current velocities (u and v) is difficult to obtain, except for particular,

18 Chapter 3

simplified cases. The shallow-water long-wave theory is used widely for the numerical modeling of tsunami propagation.

When a tsunami enters a bay, harbor resonance is induced. Every bay has its own proper (or resonant) period, according to the geometry of the bay. For example, the resonant period in a rectangular bay with a constant depth is described by the following equation:

$$T_* = \frac{4l}{\sqrt{gh}} \tag{3.12}$$

Where T_* is the resonance period in a rectangular bay, l is the length of the bay, g is the gravity acceleration, and h is the water depth at the mouth of the bay. Resonance periods in bays with other shapes can be calculated by Eq. (3.11) with some modification. If the period of the incoming wave is close to the resonance period, an amplification of the wave takes place (harbor resonance). The amplification factors of harbor resonance in bays with various shapes can be calculated analytically (e.g., Kajiura, 1982; Zelt, 1986). According to Kajiura (1982), the amplification factor is larger in V-shaped bays than in rectangular bays, and the factor is larger in bays with a linearly sloping bottom than in bays with a constant depth. Thus, harbor resonance is most significant in V-shaped bays with a linearly sloping depth. The wave height of a wave entering in a V-shaped bay can be amplified four to five times within the bay. This is consistent with observation data from actual events, such as tsunamis on the Pacific coast of north-east Japan, where rias are well developed.

Inundation of coastal areas by tsunamis takes place in various forms, such as a gradual increase in sea level by nonbreaking waves, tidal bores (like a wall of water), and a surge of breakers. The tsunami height near the coast can reach several tens of meters in the most severe cases. The duration of the water-level rise increases, and the inundation area widens in the just-mentioned order, due to the long-wave characteristics of tsunamis. On a coastal flat, the horizontal extent of inundation perpendicular to the coastline may exceed several hundred meters or even several kilometers. The transgression (run-up) and regression (backwash) of a tsunami are repeated many times with time intervals of several to several tens of minutes. As a result of strong, enduring, repeated currents, the impact on the coastal environment is significant, affecting human life and property, as well as eroding and depositing sediment.

2.3 Quantification of tsunamis

Standardized physical parameters are necessary to evaluate and compare tsunami events. Defining the tsunami height, for example, is obvious, while defining tsunami magnitude may be somewhat controversial. The tsunami magnitude evaluates the size of a tsunami, and the tsunami intensity evaluates the consequences of the tsunami event from location to location.

2.3.1 Tsunami height

The wave height of a tsunami is the amplitude of a tsunami, which is defined by the vertical difference between the wave crest and the wave trough (Fig. 3.1). The tsunami height is defined by the difference between the reference water level (e.g., mean sea level) and the crest or trough of the tsunami (Fig. 3.1). These heights are determined by the records of tidal or tsunami gauges and field investigations after a tsunami event. Note that wave height is not same as tsunami height.

2.3.2 Tsunami period

The period of a tsunami is equal to the time required for two successive wave crests to pass a fixed point (Fig. 3.1). Practically, tsunami periods are identified using records from tidal or tsunami gauges installed in particular coastal areas.

2.3.3 Run-up height

When a tsunami runs up on land, the height of the trace or run-up is identified by field measurement or interview with eye witnesses. The elevation of a discolored line on a building or of drift sand on a slope is called the "trace height." The run-up height is the elevation of the most landward point of the inundated area relative to the reference sea level (Fig. 3.1). Note that trace height is generally equivalent to run-up height. Since the sea level changes throughout a day or over days, measured tsunami and run-up heights are corrected according to the tidal data on the corresponding locations.

2.3.4 Tsunami magnitude

Like earthquake magnitude, tsunami magnitude is an objective physical parameter that measures energy radiated by the tsunami source and does not reflect the consequences (impact or damage) of the tsunami. Several studies have proposed definitions of tsunami magnitude and the tsunami intensity scale. In case of an earthquake-induced tsunami, the magnitude of the tsunami corresponds to the dimensions of sea-bottom displacement. The dimensions are physical parameters of the fault plane and are in proportion to the magnitude of the earthquake. According to empirical data from observations of earthquake tsunamis, tsunami magnitude m can be defined in an equation as follows (Iida, 1963):

$$m = 2.61M - 18.44 \tag{3.13}$$

where M is the magnitude of an earthquake.

The magnitude of a tsunami can also be defined by the tsunami height observed on the coast or measured by a tidal or tsunami gauge. According to the results of previous studies, definitions of tsunami magnitude M_t by tsunami height have the general form:

$$M_t = a\log H + b\log \Delta + D \tag{3.14}$$

20   Chapter 3

where H is the maximum single amplitude of a tsunami wave (in meters), Δ is the distance from the earthquake epicenter to the observation site along the shortest oceanic path (in kilometers), and a, b and D are constants (Papadopoulos, 2003).

2.3.5 Tsunami intensity

Tsunami intensity is a rather subjective estimate of the macroscopic effects of a tsunami event. Consequently, a tsunami event can be characterized by different tsunami intensities at different locations of the affected area. If we intend to quantify the impact of tsunamis as we do with earthquakes, the intensity of a tsunami should be independent of any physical parameter and should estimate or evaluate the degree of tsunami impact, such as loss of human life, damage to property, and effects on the natural environment. However, there has been confusion between the definition of the magnitude and the intensity of a tsunami. Some tsunami intensity scales are defined according to tsunami height with possible impact, thus providing a definition of tsunami magnitude rather than an intensity scale. In order to avoid such confusion and establish a pure tsunami intensity scale, Papadopoulos and Imamura (2001) proposed an improved and detailed 12-grade tsunami intensity scale.

3. Tsunami sedimentology

A Greek archaeologist, Marinatos (1939), may have been the first person to describe tsunami deposits in his work on deposits related to a Minoan volcanic event in the late Bronze Age, ~3700 years ago. He found that the floors of Minoan ruins on Crete were covered by a thick layer of marine sand, on top of which lay a layer of seaborne pumice from Thera (Santorini), and he hypothesized that a destructive tsunami caused by the Thera eruption struck the Aegean Sea coast of Crete. Geologists began to study tsunami sedimentation in the 1960s after the 1960 Chilean earthquake tsunami (Kon'no, 1961; Wright and Mella, 1963). The scientific interest in tsunami sedimentation has been increasing since the 1980s (e.g., Atwater, 1987; Bourgeois and Reinhart, 1989; Bourgeois et al., 1988; Dawson et al., 1988; Kastens, and Cita, 1981; Long et al., 1989; Minoura et al., 1987; Reinhart and Bourgeois, 1987). Substantial investigations have identified tsunami deposits mostly in coastal regions along active margins (Fig. 3.2). Considerable knowledge of tsunami sedimentation and of the characteristics of tsunami deposits has been obtained. Here, we classify tsunami events into two categories according to observation data.

1. Observed tsunamis, of which we have a great deal of data from scientific, instrumental observation. Most modern (from the 1800s to today) tsunami events are included in this category.

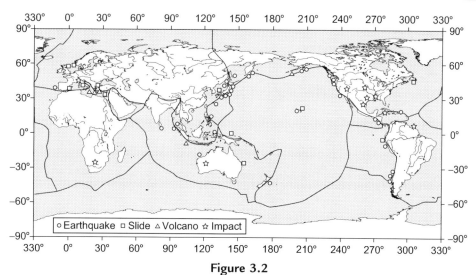

Figure 3.2
Reference map of tsunami deposits. *Modified after Bourgeois, J., Minoura, K., 1997. Paleotsunami studies—contribution to mitigation and risk assessment. In: Gusiakov, V.K. (Ed.), Tsunami Mitigation and Risk Assessment, Report of the International Workshop, 21—24 August 1996, Russia Petropavlovsk-Kamchatsky. 1—4.*

2. Unobserved tsunamis, of which we have less reliable data on magnitude and height. In some cases, even the existence of tsunami events is ambiguous. Almost all historic and prehistoric tsunami events are included in this category.

Many geologists have studied sediments deposited by unobserved tsunamis, because ancient sedimentary events are important subjects of geological study. Studies on the tsunami deposits associated with giant submarine slides, volcanic events involved in the possible disruption of ancient civilizations, and the large meteorite impact in the late Cretaceous have revealed the nature and interesting consequences of these events. Although geological studies of tsunami layers preserved in stratigraphic sections are informative, tsunami sedimentation cannot be interpreted merely on the basis of the study of unobserved tsunamis. Detailed scientific investigations of sediment transport by observed tsunamis for which the causes and consequences are well known are essential for an understanding of the triggers and thresholds of sediment transport and deposition. Furthermore, sedimentological interpretations of deposits left by observed tsunamis tend to be noncontroversial; thus, the results of field investigations of observed tsunamis are fundamental for studies on tsunami sedimentation.

This section describes the transport process of sediment and reviews the sedimentological characteristics of tsunami deposits in subaqueous and subaerial environments, on the basis of previous studies.

3.1 The mechanics of sediment transport

Understanding the mechanical process of sediment transport is necessary for discussing tsunami sedimentation. Here, we briefly summarize some basic terms and data related to sediment transport by flowing fluid.

3.1.1 Shear stress and tractive force

Due to long-wave characteristics, the horizontal movement of a tsunami is uniform throughout the water column. This means that the current velocity of the tsunami does not change with depth, and thus the current caused by the tsunami propagation may interact with the sea bottom. The current velocity immediately above the sea bottom is retarded due to the friction between the moving fluid and the constituting materials of the sea floor (Fig. 3.3). This forms a thin boundary layer of the fluid adjacent to the sea bottom. The boundary layer behaves as a laminar flow as long as the velocity gradient within the boundary layer is small. In the laminar flow, the directions of the current do not change at any point and there are no disturbances within the boundary layer. The laminar flow of a viscous fluid on the sea floor with a finite surface roughness induces a shear stress. The shear stress is a deforming or potentially deforming force that is set up within the boundary layer. The velocity gradient within the boundary layer, dU/dz, and the dynamic viscosity of fluid, η, are related to the shear stress, τ_0.

$$\tau_0 = \eta \frac{dU}{dz} \qquad (3.15)$$

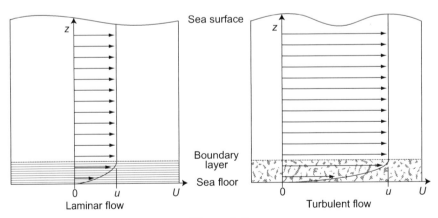

Figure 3.3
Velocity profile of a long wave. Left: The gradient of the current velocity near the sea floor is moderate and the current state of the boundary layer is laminar. Right: The velocity gradient is steep and the boundary layer becomes turbulent.

The shear stress can be comparable to the tractive force, τ, which is a more convenient expression for quantitative evaluation. The following forms are typical expressions of the tractive force:

$$\tau = \rho u^2_{\backslash hbox*}$$ (3.16)

or

$$\tau = \rho C_f U^2$$ (3.17)

Where ρ is the density of water, $u_* \equiv \sqrt{\tau_0/\rho}$ is the friction speed, $\tau_0 = \rho g h I$ is the shear stress of the sea floor, g is the gravity acceleration, h is the water depth, I is the energy gradient, C_f is the coefficient of frictional resistance, and U is the depth-averaged current velocity. Energy gradient I is a function of the current velocity and the surface gradient of the sea. A rough estimation of the tractive force on the deep-sea floor induced by tsunami propagation is possible, according to the depth-averaged current velocity provided by Eq. (3.6). Although the tractive force grows in proportion to the second power of current velocity, its effect on a typical tsunami is possibly minor in the deep sea.

3.1.2 Critical condition of sediment transport

The tractive force (or shear stress) controls the transport and diffusion of sediment on the sea bottom. A major part of the sediment on the sea bottom may resist movement by the tractive force as long as it does not exceed particular thresholds of transport. The resistance to tractive force comes from the gravitational force acting on sedimentary particles, adherence by pore-water pressure, and the electrostatic force of the particles (Friedman et al., 1992). Many analytical and empirical studies have revealed the relationship between current velocity of the boundary layer, particle size, and initial transport of the sediment. The results were compiled into diagrams such as the Hjulström diagram (Fig. 3.4; e.g., Allen, 1986; Friedman et al., 1992; Sengupta, 1994; Yalin, 1977), which illustrate the threshold of particle motion. According to the Hjulström diagram, sedimentary particles of fine sand size (-2φ; 0.25 mm) have a minimum transport velocity (corresponding to the incipient motion of particles) of ~ 2 cm/s and a critical erosion velocity (corresponding to mass transport) of ~ 20 cm/s, while clay- to silt-sized particles have a much higher resistance to currents. The distribution of sediment on the sea bottom is controlled by the physical properties of the particles, the geological configuration of the source on land, and the regime of the oceanic current. Due to a higher terminal falling velocity of the particles (see further), sand-sized sediment commonly covers the shallower sea bottom, such as near the coast or around the mouth of a river. On the other hand, clay-to silt-sized sediment has a lower terminal falling velocity and covers the relatively deeper sea bottom. Owing to the deeper distribution and the larger resistance to tractive forces, it is possible that mud on the deep-sea floor is hard to move even at the time of a

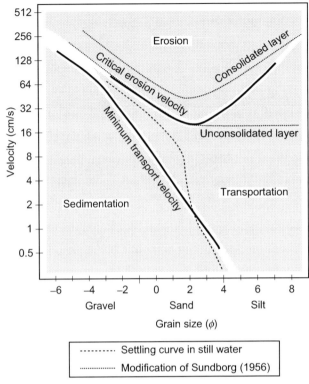

Figure 3.4
The Hjulström diagram from 1935, describing the relationship between the initial motion of sedimentary particles and current velocity, depending on the grain size of the particles. *Modified after Friedman, G.M., Sanders, J.E., Kopaska-Merkel, D.C., 1992. Principles of Sedimentary Deposits. Macmillan Publishing Company, New York, pp. 717.*

tsunami. On the other hand, sand-sized particles are sensitive to transport by currents. They are easily entrained into the current and also settle again easily.

3.1.3 Turbulence and terminal falling velocity

The linear long-wave approximation of tsunamis does no longer apply when a tsunami arrives in shallow coastal areas (approximately <200 m), and Eq. (3.6) is then no longer applicable. The shallow-water long-wave theory (Eqs. 3.9–3.11) should rather be used in this situation. Calculation of the current velocity becomes complicated and has commonly to be treated by numerical modeling. Anyway, the current velocity and tractive force probably grow in response to a decrease in water depth. The boundary layer becomes a turbulent flow, which is characterized by eddies and vortices. The degree of turbulence is expressed by the Reynolds number, which is a dimensionless parameter measuring the ratio of inertial to viscous forces acting in the boundary layer:

$$\Re = \frac{UL}{\nu} \tag{3.18}$$

where U is the mean current velocity, L is the characteristic length of the current such as water depth or particle size, and $\nu = \eta/\rho$ is the kinematic viscosity of the fluid. The transition from laminar to turbulent occurs around $\Re \sim 10{,}000$. Swirls offer upward components to the current velocity. The tractive force eventually exceeds the threshold at a particular place and a major part of the sediment on the sea bottom becomes disturbed and entrained into the turbulent flow. The mechanical processes involved in the movement of the sedimentary particles are (1) lift force on the particles, (2) sudden pressure change due to turbulence or an abrupt topographic change, (3) direct impact of fluid on the particles, and (4) support of the particles by the vertical component of the turbulent flow (Friedman et al., 1992). At the time of tsunami incursion near a coast, the vertical component of the turbulent flow may play an important role in the entrainment and transport of sediment.

Settling of the sedimentary particles is generally driven by gravity (free falling). The free falling is characterized by the terminal falling velocity, which is a physical property of the particles. Each particle has a specific terminal falling velocity, which depends on its diameter, shape, and relative density. In the case of free falling of a single spherical particle in a stationary fluid, the terminal falling velocity, w, is described by the following equation:

$$w = \frac{2}{9} \frac{a^2(\sigma - \rho)g}{\eta} \tag{3.19}$$

Where a is the particle radius, σ and ρ are the particle and fluid densities, respectively, g is the gravity acceleration, and η is the dynamic viscosity of the fluid. Under natural conditions, the shape of the particles, the mutual interference of the particles, the existence of walls, the fluctuations in current velocity and turbulence, and the coefficient of the drag affect the free falling of sedimentary particles and change the terminal falling velocity.

3.1.4 Mode of particle motion: bed load and suspension load

Sedimentary particles transported by a current are commonly classified into two categories that are based on the mode of motion: bed load and suspended load. The bed load is transported in rapid interactions with the sea floor. In contrast, suspended load is totally entrained into and transported within the flow all over its vertical extent.

The various types of motion alternate according to the four aforementioned mechanical processes (in other words, according to variations in the tractive force of the current). If the tractive force is not strong enough, most of the sedimentary particles will be present near the sea floor. When the tractive force becomes stronger, the turbulence of the flow will increase and more particles will be entrained in the flow. Once the current loses its

26　Chapter 3

energy, the particles also lose their kinetic energy and will settle on the bottom due to gravitation. Physical particle properties, such as size, density, and shape also control the motion of the particles. It is possible that larger and heavier particles have a larger resistance to the tractive force than smaller and lighter particles. In the case of tsunamis, especially in near-shore and onshore areas, suspended load may dominate the transport process.

3.1.5 Quantity of sedimentary particles in a tsunami-induced current

The quantitative evaluation of the distribution and concentration of the sedimentary particles in a tsunami is certainly a complicated problem, since tsunami run-ups are highly developed turbulent flows, and the physical properties of the sediment carried by them vary from particle to particle. In addition, instrumental observation of the sediment transport during an actual tsunami run-up is practically impossible. Thus, the study of the sediment transport by tsunamis requires much theoretical, empirical, and numerical analysis. Such studies are still developing and recent results provide numerical reconstructions of sediment transport and deposition by tsunami events (e.g., Asai et al., 1998; Takahashi, 1998).

The results of numerical modeling can be verified by sedimentological studies of tsunami deposits. Such an analysis may yield a quantitative relationship between the distribution and thickness of tsunami layers and the hydrodynamic properties of the responsible tsunamis.

3.2 Characteristics of tsunami deposits

Tsunami sedimentation can take place in both subaqueous and subaerial environments. The subaqueous environment can be divided into two realms: the submarine (sea bottom) and the lacustrine (enclosed basin) environment. The subaerial environment may correspond to the onshore environment, such as coastal flats. The lacustrine environment may occupy an intermediate position between submarine and onshore environments. Since the behavior of tsunamis depends strongly on bathymetric and topographic configurations, tsunami-induced currents result in diverse sedimentation patterns that depend on differences in these environments. Several workers have presented schematic models of subaqueous and subaerial tsunami sedimentation, sometimes, based on immediate observation of recent tsunami events in lacustrine and onshore environments; others described possible sedimentation processes on the basis of the interpretations of ancient tsunami layers in a submarine environment. The overall picture of tsunami sedimentation in the various environments is illustrated in Fig. 3.5.

Here, the sedimentation of tsunamis is described successively for the submarine, lacustrine, and onshore environments.

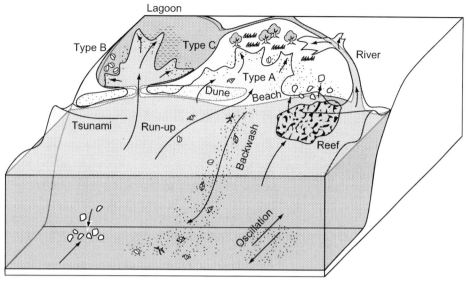

Figure 3.5
Schematic model of tsunami sedimentation. Type A sedimentation: Tsunami run-up transports materials landward from beaches and dunes; meanwhile tsunami backwash brings terrestrial materials seaward. Type B sedimentation: Molluscs are piled on the lake margin and their shells form mounds that are several tens of centimeters high. Type C sedimentation: A large volume of seawater rushes into intertidal ponds through channels and the pond water is whitened transiently with precipitated microcrystalline carbonates. Sand grains brought by the seawater from beaches and dunes accumulate on the bottom of the pond, forming a sand layer that is several centimeters thick. This type of sedimentation also causes a geochemical change in the sedimentary sequence. *Modified after Minoura, K., Nakaya, S., 1991. Traces of tsunami preserved in inter-tidal lacustrine and marsh deposits: some examples from northeast Japan. J. Geol. 99, 265–287.*

3.2.1 Submarine tsunami deposits

Observation of sea bottom changes before and after a tsunami provides data on the threshold of sediment transport. Okamura et al. (2004) presented a detailed before-and-after comparison of the condition of the sea-bottom sediment in a case study of a recent tsunami event. The M8.0 Tokachi-Oki earthquake from September 26, 2003 triggered a tsunami that struck the southern coast of Hokkaido in northern Japan, resulting in average tsunami heights from 1 to 3 m along the southern coast of the island and a maximum tsunami height of ∼4 m at several locations. According to sedimentological analysis of submarine core samples, it was reported that muddy sediment on the surface of the sea bottom at a water depth of 54 m had disappeared after the tsunami. It was also reported that the diatom assemblages from 50 to 80 m water depth had changed significantly, but that there was less notable change in the sedimentary facies.

The numerical simulation of the Tokachi-Oki earthquake tsunami showed that the current velocity was 20–30 cm/s in these areas.

Most submarine tsunami deposits have been identified as an intercalation of typically coarser sedimentary particles within overlying and underlying muddy layers (Albertao and Martins, 1996; Fujiwara et al., 2000; Takashimizu and Masuda, 2000). The sedimentation of the muddy layers implies a low-energy background environment, such as a restricted bay or deep-sea floor. Tsunamis can disturb the prevailing steady state, and high-energy sedimentation may be recorded as an obvious change in the sedimentary facies. Allochthonous material included in the event layer, such as a concentration of mollusc shells and plant fragments, can be interpreted as a result of the severe erosion and transport of coastal sediment into basins and the deep-sea environment, suggesting a tsunami origin (Albertao and Martins, 1996; Fujiwara et al., 2000; Takashimizu and Masuda, 2000; Takayama et al., 2000).

Tsunami deposits from submarine environments have been found in palaeobays and basins at depths of 10–20 m (Fujiwara et al., 2000; Massari and D'Alessandro, 2000) or deeper, to around 200 m (Hassler et al., 2000; Lawton et al., 2005; Shiki and Yamazaki, 1996). In the western Mediterranean Sea, a homogenite that is considered to have formed by the tsunami event of the Thera eruption in the late Bronze Age extends up to 4000 m depth (Cita et al., 1996; Kastens and Cita, 1981). Submarine tsunami deposits associated with the K/T boundary impact event were found on mid to outer shelves and on the slope of the platform edge, ranging from around 100 to 3000 m depth (Albertao and Martins, 1996; Bourgeois et al., 1988; Tada et al., 2002; Takayama et al., 2000). Such depths at which tsunami sedimentation takes place may vary, depending on the magnitude of the tsunami. At the time of extremely large-scale tsunami events, the effect of the tsunami can reach the deep-sea bottom and cause transport of sediment.

The boundary between the tsunami and the underlying layer frequently shows a sharp erosional contact, indicating a high tractive force at the time of sedimentation (Albertao and Martins, 1996; Bourgeois et al., 1988; Fujiwara et al., 2000; Lawton et al., 2005; Massari and D'Alessandro, 2000; Takashimizu and Masuda, 2000). Run-up and backwash, which cause a gradual but significant change in current direction, are most characteristic features of tsunamis in coastal areas. Cross-stratification (cross-bedding or lamination), low-angle wedge-shaped lamination, and imbrication of large-sized sedimentary particles such as mollusc shells are considered to be evidence of the relatively long-period oscillation of the current directions of tsunamis (e.g., Fujiwara et al., 2000; Hassler et al., 2000; Lawton et al., 2005; Massari and D'Alessandro, 2000; Tada et al., 2002; Takashimizu and Masuda, 2000). The repetition of waves and the gradual decrease in current velocity can be reflected in graded or laminated bedding (e.g., Bourgeois et al., 1988; Fujiwara et al., 2000; Massari and D'Alessandro, 2000; Takashimizu and

Masuda, 2000) and in thinner sublayers (Lawton et al., 2005; Tada et al., 2002). Antidunes and parallel lamination in tsunami layers, corresponding to an upper-flow regime, are interpreted as a predominance of a particular current direction such as run-up or backwash (Bourgeois et al., 1988; Fujiwara et al., 2000, 2003; Tada et al., 2002).

The suspension of sedimentary particles dominates the transport process at the time of a tsunami event. Free falling of sedimentary particles from suspension may form a fining-upward sequence in a tsunami layer (Albertao and Martins, 1996; Bourgeois et al., 1988; Fujiwara et al., 2000; Tada et al., 2002; Takashimizu and Masuda, 2000; Takayama et al., 2000). Tsunami layers are often identified as a stack of normal and reversed graded layers, showing a complex variation of the particles in grain size (like saw teeth), which may reflect the variation in current velocity and the transport process (Fujiwara et al., 2003). Clay- and silt-sized particles suspended within the water may begin to settle down during a relatively calm interval between wave trains. Thus, mud layers often cover the previously deposited coarser sediment (Takashimizu and Masuda, 2000).

Submarine tsunami deposits typically show evidence of significant hydrodynamic energy and oscillations of run-up and backwash. Observation and measurement of the behavior and sedimentation process of tsunamis on the sea floor are difficult, and there are few examples of modern tsunami events. Therefore, the knowledge of submarine tsunami sedimentation is still limited, and the accuracy of the interpretations of ancient submarine tsunami deposits remains uncertain. Future developments in tsunamiite research may change the current interpretations of submarine tsunamiites.

3.2.2 Lacustrine tsunami deposits

Lacustrine environments offer a much better opportunity for observing the sedimentation process of actual tsunamis than submarine environments. Actual descriptions of tsunami sedimentation, which are beyond controversy, are possible. Minoura and Nakaya (1991) directly observed tsunami sedimentation in intertidal lagoons, when the 1983 Japan Sea earthquake triggered a large-scale tsunami, which struck the western coast of north-east Japan with a maximum tsunami height of up to 15 m. They reported in detail the geological effect of the earthquake and tsunami and described the sedimentation processes that had taken place in the lagoons (Type B and C sedimentation in Fig. 3.5).

It seems that some parts of the sedimentation process are comparable to sedimentation in the submarine environment. A significant difference is the chemical interaction between sea and fresh water.

Characteristic sand layers in intertidal lakes and lagoons have been investigated worldwide in coastal areas and identified as traces of historical or prehistorical tsunami incursions. A sand layer of more than 1 m thickness in shallow coastal lagoons on the Scilly Isles has been correlated with the incursion of the 1755 Lisbon earthquake tsunami

(Foster et al., 1993). Distinctive deposits in Norwegian coastal lakes are interpreted as evidence of a tsunami triggered by the second Storegga slide ~7000 years BP (Bondevik et al., 1997). Periodical occurrences of historical earthquake tsunamis have been correlated with sand layers in lakes on the Pacific coast of central Japan (Hirose et al., 2002; Nanayama et al., 2002; Okahashi et al., 2002; Okamura et al., 2000; Tsuji et al., 2001, 2002) and north-east Japan (Minoura et al., 1994). Bottom sediment composed of graded pebbly sand was described as a record of earthquake-induced tsunamis at Owens Lake, south-east California (Smoot et al., 2000).

Tsunami deposits in lacustrine environments are commonly identified as a distinct sand layer intercalated within silty to clayey sediment, which suggests a low-energy steady state of the background sedimentary environments. The sedimentological features of the sand layer represent the variation in hydraulic profile of tsunamis such as tsunami deposits from other environments. Some examples show erosion of muddy sediment just beneath the sand layer, suggesting the erosion of the lake bottom by traction flow during the tsunami incursion (Bondevik et al., 1997), while the fine material reflects free falling of sediment under relatively calm conditions (Minoura et al., 1994). The sand layer shows a lateral decrease in thickness and grain size (Bondevik et al., 1997; Foster et al., 1993), which might indicate the direction of sediment transport such as a landward flow during tsunami run-up. The sand layer generally fines upward, suggesting a gradual decrease in tsunami energy (Bondevik et al., 1997). The occasionally present couplets of sand and silt layers can be interpreted as evidence of multiple tsunami incursions (Nanayama et al., 2002).

Traces of tsunamis in coastal lakes and lagoons are expected to have a high preservation potential because of the basically calm and stable sedimentation environment. On other hand, many coastal flats are the fundamental areas of erosion and the major part of tsunami deposits is easily disturbed and removed due to the effect of wind, precipitation, vegetation, and other processes. This means that there is a much greater chance of discovering tsunami deposits in coastal lakes and lagoons than in coastal flats. It is possible to obtain evidence of historical and older tsunami incursions from borehole sampling in coastal lakes and lagoons. Of course, the exact identification of a tsunami event depends on accurate stratigraphic observation and age determination. However, not all coastal lakes are suitable for this kind of investigation. It is suggested that smaller ponds are the most appropriate for core sampling of tsunami deposits, since larger lakes are often important places for human activities, and the bottoms of the lakes are likely disturbed by artificial topographic modifications (Tsuji et al., 2002). It is also pointed out that not all tsunami events form a characteristic sand layer in lakes and lagoons, and a sand layer may also result from another event. For example, a fine sand layer in the sedimentary succession of coastal lagoons can, according to Foster et al. (1993), have a wind-blown origin from Holocene coastal sand dunes. Some possible tsunami events are not recorded by sand layers, and on the other hand, the sand layers are often associated with ambiguous tsunami events.

3.2.3 Onshore tsunami deposits

Tsunami deposits on the Earth's surface, such as on coastal flats and marshes, have been well investigated since the 1960s, because coastal flats are best accessible for investigating the effect of tsunamis, and accounts by eye witnesses and data from instrumental measurement in coastal flats support the investigation of tsunami sedimentation. Some examples from studies of onshore tsunami deposits by observed tsunami events are described in this section.

3. The 1960 Chilean earthquake tsunami inundated the coastal plain of south-central Chile with run-up heights of 5–10 m; it caused widespread erosion and deposition. The tsunami breached a sandpit at the mouth of a river and a veneer of sand of 6–30 cm thickness covered the alluvial soil (Wright and Mella, 1963). According to Bourgeois and Reinhart (1989), the tsunami deposit was widely distributed over the supratidal surfaces. It thinned from more than 1 m to 0–2 cm over a distance of 0.6 km and fined dramatically away from the sediment source, from gravel to predominantly fine sand.

4. The 1960 Chilean earthquake tsunami also struck the north-eastern coast of Japan with a wave height of up to 5 m. Kon'no (1961) identified tsunami deposits at many coastal localities in north-east Japan, including some deposits more than 100 m inland from the beach ridge. The deposits were composed mostly of graded beds of silt and fine sand. The maximum thickness of the tsunami deposit was 25 cm.

5. The 1983 Japan Sea earthquake triggered a large-scale tsunami with a maximum height of ~15 m on the north-east coast of the Japan Sea. Large quantities of beach and dune sands were eroded by run-up and backwash of the tsunami and were finally deposited inland and on the shore face. This type of sedimentation formed sand sheets within the coastal sedimentary succession and also modified the topography of the coastal areas (Type A sedimentation in Fig. 3.5; Minoura and Nakaya, 1991).

6. Deposits from Flores Island by the 1992 Flores (Indonesia) earthquake tsunami were identified as continuous or discontinuous sheets of sediment composed of medium gray sand near the coast and of fine gray sand farther inland. The deposits range in thickness from 0.01 to 0.5 m (Shi et al., 1995).

7. The 1993 south-west Hokkaido earthquake tsunami deposited sand sheets on gentle and steep slopes and on flat fields. The deposits consisted of beach sand and had similar characteristics as those of the Flores earthquake tsunami described earlier (wide distribution along the coast, variation in thickness and mean grain size, and graded bedding) (Nanayama et al., 2000; Nishimura and Miyaji, 1995; Sato et al., 1995).

8. Minoura et al. (1996) discovered a deposit in the Ust'-Kamchatka area from the 1923 Kamchatka earthquake tsunami. The tsunami deposit consisted of a thin (2–3 cm) layer of well-sorted coarse and fine sand, which was traceable for 3 km or more inland from the present coastline.

32   Chapter 3

Many historic and prehistoric onshore tsunami deposits show similar characteristics. Both are generally composed of sand layers (Atwater, 1987; Dawson et al., 1988; Minoura et al., 2000). Vertical grain-size variation within the sand layer sometimes forms as graded bedding (Dawson et al., 1988, 1991). Some ancient tsunami deposits cover extensive coastal plains (Long et al., 1989, 1990) and can be traced up to several kilometers in one case (Minoura and Nakaya, 1991). Their thicknesses typically decrease inland (Atwater, 1987), so that they form a gradually tapering wedge (Dawson et al., 1988); and the mean grain size of these deposits also decreases with distance from the sea (Benson et al., 1997).

The earlier descriptions do not include all characteristics of onshore tsunami deposits. The aforementioned are considered "typical" characteristics of tsunami deposits. In fact, the variation is quite wide, and the diversity of onshore tsunami deposits is described in Section 3.3.

3.3 A review of onshore tsunami sedimentation

Tsunami deposits have a quite wide variation in their characteristics, including distribution, grain-size distribution, and sedimentary structures. The hydraulic profile of a tsunami, local bathymetry and topography, sources of sediment, and transport processes are jointly responsible for the diversity in tsunami deposits. Here, we review in greater detail onshore tsunami sedimentation based mostly on the results of studies on observed tsunami events. Hydraulic explanations of the sedimentation are provided as far as reasonably possible.

3.3.1 Sedimentological feature associated with tsunami run-up

Evidence of a tsunami incursion appears commonly as extensive coverage of coastal flats by sand layers. The tsunami layer typically covers coastal flats to several hundreds of meters from the coastline, and it often reaches even several kilometers (Minoura et al., 1996). The thickness of the tsunami layer varies from a few millimeters to several tens of centimeters. The distribution of the tsunami layer is either continuous or discontinuous, depending sensitively on the local topography.

The distribution of onshore tsunami deposits is not confined to broad coastal flats, and neither is the geometry of tsunami deposits confined to a layer of sand. For example, a dome-shaped accumulation of marine sands of more than 10 m high exists at the head of Iruma Bay, on the Pacific coast of central Japan (Fig. 3.6). The formation of this sand mound is ascribed to the incursion of the 1854 Ansei-Tokai earthquake tsunami. This peculiar sediment form is interpreted to be a result of an extreme amplification of waves due to harbor resonance within the V-shaped bay.

Figure 3.6
The 1854 Ansei-Tokai earthquake tsunami completely destroyed the settlement of Iruma and deposited a huge sand dome. Afterward, the survivors rebuilt their houses on the dome. The height of the dome reaches 16 m and the total volume is estimated to be 700,000 m^3 (Asai et al., 1998). (A) Bird's-eye photograph of Iruma. (B) Photograph of Iruma taken from the ocean side.

As the run-up of tsunami spreads landward and exhausts the kinetic energy, the current velocity and inundation depth decrease. The change can be gradual or abrupt, depending on local topographic configurations and the specific characteristics of the tsunami. Entrainment and suspension of sedimentary particles in the current are dominated by tractive force, while the physical properties of the particles offer resistance to the tractive force. The vertical component of the turbulence and the settling velocity of the particles, which is determined largely by their size, shape, and density, may control the deposition of the particles from suspension. The decrease in current velocity reduces the degree of turbulence and the magnitude of the tractive force, resulting in the settling of the particles. As for the transport of sediment, the quantity of particles within a current is proportional to the tractive force and the volume of water. The transport of coarser or heavier particles may require much support of the tractive force; on the other hand, the transport of finer or lighter particles may require less support of the tractive force. Thus, thickness and particle composition gradually change within a tsunami layer. If the change in current energy is gradually landward, a progressive decrease in thickness and particle-size distribution of the tsunami layer is found. Many tsunami deposits on coastal flats show a landward decrease in thickness, forming wedges (Benson et al., 1997; Gelfenbaum and Jaffe, 2003; Goff et al., 2004; Minoura et al., 1997; Nanayama et al., 2000; Nishimura and Miyaji, 1995; Sato et al., 1995), although some examples show a relatively constant thickness of tsunami deposits over their entire extent (Benson et al., 1997; Gelfenbaum and Jaffe, 2003; Minoura et al., 1996). Typical onshore tsunami deposits show a landward fining trend in particle size (Minoura et al., 1996, 1997; Nishimura and Miyaji, 1995; Shi et al., 1995), while other examples show a fairly complex trend of landward fining (Gelfenbaum and Jaffe, 2003). Note that the landward thinning and fining do not always appear simultaneously. Fig. 3.7 shows the results of a grain-size analysis of the tsunami deposit by the 1923 Kamchatka earthquake tsunami (Minoura et al., 1996). It is obvious that the mean diameter of the sedimentary particles gradually decreases landward. The lateral changes in thickness and particle size of the tsunami layer are good indicators of the direction of the current. Therefore, the landward thinning and fining of a tsunami layer are the major criterion for the identification of a tsunami run-up.

The particle size of a tsunami layer often decreases upward, for example, from coarse sand at the base to fine sand at the top (Benson et al., 1997; Gelfenbaum and Jaffe, 2003; Shi et al., 1995). This is considered to be a result of the gradual settling from suspension. At the time of sediment transport, coarser particles likely exist near the bottom, while finer ones can be distributed throughout the water column. A structureless (massive or homogeneous) tsunami layer may reflect the severe mixing of sediment and water due to high turbulence and rapid deposition.

According to data from field investigations of recent tsunami events, the typical current velocity of tsunami run-up near the coast is around several meters per second, and the inundation depth is several tens to hundreds of centimeters. Such currents are obviously beyond the threshold of sediment transport (Fig. 3.4) and have sufficient strength for the removal of the material from the preexisting surface. The boundary between a tsunamiite and the underlying layer is frequently (or typically) erosional (Benson et al., 1997; Clague and Bobrowsky, 1994; Gelfenbaum and Jaffe, 2003; Shi et al., 1995). Ingredients in the

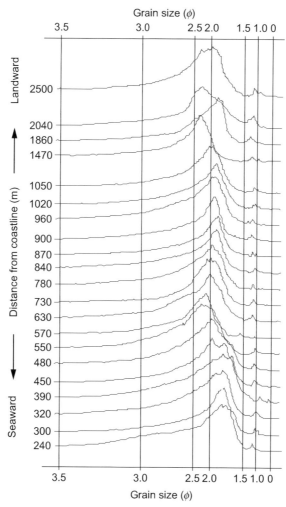

Figure 3.7
Landward fining of the tsunami deposit by the 1923 Kamchatka earthquake tsunami. *Modified after Minoura, K., Gusiakov, V.G., Kurbatov, A., Takeuti, S., Svendsen, J.I., Bondevik, S., Oda, T., 1996. Tsunami sedimentation associated with the 1923 Kamchatka earthquake. Sediment. Geol. 106, 145–154.*

tsunami layer from local soil, such as rip-up clasts, can be interpreted as being a result of considerable erosion (Gelfenbaum and Jaffe, 2003; Minoura et al., 1997).

3.3.2 Source and particle composition of tsunami deposits

The source of a tsunami deposit is associated with both the local/onshore area and the shallow sea bottom, depending on the magnitude of the tsunami, the local topography, and the distribution of sediment in the coastal areas. Some examples suggest that a major part of the tsunami deposits is derived from near-offshore areas (Gelfenbaum and Jaffe, 2003;

36   Chapter 3

Nanayama et al., 2000). Other examples show that tsunami deposits are largely derived from the nearest source, such as the local coastal substratum, without transport over a long distance (Sato et al., 1995; Shi et al., 1995). According to the results of Gelfenbaum and Jaffe (2003), the volume of deposition is twice the volume of erosion in near-shore areas, while Sato et al. (1995) suggested that the volume of erosion could account for the volume of tsunami deposits on land.

Tsunamis transport the materials that build up the sea bottom or the coastal surface area, including both clastic and biogenic sedimentary particles, and then accumulate them all as a tsunami deposit. Thus, fossils found in tsunami deposits such as coral blocks, mollusc shells, benthic foraminifers, and diatoms provide effective data for reconstruction of the source of a tsunami deposit (Benson et al., 1997; Clague and Bobrowsky, 1994; Minoura et al., 1996, 1997). Species of these fossils are associated with particular environments such as a freshwater, brackish, or marine with a specific habitat depth. Kon'no (1961) identified several kinds of marine microfossils in the deposit left by the 1960 Chilean earthquake tsunami and clarified that the sediment derived from the sea floor in near-shore areas. He concluded that the tsunami deposits did not originate in the deep sea. In the case of the 1993 south-west Hokkaido earthquake tsunami, Nanayama et al. (1998, 2000) suggested that the origin of the tsunami deposit was the sea floor between 50 and 60 m depth, based on a palaeontological analysis of the foraminifers. The depths of sediment sources can obviously vary, depending on the magnitude of the tsunami, the local bathymetry and topography, and the distribution of sediment.

The Hjulström diagram (Fig. 3.4) shows that sand-sized particles are the most sensitive to changes in the properties of the current. This is why onshore tsunami deposits commonly consist of sand with a low mud content. Shi et al. (1995) noted that most clay- and silt-sized particles are carried back into the sea by the backwash. Two sorting processes control the particle composition of tsunami deposits: (1) clay- and silt-sized particles have a higher resistance to a tractive force than sand-sized particles, and the finer particles cover the relatively deeper bottom, so that the finer particles are less likely to be entrained in the current; (2) clay- and silt-sized particles have a lower settling velocity, so that they easily stay in suspension. The backwash of a tsunami may return almost all of the finer particles to the sea. As a result of these sorting processes, the mud content of a typical tsunami deposit is very low.

The sediment source and characteristics of a tsunami control the components of a tsunami deposit. Although most examples are from historic or prehistoric events, onshore tsunami deposits often comprise gravel- to boulder-sized particles (Kato, 1987; Kato et al., 1988; Moore, 2000; Moore et al., 1994; Nott, 2000, 2004). Recently, an extensive accumulation

Figure 3.8
Tsunami boulders at Laem Pakarang, south-west Thailand. The boulders were transported and deposited by the December 26, 2004 Indian Ocean tsunami and are composed of coral blocks that have a corresponding habitat depth of around 10 m. The scale in the photograph B indicates 1 m.

38 Chapter 3

of coral boulders in intertidal zones was observed after the 2004 Indian Ocean tsunami (Fig. 3.8). The coral boulders cover a large part of the intertidal zone, about 600 m along the coastline and 400 m from the coastline at high-tide level. The tsunami-transported boulders reach sizes of several meters and weights of up to 100 tonnes (Nott, 2004). A landward fining trend appears even in tsunami deposits composed of gravels or, in lithified form, conglomerates (Moore, 2000).

The frequency curves of the grain-size particles in tsunami deposits show relatively well to poorly sorted distributions. According to Kon'no (1961), particle sorting in a tsunami deposit is generally poor. Some tsunami deposits consist of well-sorted sands (e.g., Minoura et al., 1996, 1997; Nanayama et al., 2000; Sato et al., 1995), and others vary in their degree of sorting (Shi et al., 1995). The distribution of the particle sizes of tsunami deposits can be unimodal or multimodal, but distributions without distinct modal peaks (multimodal distribution) also occur (Minoura et al., 1996; Shi et al., 1995). The multimodal grain-size distribution in tsunami deposits is considered to reflect the nature of the source material and the processes of tsunami sedimentation (Shi et al., 1995). For example, a bimodal-grain size distribution of a tsunami deposit is attributed to the activity of two different transport processes within a single flow, that is, suspended load and bed load (Minoura et al., 1996).

3.3.3 Sedimentological features associated with tsunami backwash

At its final phase of a run-up, a tsunami uses almost all of its kinetic energy, which is converted into the potential energy of the water mass. The potential energy is immediately reconverted into kinetic energy; thus, a backwash is induced. The backwash of a tsunami erodes the surface, including previously formed tsunami layers, local soil, and other material and transports them during backwash. The direction of the backwash is not always opposite to that of the run-up, as the backwash often concentrates on topographic depressions, such as channels and rivers, causing severe erosion (Kon'no, 1961). For example, at the time of the 1993 south-west Hokkaido earthquake tsunami, the concentrated backwash was responsible for erosion of about 1500 m^3 of gravel in coastal terraces, resulting in a bowl-shaped depression of 0.5−1.5 m depth (Sato et al., 1995).

Intensity of tsunami backwash varies according to the topography of the current path. This results occasionally in the stagnation of water, thus some of the material transported by the backwash is deposited on the surface. Material redeposited by the backwash forms thus also a component of a tsunami deposit, commonly, a poorly sorted sandy mud (Sato et al., 1995). As the grain size of run-up deposits directly reflects the granulometry of the source area, so that these deposits show poor sorting (ranging from fine sand to large cobbles), whereas tsunami deposits formed by backwash are commonly composed of finer particles such as sandy mud. The sand layer by run-up and the muddy layer by backwash make a couplet. This couplet represents the entire tsunami incursion (Benson et al., 1997).

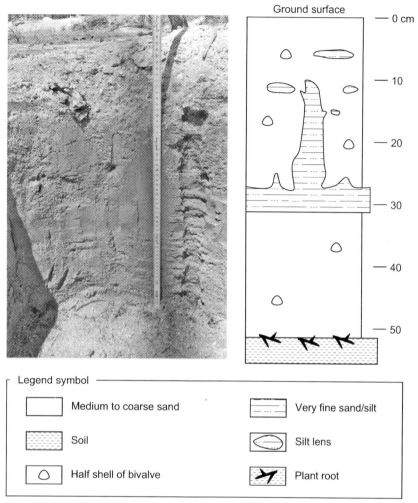

Figure 3.9
Sedimentological section of the tsunami deposit at Bang Sak, south-west Thailand. The December 26, 2004 Indian Ocean tsunami deposited 50–60 cm of medium to coarse beach sand with an intercalation of silt layer. The upper boundary of the silt layer shows significant liquefaction.

Fig. 3.9 shows a stratigraphical section of a tsunami deposit formed by the 2004 Indian Ocean tsunami. A couplet of a sand and silt layer can be identified in the sequence. Sand layers by tsunami run-up are often covered by a mud cap, which is associated with the stagnation of water under relatively calm conditions (Gelfenbaum and Jaffe, 2003; Sato et al., 1995).

Bent plants indicate the direction of the backwash of a tsunami (Nishimura and Miyaji, 1995). The direction of the current can also be reconstructed from the imbrication of

40   Chapter 3

large-sized sedimentary particles such as cobbles (Nanayama et al., 2000; Nishimura and Miyaji, 1995; Sato et al., 1995).

3.3.4 Sedimentological features associated with repetition and waning of waves

The presence of successive tsunami layers with fining-upward sequences is evidence of the repetition of tsunami incursions. According to the investigation immediately after the 1960 Chilean earthquake tsunami in north-east Japan, the tsunami deposited, indeed, graded sand layers. The sand units are interpreted as having been deposited by multiple run-ups by successive waves, and the observed period of the waves suggests that the beds were deposited with 30—40 min intervals (Kon'no, 1961).

The hydraulic energy of a tsunami generally decreases with each new incursion; thus, the thickness and the mean particle diameter of the sandy run-up deposits decrease upward, resulting in multiple fining-upward sequences (e.g., Gelfenbaum and Jaffe, 2003; Nishimura and Miyaji, 1995; Shi et al., 1995). The repetition and waning of wave successions are well illustrated by the saw-toothed vertical fluctuation in particle-size distribution within a tsunami layer (Shi et al., 1995). Note that the number of graded layers does not always correspond to the actual number of tsunami incursions. The run-up of a wave does not always leave sediment, and some of the tsunami deposits are reworked just after their sedimentation, due to the following backwash or run-up (Gelfenbaum and Jaffe, 2003).

3.4 The occurrences of tsunamis and tsunamiites

Studies on tsunami deposits have long focused on the identification of ancient tsunami events. Previous studies have mapped worldwide numerous tsunami deposits in coastal sedimentary successions (Fig. 3.2), and many of them have been correlated with historic and prehistoric tsunami events. In recent years, several issues have been discussed from the viewpoint of the further development of tsunami sedimentology and of the contribution to the risk assessment of tsunami disasters. For example, Bourgeois and Minoura (1997) identified three key problems in tsunami deposit study: (1) positive identification of tsunami deposits, (2) dating and correlation of events, and (3) quantification of palaeotsunami events.

The first two key problems deal with the reconstruction of the history of tsunami events. Incursions of tsunamis do not always form tsunami deposits. A negative identification of a tsunami deposit does not mean the nonoccurrence of a tsunami. In addition, the geographical extent of a tsunami deposit is frequently reduced due to erosion. Thus, an inundation map of a tsunami event based on tsunami deposits probably indicates the minimum area of the inundation.

The distinction between tsunami deposits and deposits from other agents is one of the most important subjects. Distinguishing between tsunami deposits (tsunamiites) and storm deposits (tempestites) is particularly difficult. Accurate recognition of a tsunami deposit assures the reliability of the history of tsunami events. If an event deposit is successfully identified as a tsunami deposit, the study should subsequently focus on the age and correlation with the cause (trigger event). There are insufficient adequate sampling methods and measurement techniques, however, so that dating of a tsunami deposit commonly poses a problem. Tsunami deposits often miss a recognizable corresponding cause, and the determined age of the tsunami event frequently is inconsistent with the age of the possible trigger event provided by historical records.

The third key problem (quantification of palaeotsunami events) is associated with the problem of estimating of the magnitude and height of ancient tsunamis. Since the characteristics of a tsunami are thought to control the extent of sediment transport, the distribution and thickness of tsunami deposits are expected to be proportional with the magnitude and height of the tsunami. This may directly contribute to the risk-assessment programs of future tsunami disasters. In this context, problems concerning the identification and the quantification of tsunami are discussed.

3.4.1 Distinction between tsunami and storm deposits

A review of the sedimentological characteristics of tsunami deposits illustrates that the characteristics of tsunami deposits such as extent, thickness, grain-size distribution, material, and stratification are fairly variable, as they depend on the properties of the tsunami and the source of the sediment. It seems very difficult to identify unobserved tsunami deposits based only on grain-size distribution or sedimentary structures at a single site (Sato et al., 1995). Tsunami deposits are not identifiable on the basis of exclusive and diagnostic criteria because other kinds of deposits share some of their characteristics (Pinegina and Bourgeois, 2001). Storm surges, gravity flows, tidal waves, and sea-level changes are possible alternatives. Particularly storm surges are a major candidate for an alternative interpretation because of the area that they affect and the order of wave energy can be comparable with those of tsunamis and this holds also for the resulting deposits.

Distinguishing between tsunami and storm deposits has therefore been a study object for a long time already (see, among others, Dawson et al., 1991). Dawson and Shi (2000) presented a review on this problem. Since there are relatively only few examples available for comparison between tsunami and storm deposits, interpretations are often still controversial. For example, Kon'no (1961) noted that there is no apparent difference between tsunami and storm sedimentation, except for the more extensive distribution of the deposits in the affected areas. Dawson et al. (1991) suggested that erosion is more significant than deposition during storms and suggested that there is a low probability of deposition by storms. The behavior of storm surges may vary from event to event, and it is affected by local bathymetric and topographic configurations. This complicates the problem. Differentiation between tsunami and storm deposits is therefore often performed on the basis of the "typical" characteristics of tsunami deposits. Some studies excluded storm as a cause for the deposition of specific layers because the recurrence intervals of major historical storm surges in the study area were not consistent with the occurrence of the layers under study.

Recently, onshore tsunami and storm sediments have been compared on the basis of both observed and unobserved events. In the case study of the 1993 south-west Hokkaido earthquake tsunami and the 1959 Miyakojima typhoon in northern Japan, Nanayama et al. (2000) showed that the tsunami deposits were composed of four layers of sand by landward- and seaward-directed flow; the storm deposits, on the other hand, comprised only one layer from a flow that was exclusively landward. They reported that the tsunami and storm deposits were similar in that their thickness decreased landward in both cases. According to a palaeontological analysis, the tsunami deposits were derived from the sea bottom at the depths of 50—60 m, whereas the storm deposits were composed of well-sorted marine sand derived from coastal areas. Their grain-size analyses illustrated that the storm deposits were better sorted than any of the tsunami deposits.

Goff et al. (2004) reported significant differences between ancient tsunami and modern storm deposits in New Zealand. The storm deposits are not extensive and show no relationship between the inundation distance and their thickness. They noted that the sorting of the storm deposit was better than that of the tsunami deposit and that the mean grain size of the storm deposit was larger than that of the tsunami deposit.

The widespread deposition of sand by a major tropical cyclone has been described by Nott (2004). Tropical cyclone Vance in 1999 left a sheet of sand up to 1—2 m thickness. The storm deposit was composed of tabular medium- to coarse-grained cross-bedded sands, shells of various sizes, and coral fragments. Most remarkable is that the sand sheet extends inland for ~300 m. The major storm did not affect a boulder accumulation, which is interpreted as having been deposited by prehistoric tsunamis.

There is no doubt that major storms are capable of forming sand layers that somewhat resemble tsunami deposits. The wave energy of major storm surges seems to be locally comparable to tsunamis. Although the distinction between tsunami and storm deposits is difficult, the characteristics of tsunamis offer possible criteria for the identification. Tsunamis are characterized by long-wave properties. This is responsible for the relatively long period of wave oscillations with strong, enduring currents around coastal areas. It results in repetitive run-up and backwash, as well as in larger run-up distances and heights. On the other hand, storm surges do not have long-wave properties; thus, the periodicity of wave repetition is shorter and the run-up distance and height are smaller

(Nott, 2004; Pinegina and Bourgeois, 2001). Differences in hydrodynamic characteristics between tsunamis and storms are important factors in the distinction between tsunamiites and tempestites.

3.4.2 Quantitative evaluation of ancient tsunamis

Several authors have recently mentioned the possibility of a quantitative evaluation of tsunami events from the viewpoint of risk assessment (e.g., Bourgeois and Minoura, 1997; Pinegina and Bourgeois, 2001). The magnitude and the height of a tsunami, for example, are objective, effective data for evaluating unobserved tsunamis. The ages of ancient tsunamis provide the recurrence intervals of tsunamis and the magnitude or the height of the tsunami allows evaluation of the possible impact.

The scale of a tsunami, which is frequently represented by its vertical run-up, is an important factor controlling the thickness and distribution of tsunami deposits (Sato et al., 1995). Local topography and the sediment source may also affect the thickness of tsunami deposits. There is no clear relationship between the tsunami height and the thickness of the tsunami deposit. In the case of the 1993 south-west Hokkaido earthquake tsunami in northern Japan, the tsunami height of 10–30 m corresponded to a thin sand layer with a thickness of 2–3 cm at a certain place, whereas a 20-cm-thick tsunami deposit at another site corresponded to a vertical run-up of 15 m (Sato et al., 1995). Gelfenbaum and Jaffe (2003) reported, however, that the thickness of tsunami deposits can be related to along-shore tsunami heights on a regional scale.

The second point to consider is the relationship between the run-up height or distance and the highest altitude (or landward limit) of a tsunami deposit. In the case of the 1993 south-west Hokkaido earthquake tsunami, the maximum run-up height was about 1 m higher than that of the deposits (Nishimura and Miyaji, 1995). Gelfenbaum and Jaffe (2003) reported that the limit of tsunami inundation (720 m) is about 40 m further landward than the limit of the tsunami deposit (680 m). It seems that these results vary from case to case, probably depending on the topographic setting and the profile of the tsunami. Since tsunami sedimentation is a relatively rare event, the number of such field measurements is still small. Therefore, it is currently difficult to provide a clear relationship between the tsunami height and the geometry of the tsunami layer based only on field results.

The earlier review of sediment transport implies that the extent and thickness of a tsunami deposit depend not only on tsunami or run-up height but also on inertial force, loading capacity, surface gradient and roughness, and other factors. We therefore give an example of a hydraulic experiment demonstrating the quantitative relationship between the character of a tsunami and the depositional pattern.

44 Chapter 3

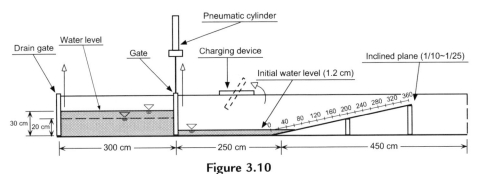

Figure 3.10
Setup of the experimental channel.

Fig. 3.10 shows the configuration of the experiment. We performed one-dimensional channel experiments regarding tsunami run-up and sedimentation on an artificial substratum (inclined plane). The aim of the experiment was to establish a quantitative relationship between the hydraulic quantities of the tsunami and the geometry of the tsunami deposit. The experiment was designed to simulate the transport of sediment as suspension load. We examined in detail the influence of the backwash for analysis of tsunami sedimentation. By adjusting the water level of the tank, we changed the magnitude of the tsunami and repeated the experiment three or five times for each setting. Fig. 3.11 shows the distribution of the sediment weight over the inclined plane under three different settings. The dash line marked by the squares indicates the weight of the sediment deposited by the run-up and backwash (weight A). The solid line marked by the circles indicates the weight deposited only by run-up (weight B). The weight of sediment reworked by backwash can be calculated by subtracting weight A from weight B. The hatched area in Fig. 3.11 indicates the weight of the reworked sediment. It seems that the total weight of the sediment clearly depends on the magnitude of the tsunami and this is especially clear under conditions with no backwash. The backwash induced downward transport of sediment on the inclined plane and deposited it on the lower part of the plane. The reworked sediment covered approximately the lower one-third of the inundated area, and a major part of the reworked sediment flowed out of the inclined plane. The result of the experiment shows the formation of the tsunami layer by backwash (Sato et al., 1995).

Fig. 3.12 shows the relationship between the water flux and the weight of the sample sediment. The water flux is defined as the product of inundation depth, width of the channel, and depth-averaged current velocity. The current velocity reached up to 1.5 m/s in this experiment. The scale of the weight of the sample sediment in the figure is converted into kilograms per square meter. The result indicates a clear and positive relationship

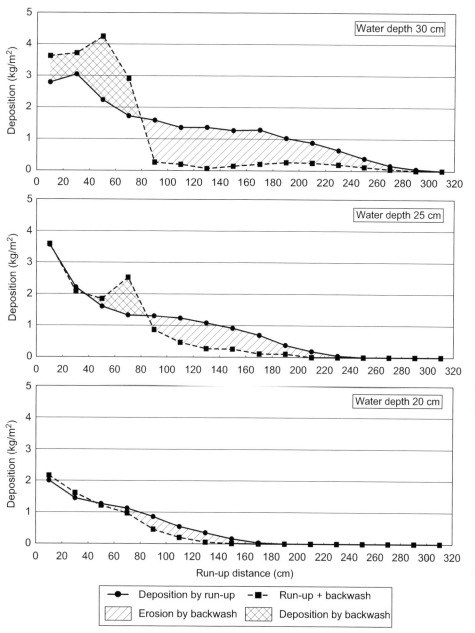

Figure 3.11
Deposition pattern of the samples formed under six different experimental conditions. The impermeable surface was used with a fixed inclination of 1/10. The water depth of the tank was set to 30, 25, and 20 cm in order to change the strength of the current. The deposition pattern shows landward thinning ("wedges"), which is also found for typical onshore tsunami deposits.

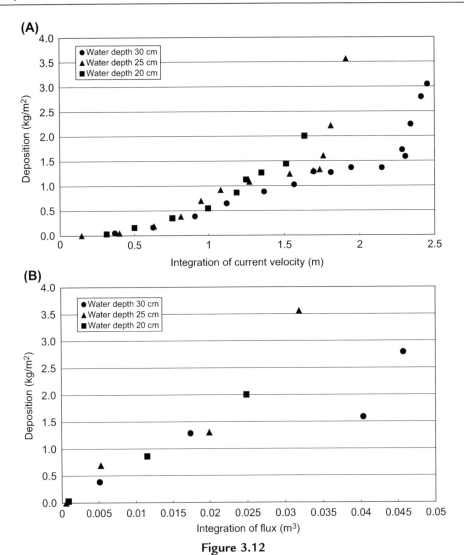

Figure 3.12
Experimental results. (A) Relationship between deposition and current velocity. (B) Relationship between deposition and water flux. The result implies a linear increase in deposition with a corresponding range of the integrated flux from 0 to 0.025 m^3.

between the water flux and the weight of the sample sediment. The relationship is expressed in Fig. 3.12. For example, when 0.01 m^3 water passes through a certain part of the inclined plane, the weight of the deposited sample sediment will be up to ∼0.5 kg/m^2. We believe that such a quantitative evaluation of ancient tsunamis is reliable as long as the configuration of the experiment is appropriate.

4. Concluding remarks

The nature of tsunamis is responsible for the diversity of tsunami deposits. The propagation of tsunamis takes place over vast areas of the ocean from deep sea to land, and the behavior of tsunamis varies intricately in response to bottom and coastal topography. Thus, tsunamis induce various types of sedimentation in submarine, lacustrine, and onshore environments. Previous studies on tsunami sedimentology have documented the characteristics of numerous tsunami deposits and have illustrated their diversity. The hydrodynamic characteristics of a tsunami play an important role in the variation of tsunami sedimentation. Therefore, tsunami deposits should be identified based on the careful consideration of the hydrodynamics and hydraulics of a tsunami and on the local topographic configuration.

A major purpose of tsunami sedimentology is to identify historic and prehistoric tsunami events and to reconstruct these tsunami incursions. Ancient geological events can be interpreted from tsunami deposits. The development of coastal topography can also be understood from the viewpoint of tsunami sedimentation. Estimation of the recurrence intervals and magnitudes of ancient tsunamis is another important purpose. It is suggested that the subject be addressed from the standpoint of risk assessment and disaster prevention. Such aspects may increase the significance of tsunami sedimentology.

The key problems in tsunami sedimentology are the identification of tsunami deposits, the dating and correlation of events, and the quantification of ancient tsunami events, as Bourgeois and Minoura (1997) pointed out. Drawing up criteria for identification is a particularly important issue. Each problem can be expected to be resolved only if close cooperation among the various fields of research is realized, such as sedimentology, palaeontology, archaeology, geomorphology, hydrodynamics, and hydraulics.

References

Albertao, G.A., Martins Jr., P.P., 1996. A possible tsunami deposit at the Cretaceous-Tertiary boundary in Pernambuco, northeastern Brazil. Sediment. Geol. 104, 189–201.

Allen, J.R.L. (Ed.), 1986. Sedimentary Structures: Their Character and Physical Basis Developments in Sedimentology, vol. 39. Elsevier Science Publishers, Netherlands, pp. 593–663 (+).

Alvarez, L.W., Alvarez, W., Asaro, F., Michel, H.V., 1980. Extraterrestrial cause for the Cretaceous-Tertiary extinction. Science 208, 1095–1108.

Alvarez, W., Smit, J., Lowrie, W., Asaro, F., Margolis, S.V., Claeys, P., Kastner, M., Hildebrand, A.R., 1992. Proximal impact deposits at the cretaceous-tertiary boundary in the Gulf of Mexico: a restudy of DSDP leg 77 sites 536 and 540. Geology 20, 697–700.

Asai, D., Imamura, F., Shuto, N., Takahashi, T., 1998. The wave height of the 1854 Ansei-Tokai earthquake tsunami and sediment transportation in Iruma, Izu Peninsula, Japan. Proc. Coastal Eng. JSCE 45, 371–375.

Atwater, B.F., 1987. Evidence for great Holocene earthquakes along the outer coast of Washington State. Science 236, 942–944.

48   Chapter 3

Bardet, J.P., Synolakis, C.E., Davies, H.L., Imamura, F., Okal, E.A., 2003. Landslide tsunamis: recent findings and research directions. Pure Appl. Geophys. 160, 1793–1809.

Beget, J.E., 2000. Volcanic tsunamis. In: Sigurdsson, H. (Ed.), Encyclopedia of Volcanoes. Academic Press, San Diego, CA, p. 1417.

Benson, B.E., Grimm, K.A., Clague, J.J., 1997. Tsunami deposits beneath tidal marshes on northwestern Vancouver Island, British Columbia. Quat. Res. 48, 192–204.

Bondevik, S., Svendsen, J.I., Mangerud, J., 1997. Tsunami sedimentary facies deposited by the Storegga tsunami in shallow marine basins and coastal lakes, western Norway. Sedimentology 44, 1115–1131.

Bourgeois, J., Minoura, K., 1997. Paleotsunami studies—contribution to mitigation and risk assessment. In: Gusiakov, V.K. (Ed.), "Tsunami Mitigation and Risk Assessment, Report of the International Workshop" 21–24 August 1996, Russia Petropavlovsk-Kamchatsky, pp. 1–4.

Bourgeois, J., Reinhart, M.A., 1989. Onshore Erosion and Deposition by the 1960, Tsunami at the Rio Lingue Estuary, South-Central Chile, p. 1331. EOS 70.

Bourgeois, J., Hansen, T.A., Wiberg, P.L., Kauffman, E.G., 1988. A Tsunami deposit at the Cretaceous-Tertiary boundary in Texas. Science 241, 567–570.

Bugge, T., Befring, S., Belderson, R.H., Eidvin, T., Jansen, E., Kenyon, N.H., Holtedahl, H., Sejrup, H.P., 1987. A giant three-stage submarine slide off Norway. Geo Mar. Lett. 7, 191–198.

Cita, M.B., Camerlenghi, A., Rimoldi, B., 1996. Deep-sea tsunami deposits in the eastern Mediterranean: new evidence and depositional models. Sediment. Geol. 104, 155–173.

Clague, J.J., Bobrowsky, P., 1994. Evidence for a large earthquake and tsunami 100–400 years ago on western Vancouver island, British Columbia. Quat. Res. 41, 176–184.

Dawson, A.G., Shi, S., 2000. Tsunami deposits. Pure Appl. Geophys. 157, 875–897.

Dawson, A.G., Long, D., Smith, D.E., 1988. The Storegga slides: evidence from eastern Scotland for a possible tsunami. Mar. Geol. 82, 271–276.

Dawson, A.G., Foster, I.D.L., Shi, S., Smith, D.E., Long, D., 1991. The identification of tsunami deposits in coastal sediment sequences. Sci. Tsunami Hazards 9 (1), 73–82.

Foster, I.D.L., Dawson, A.G., Dawson, S., Lees, J.A., Mansfield, L., 1993. Tsunami sedimentation sequences in the Scilly Isles, south-west England. Sci. Tsunami Hazards 11 (1), 35–45.

Friedman, G.M., Sanders, J.E., Kopaska-Merkel, D.C., 1992. Principles of Sedimentary Deposits. Macmillan Publishing Company, New York, p. 717.

Fujimoto, K., Imamura, F., 1997. Generation of tsunami by the K/T impact. Proc. Coastal Eng. JSCE 44, 315–319 (in Japanese).

Fujiwara, O., Masuda, F., Sakai, T., Irizuki, T., Fuse, K., 2000. Tsunami deposits in Holocene bay mud in southern Kanto region, Pacific coast of central Japan. Sediment. Geol. 135, 219–230.

Fujiwara, O., Kamataki, T., Tamura, T., 2003. Grain-size distribution of tsunami deposits reflecting the tsunami waveform: an example from a Holocene drowned valley on the Southern Boso Peninsula, East Japan. Quat. Res. **42** 67–81 (in Japanese).

Gelfenbaum, G., Jaffe, B., 2003. Erosion and sedimentation from the 17 July, 1998 Papua New Guinea tsunami. Pure Appl. Geophys. 160, 1969–1999.

Goff, J., McFadgen, B.G., Chague-Goff, C., 2004. Sedimentary differences between the 2002 easter storm and the 15th-century Okoropunga tsunami, southeastern north island, New Zealand. Mar. Geol. 204, 235–250.

Haq, B.U., Hardenbol, J., Vail, P.R., 1981. Chronology of fluctuating sea levels since the Triassic. Science 235, 1156–1165.

Hassler, S.W., Robey, H.F., Simonson, B.M., 2000. Bedforms produced by impact-generated tsunami, ∼2.6 Ga Hamersley basin, Western Australia. Sediment. Geol. 135, 283–294.

Hildebrand, A.R., Penfield, G.T., Kring, D.A., Pilkington, M., Camargo, Z.A., Jacobsen, S.B., Boynton, W.V., 1991. Chicxulub crater: a possible cretaceous/tertiary boundary impact crater on the Yucatan Peninsula, Mexico. Geology 19, 867–871.

Hirose, K., Goto, T., Mitamura, M., Okahashi, H., Yoshikawa, S., 2002. Event deposits and environment changes detected in marsh deposits at Aisa, Toba City. Chikyu Mon. 280, 692–697 (in Japanese).

Iida, K., 1963. Magnitude, energy and generation mechanisms of tsunamis and a catalogue of earthquakes associated with tsunamis. In: Proceedings of Tsunami Meeting, 10th Pacific Scientific Congress 1961, IUGG Monograph 24, pp. 7—18.

Imamura, F., 1996. Numerical simulation and visualization of tsunamis. J. Jpn. Soc. Fluid Mech. 15, 376—383 (in Japanese).

Jansen, E., Befring, S., Bugge, T., Eidvin, T., Holtedahl, H., Sejrup, H.P., 1987. Large submarine slides on the Norwegian continental margin: sediments, transport and timing. Mar. Geol. 78, 77—107.

Kajiura, K., 1982. Approximate Prediction of Wave Gain in Bays. Study on Tsunamis Part 1. Fire and Marine Insurance Rating Association of Japan, pp. 27—42 (in Japanese).

Kastens, K.A., Cita, M.B., 1981. Tsunami-induced sediment transport in the abyssal Mediterranean Sea. Geol. Soc. Am. Bull. 92, 845—857.

Kato, Y., 1987. Run-up height of Yaeyama seismic tsunami (1771). Zishin 2 (40), 377—381 (in Japanese with English abstract).

Kato, Y., Hidaka, K., Kawano, Y., Shinjo, R., 1988. Yaeyama seismic tsunami (1771) at Tarama island, the Ryukyu islands: 1. Movement of reef blocks and a run-up height. Earth Sci. 42, 84—90 (in Japanese with English abstract).

Keating, B.H., McGuire, W.J., 2000. Island edifice failures and associated hazards. Pure Appl. Geophys. 157, 899—955.

Kiyokawa, S., Tada, R., Iturralde-Vinent, M.A., Tajika, E., Yamamoto, S., Oji, T., Nakano, Y., Goto, K., Takayama, H., Delgado, D.G., Otero, C.D., Rojas-Consuegra, R., et al., 2002. Cretaceous-Tertiary boundary sequence in the Cacarajicara Formation, western Cuba: an impact-related, high-energy, gravity-flow deposit. Geol. Soc. Am. Spec. Pap. 356, 125—144.

Kon'no, E. (Ed.), 1961. Geological Observation of the Sanriku Coastal Region Damaged by the Tsunami Due to the Chile Earthquake in 1960 52 Contribution of Institute of Geology and Paleontology. Tohoku University, pp. 1—45 (in Japanese with English abstract).

Lawton, T.F., Shipley, K.W., Aschoff, J.L., Giles, K.A., Vega, F.J., 2005. Basinward transport of Chicxulub ejecta by tsunami-induced backflow, La Popa basin, northeastern Mexico, and its implications for distribution of impact-related deposits flanking the Gulf of Mexico. Geology 33, 81—84.

Long, D., Smith, D.E., Dawson, A.G., 1989. A Holocene tsunami deposit in eastern Scotland. J. Quat. Sci. 4, 61—66.

Long, D., Dawson, A.G., Smith, D.E., 1990. Tsunami risk in northwestern Europe: a Holocene example. Terra. Nova 1, 532—537.

Maramai, A., Graziani, L., Tinti, S., 2005a. Tsunamis in the Aeolian islands (southern Italy): a review. Mar. Geol. 215, 11—21.

Maramai, A., Graziani, L., Alessio, G., Burrato, P., Colini, L., Cucci, L., Nappi, R., Nardi, A., Vilardo, G., 2005b. Near- and far-field survey report of the 30 December 2002 Stromboli (Southern Italy) tsunami. Mar. Geol. 215, 93—106.

Marinatos, S., 1939. The volcanic destruction of Minoan Crete. Antiquity 13, 425—439.

Massari, F., D'Alessandro, A., 2000. Tsunami-related scour-and-drape undulations in Middle Pliocene restricted-bay carbonate deposits (Salento, south Italy). Sediment. Geol. 135, 265—281.

Matsui, T., Imamura, F., Tajika, E., Nakano, Y., Fujisawa, Y., 2002. Generation and propagation of a tsunami from the Cretaceous-Tertiary impact event. In: Koeberl, C., MacLeod, K.G. (Eds.), Catastrophic Events and Mass Extinctions, Impacts and Beyond, vol. 356. Geol. Soc. Am. Spec. Pap, pp. 69—77.

Minoura, K., Nakaya, S., 1991. Traces of tsunami preserved in inter-tidal lacustrine and marsh deposits: some examples from northeast Japan. J. Geol. 99, 265—287.

Minoura, K., Nakaya, S., Sato, H., 1987. Traces of tsunamis recorded in lake deposits—an example from Jusan, Shiura-mura, Aomori. Zishin 2 (40), 183—196 (in Japanese with English abstract).

Minoura, K., Nakaya, S., Uchida, M., 1994. Tsunami deposits in a lacustrine sequence of the Sanriku coast, northeast Japan. Sediment. Geol. 89, 25—31.

50   Chapter 3

Minoura, K., Gusiakov, V.G., Kurbatov, A., Takeuti, S., Svendsen, J.I., Bondevik, S., Oda, T., 1996. Tsunami sedimentation associated with the 1923 Kamchatka earthquake. Sediment. Geol. 106, 145–154.

Minoura, K., Imamura, F., Takahashi, T., Shuto, N., 1997. Sequence of sedimentation process caused by the 1992 Flores tsunami: evidence from Babi island. Geology 25, 523–526.

Minoura, K., Imamura, F., Kuran, U., Nakamura, T., Papadopoulos, G.A., Takahashi, T., Yalciner, A.C., 2000. Discovery of Minoan tsunami deposits. Geology 28, 59–62.

Moore, A.L., 2000. Landward fining in onshore gravel as evidence for a late Pleistocene tsunami on Molokai, Hawaii. Geology 28, 247–250.

Moore, J.G., Bryan, W.B., Ludwig, K.R., 1994. Chaotic deposition by a giant wave, Molokai, Hawaii. Geol. Soc. Am. Bull. 106, 962–967.

Nanayama, H., Satake, K., Shimokawa, K., Shigeno, K., Koitabashi, S., Miyasaka, S., Ishii, M., 1998. Sedimentary facies and sedimentation process of invading tsunami deposit—example from the 1993 southwest Hokkaido earthquake tsunami. Kaiyo Monthly 15, 140–146 (in Japanese).

Nanayama, F., Shigeno, K., Satake, K., Shimokawa, K., Koitabashi, S., Miyasaka, S., Ishii, M., 2000. Sedimentary differences between the 1993 Hokkaido-nansei-oki tsunami and the 1959 Miyakojima typhoon at Taisei, southwestern Hokkaido, northern Japan. Sediment. Geol. 135, 255–264.

Nanayama, H., Kaga, A., Kinoshita, H., Yokoyama, Y., Satake, K., Nakata, T., Sugiyama, Y., Tsukuda, E., 2002. Traces of the Nankai earthquake tsunami discovered at Tomogashima, Kidan strait. Kaiyo Mon. 28, 123–131 (in Japanese).

Nishimura, Y., Miyaji, N., 1995. Tsunami deposits from the 1993 southwest Hokkaido earthquake and the 1640 Komagatake eruption, northern Japan. Pure Appl. Geophys. 144, 719–733.

Nomanbhoy, N., Satake, K., 1995. Generation mechanism of tsunamis from the 1883 Krakatau eruption. Geophys. Res. Lett. 22, 509–512.

Nott, J., 2000. Records of prehistoric tsunamis from boulder deposits—evidence from Australia. Sci. Tsunami Hazards 18, 3–14.

Nott, J., 2004. The tsunami hypothesis—comparisons of the field evidence against the effects, on the Western Australian coast, of some of the most powerful storms on Earth. Mar. Geol. 208, 1–12.

Okahashi, H., Akimoto, K., Mitamura, M., Hirose, K., Yasuhara, M., Yoshikawa, S., 2002. Event deposits detected in marsh deposits at Aisa, Toba City, Mie Prefecture—identification of tsunami deposits using foraminifera. Chikyu Mon. 280, 698–703 (in Japanese).

Okamura, M., Matsuoka, H., Tsukuda, E., Tsuji, Y., 2000. Monitoring of tectonics and historical tsunamis during the past 10,000 years by coastal lacustrine deposits. Chikyu Mon. 28, 162–168 (in Japanese).

Okamura, Y., Satake, K., Katayama, H., Noda, A., Sagayama, T., Suga, K., Uchida, Y., 2004. Effect of the earthquake and tsunami on the sea bottom. In: Hirata, N. (Ed.), Report of the Emergency Survey and Study on the 2003 Tokachi-Oki Earthquake. Earthquake Research Institute of Tokyo University, Tokyo, p. 7 (in Japanese).

Papadopoulos, G.A., 2003. Quantification of tsunamis: a review. In: Yalciner, A.C., Pelinovsky, E., Synolakis, C.E., Okal, E. (Eds.), Submarine Landslides and Tsunamis. Kluwer, Deventer, pp. 285–291.

Papadopoulos, G.A., Imamura, F., 2001. A proposal for a new tsunami intensity scale. Proc. Int. Tsunami Sym. 5 (1), 569–577.

Pinegina, T.K., Bourgeois, J., 2001. Historical and paleo-tsunami deposits on Kamchatka, Russia: long-term chronologies and long-distance correlations. Nat. Hazards Earth Syst. Sci. 1, 177–185.

Reinhart, M.A., Bourgeois, J., 1987. Distribution of Anomalous Sand at Willapa Bay, Washington: Evidence for Large-Scale Landward-Directed Processes, p. 1469. Eos 68.

Rothwell, R.G., Thomson, J., Kahler, G., 1998. Low-sea-level emplacement of a very large Late Pleistocene 'megaturbidite' in the western Mediterranean Sea. Nature 392, 377–380.

Sato, H., Shimamoto, T., Tsutsumi, A., Kawamoto, E., 1995. Onshore tsunami deposits caused by the 1993 Southwest Hokkaido and 1983 Japan Sea earthquakes. Pure Appl. Geophys. 144, 693–717.

Self, S., Rampino, M.R., 1981. The 1883 eruption of Krakatau. Nature 294, 699–704.

Sengupta, S.M., 1994. Introduction to Sedimentology. Balkema Publishers, Brookfield, p. 314.

Shi, S., Dawson, A.G., Smith, D.E., 1995. Coastal sedimentation associated with the December 12th, 1992 tsunami in Flores, Indonesia. Pure Appl. Geophys. 144, 525−536.

Shiki, T., Yamazaki, T., 1996. Tsunami-induced conglomerates in Miocene upper bathyal deposits, Chita Peninsula, central Japan. Sediment. Geol. 104, 175−188.

Sigurdsson, H., Carey, S., Mandeville, C., Bronto, S., 1991. Pyroclastic Flows of the 1883 Krakatau Eruption, pp. 377−392. EOS 72.

Smit, J., Montanari, A., Swinburne, N.H.M., Alvarez, W., Hildebrand, A.R., Margolis, S.V., Claeys, P., Lowrie, W., Asaro, F., 1992. Tektite-bearing, deep-water clastic unit at the Cretaceous-Tertiary boundary in northeastern Mexico. Geology 20, 99−103.

Smit, J., Roep, T.B., Alvarez, W., Montanari, A., Claeys, P., Grajales-Nishimura, J.M., Bermudez, J., 1996. Coarse-grained, clastic sandstone complex at the K/T boundary around the Gulf of Mexico: deposition by tsunami waves induced by the Chicxulub impact? Geol. Soc. Am. Spec. Pap. 307, 151−182.

Smoot, J.P., Litwin, R.J., Bischoff, J.L., Lund, S.J., 2000. Sedimentary record of the 1872 earthquake and "tsunami" at Owens Lake, southeast California. Sediment. Geol. 135, 241−254.

Stinnesbeck, W., Keller, G., 1996. K/T boundary coarse-grained siliciclastic deposits in northeastern Mexico and northeastern Brazil: evidence for mega-tsunami or sea-level changes? Geol. Soc. Am. Spec. Pap. 307, 197−209.

Sundborg, Å., 1956. The River Klarälven: a study of fluvial process. Geogr. Ann. 38, 127−316.

Tada, R., Nakano, Y., Iturralde-Vinent, M.A., Yamamoto, S., Kamata, T., Tajika, E., Toyoda, K., Kiyokawa, S., Delgado, D.G., Oji, T., Goto, K., Takayama, H., et al., 2002. Complex tsunami waves suggested by the Cretaceous-Tertiary boundary deposit at the Moncada section, western Cuba. Geol. Soc. Am. Spec. Pap. 356, 109−123.

(Ph.D. thesis). In: Takahashi, T. (Ed.), 1998. Study on Sediment Transportation by Tsunami Tohoku University 184 (in Japanese).

Takashimizu, T., Masuda, F., 2000. Depositional facies and sedimentary successions of earthquake-induced tsunami deposits in Upper Pleistocene incised valley fills, central Japan. Sediment. Geol. 135, 231−239.

Takayama, H., Tada, R., Matsui, T., Iturralde-Vinent, M.A., Oji, T., Tajika, E., Kiyokawa, S., Garcia, D., Okada, H., Hasegawa, T., Toyoda, K., 2000. Origin of the Penalver Formation in northwestern Cuba and its relation to K/T boundary impact event. Sediment. Geol. 135, 295−320.

Tappin, D.R., Matsumoto, T., Watts, P., Satake, K., McMurtry, G.M., Matsuyama, M., Lafoy, Y., Tsuji, Y., Kanamatsu, T., Lus, W., Iwabachi, Y., Yeh, H., et al., 1999. Sediment Slump Likely Caused the 1998 Papua New Guinea Tsunami, p. 340. EOS 80 329. 334.

Tsuji, Y., Goto, T., Okamura, M., Matsuoka, H., Hang, S., 2001. Traces of Historical and Pre-historical Tsunamis in the Lake-Bottom Sedimentary Sequences at Oh-Ike, Owase Sukari-Ura, Mie Prefecture, West Japan. Tsunami Engineering Technical Report 18 Disaster Control Research Center of Tohoku University, pp. 11−14 (in Japanese).

Tsuji, Y., Okamura, M., Matsuoka, H., Goto, T., Hang, S., 2002. Traces of historical and pre-historical tsunamis in the lake-bottom sedimentary sequences at Oh-Ike (Owase city) and Suwa-Ike (Kii-Nagashima town), Mie Prefecture. Chikyu Mon. 280, 743−747 (in Japanese).

Wright, C., Mella, A., 1963. Modifications to the soil pattern of south-central Chile resulting from seismic and associated phenomena during the period May to August 1960. Bull. Seismol. Soc. Am. 53, 1367−1402.

Yalin, M.S. (Ed.), 1977. Mechanics of Sediment Transportation, second ed. Pergamon Press, Oxford, p. 295.

Zelt, J.A., 1986. Tsunamis: The Response of Harbours with Sloping Boundaries to Long Wave Excitation. Report, Division of Engineering and Applied Science California Institute of Technology, Pasadena, California, 318. KH-R-47.

CHAPTER 4

Bedforms and sedimentary structures characterizing tsunami deposits

O. Fujiwara

Geological Survey of Japan, National Institute of Advanced Industrial Science and Technology (AIST), Tsukuba, Ibaraki, Japan

1. Introduction

Tsunami deposits show a wide variety of bedforms and sedimentary structures; their nature strongly depends on the physical setting of the depositional sites and on the nature of the responsible events, such as current strength and the surface geology. However, several common features are recognized in most tsunami deposits, reflecting the physical properties of tsunami waves. The extremely large wavelength and period compared with those of storm waves are responsible for the common features of tsunami deposits. At the same time, these common features reflecting the tsunami waveform may serve as criteria for distinguishing tsunami deposits from storm deposits. This distinction has been an embarrassing problem for researchers from the very beginning of "tsunami geology."

This chapter discusses the bedforms and sedimentary structures of tsunami deposits that reflect the physical properties of the responsible tsunami waves and provides a clue for solving the problem of correct recognition. The common features of tsunami deposits are described in the following on the basis of some examples from an ideal environment for deposition and preservation. This chapter refers to tsunami deposits in coastal lowlands, including shallow bays, lagoons, and ponds. These areas, which are separated from the open sea, usually have a sufficient water depth and accommodation space for the deposition of tsunami deposits with well-developed sedimentary structures. The relatively high-sediment flux in these shallow-water areas provides favorable preservation conditions for tsunami deposits by rapid burial. Sedimentary structures from deep-sea tsunamis and from meteoric and volcanic impact tsunamis are discussed in Chapters 13, 14, 15, and 17.

Tsunamiites. https://doi.org/10.1016/B978-0-12-823939-1.00004-5
Copyright © 2008 Elsevier B.V. All rights reserved.

2. Differences of waveforms between tsunami- and storm-induced waves

A definite difference of physical properties between tsunamis and other waves is the extremely large wavelength, as described in Chapter 20. The wavelengths are up to some 100 km, and periods are 10 min to 1 h, which is 100 times more than those of wind-induced waves. The wavelength and period of tsunamis are controlled by the water depth and the size of the areas where they originated.

Schematic cross sections of tsunamis and wind waves are shown in Fig. 4.1. In tsunami waves, a huge volume of water comes behind the sheer wave head. Large waves composing a wave train arrive one after another at the coastal area when a large tsunami occurs. Each wave generates high-energy flows on and along the coast, with durations of several to 10 min. The repeated arrival of tsunami waves generates a cycle of run-up and backwash flows on the coastal area, thus resulting in a succession of sediments reflecting the succession of arriving waves.

Another remarkable character of tsunamis is the group of large waves during the middle stage of their wave train. In many cases, the second or the third wave is the largest. This phenomenon mainly reflects the propagation process of the tsunamis, as well as the faulting process in the area of origin (Watanabe, 1998). This characteristic wave-size distribution in the tsunami wave train is, in some fortunate cases, recorded in the sedimentary succession, as will be detailed in Section 5.

The period of storm waves is generally several seconds in small bays, and 10–20 s at most, even on open coasts. Traces of individual waves are scarcely preserved in storm deposits, because of frequent erosion. Storm waves form a continuous wave group, which generates a depositional succession represented by a fining-upward sequence reflecting the waning stage of the wave group (Cheel and Leckie, 1993; Walker et al., 1983).

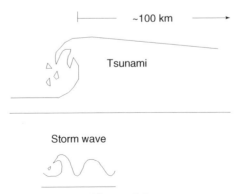

Figure 4.1
Comparison of waveforms between tsunami and wind wave. Tsunamis have extremely longer wavelength and wave period compared with the wind waves represented by storm waves.

The waveform of storm surges is a composite of the slow, tide-induced water-level rise mainly driven by the atmospheric pressure and wind force and of the short-period wind waves. The tide-induced water-level rise, <1 mm/s, in general is insufficiently powerful to transport coarse sedimentary particles on the sea floor; the major force responsible for sediment transport over the sea floor during storms is formed by the short-period wind waves. Subaqueous storm surge deposits have therefore, at least theoretically, a similar character as storm deposits (Fujiwara and Kamataki, 2008).

3. Bedforms and sedimentary structures reflecting the tsunami waveform

Considering the aforementioned physical properties of tsunamis, in particular the extremely large wavelength and period, tsunami deposits necessarily reflect deposition by successive individual currents, not from continuous short-period wave groups. The bedforms, sedimentary structures, and thicknesses of tsunami deposits vary widely, according to the hydrodynamic conditions at the depositional sites (Kon'no, 1961; Sato et al., 1995). The hydrodynamic conditions of tsunami currents, such as direction, speed, density, and dominant grain size, change continuously, corresponding to the geologic and geomorphic conditions at and around the depositional sites. The changes of flow conditions with time are recorded in the vertical succession of sedimentary structures and bedforms of a tsunami deposit at each depositional site.

Tsunami deposits, both on land (washover) and under water (subaqueous), can roughly be divided into two types: the single-bed and the multiple-bed type. The former is represented by a single sediment sheet. Many of washover tsunami deposits belong to this type. The latter type shows a cyclic succession of several sediment sheets and is known from both washover and subaqueous tsunami deposits. The preservation potential of tsunami deposits is also strongly controlled by geologic conditions such as the sedimentation rate and the roughness of the sedimentary surface around the depositional sites. Even by the same tsunami, both types of tsunami deposits can be generated at different sites. Thick multiple-bed deposits are best preserved in depressions in the sedimentary surface, as they have there the best chance to escape the subsequent erosion.

4. Single-bed deposits

Many tsunami deposits have been described as thin (as a rule <10 cm), single sand sheets within the muddy (or peaty) sediments of coastal lowlands, marshes, and lagoons (Bourgeois and Johnson, 2001; Clague et al., 2000; Kelsey et al., 2005; Minoura et al., 1994; Okahashi et al., 2005; Pinegina et al., 2003; Tsukuda et al., 1999; Williams et al., 2005). These tsunami deposits were generally identified as light-colored, coarse-grained beds in outcrops and cored samples. Marine fossils, a basal erosion surface, and rip-up clasts within the beds can often support the identification.

Single-bed deposits generally show only one-graded interval, suggesting deposition from one sediment flow. They generally show a wedge shape, decreasing in thickness and grain size landward. This lateral change of these tsunami deposits reflects the deposition from a decelerating sediment flow, which is caused mainly by friction with the sedimentary surface (Clague and Bobrowsky, 1994; Sato et al., 1995). Reversed grading and normal grading are common (Benson et al., 1997; Clague et al., 2000). Reversed grading, due to the so-called traction carpet (Lowe, 1982, 1988), suggests the deposition from currents with a high grain content and a high current power. Imbricated gravel-sized clasts (Nanayama and Shigeno, 2004), cross-lamination, and current ripples (Choowong et al., 2005; Sato et al., 1995) are also common in tsunami deposits. Landward-directed currents can often be reconstructed from these bedforms and sedimentary structures, though backwash flows can also be reconstructed in some cases (Nanayama et al., 2000; Sato et al., 1995). Muddy films with various bioclasts, such as branches and seeds, often drape these tsunami deposits (Clague et al., 1994; Sato et al., 1995). The films are due to suspension fallout, resulting in a kind of mud drape, and debris left behind by the tsunami water.

The distinction of single-bed tsunami deposits from other surge deposits, such as washover deposits, abandoned-channel deposits, and river-flood deposits, is still a problem, because reliable criteria for the identification have not been established yet. The extremely long run-up distance of some tsunami deposits (Hirakawa et al., 2005; Nanayama et al., 2003) compared to storm surge deposits is still the only positive criterion, because it indicates that the responsible waves had a very large wavelength.

5. Multiple-bed deposits

Tsunami deposits composed of a stack of several sharp-based and graded sandy layers have been increasingly reported from several localities (e.g., Bondevik et al., 1997; Clague et al., 2000; Dawson et al., 1991; Foster et al., 1991; Kamataki and Nishimura, 2005; Kelsey et al., 2005; Nanayama et al., 2000; Williams and Hutchinson, 2000). Each sandy layer is commonly several to tens of centimeters thick. The thickness of this type of tsunami deposits generally exceeds that of the single-bed type, reaching over 1 m in some cases (Fujiwara and Kamataki, 2007). The bedforms and sedimentary structures of each individual layer are closely similar to those of the single-bed deposits.

Three major indications reflecting the extremely large wavelength and period of tsunamis are observed in multiple-bed deposits: (1) a succession of sand sheets capped by mud drapes, (2) repeated reversals of the current direction, and (3) a fining and thinning upward stack of sand sheets. A combination of these three indications forms the best criterion so far for the recognition of the tsunami depositional process.

5.1 Succession of sand sheets capped by mud drapes

A tsunami consists of a chain of high-energy flows, as mentioned earlier. The individual links of the chain can be compared with the run-up and backwash flows of the tsunami waves, and the knots of the chain can be compared with transition stages between run-up and backwash flows. The chain of tsunami waves results in the deposition of three types of depositional units, namely a run-up unit, a backwash flow unit, and a stagnation unit. Run-up and backwash units have generally an erosional basis, they exhibit reversed and normal grading, and they are covered by a mud drape or by a layer with concentrations of gyttja or plant debris (Fujiwara et al., 2003; Kelsey et al., 2005; Okahashi et al., 2005; Tsukuda et al., 1999; Tuttle et al., 2004).

Mud and gyttja drapes or layers result from suspension fallout and reflect the long stages of stagnant muddy water between the successive currents. These muddy layers are commonly sharp and clear (Fujiwara et al., 2003, 2005; Kon'no, 1961; Tuttle et al., 2004) and sometimes include plant debris and large branches (Fig. 4.2A and B). However, only dim films are observed in some examples (Kelsey et al., 2005).

Multiple-bed tsunami deposits can often be interpreted as a composite of several single-bed tsunami deposits that are found immediately on the top of each other due to the erosion of other material that was deposited before the next tsunami layer was deposited. Each run-up and backwash unit is very similar to a single-bed tsunami deposit, though they show opposite current directions. Sedimentary structures and bedforms resulting from the upper flow regime are common in run-up and backwash units. Coarse-grained dunes, antidunes, and plane beds have been frequently observed (Clague et al., 2000; Fujiwara and Kamataki, 2007; Fujiwara et al., 2003, 2005; Nanayama et al., 2001). Hummocky cross-stratification (HCS: Cheel and Leckie, 1993; Harms et al., 1975) is a major component of subaqueous tsunami deposits (Fig. 4.2B).

5.2 Repeated reversal of current directions

Repeated reversal of current directions, landward and seaward, is a typical character of tsunami deposits (Fig. 4.2C). Imbricated gravel and shells (Fujiwara et al., 2003; Goff et al., 2001; Nanayama et al., 2000) and herringbone structures (Takashimizu and Masuda, 2000) are useful sedimentary structures for reconstructing the reversal of currents. Tsunami deposits composed of a couplet of a lower run-up unit and an overlying backwash unit have been reported from some localities (Choowong et al., 2005; Nanayama et al., 2000; Nishimura and Miyaji, 1995). They show one cycle of run-up and backwash flows.

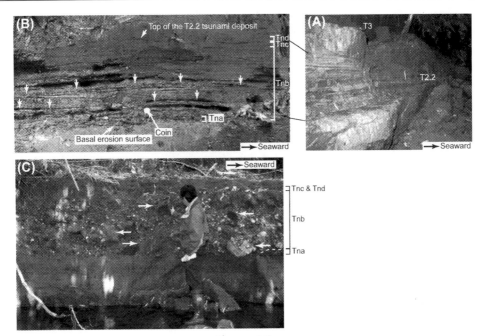

Figure 4.2
Representative Holocene subaqueous tsunami deposits in east Japan. (A) Holocene tsunami deposits (T2.2 and T3) outcropping along the river cliff in the southern Boso Peninsula, Pacific coast of Japan. They were deposited in the muddy bay center with the 10–20-m water depth during 7500–7200 cal BP (Fujiwara et al., 2003). Scale (a pencil on the T2.2 Tsunami deposit) is about 15-cm long. (B) Close-up of the T2.2 Tsunami deposit. It is mainly composed of a stack of several HCS sand sheets. Clearly preserved top surfaces of HCS are shown by the *arrows*. Black-colored muddy layers with plant debris concentration, showing the stagnation of the currents, drape the HCS sand sheets in many cases. Units Tna to Tnd are depositional units discussed in the text and shown in Fig. 4.3. Scale (a coin) is about 2.5 cm in diameter. (C) T2.2 Tsunami deposit composed of various size lithic and bio clasts. Vertical stack of several gravel sheets showing the inverse- and normal grading comprises the main part of the tsunami deposit. Cyclical turnover of the palaeocurrents is reconstructed from the imbrications of gravels (*arrows*). It was deposited in the muddy bay mouth with the 10–20-m water depth and yields abundant molluscan shells.

Complete sets of run-up and backwash units are rarely preserved, due to wave erosion. Either a run-up or a backwash unit is found alone in many cases. Thickness and grain size are generally different in the run-up and backwash deposits, due to the differences in current power and direction (Choowong et al., 2005; Fujiwara et al., 2005; Nanayama et al., 2000; Sato et al., 1995).

5.3 Fining and thinning upwards series of sand sheets

The vertical succession of sand sheets in the multiple-bed tsunami deposits reflects the changes in physical conditions of tsunami waves. In some cases, a series of ever finer and thinner sand sheets are piled up. This succession indicates the repeated arrival of successively less powerful waves. In other cases, the multiple-bed tsunami sand sheets have a more complex stacking pattern. Relatively coarse-grained and thick sand sheets are sometimes intercalated in the middle of the deposits. Outsized clasts, up to boulder size, are included in the coarse-grained sheets (Fig. 4.2C), which suggests deposition from relatively large waves in a tsunami wave train. This stacking pattern corresponds well to the tsunami waveform mentioned in Section 2. Many tsunamis have the largest waves in their middle stage.

Both washover and subaqueous tsunami deposits potentially record the waveform of the responsible tsunamis. Fig. 4.2 is an example of a subaqueous tsunami deposit from a Holocene drowned valley in east Japan. Another clear subaqueous example is the tsunami deposit from the Storegga slide in the North Atlantic, ~8000 BP, reported from coastal lakes in Norway (Bondevik et al., 1997). The tsunami deposit consists of an alternation of graded sand covered by plant debris deposited by at least four successive tsunami waves of diminishing height. An excellent example of a washover tsunami deposit has been reported from east Japan: the 1703 CE Kanto tsunami deposit from the coastal plain near Tokyo shows sediments from at least six successive waves and has the thickest and coarsest sand sheet in its middle part (Fujiwara et al., 2005). The 1994 CE Java tsunami deposit (Dawson et al., 1996), the 1755 CE Lisbon tsunami deposit in England (Foster et al., 1991), and the 1992 CE Flores tsunami deposits in Indonesia (Shi et al., 1995), which all represent the washover type, also consist of a succession of several fining-upward sand sheets.

The numbers and the amplitude distribution of waves seem, in the case of some recent tsunami deposits, to agree well with the responsible tsunamis (which have been actually observed). Tuttle et al. (2004) reported that the deposit from the 1929 CE Grand Banks earthquake, up to 25-cm thick, is composed of a stack of three units separated by peaty laminations. Three large waves have, indeed, been reported from the depositional sites at the 1929 tsunami (Tuttle et al., 2004). A multiple-bed tsunami deposit generated by the 1964 CE Alaska earthquake, up to 25-cm thick, has been reported from the marsh (Clague et al., 2000) and tidal delta front (Carver and McCalpin, 1996) along the Pacific coast of North America. The tsunami deposit consists of three (Clague et al., 2000) and four (Carver and McCalpin, 1996) beds of coarse to fine sand and mud, each of which fines upward. The 1964 CE tsunami had three (Clague et al., 2000) and six (Carver and McCalpin, 1996) main waves in the areas of the respective depositional sites. The 1993 CE Hokkaido-Nansei-Oki tsunami generated a 50-cm-thick gravelly sand bed on the southwestern coast of Hokkaido in northern Japan (Nanayama et al., 2000). This tsunami deposit can be divided into four layers that can probably be correlated with the run-up and backwash flows of the two main tsunami waves.

6. Depositional model in shallow water

A depositional model for tsunamis was constructed based primarily on observations from subaqueous multiple-bed tsunami deposits (Fig. 4.3). The repeated occurrence of run-up and backwash with long intervals in between is reconstructed from the succession of sediment sheets divided from one another by mud drapes. Each sediment sheet represented by an HCS sequence reflects the deposition from a waning high-density flow. The successive deposition by waves with an extremely long period cannot be explained by short-period storm waves.

Cyclic alternations of opposite current directions are indicated by the imbrication of gravel and by the pattern of lamination in each sand sheet. This structure differs from that of subaqueous gravity-induced debris flows, which show repeated seaward-directed current directions (Sohn, 2000).

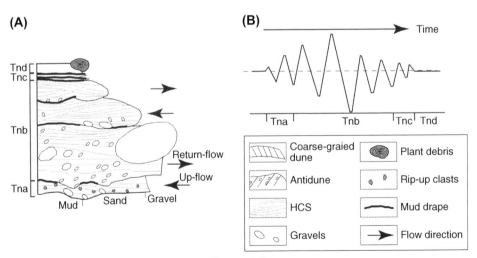

Figure 4.3
Depositional model and waveform of tsunamis. (A) Schematic succession of sediment sheets in tsunami deposits. Depositional Units Tna to Tnd correspond with the four wave groups in Fig. 4.3B, respectively. *Modified from Fujiwara and Kamataki (2007).* (B) Schematic tsunami waveform including the large wave group in its middle stage.

The changes of wave amplitude in a tsunami wave train with time are shown by the sequences of Unit Tna to Unit Tnd (Fig. 4.3). The relatively small waves of the early stage of tsunamis are represented by Unit Tna. Unit Tnb—coarse sand sheets with outsized clasts in the middle part of a tsunami deposit—corresponds to the large waves of the middle stage of the tsunami wave train. The occurrence of these large waves is closely related to the reflection of long waves on the shelves and on the continental coasts. Such large waves can hardly be expected from storms because storms hardly affect water that is deeper than continental shelves. The fining- and thinning-upward sequence of the upper half of Unit Tnb and of Unit Tnc corresponds to the waning stage of a tsunami. Unit Tnc comprises a fine alternation of sand and mud layers with plant debris. Unit Tnd—a muddy layer rich in plant debris building the uppermost part of a tsunami deposit—results from suspension fallout under low-energy condition subsequent to the tsunami.

The order in the succession of Unit Tna to Unit Tnd is always the same, but the bedforms and sedimentary structures in each sediment sheet depend on the physical properties of the responsible flows, such as viscosity, speed, and mean grain size. This depositional model for tsunamis is widely applied, because it is based on the waveform, the unique physical property of tsunamis.

A complete succession of the Unit Tna to Unit Tnd is not always present, because the preservation of tsunami deposits is limited by the physical conditions of the depositional sites, such as insufficient sediment supply to cover the tsunami deposits and thus protect them from erosion, or the erosion by the next waves of the same tsunami. The successions preserved in the sedimentary record are therefore often incomplete, depending on the circumstances. Synthesis of data from various depositional sites is consequently necessary for a precise reconstruction of the tsunami waveform.

7. Conclusions

The bedforms and sedimentary structures in tsunami deposits vary widely, as they depend not only on the physical properties of the responsible tsunami waves, such as speed, density, and viscosity, but also on the dominant grain size of the transported material. They are also strongly controlled by the geologic and geomorphic conditions of the depositional sites. Detailed observations of the bedforms and sedimentary structures in tsunami deposits reveal, however, that tsunami deposits have several characteristics in common. This results in the following conclusions regarding the depositional processes.

1. Bedforms and sedimentary structures result from high-energy and high-density sediment flows; characteristic are antidunes, HCS, and coarse-grained dunes.
2. Tsunami deposits are divided into two types, the single-bed and the multiple-bed type. The former is characterized by a single sediment sheet deposited from one

sediment flow; it generally shows a wedge shape decreasing in thickness and grain size landward. The multiple-bed type deposits are commonly composed of a number of stacked single-bed deposits.

3. Multiple-bed deposits have properties that indicate an extremely large wavelength and period of the tsunami: (a) a succession of sand sheets capped by mud drapes, (b) repeated alternation of current directions, and (c) the fining and thinning upward of the succession of sand sheets.

4. Four depositional units reflect changes in wave amplitude: Tna to Tnd in ascending order; these occur sometimes in multiple-bed deposits.

References

Benson, B.E., Grimm, K.A., Clague, J.J., 1997. Tsunami deposits beneath tidal marshes on northwestern Vancouver Island, British Columbia. Quat. Res. 48, 192–204.

Bondevik, S., Svendsen, J.I., Mangerud, J., 1997. Tsunami sedimentary facies deposited by the Storegga tsunami in shallow marine basins and coastal lakes, western Norway. Sedimentology 44, 1115–1131.

Bourgeois, J., Johnson, S.Y., 2001. Geologic evidence of earthquakes at the Snohomish delta, Washington, in the past 1200 yr. Geol. Soc. Am. Bull. 113, 482–494.

Carver, G.A., McCalpin, J.P., 1996. Paleoseismology of compressional tectonic environments. In: McCalpin, J.P. (Ed.), Paleoseismology. Academic Press, New York, pp. 183–270.

Cheel, R.J., Leckie, D.A., 1993. Hummocky cross-stratification. Sedimentol. Rev. 1, 103–122.

Choowong, M., Charusiri, P., Murakoshi, N., Hisada, K., Daorerk, V., Charoentitirat, T., Chutakositkanon, V., Jankaew, K., Kanjanapayont, P., Chulalongkorn University Tsunami Research Group, 2005. Initial report of tsunami deposits in Phuket and adjacent areas, Thailand induced by the earthquake off Sumatra December 26, 2004. J. Geol. Soc. Jpn. 111, xvii–xviii (in Japanese).

Clague, J.J., Bobrowsky, P.T., 1994. Evidence for a large earthquake and tsunami 100–400 years ago on western Vancouver Island, British Colombia. Quat. Res. 41, 176–184.

Clague, J.J., Bobrowsky, P.T., Hamilton, T.S., 1994. A sand sheet deposited by the 1964 Alaska tsunami at Port Alberni, British Columbia. Estuar. Coast Shelf Sci. 38, 413–421.

Clague, J.J., Bobrowsky, P.T., Hutchinson, I., 2000. A review of geological records of large tsunamis at Vancouver Island, British Columbia, and implications for hazard. Quarter. Sci. Rev. 19, 849–863.

Dawson, A.G., Foster, I.D.L., Shi, S., Smith, D.E., Lond, D., 1991. The identification of tsunami deposits in coastal sediment sequences. Sci. Tsunami Hazards 9, 73–82.

Dawson, A.G., Shi, S., Dawson, S., Takahashi, T., Shuto, N., 1996. Coastal sedimentation associated with the June 2nd and 3rd, 1994 tsunami in Rajegwesi, Java. Quat. Sci. Rev. 15, 901–912.

Foster, I.D.L., Albon, A.J., Bardell, K.M., Fletcher, J.L., Jardine, T.C., Mothers, R.J., Pritchard, M.A., Tirner, S.E.S., 1991. High energy coastal sedimentary deposits and evaluation of depositional processes in southwest England. Earth Surf. Process. Landforms 16, 341–356.

Fujiwara, O., Kamataki, T., 2007. Identification of tsunami deposits considering the tsunami waveform: an example of subaqueous tsunami deposits in Holocene shallow bay on southern Boso Peninsula, central Japan. Sediment. Geol. 200, 295–313.

Fujiwara, O., Kamataki, T., 2008. Tsunami depositional processes reflecting the waveform in a small bay—interpretation from the grain-size distribution and sedimentary structures. In: Shiki, T., Tsuji, Y., Yamazaki, T., Minoura, K. (Eds.), Tsunamiites—Features and Implications. Elsevier, Amsterdam, pp. 133–152.

Fujiwara, O., Kamataki, T., Tamura, T., 2003. Grain-size distribution of tsunami deposits reflecting the tsunami waveform—an example from the Holocene drowned valley on the southern Boso Peninsula, east Japan. Quat. Res. (Daiyonki-kenkyu) 42, 67–81 (in Japanese with English abstract).

Fujiwara, O., Hirakawa, K., Irizuki, T., Kamataki, T., Uchida, J., Abe, K., Hasegawa, S., Takada, K., Haraguchi, T., 2005. Depositional structures of historical Kanto earthquake tsunami deposits from SW coast of Boso Peninsula, Central Japan. Abstract. In: Japan Earth and Planetary Science Joint Meeting. J027–P023. (in Japanese with English abstract).

Goff, J.R., Chagué-Goff, C., Nichol, S., 2001. Paleotsunami deposits: a New Zealand perspective. Sediment. Geol. 143, 1–6.

Harms, J.C., Southard, J.B., Spearing, D.R., Walker, R.G., 1975. Depositional environments as interpreted from primary sedimentary structures and stratification sequences. SEPM Short. Course 2, 161pp.

Hirakawa, K., Nakamura, Y., Nishimura, Y., 2005. Mega-tsunamis since last 6500 years along the Pacific coast of Hokkaido. Gekkan Chikyu Suppl. 49, 173–180 (in Japanese).

Kamataki, T., Nishimura, Y., 2005. Field survey of the 2004 Off-Sumatra earthquake tsunami around the Banda Aceh, northern Sumatra, Indonesia. J. Geogr. 114, 78–82 (in Japanese with English abstract).

Kelsey, H.M., Nelson, A.R., Hemphill-Haley, E., 2005. Tsunami history of an Oregon coastal lake reveals a 4600 yr record of great earthquakes on the Cascadia subduction zone. Geol. Soc. Am. Bull. 117, 1009–1032.

Kon'no, E. (Ed.), 1961. Geological observations of the Sanriku coastal region damaged by the tsunami due to the Chile Earthquake in 1960. Contributions from the Institute of Geology and Paleontology, Tohoku University, 52, 40p. with 13 Plates. (in Japanese with English abstract).

Lowe, R.D., 1982. Sediment gravity flows: depositional models with special reference to the deposits of high-density turbidity currents. J. Sediment. Petrol. 52, 279–297.

Lowe, R.D., 1988. Suspended-load fallout rate as independent variable in the analysis of current structures. Sedimentology 35, 765–776.

Minoura, K., Nakaya, S., Uchida, M., 1994. Tsunami deposits in a lacustrine sequence of the Sanriku coast, northeast Japan. Sediment. Geol. 89, 25–31.

Nanayama, F., Satake, K., Furukawa, R., Shimokawa, K., Atwater, B.F., Shigeno, K., Yamaki, S., 2003. Unusually large earthquakes inferred from tsunami deposits along the Kuril trench. Nature 424, 660–663.

Nanayama, F., Shigeno, K., 2004. An overview of onshore tsunami deposits in coastal lowland and our sedimentological criteria to recognize them. Mem. Geol. Soc. Jpn. 58, 19–34 (in Japanese with English abstract).

Nanayama, F., Shigeno, K., Makino, A., Satake, K., Furukawa, R., 2001. Evaluation of tsunami inundation limits from distribution of tsunami event deposits along the Kuril subduction zone, eastern Hokkaido, northern Japan: case studies of lake Choboshi-ko, lake Tokotan-numa, lake Pashukuru-numa, Kinashibetsu marsh and lake Yudo-numa. Annual report on active fault and paleoearthquake researches. Geol. Survey Jpn./AIST 1, 251–272 (in Japanese with English abstract).

Nanayama, F., Shigeno, K., Satake, K., Shimokawa, S., Koitabashi, S., Miyasaka, S., Ishii, M., 2000. Sedimentary differences between the 1993 Hokkaido-nansei-oki tunami and the 1959 Miyakojima typhoon at Taisei, southwestern Hokkaido, northern Japan. Sediment. Geol. 135, 255–264.

Nishimura, Y., Miyaji, N., 1995. Tsunami deposits from the 1993 southwest Hokkaido earthquake and the 1640 Hokkaido Komagatake eruption, northern Japan. Pure Appl. Geophys. 144, 719–733.

Okahashi, H., Yasuhara, M., Mitamura, M., Hirose, K., Yoshikawa, S., 2005. Event deposits associated with tsunamis and their sedimentary structure in Holocene marsh deposits on the east coast of the Shima Peninsula, central Japan. J. Geosci. Osaka City Univ. 48, 143–158.

Pinegina, T.K., Bourgeois, J., Bazanova, L.I., Melekestsev, I.V., Braitseva, O.A., 2003. A millennial-scale record of Holocene tsunamis on the Kronotskiy bay coast, Kamchatka, Russia. Quat. Res. 59, 36–47.

Sato, H., Shimamoto, T., Tsutsumi, A., Kawamoto, E., 1995. Onshore tsunami deposits caused by the 1993 southwest Hokkaido and 1983 Japan sea earthquakes. Pure Appl. Geophys. 144, 693–717.

64   Chapter 4

Shi, S., Dawson, A.G., Smith, D.E., 1995. Coastal sedimentation associated with the December 12th, 1992 tsunami in Flores, Indonesia. Pure Appl. Geophys. 144, 525–536.

Sohn, Y.K., 2000. Depositional processes of submarine debris flows in the Miocene fan deltas, Pohang Basin, SE Korea with reference to flow transformation. J. Sediment. Res. 70, 491–503.

Takashimizu, Y., Masuda, F., 2000. Depositional facies and sedimentary successions of earthquake-induced tsunami deposits in Upper Pliocene incised valley fills, central Japan. Sediment. Geol. 135, 231–240.

Tsukuda, E., Okamura, M., Matsuoka, H., 1999. Earthquake history of the Nankai Trough recorded in lake sediments. Gekkan Chikyu Suppl 24, 64–69 (in Japanese with English abstract).

Tuttle, M.P., Anderson, T., Jeter, H., 2004. Distinguishing tsunami from storm deposits in eastern North America: the 1929 Grand Banks tsunami versus the 1991 Halloween storm. Seismol Res. Lett. 75, 117–131.

Walker, R.G., Duke, W.L., Leckie, D.A., 1983. Hummocky stratification: significance of its variable bedding sequences: Discussion. Geol. Soc. Am. Bull. 94, 1245–1249.

Watanabe, H. 1998. Comprehensive list of tsunamis to hit the Japanese Island, second ed. University of Tokyo Press, Tokyo, p. 238 (in Japanese).

Williams, H., Hutchinson, I., 2000. Stratigraphic and microfossil evidence for late Holocene tsunamis at Swantown marsh, Whidbey Island, Washington. Quat. Res. 54, 218–227.

Williams, H., Hutchinson, I., Nelson, A.R., 2005. Multiple sources for late-Holocene tsunamis at Discovery bay, Washington state, USA. Holocene 15, 60–73.

CHAPTER 5

Tsunami depositional processes reflecting the waveform in a small bay: interpretation from the grain-size distribution and sedimentary structures

O. Fujiwara[1], T. Kamataki[2]

[1]*Geological Survey of Japan, National Institute of Advanced Industrial Science and Technology (AIST), Tsukuba, Ibaraki, Japan;* [2]*Center for Regional Revitalization in Research and Education, Akita University, Akita, Japan*

1. Introduction

Discrimination from other event deposits such as storm deposits is a fundamental problem regarding tsunami deposits. Unfortunately, this problem has remained unsettled, mainly owing to the lack of data on sedimentary structures and grain-size distribution in tsunami deposits.

Sedimentary structures have been reported from both modern and historic run-up tsunami deposits, including run-up into a river mouth and lagoon (Dawson et al., 1991; Foster et al., 1991; Nanayama et al., 2000). Despite this, the differences between tsunami deposits and storm deposits (including storm-surge deposits) have remained unclear.

Vertical and horizontal changes in grain-size distribution are likely indicators of temporal and spatial transitions within the current that transported and deposited the sediments. In fact, the grain-size distribution of run-up tsunami deposits has been analyzed by, among others, Minoura and Nakaya (1990), Dawson et al. (1991), Foster et al. (1991), Nishimura and Miyaji (1995), Shi et al. (1995), Clague et al. (2000), and Nanayama et al. (2000, 2001). These studies have demonstrated that tsunami deposits generally exhibit fining upward and become finer and better-sorted landward. Many of these studies analyzed a tsunami deposit as an entity, however, and did not make clear the differences between tsunami and storm deposits.

Tsunamiites. https://doi.org/10.1016/B978-0-12-823939-1.00005-7
Copyright © 2008 Elsevier B.V. All rights reserved.

Fujiwara et al. (2002, 2003) reported in detail the genetic processes responsible for the sedimentary structures and vertical changes in grain-size distribution within some tsunami deposits from a Holocene drowned valley on the southern Boso Peninsula, central Japan. They paid attention to the differences in waveform (time series of amplitude and wave period) between tsunamis and storms. For example, the wave period of a tsunami is 100 times longer than that of storms. These differences are reflected by the sedimentary structures and grain-size distribution of tsunami and storm deposits. We discuss how to differentiate between tsunami and storm deposits using additional data to those presented by Fujiwara et al. (2002, 2003).

2. Regional setting

2.1 The Paleo-Tomoe Bay and its Holocene deposits

The paleo-Tomoe Bay was a Holocene drowned valley located on the south-western coast of the Boso Peninsula, east Japan (Fig. 5.1). This bay extended E–W for 2.6 km and was 0.3–0.6 km wide during the highstand in sea level of ~7200 cal. BP (Fig. 5.1C). The southern Boso Peninsula, including the study area, faces a plate convergent boundary

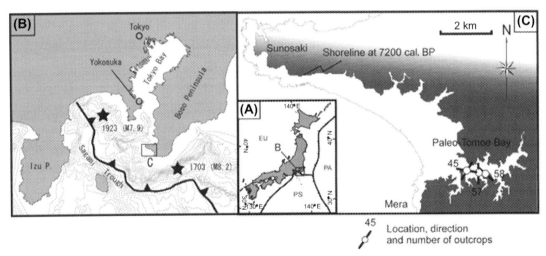

Figure 5.1
Regional setting of the study area. (A) Tectonic setting. *EU*, Eurasian plate; *PA*, Pacific plate; *PS*, Philippine Sea plate. (B) Coastal and sea-floor topography around the Sagami Trough. Bathymetric contour lines (200 m intervals) calculated from Asada (2000). The epicenters of the historical Kanto earthquakes (★) after Watanabe (1998). (C) Outcrop localities. Shoreline of the paleo-Tomoe Bay (~7200 cal. BP) was modified from the 1/25,000 topographical maps "Tateyama", "Mera" and "Chikura" of the Geospatial Information Authority of Japan and Shishikura (2001).

(Sagami Trough) (Fig. 5.1A and B) and is one of the most active coseismic uplift areas in the world. Series of Holocene marine terraces (Nakata et al., 1980) and beach ridges (Shishikura et al., 2001) indicate repeated significant submarine earthquakes at the plate boundary around the Sagami Trough.

The Holocene sediment in the paleo-Tomoe Bay overlies tilted Mio-Pliocene basement rocks and is composed of silt and sandy silt (Fig. 5.2). Deposition under low-energy conditions is indicated by the abundant in situ mollusk fossils, the good preservation of delicate shells, and the heavy bioturbation in the bay muds. Depositional ages of these bay muds have been determined by detailed ^{14}C dating of in situ or indigenous mollusk fossils (Fujiwara and Kamataki, 2003; Fujiwara et al., 1999a,b, 2000, 2003). The bay muds cropping out in the river cliff of the paleo-Tomoe Bay were deposited between 8500 and 6900 cal. BP (Fig. 5.2). Dominant species of the in situ or indigenous mollusks in the bay muds are *Dosinella penicillata*, *Fulvia mutica*, *Theora fragilis*, and *Protothaca* (*Neochione*) *jedoensis*. A paleodepth of 10−20 m is indicated from these mollusks.

Fujiwara et al. (1997, 1999a,b, 2000, 2003) studied the Holocene deposit of the paleo-Tomoe Bay and found seven tsunami deposits, coded T2 to T3.3 (Fig. 5.2).

2.2 Storm waves and tides around the Southern Boso Peninsula

Typhoons hit the southern Boso Peninsula almost every year. The storm wave height and period are 5−9 m and 13−17 s on the Pacific coast and about 1 m and 2−6 s in the Tokyo Bay, respectively (Masuda and Makino, 1987). Waves around the paleo-Tomoe Bay located near the mouth of the Tokyo Bay (Fig. 5.1) have intermediate height and period. The maximum tide range around the paleo-Tomoe Bay is about 170 cm (Japan Meteorological Agency, 2001). The sea-level rise by the largest storm surge around the Tokyo Bay during the 20th century was 2.1−2.2 m (Japan Meteorological Agency, 2001).

3. Sedimentary facies of the tsunami deposits

The seven tsunami deposits in the paleo-Tomoe Bay, 10- to 40-cm thick, form sheets of loosely packed sand with abundant mollusk shells, gravels, and rip-up clasts. They erosionally overlie the bay muds (Fig. 5.2). All seven consist of a stack of four units, determined by the depositional facies described as follows (Fujiwara et al., 2002, 2003).

3.1 Unit Tna

Unit Tna, less than 15-cm thick, is composed of fine to coarse sand that contains abundant rip-up clasts and erosionally overlies bay muds with abundant mollusks including bivalves in living position (Figs. 5.3 and 5.4). This unit contains sedimentary structures formed under the upper-flow regime, including plane beds, antidunes (Araya and Masuda, 2001;

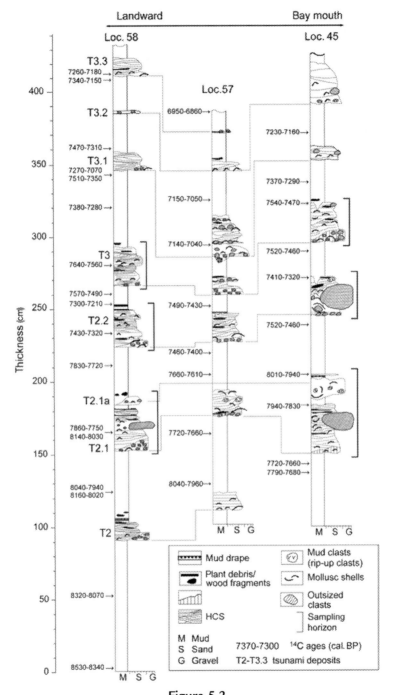

Figure 5.2
Columnar sections and calibrated ^{14}C ages from the paleo-Tomoe Bay (Fujiwara et al., 2003). Some of age data were originally published by Fujiwara et al. (1999a) and Fujiwara and Kamataki (2003).

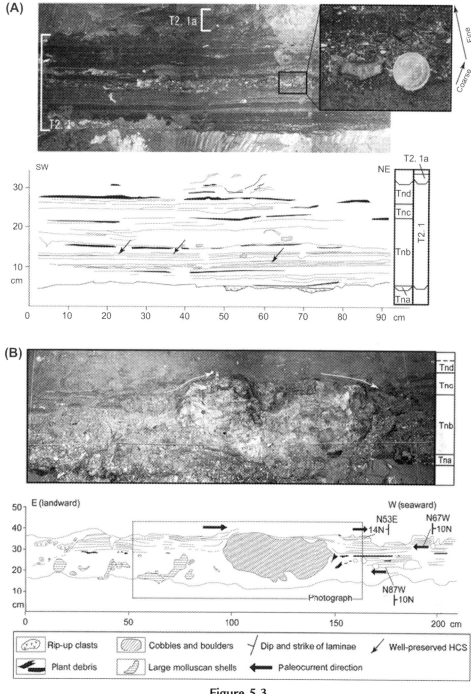

Figure 5.3
Sedimentary structures of the T2.1 and T2.2 tsunami deposits (Fujiwara et al., 2003). The tsunami deposits are composed of four depositional units (Tna to Tnd). (A) Hummocky cross-stratification (HCS) and plant debris—rich mud drapes in the T2.1 tsunami deposit. Section perpendicular to the long axis of the bay at loc. 45. (B) T2.2 tsunami deposit in section parallel to the long axis of the bay at loc. 45. Grid interval 50 cm. Coarse subunits including abundant white mollusk shells and outsized clasts are intercalated in the middle part of Unit Tnb. Paleocurrents in both seaward and landward directions reconstructed from dip of laminae (see sketch) and configuration of laminae over boulders (see photograph).

70 Chapter 5

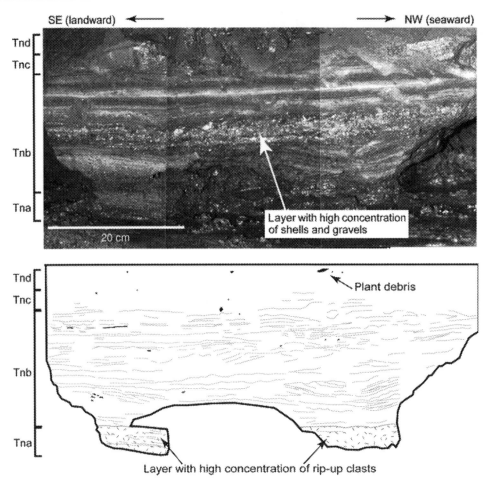

Figure 5.4
Sedimentary structures of the T3 tsunami deposit (Fujiwara, 2004). Section parallel to the long axis of the bay at loc. 58. Hummocky cross-stratification (HCS) and a coarse subunit are present in Unit Tnb.

Cheel, 1990), and prograding coarse-grained dunes. A landward paleocurrent has been inferred from these sedimentary structures, as well as from the imbrication of shells and gravels. The concentration of shells is relatively low in this unit, which has often been eroded by overlying units. It exhibits upward fining and is sometimes draped by plant debris and a thin mud layer.

3.2 Unit Tnb

Unit Tnb, less than 25-cm thick, is composed of sand with abundant mollusk shells and gravels and overlies Unit Tna with an erosional base (Figs. 5.3 and 5.4). This unit consists of a stack of several subunits including hummocky cross-stratification (HCS: Cheel and

Leckie, 1993; Harms et al., 1975) (Figs. 5.3 and 5.4). Each subunit, a few to 10-cm thick, has an erosional base and is composed of a lower division (layers) that shows upward coarsening and an upper division that is normally graded. The subunits are often draped by clay-rich layers of plant debris.

One or two subunits intercalated in the middle part of Unit Tnb are extremely coarse and include subrounded boulders (outsized clasts) up to 120 cm in diameter that were derived from the Tertiary basement rocks. The boulders show many holes made by boring shells. Shell concentrations are high in these coarse subunits. Finer subunits occur in the upper part of Unit Tnb. Paleocurrent directions are indicated by the dip of wedge-shaped cross-stratification, by imbrication of shells and gravels and by the deformation pattern of laminae around the outsized clasts (Fig. 5.3B). The currents were directed partly landward and partly seaward.

3.3 Unit Tnc

Unit Tnc, less than 10-cm thick, overlies Unit Tnb, but the transition is gradual. The unit is composed of alternating fine laminae of plant debris and sandy silt layers. The unit exhibits an overall fining-upward trend (Figs. 5.3 and 5.4).

3.4 Unit Tnd

Unit Tnd, up to 15-cm thick, overlies Unit Tnc gradually and is composed of sandy silt or silt beds with seams of plant and wood fragments (Figs. 5.3 and 5.4).

4. Grain-size distribution of the tsunami deposits

Vertical and horizontal changes of the median grain size, sorting, and mud content in three tsunami deposits, T2.1, T2.2, and T3, have been presented by Fujiwara et al. (2003). They analyzed samples from both the bay mouth (loc. 45) and the bay center (loc. 58; 750 m bay inward of loc. 45) of the paleo-Tomoe Bay (Figs. 5.1C and 5.2). Each sample was taken from a 1-cm-thick layer with mutual distances of 1 or 2 cm in the three tsunami deposits. All three tsunami deposits show a comparable grain-size distribution, which is considered to reflect the waveform of the tsunami wave train. We discuss the relationship between the grain-size distribution and the tsunami waveform in detail for the T3 tsunami deposit.

4.1 Sampling and methodology

The grain size has been determined for grains coarser than the 250 mesh (0.063 mm) for 24 samples taken over a distance of 24 cm (loc. 45) and 33 samples taken over a distance of 33 cm (loc. 58) from the T3 tsunami deposit.

72   Chapter 5

A settling tube system of Kyoto University was used for the purpose. The tube was 14 cm in diameter and the settling distance was 150 cm. The cumulative sediment weight and settling time were automatically registered by a computer. The measured data were converted to grain-size distribution. Most samples show a unimodal and log-normal distribution (Fig. 5.5).

The median, mode, sorting, skewness, and kurtosis of the samples were calculated from the grain-size distribution data using the moment method (Friedman, 1961). The median size and sorting could not be calculated for samples from the silt beds overlying and underlying the tsunami deposits, because of their low content of sand grains. The mud content was calculated from the weight ratio of grains finer than the 250 mesh.

4.2 The T3 tsunami deposit at location 58

Unit Tna (Fig. 5.5) is characterized by an extremely high mud content and moderately well-sorted fine sand. Unit Tnb has a low mud content and is well sorted. The median size, sorting, and mud content show regular vertical changes within the subunits of Unit Tnb (Fig. 5.5). Each subunit is composed of an inversely graded lower interval and a normally graded upper interval. Sorting is best in the middle part of the subunits. The mud content decreases upward to 5% in the inversely graded lower division and increases up to 20% in the uppermost part of the normally graded interval draped by the plant-debris layer.

The vertical changes of the grain-size parameters in Unit Tnb exhibit a saw-toothed curve reflecting the succession of subunits (Fig. 5.5). This saw-toothed curve generally has an extremely large (coarse and thick) tooth in the middle part and a stack of smaller teeth toward the upper part.

Unit Tnc differs from Unit Tnb by a small median grain size and a high mud content. Unit Tnc overlies Unit Tnb as a rule with a sharp boundary and shows a cyclic change of grain-size parameters.

Unit Tnd has a smaller median grain size and a higher mud content than Unit Tnc. The values are close to those of the underlying bay muds. Unit Tnd generally shows upward fining.

The skewness initially increases upward in Unit Tnb but becomes negative again in the uppermost part of the unit (Fig. 5.5). It is nearly symmetrical at the base of Unit Tnc and becomes negative upward in Unit Tnd. The mode shows closely similar values and vertical changes as the median in the tsunami deposit (Fig. 5.5). Most samples are extremely leptokurtic (3−6) throughout the sequence of this tsunami deposit.

Tsunami depositional processes reflecting the waveform   73

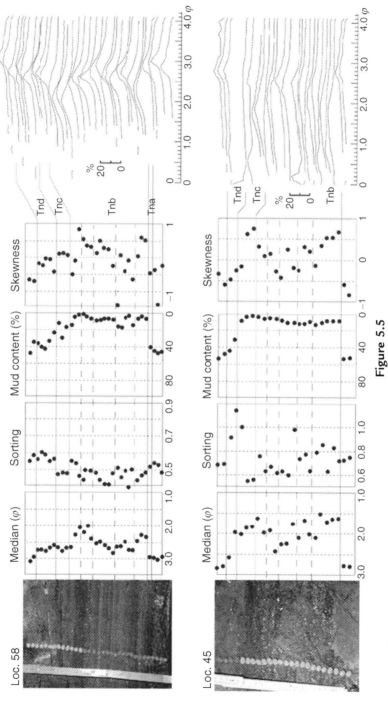

Figure 5.5

Grain-size distribution in the T3 tsunami deposit. Samples (1-cm-thick strata) were taken from the horizons of colored pins at 1 or 2 cm interval. Loc. 58 is about 750 m landward of loc. 45 (Fig. 9.1C). *Modified from Fujiwara et al. (2003) with additional data.*

74 Chapter 5

4.3 The T3 tsunami deposit at location 45

The T3 tsunami deposit at loc. 45 (Fig. 5.5) shows a larger median grain size, worse sorting, and smaller kurtosis (2−4) than at loc. 58, which is situated about 750 m more landward (Figs. 5.1C and 5.2). The saw-toothed curves of the median, sorting, and mud content are closely similar to those at loc. 58. The skewness shows a similar vertical change as at loc. 58. The mode shows closely similar values and vertical changes as the median throughout the sequence of the tsunami deposit.

5. Discussion

5.1 Tsunami waveform

Tsunamis originating in a deep sea have a unique waveform. The waveform of the 1923 CE Kanto tsunami ($M = 7.9$), a representative of the great tsunami originating in the Sagami Trough (Fig. 5.1B), is shown in Fig. 5.6. It was recorded at Yokosuka, which is shielded behind the Miura Peninsula, opposite the source area (Fig. 5.1B). This waveform has two remarkable features that are commonly observed in tsunamis, though the amplitude is relatively small due to the geomorphological situation at Yokosuka.

First, this tsunami had a very long wave period of 10−15 min. The wave period of tsunamis is defined by the following equation:

$$T = \frac{2R}{\sqrt{gh}} \tag{5.1}$$

where T is the wave period, R is the radius of the source area, g is the gravity-induced acceleration, and h is the water depth at the source area. In the case of an $M8$ earthquake around the Sagami Trough, R is about 50 km and h is about 2000 m, so T is about 714 s (about 12 min). Tsunamis have extraordinary longer wave periods than do storms. Around the Kanto region, the wave periods during storms are in general several seconds in the embayments and 10−20 s on the open coasts (Masuda and Makino, 1987).

The second feature of the Kanto tsunami was the delayed arrival of the largest waves. The wave group with larger amplitudes (waves 4−8) arrived after the small-amplitude waves. This phenomenon is well known as the edge-wave effect (Watanabe, 1998). Long waves that propagate toward the coast from offshore reflect from barriers such as continental shelves. The interference of reflected waves generates larger waves.

The wave height of the 1923 CE Kanto tsunami was very large on the coast facing the source area. In particular, huge waves with 10- to 15-min periods and 7- to 8-m heights hit the coasts of the paleo-Tomoe Bay during the 1923 CE Kanto tsunami (Japan Society of Civil Engineers, 1926).

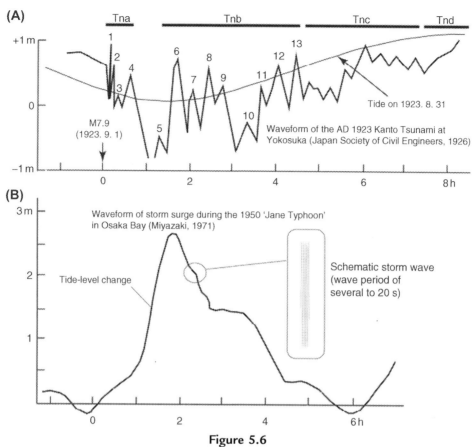

Figure 5.6
Comparison of waveforms of a tsunami (A) and a storm surge (B). The 1923 CE Kanto tsunami had a wave period of the order of 10 min and large waves during its middle stage. Four tsunami depositional units (Tna to Tnd) are attributed to early, middle, later, and after stages of the tsunami, respectively. The waveform of the storm surge is composed of a slow tide-level rise and superimposed short-period storm waves (Miyazaki, 1971). Typhoon Jane was the most severe that hit the Japanese islands in the 20th century. Quoted from Fujiwara et al. (2003).

5.2 Relationship between grain-size distribution and tsunami waveform

The vertical succession of Units Tna to Tnd and the saw-toothed grain-size distribution in each depositional unit are a consequence of the aforementioned tsunami waveform.

Unit Tna is thought to have formed by waves in the early stage of the tsunami. Dominant landward-directed paleocurrents are evidenced by the sedimentary structures. It is uncertain, however, if this unit was deposited by the very first wave, because this unit eroded underlying deposits.

Unit Tnb is subdivided into some subunits that evidence periodic changes in current power and direction. Each subunit is composed of an inversely graded lower interval and a normally graded upper interval including an HCS sequence (Figs. 5.3 and 5.4) and is covered by a mud drape. This cyclical sedimentary succession is discussed here in relatively much detail because of its implications for the interpretation of the saw-tooth-like stacking pattern of its grain-size distribution.

The depositional processes considered responsible for the genesis of each subunit that contributes to the saw-toothed curve are shown in Fig. 5.7. The two curves in Fig. 5.7 indicate water level and current—velocity changes in the tsunami wave.

The condition of a tsunami wave in a shallow bay is subdivided into three stages, namely A, B, and C.

Stage A indicates the arrival of a tsunami wave with very long wavelength (generally over 100 km). During this stage, a unidirectional current continues for several minutes. The tsunami current is reflected by both the bay bottom and the coast and forms turbulent waves in the run-up process. These waves become superimposed on the unidirectional flow, thus forming a combined flow that transports a high concentration of sedimentary particles that are eventually deposited as an HCS sequence (Cheel, 1991; Duke, 1990; Duke et al., 1991). The inversely graded structure represents a traction carpet (Lowe, 1982, 1988) and/or a transport lag (Hand, 1997) and indicates the deposition from high-density currents. Winnowing by the combined flow results in a well-sorted sand unit.

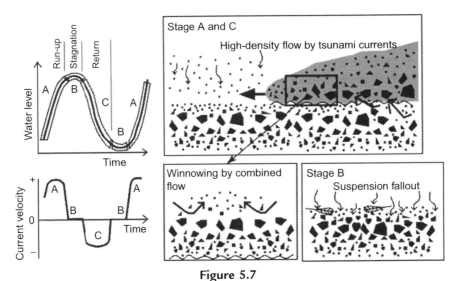

Figure 5.7
Genesis of the hummocky cross-stratified (HCS) sand unit covered by a mud drape in Unit Tnb. The HCS sand units with inverse grading are deposited from a high-density combined flow by tsunami currents in stage A (run-up current) and C (backwash current). Mud drapes formed by settling from suspension cover the HCS sand unit during stage B (stage of stagnance between two tsunami waves). *Modified from Fujiwara et al. (2003)*

A normally graded division was deposited under the waning combined flow. The upward increasingly positive skewness in the middle part of Unit Tnb indicates a relative increase in coarse grains as a consequence of winnowing. The upward increasingly negative skewness in the uppermost part of Unit Tnb indicates dominant suspension settling from the waning combined flow.

Stage B indicates the stagnation of the tsunami current at the transition from upflow to return flow (backwash). This stage of stagnance is the most important "key" for the identification of tsunami deposits. At this full-tide stage, mud grains and plant debris settle to drape the HCS sand unit. Mud grains, here defined as particles finer than 0.063 mm in diameter, settle at a rate of the order of 10^{-1} mm/s in clear, still water (Gibbs et al., 1971; Rubey, 1933) and more slowly in water with a high mud content. Therefore, a fairly long stagnant stage is indispensable to form the mud drapes. This stage of stagnance can last long if it is generated by a great tsunami, such as the 1923 CE Kanto tsunami, which had a wave period of the order of 10 min.

Stage C represents the return flow. A combination of backwash current and reflected waves again generates deposits that are similar to those of stage A. A mud drape settles during the following stagnation stage at low tide. The basal part of a new inversely graded interval with high mud content is formed by mixing of the remaining suspension and rip-up clasts derived from mud drapes on the underlying subunit.

As mentioned earlier, the repeated arrival of high-density currents alternating with long stagnant intervals (which allow enough settling from suspension to drape the subunit) is indispensable for the formation of the saw-toothed stacking pattern in Unit Tnb.

Assuming that the coarser a subunit is, the higher the energy of the wave is by which the subunit is deposited, the coarsest subunit—which is found intercalated in the middle part of Unit Tnb—is inferred to result from the largest wave in the wave train. This stacking pattern is in accordance with the tsunami waveform, which has its largest wave in the middle of the wave train. The succession of better-sorted, finer, and thinner subunits on the top of the coarsest subunit is thought to be formed by waning waves in the later stage of the tsunami.

Unit Tnc also exhibits cyclic alternations of laminae composed of (1) sandy silt and (2) plant debris, which implies repeated waves with long stagnant intervals that allowed the sedimentation of abundant plant debris. All laminae in Unit Tnc are finer and thinner than those in Unit Tnb, and they are thought to be related to the small waves during the final stage of the tsunami. This poorly sorted unit with a high content of mud suggests that it is composed of sedimentary particles from various sizes settling from suspension.

Unit Tnd was formed during the subsequent foundering of wood and plant fragments that were mainly derived from the land by the tsunami's backwash current. The upward increasingly negative skewness in Unit Tnc and Tnd indicates selective transport of fine grains and dominant suspension settling during a waning current.

The vertical pattern of sedimentary structures and the saw-tooth-like grain-size distribution in tsunami deposits are genetically related to the tsunami waveform (Figs. 5.6A and 5.8). The four units contrast sharply with each other and correspond to the developments from the beginning to the end of a tsunami wave series.

5.3 Discriminating tsunami deposits from storm deposits

Storm deposits are well known to include HCS, which is an indicator of oscillatory currents in the upper-flow regime, that is, combined flows (Cheel, 1990; Southard et al., 1990; Yokokawa et al., 1999). Storm deposits are generally understood to consist of well-sorted sand with a fining-upward sequence that reflects the waning storm waves (e.g., Cheel and Leckie, 1993; Duke et al., 1991; Meldahl, 1993; Saito, 1989; Swift et al., 1983). The vertical changes in grain-size distribution in a storm bed have, however, hardly been studied in any detail. The aforementioned characteristics of tsunami deposits, composed of a cyclical alternation of poorly sorted deposits ranging from boulders to mud, and with some units and subunits capped by mud drapes, have not been reported from storm deposits.

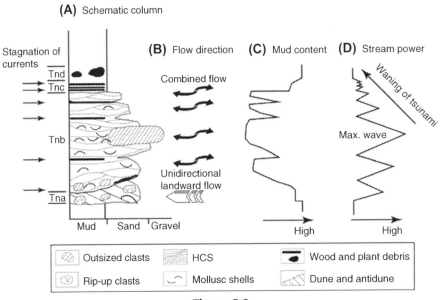

Figure 5.8

Relationship between the depositional successions and current conditions in a tsunami wave train. Unit Tna: Landward-directed paleocurrent and high mud content; early stage of a tsunami. Unit Tnb: Combined flow, maximum stream power, low mud content, saw-toothed stack of sub-units, outsized clasts; large waves in the middle stage of a tsunami. Unit Tnc: Cyclic deposition of sandy silt and plant-debris layers; small waves in the later stage of a tsunami. Unit Tnd: Silt bed with intercalated plant and wood-debris layer; return to low-energy environment after a tsunami.

These differences in texture and structure between tsunami and storm deposits can be inferred by comparing their waveforms (Fig. 5.6). The wave period of the storms in a small bay along the Pacific coast of Japan is a few to 10 s or $\sim 1\%$ of that of tsunamis. The periods of storm waves are too short to form mud drapes at the intervals of successive waves.

The knowledge regarding the grain-size distribution in storm-surge deposits is also limited. The waveform of a storm surge consists of a tidal rise of the order of 100 min, superimposed upon the storm waves with a wave period of a few to 10 s (Fig. 5.6B). The slow tide-level rise can neither move the gravel particles nor form the HCS in the coarse sand sheet, and short-period storm waves would not allow deposition of mud drapes or plant debris.

It is useful to keep in mind that a tide-level rise caused by a tsunami occurs in an instant. A great tsunami such as the 1923 CE Kanto tsunami (maximum wave height about 8 m in the study area) could generate deposits, including HCS, in a bay and also form mud drapes during the intervals between successive oscillatory currents with a long wave period.

Edge waves can be formed in large storm surges that move deeper seawater onto the continental shelf. However, large-scale edge waves like those mentioned earlier have not been reported from storms.

5.4 Tsunami deposits with a saw-toothed grain-size distribution

Vertical changes in grain-size distribution that exhibit a saw-toothed curve have been observed in both modern and historical washover tsunami deposits. Grain-size distributions of 37 samples from the 1992 CE Flores tsunami sand sheet, a 30-cm-thick deposit on a lowland along a small river, were analyzed by Shi et al. (1995). This tsunami deposit is composed of a succession of several units. The coarsest unit is found in the middle of the tsunami deposit, and finer units overlie this coarsest unit. The tsunami deposit becomes better-sorted and finer-grained upward. Saw-toothed vertical changes in grain-size distribution were also obtained from the 1755 CE Lisbon tsunami deposits reserved in lagoons in southern Portugal (Dawson et al., 1995) and southern England (Foster et al., 1991). The thicknesses of the analyzed tsunami sand sheets were 22 cm (Portugal) and 32 cm (England), and in both cases the grain-size distribution was analyzed at 1-cm intervals. The 1964 CE Alaska tsunami deposit and the 1700 CE Cascadian tsunami deposit are also composed of several units, even though these tsunami deposits are relatively thin (only a few centimeters) (Benson et al., 1997; Clague et al., 2000). These latter workers initially thought that these tsunami sand sheets consisted of a stack of normally graded laminae. Instead, some laminae have been found to be composed of an inversely graded lower division and a normally graded upper division (Fig. 6 of Benson et al., 1997 and Fig. 6 of Clague et al., 2000).

The relationship between the saw-toothed curve for the grain-size distribution in tsunami deposits and the tsunami wave form has been studied for the 1992 CE Flores tsunami deposit (Shi et al., 1995), the 1964 CE Alaska tsunami deposit (Benson et al., 1997; Clague et al., 2000), and the 1755 CE Lisbon tsunami deposit (Foster et al., 1991). A relationship between the coarsest unit in a tsunami deposit and a relative large wave in the tsunami wave train has been suggested for the 1993 CE Hokkaido-Nansei-Oki tsunami deposit (Nanayama et al., 2000) and the 1755 CE Lisbon tsunami deposit (Foster et al., 1991).

A similar interpretation of the depositional facies and grain-size distribution has been suggested for the Storegga tsunami deposit that formed due to a huge submarine slide in the Norwegian Sea ~8000 years BP (Bondevik et al., 1997; Dawson et al., 1991). The Storegga tsunami deposit is divided into five units, named wave 1–5 in an ascending order, that are capped by thin plant-debris layers (Dawson et al., 1991). Because units 1 and 4 contain relatively coarse grains, Dawson et al. (1991) assumed that the first and fourth waves in the tsunami wave train were relatively large.

These past studies have pointed out similarities between the saw-toothed vertical changes in grain-size distribution and tsunami waveforms, but have not discussed details of the genetic processes underlying the saw-toothed curves. Our new interpretation of the relationship between grain-size distribution, run-up distance, and tsunami waveform is schematically shown in Fig. 5.9. For simplicity, each subunit bounded by mud drapes corresponds to a run-up current. The tsunami waveform is recorded as a saw-toothed curve of the vertical grain-size distribution. The largest waves behind the first wave (edge waves) form the coarse and thick units. The larger waves in a tsunami wave train produce a sand sheet that

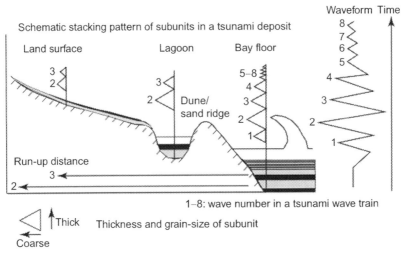

Figure 5.9
Relationship between grain-size distribution, run-up distance, and tsunami waveform (Fujiwara et al., 2003). The grain-size distribution and vertical succession of the tsunami deposits vary with the depositional conditions. A better continued record of waveform is expected in the bay where hydraulically it gets connected to the open sea. A part of large waves leave their traces on land as tsunami deposits.

has run-up farther inland. The tsunami sand sheets in the lagoon and on land were formed only by large waves that washed over the dunes and sand ridges. A more continuous waveform would be expected in a bay connected directly with the open sea, instead of a lagoon protected by barriers.

6. Conclusions

The following insights into the genesis of the sedimentary structures and the processes responsible for the grain-size distribution of the tsunami deposits have been gained from this study.

1. Vertical changes in the grain-size distribution and in the type of sedimentary structures in the tsunami deposits are clarified. The tsunami deposits are composed of four units (Tna to Tnd in an ascending order). Unit Tna erosionally covers lower bay mud and consists of poorly sorted sand with a high mud content. It includes plane bed and antidune structures. Unit Tnb consists of stacked HCS sand units. Each HCS sand unit includes traction carpet or transport-lag structures and is often covered by a mud drape. It contains one or two extremely coarse subunits with outsized clasts in its middle part. Unit Tnc shows a fine alternation of poorly sorted sandy silt layers and plant-debris laminae. Unit Tnd is composed of a silt bed with intercalated wood and plant-fragment layers.

2. The genesis of the tsunami deposits is reconstructed by integrating the vertical variation in depositional structures and grain-size distribution. The sedimentary structures of Unit Tna indicate that it was deposited by a landward-directed current in the upper-flow regime. Unit Tnb, which is composed of an alternation of HCS sand units and mud drapes, indicates a cyclic arrival of high-density currents with long stagnant stages in between, permitting mud to settle from suspension. Unit Tnc also shows cyclic deposition by high-energy waves separated by stagnant stages. Unit Tnd exhibits a return to a low-energy bay environment.

3. The succession of Units Tna to Tnd corresponds to a tsunami waveform. Unit Tna is attributed to an upflow current in the early stage of a tsunami. The coarse-grained Unit Tnb with outsized clasts was deposited by the larger waves of the middle stage of the tsunami. The poorly sorted and thin-laminated Unit Tnc is attributed to relatively small waves in the later stage of the tsunami. Unit Tnd is interpreted as the subsequent foundering of wood and plant debris derived from the land by the backwash current of the tsunami. This cyclic deposition alternating with the long stagnant stages cannot be explained by storm waves with wave periods of a few to some 10 s in a small bay. However, tsunamis have wave periods of the order of 10 min, and these can form such cyclic structures. The extremely coarse subunits in Unit Tnb are thought to have been formed by the largest of the successive waves. This unit indicates the delayed arrival of the largest waves from the wave train. The delayed largest waves are explained as an edge-wave effect and are diagnostic for tsunamis generated in a deep sea.

82   Chapter 5

4. The aforementioned sedimentary characteristics (such as the saw-toothed vertical grain-size distribution and the intercalated coarsest subunits) that allow to reconstruct the tsunami waveform are also observed in modern run-up tsunami deposits. Detailed analysis of the vertical changes in grain-size distribution is an effective approach for the identification of tsunami deposits.

Acknowledgments

The authors are grateful to Prof. Fujio Masuda of Kyoto University for his advice. Mr. Keisuke Fuse of Daiwa Geological Laboratory Co. Ltd. supported the fieldwork and helped the authors by many valuable discussions. Sincere gratitude is also expressed to Dr. Toru Tamura of the GSJ, AIST for his support regarding the grain-size analyses.

References

Araya, T., Masuda, F., 2001. Sedimentary structures of antidunes. An overview. J. Sed. Soc. Jpn. 53, 1–15.

Asada, A., 2000. 500 m mesh bathymetry data around Japan and visual edit program. J. Jpn. Soc. Mar. Surv. Technol. 12, 21–33 (in Japanese with English abstract).

Benson, B.E., Grimm, K.A., Clague, J.J., 1997. Tsunami deposits beneath tidal marshes on northwestern Vancouver Island, British Columbia. Quat. Res. 48, 192–204.

Bondevik, S., Svendsen, J.I., Mangerud, J., 1997. Tsunami sedimentary facies deposited by the Storegga tsunami in shallow marine basins and coastal lakes, western Norway. Sedimentology 44, 1115–1131.

Cheel, R.J., 1990. Horizontal lamination and the sequence of bed phases and stratification under upper-flow-regime conditions. Sedimentology 37, 517–529.

Cheel, R.J., 1991. Grain fabric in hummocky cross-stratified storm bed: generic implications. J. Sediment. Petrol. 61, 102–110.

Cheel, R.J., Leckie, D.A., 1993. Hummocky cross-stratification. Sedimentol. Rev. 1, 103–122.

Clague, J.J., Bobrowsky, P.T., Hutchinson, I., 2000. A review of geological records of large tsunamis at Vancouver Island, British Columbia, and implications of hazard. Quat. Sci. Rev. 19, 849–863.

Dawson, A.G., Foster, I.D.L., Shi, S., Smith, D.E., Lond, D., 1991. The identification of tsunami deposits in coastal sediment sequences. Sci. Tsunami Hazards 9, 73–82.

Dawson, A.G., Hindson, R., Andrade, C., Freitas, C., Parish, R., Bateman, M., 1995. Tsunami sedimentation associated with the Lisbon earthquake of 1 November AD 1755: Boca do Rio, Algarve, Portugal. Holocene 5, 209–215.

Duke, W.L., 1990. Geostrophic circulation or shallow marine turbidity currents? The dilemma of paleoflow patterns in storm-influenced prograding shoreline systems. J. Sediment. Petrol. 60, 870–883.

Duke, W.L., Arnott, R.W.C., Cheel, R.J., 1991. Shelf sandstones and hummocky cross-stratification: new insight on a stormy debate. Geology 19, 625–628.

Foster, I.D.L., Albon, A.J., Bardell, K.M., Fletcher, J.L., Jardine, T.C., Mothers, R.J., Pritchard, M.A., Tirner, S.E.S., 1991. High energy coastal sedimentary deposits and evaluation of depositional processes in southwest England. Earth Surf. Process. Landforms 16, 341–356.

Friedman, G.M., 1961. Distinction between dune, beach, and river sands from their textural characteristics. J. Sediment. Petrol. 31, 514–529.

Fujiwara, O., 2004. Tsunami depositional sequence model in shallow bay sediments - an example from Holocene drowned valleys on the southern Boso Peninsula, eastern Japan -. PhD Thesis. University of Tsukuba, 157 p.

Fujiwara, O., Kamataki, T., 2003. Significance of sedimentological time-averaging for estimation of depositional age by ^{14}C dating on molluscan shells. Quat. Res. (Daiyonki-kenkyu) 42, 27–40 (in Japanese with English abstract).

Fujiwara, O., Kamataki, T., Sakai, T., Fuse, K., Masuda, F., Tamura, T., 2002. Tsunami depositional sequence model in coastal sediments—an example from Holocene bay sediments in southern Kanto region. [Abstract]. In: The 109th Annual Meeting of the Geological Society of Japan, p. 100 (in Japanese).

Fujiwara, O., Kamataki, T., Tamura, T., 2003. Grain-size distribution of tsunami deposits reflecting the tsunami waveform—an example from the Holocene drowned valley on the southern Boso Peninsula, east Japan. Quat. Res. (Daiyonki-kenkyu) 42, 67–81 (in Japanese with English abstract).

Fujiwara, O., Masuda, F., Sakai, T., Fuse, K., Saito, A., 1997. Tsunami deposits in Holocene bay-floor mud and the uplift history of the Boso and Miura peninsulas. Quat. Res. (Daiyonki-kenkyu) 36, 73–86 (in Japanese with English abstract).

Fujiwara, O., Masuda, F., Sakai, T., Irizuki, T., Fuse, K., 1999a. Holocene tsunami deposits detected by drilling in drowned valleys of the Boso and Miura peninsulas. Quat. Res. (Daiyonki-kenkyu) 38, 41–58 (in Japanese with English abstract).

Fujiwara, O., Masuda, F., Sakai, T., Irizuki, T., Fuse, K., 1999b. Bay-floor deposits formed by great earthquakes during the past 10,000 yrs, near the Sagami trough, Japan. Quat. Res. (Daiyonki-kenkyu) 38, 489–501 (in Japanese with English abstract).

Fujiwara, O., Masuda, F., Sakai, T., Irizuki, T., Fuse, K., 2000. Tsunami deposits in Holocene bay mud in southern Kanto region, Pacific coast of central Japan. Sediment. Geol. 135, 219–230.

Gibbs, R.J., Matthews, M.D., Link, D.A., 1971. The relationship between sphere size and setting velocity. J. Sediment. Petrol. 41, 7–18.

Hand, B.M., 1997. Inversely grading resulting from coarse-sediment transport-lag. J. Sediment. Res. 67, 124–129.

Harms, J.C., Southard, J.B., Spearing, D.R., Walker, R.G., 1975. Depositional environments as interpreted from primary sedimentary structures and stratification sequences. SEPM Short Course 2, 161 p.

Japan Meteorological Agency, 2001. Tide Tables for the Year 2002, p. 290 (in Japanese).

Japan Society of Civil Engineers, 1926. Tide Level Changes and Tsunami. Report on the Disasters by the Taisho 12 (1923) Kanto Earthquake, Vol. 1; Part of the Rivers, Irrigations, Sand Arrestation, Canals and Harbors and Part of Electricity and Engineering, pp. 116–118 (in Japanese).

Lowe, R.D., 1982. Sediment gravity flows: depositional models with special reference to the deposits of high-density turbidity currents. J. Sediment. Petrol. 52, 279–297.

Lowe, R.D., 1988. Suspended-load fallout rate as independent variable in the analysis of current structures. Sedimentology 35, 765–776.

Masuda, F., Makino, Y., 1987. Paleo-wave conditions reconstructed from ripples in the Pleistocene Paleo-Tokyo Bay deposits. J. Geogr. 96, 23–45 (in Japanese with English abstract).

Meldahl, K.H., 1993. Geographic gradients in the formation of shell concentrations: Plio-Pleistocene marine deposits, Gulf of California. Palaeogeogr. Palaeoclim. Palaeoecol. 101, 1–25.

Minoura, K., Nakaya, S., 1990. Origin of inter-tidal lake and marsh environments in and around lake Jusan, Tsugaru. Mem. Geol. Soc. Jpn. 36, 71–87 (in Japanese with English abstract).

Miyazaki, M., 1971. In: Tsunami and Storm Surge. Oceanophysics. Fundamentals on Ocean Sciences 3. Tokai University Press, Kanagawa, pp. 255–323 (in Japanese).

Nakata, T., Koba, M., Imaizumi, T., Jo, W., Matsumoto, H., Suganuma, K., 1980. Holocene marine terraces and seismic crustal movements in the southern part of Boso peninsula, Kanto, Japan. Geogr. Rev. Jpn. 53, 29–44 (in Japanese with English abstract).

Nanayama, F., Shigeno, K., Satake, K., Shimokawa, S., Koitabashi, S., Miyasaka, S., Ishii, M., 2000. Sedimentary differences between the 1993 Hokkaido-nansei-oki tunami and the 1959 Miyakojima typhoon at Taisei, southwestern Hokkaido, northern Japan. Sediment. Geol. 135, 255–264.

Nanayama, F., Makino, A., Satake, K., Furukawa, R., Yokoyama, Y., Nakagawa, M., 2001. Twenty tsunami event deposits in the past 9000 years along the Kuril subduction zone identified in lake Horutori-ko,

84   Chapter 5

Kushiro City, eastern Hokkaido, Japan. Annual Report on Active Fault and Paleoearthquake Researches, Geol. Surv. Jpn., AIST, 233−249 (in Japanese with English abstract).

Nishimura, Y., Miyaji, N., 1995. Tsunami deposits from the 1993 southwest Hokkaido earthquake and the 1640 Hokkaido Komagatake eruption, northern Japan. Pure Appl. Geophys. 144, 719−733.

Rubey, W.W., 1933. Setting velocities of gravel, sand, and silt particles. Amer. J. Sci. 5th Ser. 25, 325−338.

Saito, Y., 1989. Modern storm deposits in the inner shelf and their recurrence interval, Sendai Bay, northeast Japan. In: Taira, A., Masuda, F. (Eds.), Sedimentary Facies in the Active Plate Margin. TERRA Science Publication, Tokyo, pp. 331−344.

Shi, S., Dawson, A.G., Smith, D.E., 1995. Coastal sedimentation associated with the December 12th, 1992 tsunami in flores, Indonesia. Pure Appl. Geophys. 144, 525−536.

Shishikura, M., 2001. Crustal movements in the Boso Peninsula from the analysis of height distribution of the highest Holocene paleo-shoreline. Annual report on active fault and paleoearthquake researches. Geol. Surv. Jpn., AIST 1, 273−285 (in Japanese with English abstract).

Shishikura, M., Haraguchi, T., Miyauchi, T., 2001. Timing and recurrence interval of the Taisho-type Kanto earthquake, analyzing Holocene emerged shoreline topography in the Iwai lowland, the southwestern part of the Boso peninsula, central Japan. J. Seismol Soc. Jpn. 53, 357−372 (in Japanese with English abstract).

Southard, J.B., Lambie, J.M., Federico, D.C., Pile, H.T., Weidman, C.R., 1990. Experiments on bed configurations in fine sands under bidirectional purely oscillatory flow, and the origin of hummocky cross-stratification. J. Sediment. Petrol. 60, 1−17.

Swift, D.J.P., Figueiredo, A.G., Freeland, G.L., Oertel, G.F., 1983. Hummocky cross-stratification and megaripples: a geological double standard? J. Sediment. Petrol. 53, 1295−1317.

Watanabe, H., 1998. Comprehensive list of tsunamis to hit the Japanese Island, second ed. University of Tokyo Press, Tokyo, p. 238 (in Japanese).

Yokokawa, M., Masuda, F., Sakai, T., Endo, N., Kubo, Y., 1999. Sedimentary structures generated in upper-flow-regime with sediment supply: antidune cross-stratification (HCS mimics) in a flume. In: Saito, Y., Ikehara, K., Katayama, H. (Eds.), "Land-Sea Link in Asia (Prof. K.O. Emery Commemorative International Workshop)," Proceedings of the International Workshop on Sediment Transport and Storage in Coastal Sea-Ocean System, STA (JISTEC) & Geological Survey of Japan, Tsukuba, pp. 409−414.

CHAPTER 6

Deposits of the 1992 Nicaragua tsunami

B. Higman, J. Bourgeois
Earth and Space Sciences, University of Washington, Seattle, WA, United States

1. Introduction

Following a moment magnitude 7.6—7.7 earthquake, the September 1992 Nicaragua tsunami was typified by onshore elevation and run-up of 4—6 m along about 200 km of the Nicaraguan coast (Fig. 6.1; Satake et al., 1993). The tsunami occurred near high tide (Fig. 6.2), which maximized its on-land effects. No coseismic subsidence or uplift was detected onshore, and most models of the earthquake place deformation toward the seaward edge of the subduction zone (e.g., Geist, 1999; Ide et al., 1993; Imamura et al., 1993; Kanamori and Kikuchi, 1993; Piatanesi et al., 1996; Satake, 1994, 1995).

The Nicaragua 1992 earthquake was a typical "tsunami earthquake" (Kanamori, 1972), meaning that it was anomalously efficient in its tsunami generation (Okal and Newman, 2001). Although some authors have invoked a landslide off Nicaragua to explain the high run-up (e.g., Herzfeld et al., 1997), models without landsliding can explain nearly all observed run-up and tsunami-elevation data (Tanioka and Satake, 1996; Titov and Synolakis, 1994). Note that tsunami height at the limit of inundation, commonly reported as "true run-up," is partly a function of local topography. The Nicaragua tsunami's height on land was reduced by about 1% of its penetration distance, so areas where the tsunami penetrated farther had a lower "true run-up" (Fig. 6.3). Other factors such as topographic focusing and wave sloshing affected local tsunami heights, but the data are coherent overall (Figs. 6.1 and 6.3).

The coastline of Nicaragua is a mix of rocky headlands and sandy beaches. On average, two-thirds of the southern coastline is rocky and more rugged, projecting south-westward into the Pacific; there are local salinas and lagoons, as at Playa Hermosa, Las Salinas, Playa de Popoyo, and Huehuete. The northern coastline is relatively straight, of lower relief and mostly sandy; several navigable estuaries are present.

The maximum tsunami elevation along the northern coastline was generally less than 4 m; along the central to southern coastline it was typically 4—6 m (Fig. 6.1). The tsunami apparently had only one damaging wave. The tide-gauge record from Corinto

Tsunamiites. https://doi.org/10.1016/B978-0-12-823939-1.00006-9
Copyright © 2021 Elsevier B.V. All rights reserved.

86 Chapter 6

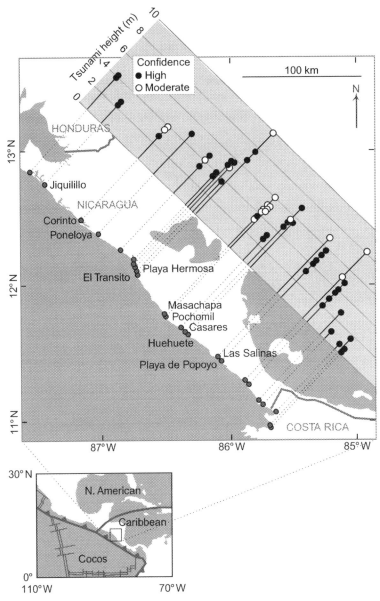

Figure 6.1
Location map and plot of 1992 Nicaragua tsunami elevations in Nicaragua and northernmost Costa Rica, as reported by Abe et al. (1993), Satake et al. (1993), and Baptista et al. (1993). Note that many of these measurements are not "true run-up" (elevation at inland limit of inundation) (see text and Fig. 6.3). Points are coded as "high" or "moderate" confidence depending on the reliability of the marker for tsunami height that was used.

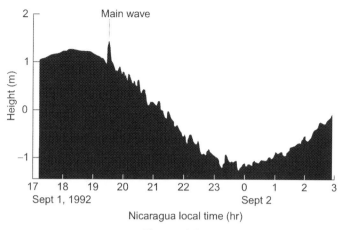

Figure 6.2
Tide-gauge record of the 1992 Nicaragua tsunami from the port of Corinto, within an embayment. Note the single relatively large positive wave, following initial withdrawal. Tsunami heights on the outer shore at this latitude were 2–4 m (Fig. 6.1).

harbor (Fig. 6.2) shows a small retreat and only one large wave (about half a meter on the tide gauge, 2–4 m at the same latitude on the open coast). Eyewitnesses along the Nicaragua coast typically described a single large wave, consistent with the tide-gauge record. Some eyewitnesses described the tsunami as two or three waves in quick succession, without any retreat of the water in between.

The tsunami occurred during the rainy season (boreal summer), so coastal water bodies were full and remained full during initial surveys. All later surveys were conducted in the dry season (boreal winter), when many of these lagoons desiccate; these dry lagoons are called salinas and are commonly mined for salt. Several teams surveyed the run-up and damage of the tsunami in the fall of 1992 (Abe et al., 1993; Baptista et al., 1993, Fig. 6.1) and interviewed local witnesses. Bourgeois participated in a reconnaissance survey of the entire coastline (Satake et al., 1993), visiting most road-accessible localities in late September 1992 and noting and photographing deposits left by the tsunami. In February 1993 (6 months after the tsunami), Bourgeois conducted a survey of a single site, Playa de Popoyo. She measured five profiles and sampled the tsunami deposit. In March 1995, a joint US–Nicaragua team including Bourgeois conducted nearshore bathymetric surveys at El Velero, Playa Hermosa, El Transito and Playa de Popoyo. In March 2003, the Popoyo site and Las Salinas were revisited, surveyed, and sampled by Higman.

2. 1992 tsunami deposits along the Nicaragua coast

In this section, we describe the general characteristics of the tsunami deposits based on 1992, 1993, 1995, and 2003 observations. Where the tsunami run-up was more than about

88 Chapter 6

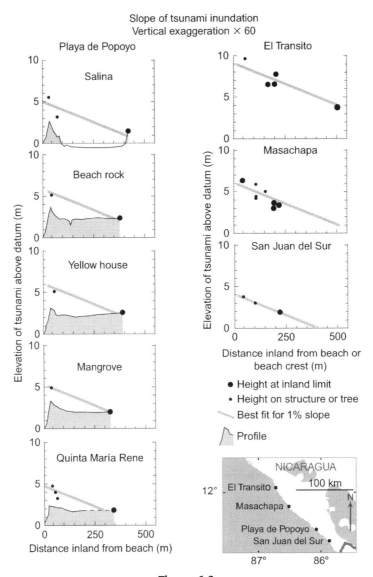

Figure 6.3
Plot of tsunami elevation and distance from shore, from Playa de Popoyo (this contribution) and El Transito, Masachapa, and San Juan del Sur (Abe et al., 1993). Tsunami heights at the limit of inundation were usually marked by floating debris. Other indicators included water marks, tree scars, damage to structures, and seaweed in vegetation. A best-fit 1% slope is depicted to illustrate that the reduction in height with distance is similar at different sites.

2 m, there was some evidence of sediment deposited by the tsunami; at sites with 4 m or more of run-up, the deposits were generally distinctive, of measurable thickness, and potentially preservable. Deposits described further as "thin" are around 1-cm thick.

Correlations between run-up height and deposit properties may be influenced by factors such as local topography and sediment source, as can be shown for the Popoyo site, discussed in more detail in Section 3. If we compare the Nicaragua deposits to those from the 1960s event in Chile (Bourgeois and Reinhart, 1993), they are less extensive, less distinctive, and less likely to be preserved.

The Nicaragua 1992 tsunami deposits are composed primarily of sand and shell debris eroded from the beach or shoreface. The deposits also include larger blocks of beachrock (naturally cemented beach sediment), plant debris such as uprooted shrubs, and anthropogenic materials such as bricks, cut blocks of tuff (a common wall material), concrete and roof tiles, as well as clothing and other artifacts.

In most localities, evidence of flow direction (flopped-over plants or fallen structures) was landward, except in some inlets where flow was parallel to local contours. Flopped-over plants indicating the backwash current (return flow toward the sea) were observed in only one case, at Las Salinas. More detailed paleocurrent observations are presented in the survey of Playa de Popoyo (see further).

2.1 Site-by-site observations

This is a summary, site by site, from north to south, of observations and interviews made principally by Bourgeois in 1992 (September 20 to October 1), with a few additions from later surveys; localities are on Fig. 6.1.

2.1.1 Jiquilillo

The Jiquilillo/Peninsula Padre Ramos site is located on an estuary, Estero Padre Ramos/ Estero San Cayetano. Open-coast run-up at this location (about 2 m) barely topped the low spit south of the estuary entrance, and there was minimal damage. There has been, historically, dramatic coastal erosion of the spit, where the town has lost two streets to erosion that predates the tsunami. Thin sand deposits left by the tsunami will be hard to distinguish from underlying beach-ridge sediment due to their similar character and potential to be reworked. In the interior of the estuary, run-up was less than 1 m, and most tide flats are mangrove-infested, so the only evidence found of the tsunami was washed-up wood and plant debris. We noted no tsunami effects on intertidal vegetation on sand flats directly landward of the mouth of the estuary. Local witnesses said that a storm in 1982 generated very high water in the estuary and did major damage to mangroves.

90 Chapter 6

2.1.2 Corinto

The town and harbor of Corinto are somewhat protected from the open sea by a rocky headland and spit; tsunami sand deposits were present but subtle at the southern end of the Corinto peninsula. In town, to the north (community of Barrio Nuevo), the tsunami overtopped and eroded an artificial sand barrier and deposited a fan delta of sand within the town. This deposit was most extensive 100 m or more landward of the barrier where flow was unobstructed, down a street perpendicular to the shoreline; the deposit texture mimicked the source of well-sorted fine sand. About 2 km north of town (community of Barrio La Boya), where there was no artificial barrier, a thin layer of sand was present in the yard of a house proximal to the beach. The interior of the estuary at Corinto is similar to Estero Padre Ramos, mangrove-infested; the tide gauge in the harbor recorded on the order of a 50-cm first-wave amplitude (Fig. 6.2). Observers reported little effect of the tsunami on tide flats.

2.1.3 Poneloya

Poneloya is a small beach resort and has a low sea wall. A thin, mostly continuous layer of sand was present on the lawns just landward of the sea wall and on floors of the first row of houses.

2.1.4 Playa Hermosa

Run-up at Playa Hermosa was less than 4 m, dramatically less than at El Transito just to the south (Fig. 6.1). At the north and south ends of the embayment, the coastal terrain is upward sloping, and there was little damage to the few, generally well-constructed houses. When observed in 1992, the beach face was eroded slightly into a low cliff; the tsunami deposit was thin, extending only tens of meters from the shoreline, and is unlikely to be preserved. In 1995, excavations were made in small salinas behind the beach ridge in the center of the embayment, ~120 m from the shoreline. The 1992 tsunami deposit (1–8 cm of fine-medium black sand) overlaid gray mud and was in 1995 overlain by about 3 cm of brown mud.

2.1.5 El Transito

El Transito was surveyed in detail by the Japanese team (Abe et al., 1993). Run-up at this site was as high as, or higher than, any other site measured, typically 6–8 m, with massive damage to the town. In 1992, Bourgeois visited only the south end of the site, where most of the terrain is natural topography, including a small stream that directed part of the tsunami flow. Where observed, a mostly continuous sheet of gray sand overlay a hard, brown soil surface to a distance inland of no more than 0.5 km. The sand was typically on the order of 1-cm thick and thinned inland. This site was subsequently heavily reworked by people.

2.1.6 Masachapa

The coastal terrain is upward sloping, and in 1992 some parts of the shoreline had a low sea wall. Reported damage was extensive, but bulldozing had already altered much of the near-field area by late September 1992. Landward of the beach, a thin layer of sand was observed on the lower parts of the slope. It is unlikely that identifiable deposits will be preserved at this location.

2.1.7 Pochomil

The terrain is relatively flat, including a gently sloping beach profile. A thin sheet of sand was deposited on the ground, some beach patios, and the floor of a hotel.

2.1.8 Casares

The coastal terrain is overall upward sloping. Some localities had low sea walls. The flow was directed up an inlet splitting the north and south parts of the town. Thin tsunami sand layers were present in both natural settings and on floored structures.

2.1.9 Huehuete

A small lake exists landward of Huehuete, separated from the sea by a low ridge of compact soil. The tsunami overrode the ridge and flowed into this lake, temporarily raising its level (based on eyewitness accounts). The level had returned to its normal rainy-season elevation by September 22; typically this lake dries up in the dry season. A quick reconnaissance in 1992 near the seaward shore of the lake revealed a tsunami sand 1- to 2-cm thick, overlying brown and black mud, and overlain by about 1 mm of mud; the deposit thinned landward. Grading of the tsunami deposit was not obvious.

2.1.10 Punta Teonoste

Higman visited a low area between rocky headlands near Punta Teonoste in 2003. He took sand samples from a location about 50 m from the beach and a few meters from a mangrove slough that in 2003 was blocked by a beach berm. The deposit is interpreted to be from the 1992 tsunami because it lies on top of cohesive sandy soil and extends well into the shrubby forest along the mangrove slough. The site is close enough to the beach and low enough that a storm-deposit interpretation cannot be ruled out.

2.1.11 Las Salinas

The tsunami washed up the side of a large, steeply sloping dune, where possible tsunami sand could not be distinguished from the dune sand. Just south of the dune, a low, flat surface was covered with one to several centimeters of gray sand. A trench revealed that this sand was underlain by dark, muddy, soft soil. The deposit was apparently crudely graded, coarsening again at the surface. This surface coarsening is interpreted as a lag left

92 Chapter 6

by the backwash, which in this case was strong enough to reorient the local stiff grasses toward the sea—the only place where Bourgeois observed such seaward-directed flop overs.

Higman visited a small salina NW of the dune in 2003. He sampled on a mudflat behind a beach ridge; scattered broken mangrove stumps may be remnants of a destroyed 1992 mangrove swamp. The 1992 tsunami deposit on the mudflat overlies an irregular and locally sandy mud surface and contains some beach gravel and even a small beach cobble. The deposit is obviously graded, and in some places very coarse sand has filled crab burrows in the underlying surface. The location chosen for sampling had a locally more planar base, suggesting it had not been very disturbed since deposition. The deposit was interpreted as a tsunami deposit because the depositional agent transported cobbles over a vegetated beach ridge with structures on it.

2.1.12 Playa de Popoyo

This site is discussed in detail further. Reconnaissance in September 1992 was focused principally on run-up heights and damage to houses, but tsunami-transported sand and debris were observed to be widespread. This site was visited again in 1993, 1995, and 2003. By 2003, several fields behind the beach ridge had been ploughed, destroying the tsunami deposit in those locations.

2.1.13 Northernmost Costa Rica

The terrain at these locales is mostly quite rugged. Run-up was typically less than 1 m, and in 1992 Bourgeois observed only transported plant debris and a displaced boat.

3. Tsunami deposits near Playa de Popoyo

Playa de Popoyo (Fig. 6.4) is a 2-km-long, straight beach, at about $11°27'$ latitude, which provides an excellent case history of the effects and deposits of a tsunami. Overtopping the beach ridge, the 1992 tsunami reached heights of about 5 m, and at the inland limit its elevation was typically 2–3 m above our datum (top of beachrock on beach; Fig. 6.3). Sixteen people lost their lives at this site and most survivors have not returned to live on the Playa. All but three houses along Playa de Popoyo were destroyed beyond repair; the tsunami removed most of them completely from their foundations. These houses were a major source of large clasts in the tsunami deposit (Fig. 6.5; Table 6.1). The other source of large clasts was beachrock, some of which crops out along the beach, except in the far north.

The topography of Playa de Popoyo is relatively simple, and the tsunami came over the beach ridge orthogonally, then drained out through low spots (Fig. 6.4). The interpretation of the tsunami routing (see Fig. 6.4) is based principally on observations of pushed-over and transported vegetation and house parts, as well as on geomorphic features noted in 1992 and early 1993. The northern part of the Playa is typified by a salina behind a single beach ridge (Fig. 6.5). The salina was covered by about 1 m of water when the tsunami arrived and was a salt flat in the dry season when surveys were conducted. Toward the south along Playa de Popoyo, the beach ridge is subdued, and topography behind the beach ridge is flat, entirely supratidal and vegetated (Yellow House, Quinta María Rene, and Mangrove profiles in Fig. 6.5). At the north and south ends of the playa, the tsunami followed topographic contours into the lagoon and the mouth of Rio El Limón. Tsunami withdrawal was focused to three low spots—the lagoon entrance, a slough south of the Mangrove profile, and Rio El Limón (Fig. 6.4).

Inundation distances at Playa de Popoyo range from about 250 to 500 m, except in the salina, which was completely inundated (Fig. 6.4); inundation was interpreted based on dead vegetation and on wrack lines observed in March 1993. Tsunami deposits extended for most of this distance, although they were thin and very fine toward the landward

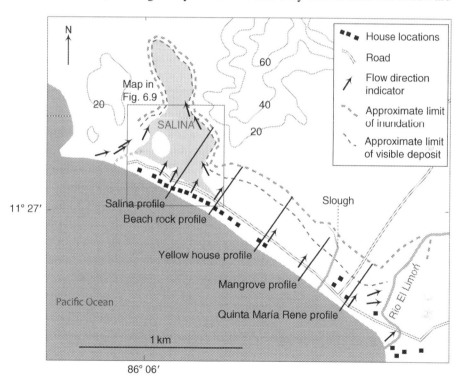

Figure 6.4
Overview map of Playa de Popoyo taken from a 1:50,000 topographic sheet showing profile (Fig. 6.5) locations and tsunami effects recorded in 1992 and 1993. Current indicators were measured in March 1993. Houses plotted north of "Yellow House" are schematic. Limits of tsunami inundation and deposit are approximate, except along profiles, and as noted around the salina in March 1993.

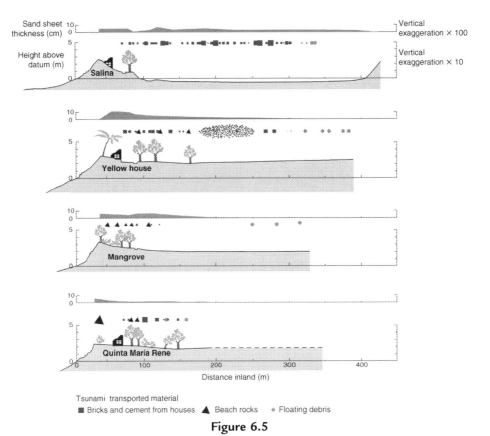

Figure 6.5
Profiles (mapped in Fig. 6.4) measured and described in March 1993 (excluding "Beachrock"). See Appendix A for notes on protocol. Tsunami sand thickness is generalized from point observations (small excavations at 10–20 points per profile). Positions of large clasts are generalized; see Table 6.1 for some more specific large-clast data.

limit (Fig. 6.5). The extent of inundation and tsunami deposits increases toward the north. In general, vegetation is denser in the southern part of Playa de Popoyo. By 1995, shrubby vegetation along the entire shoreline had increased in density and even more so by 2003.

Along the profiles that were measured in 1993 (Fig. 6.5), the tsunami deposit becomes finer and thinner landward and fines upward. In seaward proximal deposits (the term "proximal" is used here to indicate a position relatively close to the ocean, whereas "distal" is used here to indicate a more landward position), a coarse surface layer was present in several localities and interpreted as a withdrawal scour lag; outside the salina, some wind deflation was observed in 1993. Proximal deposits on the beach ridge are typically 5- to 10-cm thick and composed of very coarse sand with pebbles and

Deposits of the 1992 Nicaragua tsunami

Table 6.1: Large clasts surveyed (March 1993) and other basic observations.

Material	Distance from source[a] (m)	Dimensions (cm)	Notes
Profile 1 (south of Quinta María Rene)			
Beachrock	5	230 × 160 × 45	Minimum vertical 1.5 m
Beachrock	65	36 × 15 × 5	Begin field of house blocks
Beachrock	75	35 × 20 × 5	Last large beachrock
Wall	50	54 × 23 × 18	
Wall	50	90 × 54 × 18	
Wall	70	112 × 78 × 25	
Wall	100	47 × 44 × 11	Last large clast
Field of blocks	25–120	Many 10–20 cm diameter	
Limit of sand	~200		
Water limit	~300		
Profile 3 (Mangrove)			
Beachrock	20	100 × 75 × 16	First large block
Beachrock	30	124 × 117 × 24	
Beachrock	35	120 × 74 × 23	Tilted on tree
Beachrock	50	100 × 72 × 13	
Beachrock	60	70 × 50 × 25	Against fence
Beachrock	85	120 × 75 × 15	Clast is very scalloped
Beachrock	90		Last rocks about here
Limit of sand	~200		
Last debris	~250		Floating debris (woody)
Profile 4 (yellow house)			
Beachrock and walls		Not measured	Against tree
Wall	40	80 × 65 × 14	
Wall	50	120 × 112 × 14	
Tuff block	50	43 × 34 × 16	
Beachrock	110	75 × 65 × 20	
Wall	75	143 × 67 × 14	
Beachrock	140	89 × 84 × 17	
Field of bricks	100–200	Many clasts of 1–4 bricks	
Wall	160	87 × 71 × 14	Large wall blocks near here
Wall	170	85 × 76 × 14	
Wall	200	85 × 58 × 14	
Last wall pieces	220	Not measured	
Last bricks	240	Not measured	
Limit of sand	~300		
Last debris	~300	Less dense than bricks (Roofing, wood, ...)	

[a]The source is generally either the zone of beachrock on the beach or a house foundation. In the case of beachrock, the distance is a minimum. In the case of sand limit, the distance is from the shoreline.

shell debris. The body of the deposit, over most of the profiles, comprises 1−5 cm of coarse to fine sand, fining landward (Fig. 6.6). Some deposits in the salina exhibit flat to low-angle heavy mineral lamination (Fig. 6.7). The landward tail of the deposit is less than 1 cm of fine to very fine sand and silt. In most cases, large clasts are present for over half of the inundation distance (Figs. 6.5 and 6.8; Table 6.1).

4. Grading of the tsunami deposits

The grading in the tsunami deposits is a reflection of the transport and deposition by the tsunami. Differences in vertical grading between proximal and distal portions of the deposit may help constrain how the tsunami is recorded by its deposit. We structured our interpretation of grading in the 1992 Nicaragua tsunami deposit around several questions:

- How do vertical and horizontal grading trends relate?
- At what point(s) in the cycle of erosion, transport, and deposition does sorting occur?
- Are laterally similar layers, appearing to be correlative, deposited at the same time from a flow, or progressively in time (as in progradation)?
- How are different flow features, such as breaking fronts or surges, recorded in the deposit?

Grading within tsunami deposits is commonly described in three ways: proximal to distal, bottom to top, and along the shore. Only proximal-to-distal grading and bottom-to-top grading are treated in this analysis of the Nicaragua tsunami deposit.

The 1992 Nicaragua tsunami near Playa de Popoyo transported everything from mud to boulders, so the deposit contains any sediment that was available for transport. A textural description of such a deposit primarily reflects the sediment source rather than the tsunami itself. Therefore, a description of *change* within a deposit, such as grading, is more likely than the overall grain-size distribution to reflect the processes of erosion, transport, and deposition.

The data presented here are grain sizes inferred from terminal settling velocity distributions generated using a settling column. We prefer these data over directly measured grain-size data for this study because grains with similar settling velocity but different grain size are likely to behave more similarly than grains with similar size but different settling velocity.

4.1 Landward grading

At Popoyo, settling velocity distributions generally fine landward (Fig. 6.6), although grading is reversed between some sample pairs. Change in mean grain size is typically around 0.5 φ per 100 m, although it is much less on the Quinta María Rene profile.

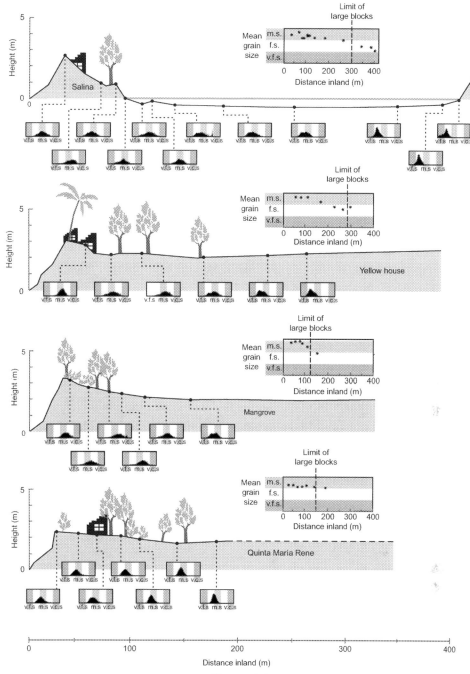

Figure 6.6
Grain-size distributions of 1992 Nicaragua tsunami deposits sampled in 1993 along four measured profiles at Playa de Popoyo (Fig. 6.5). The mean grain size is plotted versus distance above each profile. On the mean-grain-size plot, the analytical error is less than the size of the dot. See Appendix A for comments on sampling.

98 Chapter 6

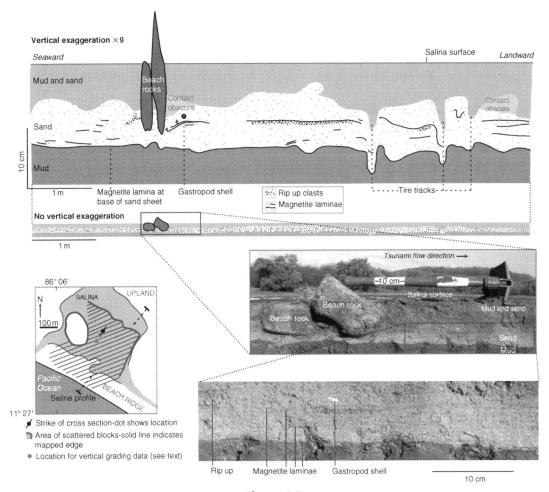

Figure 6.7

Internal structure of the 1992 Nicaragua tsunami sand sheet. Sketch and photo from a 2003 trench in the Popoyo salina, about 10 m NW of the "Salina" profile and about 250 m from the beach, where the deposit is particularly thick and undisturbed. Heavy-mineral laminae are visible at different depths in the deposit; however, one particularly prominent lamina divides the deposit approximately in half in many places. Also noted on the sketch are rip-up clasts (<1 cm diameter), which most commonly occur directly below the prominent lamina. At the distal end of the sketch, several tire tracks have pushed the deposit down into the soft underlying mud.

Large-clast grading at Playa de Popoyo is more complex. For example, at the Yellow House profile, a lozenge-shaped field of bricks was strewn in a wake-like pattern between 100 and 160 m from the source house (Fig. 6.5). The density of clast distribution was lower both seaward and landward of this field of bricks. Also along the Salina, Yellow House, and Quinta María Rene profiles, the number of large clasts decreases abruptly near the limit of their extent. Along the Salina profile, at about 300 m inland, this abrupt

Deposits of the 1992 Nicaragua tsunami 99

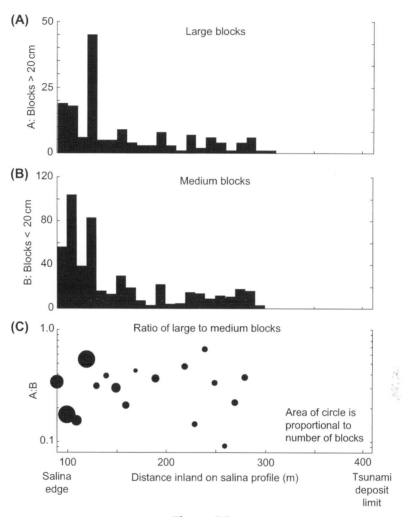

Figure 6.8
Summary of data collected on the distribution of blocks along the Popoyo "Salina" profile (Fig. 6.4). Almost all of the blocks along this profile were house fragments, including tuff blocks, reinforced concrete, and large sections of intact brick walls. The profile was divided into 10-m-long and 15-m-wide rectangles, and all solid blocks larger than a few centimeters within each section were counted and categorized by median length. Most blocks in the "medium" category (<20 cm) were individual bricks, while those in the "large" category (<20 cm) were multiple bricks connected together, or concrete, or, especially, large bricks. Blocks with very high surface areas, such as tiles and cinder blocks, were not counted. The dots mark the ratio between the number of large blocks and the number of medium blocks within each 10-m increment. Ratios for counts of less than 10 blocks are excluded.

termination in the blocks is coincident with a transition in the sand-size settling velocity distributions (Figs. 6.6 and 6.8).

Along the Salina profile, we counted the number of two different-size classes of large clasts (Fig. 6.8). Most of these clasts are house fragments, and they appeared to be all of similar density. Overall the ratio between the large (>20 cm) and medium (<20 cm) classes has no clear trend. If these clasts were graded to fine landward, this ratio would decrease landward. We also mapped two distinctive groups of bricks, patterned cinder blocks and yellow bricks, coming from one house (Fig. 6.9). The cinder blocks had a higher source within the house walls than did the yellow bricks. We found that the cinder blocks were less scattered down flow.

Both of these observations support the idea that the large clasts were distributed throughout the tsunami rather than concentrated at or near the base. The lack of grading in the large clasts (Fig. 6.8C) suggests that their transport in the tsunami was not primarily

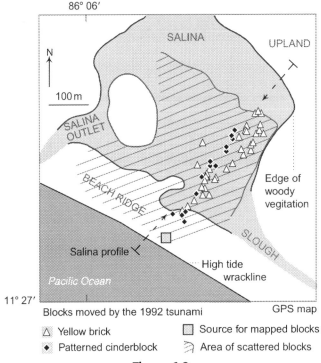

Figure 6.9
GPS map (2003) showing distribution of tsunami-transported blocks on the Popoyo salina, approximately along the "Salina" profile (Fig. 6.6). Two different distinctive brick types from a single house are plotted. The patterned cinder blocks appeared to have come from high on the wall around the bathroom, while the yellow bricks formed complete walls of the house. Solid boundaries are GPS-mapped, including the landward limit of scattered blocks, which is abrupt.

dependent on their size. We think that it is unlikely that the clasts were transported by sliding or rolling, cases where the size influences the friction with the bed and the degree of coupling to the flow. We interpret the different scattering of the two clast types (Fig. 6.9) to indicate that source height was the characteristic that distinguished the two populations in their scattering. If so, this pattern indicates that the clasts were distributed throughout the tsunami water column, not concentrated at the base of the flow during their transport. However, these patterns may not be representative of cobbles or boulders in natural settings, as the clasts in question were "entrained" from standing structures, not from the ground.

4.2 Vertical grading

Vertical grading was measured in tsunami deposits from five trenches where sets of vertically contiguous samples were collected (Fig. 6.10). Normal grading dominated in each case, but a small section with inverse grading was present at the base of all but the proximal Playa de Popoyo deposit. Generally, grading was weaker and sorting better (grain-size distributions narrower) at the top of each deposit. These similarities in grading pattern occur in spite of differences in the overall distribution of sand size among the locations sampled.

4.2.1 Playa de Popoyo

In the salina at Playa de Popoyo, sets of vertically contiguous samples were taken from trenches at two locations at distances of about 200−250 m from the beach. These trenches were along a line where the deposit was consistently over three times as thick as the usual 3 cm observed in most of the salina. We used detailed measurements of the samples to correlate grain-size distributions between the proximal and distal trenches (Fig. 6.11). The distributions are very similar between the upper part of the proximal deposit and two different portions of the base of the distal deposit. We favor the upper of these grain-size correlations because it is additionally associated with an abrupt jump in mean grain size and a layer of rip-ups in both deposits.

The similarity in grain size, grading, and relative rip-up concentration between the two points is probably a reflection of the sorting process rather than a marker for simultaneous deposition. At any given moment, the water passing over these locations is different and is carrying sediment eroded either from different places or at different times. Also, the flow conditions are unlikely to be homogeneous at any given time, as tsunami flows are always changing, and it takes time for changes to propagate through space. However, the similarity in the material deposited (grain-size distribution and rip-ups) and the similarity in the grading of the deposit (abrupt fining overlain by subtle grading) suggest that the process most responsible for the vertical variation in grain-size distribution was similar between the two locations. The sorting may occur during either erosion or deposition (or both).

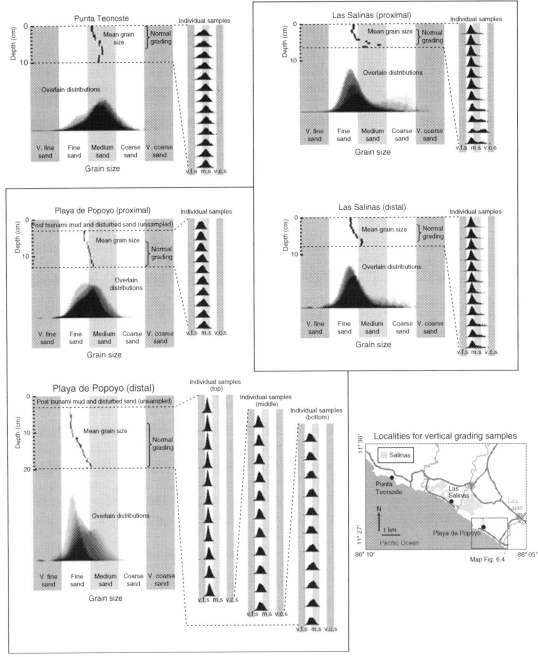

Figure 6.10

Grain-size distributions for five vertically distributed sample sets collected in 2003. For each trench, the mean grain size is plotted versus depth; all the distributions are overlain to make a composite distribution, and each distribution is plotted separately to the right. For the mean grain size versus depth plot, the height of rectangular marks indicates the depth range that was sampled, while the width is one standard deviation of at least three separate analyses of the same sample (an estimate of standard error). The only exception to this is "Las Salinas" (distal), where samples were run only once, in which case a mark that is an estimate of standard error is plotted.

Deposits of the 1992 Nicaragua tsunami 103

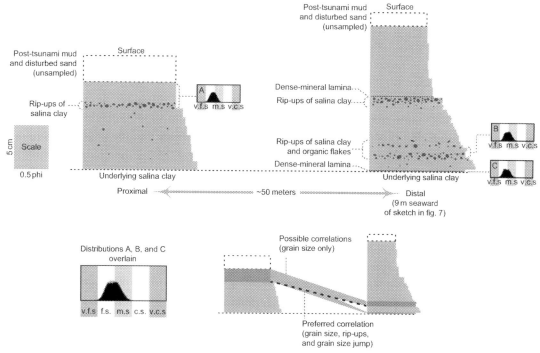

Figure 6.11

Stratigraphy and interpretation of two trenches in the salina at Playa de Popoyo (see location in Fig. 6.7). The grain-size data for these trenches is reported in Fig. 6.10. These trenches were about 100 m north of the Salina profile and were about 200 (proximal) and 250 (distal) m from the ocean. For the stratigraphic sections, the width of the section is related to the mean grain size of samples collected at different depths (see Appendix A on vertical sampling). Wider parts of the stratigraphic section have coarser sediment. The tops of both of these trenches were not sampled. A correlation is noted where the top of the proximal deposit is similar in structure and composition to the base of the distal deposit.

In the case of *erosional sorting*, the sorting reflects a biased removal of sediment from the source. The similar sections in the two deposits reflect sediment deposited from the same water at different times. They are similar because little sorting occurred during transport and deposition, so the sorting processes involved in erosion set the distribution that a given section of the flow laid down as it propagated inland.

In the case of *depositional sorting*, the sorting reflects a biased addition of sand to the deposit. The similar sections in the two deposits reflect sediment eroded from a homogenized source and then sorted during deposition. When the same process acted to the same extent on the same source sediment, the sediment deposited would be the same.

Our data are insufficient to distinguish these two cases. However, each case makes different predictions about how different parts of the tsunami flow would be recorded in the deposit. If most sorting occurred during erosion, then it was the conditions at the sediment source that controlled grading, except that it was the conditions at the point of deposition that controlled whether deposition occurred. If, instead, most sorting occurred during deposition, then the source only set the overall distribution of sediment in the deposit, and it was conditions near the point of deposition that were recorded.

4.2.2 Las Salinas

Just north of Playa de Popoyo, in a salina near Las Salinas, we measured vertical grading in two trenches, about 35 m apart, perpendicular to the beach (Fig. 6.10). Correlations of grain-size distributions, drawn and interpreted in a similar way as in the Playa de Popoyo salina, show little lateral change in the deposit, except for fining of the base of the deposit and also of patches of sediment in local depressions (Fig. 6.12).

In this salina, there were burrows and other depressions that were filled with sand before the main deposit formed (Fig. 6.12). Sediment in the depressions is coarser than the main sand sheet. The depressions varied in geometry, but some burrows were deep enough, and it is likely that any grain that reached there was deposited permanently. Thus, burrowed

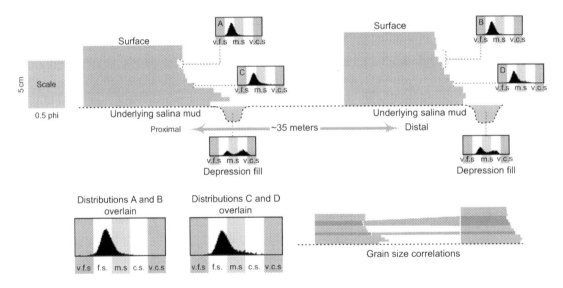

Figure 6.12
Stratigraphy and interpretation of two sections in a salina near Las Salinas (see location on insert of Fig. 6.10). Both these sections are part of a 50-m-long trench across a portion of a thicker and less disturbed deposit about 200 m from the ocean. No profile was measured here; however, it was generally similar to other measured profiles with a beach ridge separating the salina from the ocean (e.g., the Popoyo "Salina" profile; Fig. 6.5). Along this long trench, 10 depressions and burrows filled with particularly coarse sediment were sampled, and the grain-size data for depressions nearest the two sections are reported here. Grain-size correlations between the two sections show little change in the upper two-third of the deposit, but a marked lateral fining of the basal sections and depression fills.

deposits should be a record of the first water in the flow. Eyewitnesses described the tsunami approaching as a breaking bore, so the front of the wave probably had greater basal shear stress and turbulence than the water immediately behind it. High-shear and strong mixing is ideal for the entrainment and transport of coarse sediment such as what is plugging the depressions along an otherwise erosive bed.

5. Discussion

Grading in the deposit is similar at all locations studied along the coast, suggesting that this grading reflects sorting processes in the tsunami rather than in the sediment source. There *are* variations in sediment source, as reflected in variable grain-size distributions along the coast. It is unlikely that every location would show dominantly normal grading that became less strong toward the top if this grading were a result of local source conditions.

Discrete, coarse sediment–filled pockets below a tsunami sand sheet, such as the burrow fillings in this case, may provide distinct records of steep bore-fronted flows such as tsunamis and turbidity currents. Storms rarely approach as a breaking bore, so depressions plugged with coarse sediment might provide a criterion for distinguishing storm and tsunami deposits.

The tsunami deposit reflects the tsunami flow on various spatial scales, from tens of meters to kilometers or more. The wave in this area—and probably across much of Nicaragua—was simple. There was only a single large wave, and at the sites studied beach ridges limited the strength of withdrawal. Deposit grading is therefore simple and consistent from place to place. Some features, such as subtle grading at the top of the deposit, are consistent all along this section of coast. Other features, such as the abrupt jump to finer grain size overlain by rip-ups in the Playa de Popoyo salina, are consistent between closely spaced trenches, but do not extend between locations. There are also differences between trenches that are close together. These variations reflect spatial variation in the temporal flow structure of the tsunami and show that deposits have the fidelity to record both large-scale tsunami structure and local variation.

Appendix A
Field and laboratory protocols
A.1
Sea level data

Most of the tsunami elevation data were collected in late September 1992 and corrected (with tide tables) from tide level on the day of measurement to tide level at the time of the tsunami, just after high tide on September 1, 1992. In most cases, a hand level and stadia rod were used, and distances estimated, mapped, or measured by GPS.

In March 1993, five profiles were measured across the Playa de Popoyo (Figs. 6.4 and 6.5), using a tripod, transit level, and stadia rod. On the profiles, the water level at the time of the 1992 tsunami is estimated and used as datum (=0). This estimate is based on local observation of high tide over several days, which typically came to a level just below the upper beachrock step. This level also corresponds to the limit of grassy vegetation in the salina. The tsunami elevation data collected in September 1992 are plotted onto the profiles, but only some were remeasured, so these elevation data may have an error on the order of decimeters.

A.2
Proximal to distal sampling

In March 1993, four of the profiles were sampled for tsunami deposits (Fig. 6.8), associated with survey points along the profiles. Because it was dry season, the tsunami deposits were loose, and some wind deflation had taken—and was still taking—place. Some of the finer sediments were blown away during sampling. Bourgeois tried to take a representative sample of the entire deposit, top to bottom. The best samples were from the Salina profile because they were damp and capped by mud, except the most distal cases, which were exposed and dry. During the 2003 return trip to Nicaragua, Higman outlined 10-m-long and 15-m-wide rectangles along the Salina profile and counted the number of bricks in each of the two categories, one with a median diameter between 10 and 20 cm, the other with a median diameter of over 20 cm.

A.3
Vertical sampling

No vertical sampling was done during the 1993 or 1995 visits. In 2003, none of the locations where profiles had been measured in 1993 proved suitable for collecting vertical grading samples. However, in three other locations, one close to the Salina profile, Higman collected vertically distributed samples. To sample for vertical grading, we cleaned the upper surface of the deposit and cut a small tablet of sand out with a knife. When this sample was completely removed, leaving a flat surface below, another tablet was cut out as the next sample. This technique yielded samples that were contiguous and minimally mixed with adjacent samples. The top of the salina deposit, where it is mixed with overlying mud, was not sampled.

A.4
Analysis

For analysis, these samples were split to less than 0.7-g submerged weight and run through a 189-cm settling column. Cumulative submerged weight versus time data from the settling column were used to estimate equivalent grain size of silica spheres. The plotted distributions have arbitrary proportion units with constant area.

Acknowledgments

The authors would like to thank Andrew Mattox, Jose Borrero, Costas Synolokis, Harry Yeh, Becquer Fernandez, and Gordie Harkins for their participation in the posttsunami surveys and David Trippett for assistance with laboratory analysis of samples. Brian Atwater provided critique that led to substantial improvement in our figures.

References

Abe, K., Tsuji, Y., Imamura, F., Katao, H., Yoshihisa, I., Satake, K., Bourgeois, J., Noguera, E., Estrada, F., 1993. Field survey of the Nicaragua earthquake and tsunami of September 2, 1992. Bull. Earthquake Res. Inst. 68, 23—70. Tokyo Daigaku Jishin Kenkyusho Iho.

Baptista, A.M., Priest, G.R., Murty, T.S., 1993. Field survey of the 1992 Nicaragua tsunami. Mar. Geodes. 16, 169—203.

Bourgeois, J., Reinhart, M.A., 1993. Tsunami deposits from 1992 Nicaragua event: implications for interpretation of paleo-tsunami deposits, Cascadia subduction zone. EOS 74 (43 Suppl.), 350.

Geist, E.L., 1999. Local tsunamis and earthquake source parameters. In: Dmowska, R., Saltzman, B. (Eds.), "Tsunamigenic Earthquakes and Their Consequences" Advances in Geophysics, vol. 39, pp. 117—209.

Herzfeld, U.C., Von Huene, R., Kappler, W., Mayer, H., 1997. Characterization of high-resolution geomorphological processes; slumps on the slope of the Middle America Trench. In: Pawlowsky-Glahn, V. (Ed.), Proceedings of IAMG '97, the Third Annual Conference of the International Association for Mathematical Geology (IAMG), pp. 585—590.

Ide, S., Imamura, F., Yoshida, Y., Abe, K., 1993. Source characteristics of the Nicaraguan tsunami earthquake of September 2, 1992. Geophys. Res. Lett. 20, 863—866.

Imamura, F., Shuto, N., Ide, S., Yoshida, Y., Abe, K., 1993. Estimate of the tsunami source of the 1992 Nicaraguan earthquake from tsunami data. Geophys. Res. Lett. 20, 1515—1518.

Kanamori, H., 1972. Mechanism of tsunami earthquakes. Phys. Earth Planet. In. 6, 246—259.

Kanamori, H., Kikuchi, M., 1993. The 1992 Nicaragua earthquake: a slow tsunami earthquake associated with subducted sediments. Nature 361, 714—716.

Okal, E.A., Newman, A.V., 2001. Tsunami earthquakes: the quest for a regional signal. Phys. Earth Planet. In. 124, 45—70.

Piatanesi, A., Tinti, S., Gavagni, I., 1996. The slip distribution of the 1992 Nicaragua earthquake from tsunami run-up data. Geophys. Res. Lett. 23, 37—40.

Satake, K., 1994. Mechanism of the 1992 Nicaragua tsunami earthquake. Geophys. Res. Lett. 21, 2519—2522.

Satake, K., 1995. Linear and nonlinear computations of the 1992 Nicaragua earthquake tsunami. In: Satake, K., Imamura, F. (Eds.), Tsunamis; 1992—1994, Their Generation, Dynamics, and Hazard. Pure and Applied Geophysics, pp. 455—470, 144.

Satake, K., Bourgeois, J., Abe, K., Tsuji, Y., Imamura, F., Iio, Y., Katao, H., Noguera, E., Estrada, F., 1993. Tsunami field survey of the 1992 Nicaragua earthquake. EOS 74 (145), 156—157.

Tanioka, Y., Satake, K., 1996. Tsunami generation by horizontal displacement of ocean bottom. Geophys. Res. Lett. 23, 861—864.

Titov, V.V., Synolakis, C.E., 1994. A numerical study of wave runup of the September 2, 1992 Nicaraguan tsunami. In: Proc. IUGG/IOC International Tsunami Symposium, pp. 627—635.

CHAPTER 7

Sedimentary characteristics and depositional processes of onshore tsunami deposits: an example of sedimentation associated with the July 12, 1993 Hokkaido-Nansei-Oki earthquake tsunami

F. Nanayama

Geological Survey of Japan, National Institute of Advanced Industrial Science and Technology (AIST), Tsukuba, Ibaraki, Japan

1. Introduction

Numerous studies of modern and past onshore tsunami deposits have been undertaken worldwide since 1960. Sand sheets are a common feature in tsunami-devastated lowlands such as coastal marshes and lagoons. Such material has been noted for most modern tsunamis, including the 1960 Chile (Kon'no, 1961; Wright and Mella, 1963), the 1983 Nihonkai-Chubu (Minoura and Nakaya, 1991), the 1992 Flores Island (Minoura et al., 1997; Shi et al., 1995), the 1993 Hokkaido-Nansei-oki (Nanayama et al., 2000; Nishimura and Miyaji, 1995; Sato et al., 1995), the 1994 Java (Dawson et al., 1996), and the 1998 Papua New Guinea (Gelfenbaum and Jaffe, 2003) tsunamis.

Ancient tsunamis have also been inferred on the basis of sand sheets found in coastal lowlands; these include tsunamis that occurred at (or reached) Scotland (Dawson et al., 1988, 1991), Norway (Bondevik et al., 1997), Cascadia along the Pacific coast of North America (Atwater, 1987, 1992; Atwater and Moore, 1992; Benson et al., 1997; Clague and Bobrowsky, 1994; Clague et al., 1999; Hutchinson et al., 1997), northern Japan (Minoura et al., 1994), Kamchatka (Minoura et al., 1996; Pinegina and Bourgeois, 2001), and the Aegean Sea area (Minoura et al., 2000). Sand sheets are important markers of ancient tsunamis especially and may serve as the only record of some prehistoric tsunamis.

Tsunamiites. https://doi.org/10.1016/B978-0-12-823939-1.00007-0
Copyright © 2008 Elsevier B.V. All rights reserved.

110   Chapter 7

In general, (1) relatively coarser sediments such as sandy or gravelly layers within a peat succession or within lacustrine sediments, (2) the presence of fining upward sequences or graded structure, and (3) a landward thinning of wedge-shaped sediment sheets are common characteristics of onshore tsunami deposits throughout the world (e.g., Dawson et al., 1988). Minoura and Nakaya (1991) also pointed out that the seawater that floods onto land as a result of a tsunami incursion causes distinctive processes of sedimentation in coastal marshes and lagoons.

There are many other coastal phenomena, however, that leave such sand sheets. Similar sheets have been reported from long-term sea-level changes, storm surges, and river floods, but they can also result from jetted sand by liquefactions. Previous tsunami studies have commonly involved a system of eliminating all other possibilities so that a tsunami was the only reasonable explanation. Such an elimination process is possible only, however, if detailed descriptions of observed tsunami sedimentation are available so that depositional models including the general features of onshore tsunami deposits can be developed.

There are, however, only few descriptions of sedimentary structures and facies of onshore tsunami deposits, because such studies have usually been carried out by coastal engineering researchers and seismologists who are interested in hazard prediction rather than in sedimentation processes. The following workers published some of the most important reports of tsunami-related sedimentary structures and facies. Bondevik et al. (1997) described sedimentary facies and sedimentation processes of the Storegga tsunami that, 7000 year BP, reached shallow-marine basins and coastal lakes in Norway. Sato et al. (1995) reported current ripples and dunes in the onshore tsunami deposits caused by the 1982 Japan Sea and the 1993 Hokkaido-Nansei-Oki earthquakes. Also, Nanayama et al. (2000) reported many sedimentary structures controlled by the inflow and the outflow of the 1993 Hokkaido-Nansei-Oki tsunami on the Miyano coast, northern Japan, and Peters et al. (2001) summarized the sedimentary facies of the 1700 CE Cascadia Subduction Zone (CSZ) tsunami deposits in coastal marshes and lagoons, NW America.

In this chapter, the onshore tsunami deposits from the 1993 Hokkaido-NanseiOki earthquake tsunami are described. We had the opportunity to study these deposits in detail 4 years after the tsunami had occurred, in an estuary area of the Usubetsu River at Taisei, along the western side of the Oshima Peninsula (Fig. 7.1A). These relatively recent tsunamis are of particular value because eyewitnesses can confirm how tsunami sediment deposition occurred at particular locations.

2. General setting

On July 12, 1993, an earthquake of 7.8 magnitude occurred with its epicenter ~200 km northwest of Okushiri Island (42°46.8′N, 139°11.1′E) (Fig. 7.1B). This Hokkaido-Nansei-

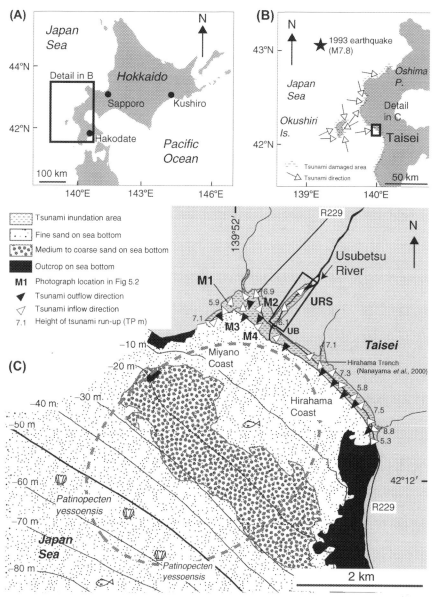

Figure 7.1

Location maps. (A) Location of the study area in south-western Hokkaido. (B) Epicenter (asterisk) of the 1993 Hokkaido-Nansei-Oki earthquake (Nakanishi et al., 1993), the tsunami-damaged area along the western coast of Oshima Peninsula, and the location of the study area at Taisei. Small arrows represent directions of advancement near the coast. (C) Onshore and offshore areas near Taisei showing the area that was inundated in 1993, and the study area at the Usubetsu River. Small arrows show the inflow (white) and the outflow (black) directions of the 1993 tsunami. The ellipse indicated by a dotted line shows the postulated provenance area of the 1993 onshore tsunami deposits (Nanayama et al., 2000). *UB*, Usubetsu-bashi Bridge, *URS*, the study section at the Usubetsu River. *(B) Modified from Shimamoto, T., Tsutsumi, A., Kawamoto, E., Miyawaki, M., Sato, H., 1995. Field survey report on tsunami disasters caused by the 1993 Southwest Hokkaido Earthquake. Pure Appl. Geophys. 144, 665–691. (C) Modified from Ganzawa, Y., Kito, N., Sadakata, N., 1995. Tsunami of the 1993 southwest off Hokkaido earthquake and refuge behavior of people: in the case of the Taisei town in Hokkaido, Japan. Earth Sci. 49, 379–390. (in Japanese, with English abstract).*

112 Chapter 7

Oki earthquake caused a tsunami that flooded not only Okushiri Island but also the parts of the western side of the Oshima Peninsula in south-western Hokkaido (Fig. 7.1B), including Taisei. The tsunami developed immediately after the earthquake had struck the south-western coast of Hokkaido, which resulted in 229 people reported missing or killed (Fig. 7.2).

Investigations by Ganzawa et al. (1995) and Tsuji et al. (1994) at Taisei indicated that two tsunami flood waves had struck the coast and caused widespread destruction. The first had a maximum height of 7.0 m, and the second wave was even higher, at 8.6 m (Tsuji et al., 1994). The alluvial plain at Taisei is 3–7 m above the mean sea level, and the 1993 tsunami reached a maximum of 200 m in distance from the shoreline (Fig. 7.1C). Several Taisei residents were drowned, and buildings were moved as much as 30 m inland (Fig. 7.2: M1 and M2).

The sea floor off this section of the coast gradually descends (gradient 20/1000) to a depth of −400 m about 2 km from the shore. The bottom sediments off Taisei between the shoreline and −200-m water depth are fine sands, and those deeper than −200 m are silt. At −20-m depth, coarse to medium-grained sand bodies have been deposited by longshore currents parallel to the shoreline (Sagayama et al., 2000, Fig. 7.1C). The shoreline of the Miyano and Hirahama coasts has an arcuate shape (Fig. 7.1C). The coastal gravels are subrounded and pebble- to cobble-sized and have a basaltic, andesitic, or granitic composition: the coastal sands are subrounded medium- to coarse-grained quartz sands supplied from the Usubetsu River.

The Usubetsu River, which has its mouth at Taisei, is about 11-km long; its maximum width is 50 m at the Usubetsu-bashi Bridge (UB in Figs. 7.1C and 7.3). The river's gradient is 9.7/1000 in its lower reaches. This river flows along the north-eastern side of an alluvial plain, where its straight channel conforms to the local geological lineaments. No delta has developed at the river mouth because of strong wave action and longshore currents (Fig. 7.1C). Its banks are protected by concrete levees between the Usubetsu-bashi Bridge and the estuary area of the Usubetsu River. The fluvial gravels and sands of the riverbed are subrounded to subangular and consist of basic volcanic and granitic rocks derived from the geological formations in the river's upper reaches (Fig. 7.1C).

3. Methods

Sediments were sampled along two survey lines on a large longitudinal bar in the Usubetsu River between the Usubetsu-bashi Bridge and a point 260 m upstream of the bridge (the white asterisk in URS, Fig. 7.2). Samples were analyzed as described further.

Figure 7.2
Damage caused by the tsunami. M1: Tsunami damage photographs taken in Miyano, Taisei. M2: Breakwater blocks moved by the inflows. M3: Erosional cliffs caused by the outflows. M4: Flattened plants showing the outflow directions. URS: Upstream side of the Usubetsu-bashi Bridge (UB), showing study area, with two survey lines (*lines* A and B) that piled up the tsunami-flooded image. *White asterisks* show tsunami heights (TP m), and *white triangles* show sites marked on the survey lines of Fig. 7.3. For locations, see Fig. 7.1C.

3.1 Field survey

Two survey lines (lines A and B) were established at approximately right angles to the modern shoreline on the surface of the longitudinal bar in the Usubetsu River (Fig. 7.3). Altitudes and distances from the shoreline were measured at all sampling points (Fig. 7.3).

114   Chapter 7

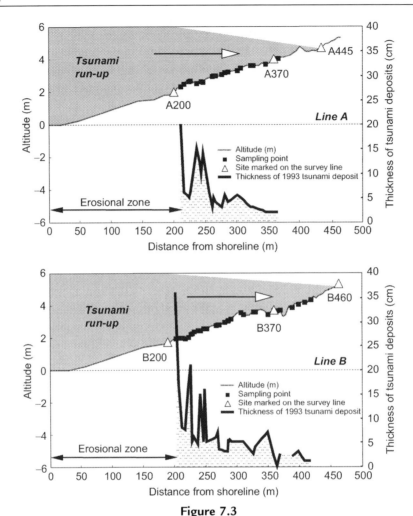

Figure 7.3
Topographic profiles and distance from the present shoreline of each survey line, showing the thickness of the 1993 onshore tsunami deposits along *lines* (A and B).

Furthermore, the tsunami deposits were traced by scoop sampling every 5 m; in addition, 50 pits smaller than 50 cm × 30 cm with a depth of 30 cm had been dug, and 10 lacker-peel samples had been obtained by spray bond; 22 oriented lunch box samples had been taken from the walls of these pits. Furthermore, the gravel fabric, sizes, and compositions had been measured in three-pit walls near the bridge.

3.2 Sedimentary description

After bringing the samples to the laboratory, the box samples were opened and three-dimensional-oriented peels were made according to the lunch box method described by Nanayama and Shigeno (1998). Then, the sedimentary structures, bed forms, grain size, and colors of each sample were described. Finally, a stereoscope was used to describe the mineral composition and shapes of the grains.

4. Results

4.1 General characteristics

Multiple layers of the 1993 tsunami deposits were found in many pit walls. The deposits, which overlie the 1993 soil, are <36-cm thick (Figs. 7.3 and 7.4), and they were found only between 200 and 430 m from the present shoreline because of intensive erosion by concentrated outflows at the narrow river mouth (Fig. 7.3).

At B210, the tsunami deposits reach their maximum thickness of 36 cm; they consist of cobbles and pebbles (maximum diameter 20 cm) and pebbly coarse sands (Fig. 7.4). These coarse deposits have a clear mound shape (Fig. 7.4). Between B216 and B250, sand sheets

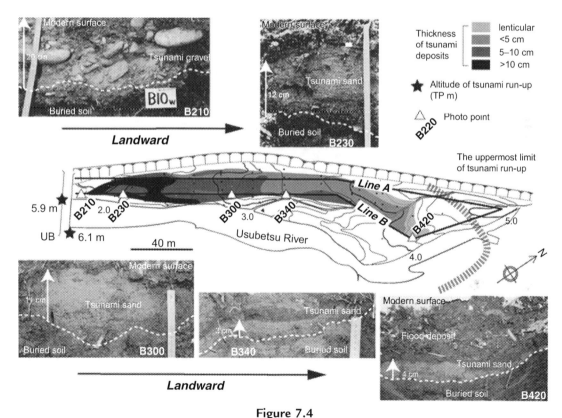

Figure 7.4
Typical occurrences of tsunami deposits, and an isopach map of the study section showing the thickness of the 1993 onshore tsunami deposits and the locations of the photographs. *UB*, Usubetsu-bashi Bridge.

116   Chapter 7

of more than 10-cm thickness are present; they are composed of coarse to fine sand with gravel. Furthermore, between B250 and B420, thin fine-sand layers occur. Above B420, the deposits are lenticular, and they pinched out at B430 (Fig. 7.4).

Generally, the thickness of the tsunami deposits appears to be determined by the topography so that the thinning takes place away from the shoreline (Figs. 7.3 and 7.4). When the tsunami-flooded area reported by Ganzawa et al. (1995) is compared with the upstream limit of the tsunami deposits, the former is 24 m further from the shoreline than the latter, but the difference in altitude of maximum run-up is only 58 cm (Fig. 7.4).

4.2 Sedimentary structures

The 1993 tsunami deposits are mainly composed of poorly sorted sands, which appeared to be mixed with marine sands and gravels that occur near the coast. The sedimentary structures include gravel fabric, current ripples, and dunes (Figs. 7.5 and 7.6) and show that the onshore tsunami deposits were produced by current energy, not by wave energy, and that the sedimentary particles were mainly transported by traction, not in suspension. These deposits can be divided into two sedimentary facies, a gravel facies and a sheet-sand facies, which will be detailed further.

4.3 Sedimentary units and facies

The sedimentary structures of the 1993 tsunami deposits provide information about the current direction, allowing us to distinguish inflow deposits from outflow deposits. Four sedimentary units have been identified: Unit 1 through Unit 4, in ascending order. The lower boundary of each unit is clearly erosional, as indicated by scouring (Figs. 7.5 and 7.6). Furthermore, a reworked unit was occasionally found on the top of Unit 4; this cover was caused by river flooding after the tsunami sedimentation (Fig. 7.7).

4.3.1 Unit 1 as the first inflow deposits

Unit 1 is found only at A225 and B210, near the Usubetsu-bashi Bridge (Fig. 7.7). It overlies the 1993 soil and fills a topographic depression. The lower and upper boundaries of the unit are erosional. The unit, consisting mainly of matrix-supported, thickly bedded and pebbly sands, contains a gravel facies that is composed of pebbles and cobbles (maximum diameter 16 cm). No grading or current structures were found in this unit.

Although no more information is available, we imply that Unit 1 was deposited during the inflow because it is the lowest unit in the tsunami sequence and has sedimentary characteristics similar to those of the gravel facies of Unit 3, as follows.

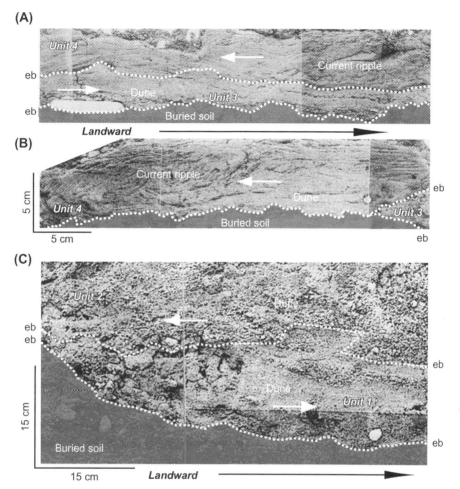

Figure 7.5
Sedimentary structures of the sheet-sand facies in the 1993 onshore tsunami deposits at B250 (A, B) and A225 (C). Samples collected by the peel method. The *dotted lines* show unit boundaries. *eb*, erosional base.

4.3.2 Unit 2 as the first outflow deposits

Unit 2 mainly consists of fine sand and was found only at A225 and B210 near the bridge (Fig. 7.7). It shows a sheet-sand facies associated with gravel and coarse sand and fills a topographic depression. Unit 2 overlies Unit 1 and the 1993 soil with an erosional lower boundary. No current-induced structures have been found in this unit.

We imply that Unit 2 was deposited during the outflow because the sedimentary characteristics are similar to those of Unit 4, as follows.

118   Chapter 7

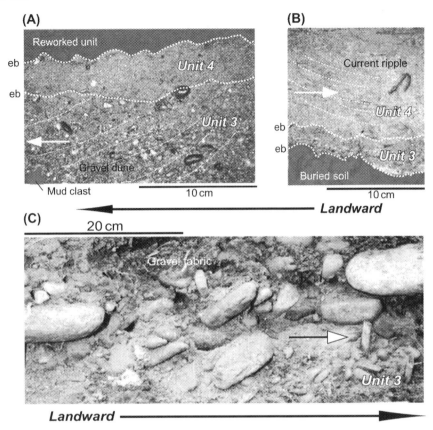

Figure 7.6
Sedimentary structures of the sheet-sand facies such as current ripples and dunes at B216 (A), A260 (B) obtained by the oriented lunch box method and gravel fabric of the gravel facies at B210 (C). The *dotted lines* show unit boundaries. *eb*, erosional base.

4.3.3 Unit 3 as the second inflow deposits

Unit 3 is distributed between B210 and B250 and between A210 and A320 (Fig. 7.7). It overlies both Units 1 and 2 and the 1993 soil with an erosional boundary and consists of coarse to fine sands with gravel and shell fragments. It contains both sedimentary facies, the gravel facies and the sheet-sand facies.

The gravel facies is found between B210 and B216 and consists of matrix-supported gravels. The matrix is composed of coarse to fine marine sand, which contains cobbles (maximum diameter 26 cm) and shell fragments such as *Patinopecten yessoensis*, which inhabit nearshore areas at depths of -50 to -60 m off Taisei (Nanayama et al., 2000, Fig. 7.1C).

A typical example of the gravel fabric showing an N30°E current direction is shown in Figs. 7.6C and 7.7. This gravel bed is mound-shaped and shows normal and inverse

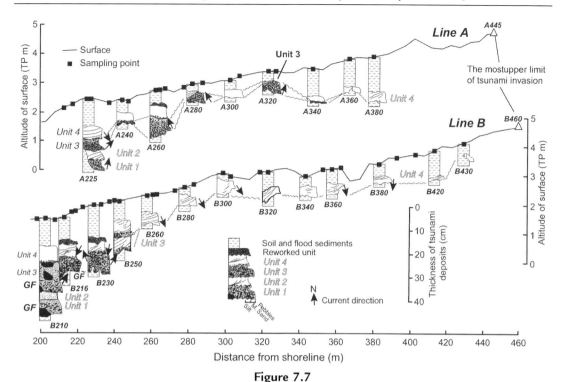

Figure 7.7
Sedimentary logs along the two survey lines (A and B), and correlations of the four sedimentary units between sampling locations. The altitudes were measured with respect to the mean sea level in Tokyo Bay (TP m). GF marked sediments belong to the gravel facies: the other sediments belong to the sheet-sand facies.

grading. It is inferred to have been deposited by a powerful turbulent current because the characteristics are very similar to sedimentary structures associated with a high-density turbidite (Lowe, 1982).

The sheet-sand facies consist of coarse to fine marine sand. It was found at A320 and B230. We identified dunes in these sand layers at B220, B230, and B245 and current ripples at A280 and A320. These sedimentary structures show that the sediments were mainly transported as bed load (traction) but not in suspension.

The change in the bed forms from dunes to current ripples indicates a decreasing current energy between 237 and 367 m from the shoreline. The deposits pinch out between A320 and A340 (Fig. 7.7).

In some places, Unit 3 is capped by a thin silt layer containing many plant fragments, indicating that the layer settled from suspension when the land was temporarily flooded.

120   Chapter 7

4.3.4 Unit 4 as the second outflow deposits

Unit 4 deposits are widely distributed between B210 and B430 and overlie Units 1 to Unit 3, as well as the 1993 soil with an erosional boundary. It shows a sheet-sand facies. This unit is mainly composed of fine marine sands, showing graded structures and current ripples that indicate a seaward current direction (Fig. 7.7). Wood and plant fragments are found at the top of the unit. We imply that Unit 4 was deposited during the outflow because of the seaward current direction interpreted from the sedimentary structures.

4.3.5 Interpretation of the four units

We have considered the four units in the context of the two known tsunami run-up processes in the Usubetsu River. Eyewitness observations of the 1993 tsunami in this area (e.g., Ganzawa et al., 1995; Tsuji et al., 1994) allow interpreting Unit 1 as a deposit of the first inflow, Unit 2 as those of the first outflow, Unit 3 as those of the second inflow, and Unit 4 as a deposit of the second outflow (Fig. 7.8).

Units 3 and 4 have a larger extent and are coarser grained than Units 1 and 2 do, because the second run-up flow was larger than the first one (Fig. 7.8). Unit 4 is more widely distributed than Unit 3 due to the erosion of the Unit 3 by the subsequent outflow (Fig. 7.8).

5. Discussion

5.1 Sedimentary characteristics and facies of the 1993 onshore tsunami deposits

Most sand layers in intertidal and marsh environments are deposited by channel migration, river floods, and storms, so some criteria are needed to distinguish tsunami sands from sands with another origin. There are, however, only few descriptions of sedimentary structures and facies within onshore tsunami deposits, because the tsunami run-up process is very difficult to understand, while it also varies in space because of the differences of the sea-bottom and nearshore topography.

During tsunami run-up, tsunamis produce sedimentary structures, including fining upward sequences or graded beds (Benson et al., 1997; Gelfenbaum and Jaffe, 2003; Nishimura and Miyaji, 1995; Tuttle et al., 2004), nongraded (massive) or multiple graded sheet sands (Benson et al., 1997; Gelfenbaum and Jaffe, 2003), sharp erosional bases (Nanayama et al., 2000) and flame structures (Minoura and Nakata, 1994), and landward-thinning wedges of marine material (Dawson et al., 1996; Minoura et al., 1996). Gelfenbaum and Jaffe (2003) reported that some deposits contain rip-up clasts of muddy soil and that locally a mud cap is present. Current ripples and dunes in onshore tsunami deposits created by the tsunami inflows and outflows have been reported by Sato et al. (1995) and Nanayama et al. (2000).

Sedimentary characteristics and depositional processes of onshore 121

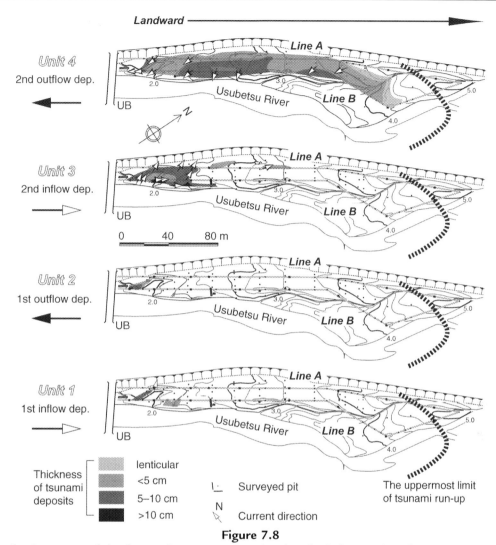

Figure 7.8
Distribution maps of the four sedimentary units, showing the inflow and outflow directions, bed thicknesses of each unit, and tsunami run-up limit.

The 1993 onshore tsunami deposits mainly consist of fine sand derived from the offshore area with a depth of less than −50 to −60 m. This is proven by the occurrence of *P. yessoensis* that inhabits nearshore areas off Taisei (Nanayama et al., 2000, Fig. 7.1C). Consistent with the findings documented in other tsunami studies, seaward deposits are coarser, being composed of coarse sands and gravels. Furthermore, the deposits thin landward, finally pinching out at the uppermost limit of the tsunami run-up (Fig. 7.7), as was also observed by Dawson et al. (1988, 1991).

122   Chapter 7

Besides the sheet-sand facies examined earlier, we also recognized another facies within the 1993 tsunami deposits, namely the gravel facies. Development of this facies in the tsunami deposits, especially in Unit 1 and Unit 3, reflects the deposition of these units from a powerful turbid current such as a "high-density turbidity current."

Benson et al. (1997) and Gelfenbaum and Jaffe (2003) described multiple graded sand beds. The genetic relationship of this multiple structure of the tsunami deposits, however, was not examined on the basis of the observations of real tsunami waves. In the present study of the 1993 onshore tsunami deposits, we also identified four sedimentary units, as stated earlier, and confirmed, based on the eyewitness accounts by many residents of Taisei in this area, that the multiple layers were produced by two tsunami inundations.

Furthermore, it must be emphasized that

1. Units 3 and 4 are more widely distributed than Units 1 and 2, as stated earlier. This coincides well with the eyewitness observation that the second run-up was larger than the first run-up. So, we suggest that the tsunami deposits mainly record the largest run-up during one tsunami event in general.
2. Unit 4 is more widely distributed than Unit 3 because the outflow eroded the inflow deposits. It might be said that the outflow commonly erodes the inflow deposits and redeposits them.
3. The bedforms of the Unit 3 change from dunes to current ripples. This implies decrease of the hydrodynamic energy during the inflow. So, we suggest, if we study the sedimentary structures of onshore tsunami deposits furthermore, then we should be able to understand the variations of current regimes better and calculate the current velocities of each current.

5.2 An ideal model of the 1993 tsunami sedimentation

The results of our field researches, in combination with eyewitness reports and seismic data, allow the following reconstruction of the sedimentary processes responsible for the 1993 onshore tsunami deposits in the estuary area of the Usubetsu River as follows (Fig. 7.9).

The 1993 tsunami, which originated at the earthquake epicenter northwest of Okushiri Island, struck and eroded the coastal shelf slope off Taisei shallower at a depth of less than −50 to −60 m. The eroded marine sands were then transported onto the land by the powerful turbid current of the tsunami run-up (Fig. 7.9A).

The tsunami inflow became turbid during their run-up into the Usubetsu River, where coarse fluvial sands and gravels were eroded from the riverbed and mixed with suspended marine sands; this mixture then transported farther inland via two mechanisms, in suspension and as bed load (Fig. 7.9B).

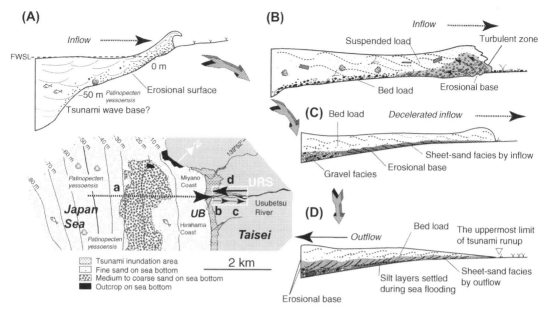

Figure 7.9
Idealized model of the depositional processes. (A) Offshore erosion by the inflow. (B) Erosion and transportation by the inflow. (C) Sedimentation by the decelerated inflow before the outflow. (D) Resedimentation by the outflow. *FWSL*, fair-weather sea level; *URS*, the study section at the Usubetsu River.

The current velocity and energy decreased during the run-up process in the Usubetsu River. The hydrodynamic energy of the inflow decreased near the Usubetsu-bashi Bridge; the mound-shaped gravel bed was settled; the current continued as a high-density turbiditic current during the initial stage of deposition. After this hydraulic jump (Allen, 1984), fine sand grains were easily transported as a bed load (traction) mechanism, and they were deposited when the run-up current of the tsunami came to an end and lost its energy (Fig. 7.9C).

After the run-up had reached its limit, the transported seawater became temporarily more or less stagnant, during which time suspended silt grains settled out as sheet sands. The seawater then flowed back to sea, and the outflow eroded the inflow deposits and redeposited them as outflow deposits, creating new sedimentary structures via traction transport (Fig. 7.9D).

6. Conclusions

The onshore tsunami deposits resulting from the July 12, 1993 Hokkaido-Nansei-Oki earthquake tsunami can be described as follows:

1. The deposits contain many sedimentary structures such as gravel fabric, current ripples, and dunes. These structures show that the sedimentary particles were mainly transported as bed load (traction).
2. Two facies can be recognized: a gravel facies and a sand-sheet facies in the 1993 onshore tsunami deposits. The deposits of the gravel facies were produced by a powerful turbulent current similar to that of a high-density turbidity current.
3. Four sedimentary units (Unit 1 through Unit 4) are distinguished. These are interpreted as resulting from the first inflow and the first outflow and from their second equivalents.
4. Units 3 and 4 have a wider extent than Units 1 and 2 because the second run-up current was larger than the first one. It is suggested that commonly tsunami deposits mainly record the largest run-up flow.
5. Unit 4 is more widely distributed than Unit 3 because the outflow eroded the inflow deposits and redeposited them as outflow deposits.
6. The Unit 3 bedforms changed in time from dunes to current ripples due to the decrease in hydrodynamic energy during the inflow. Further study of the sedimentary structures of the onshore tsunami deposits will increase the understanding of the variations in inflow regimen and may provide sufficient data to calculate the current velocities.

Acknowledgments

The author thanks Professor T. Shiki (Kyoto University), Brian F. Atwater (USGS), K. Satake and K. Shimokawa (Active Fault Research Center, AIST), K. Shigeno, S. Miyasaka, and M. Ishii (Meiji Consultant Co., Ltd.), who provided comments that helped to improve the manuscript. This research has been supported in part by Grant-in-Aid for Scientific Research from the Ministry of Education, Science and Culture of Japan (No. 16540423).

References

Allen, J.R.L., 1984. Sedimentary Structures, Their Character and Physical Basis. Elsevier, Amsterdam, p. 663.

Atwater, B.F., 1987. Evidence for great Holocene earthquakes along the outer coast of Washington state. Science 236, 942−944.

Atwater, B.F., 1992. Geologic evidence for earthquakes during the past 2000 years along the Copalis River, southern coastal Washington. J. Geophys. Res. 97, 1901−1919.

Atwater, B.F., Moore, A.L., 1992. A tsunami about 1000 years ago in Puget Sound, Washington. Science 258, 1614−1617.

Benson, B.E., Grimm, K.A., Clague, J.J., 1997. Tsunami deposits beneath tidal marshes on northwestern Vancouver Island, British Columbia. Quat. Res. 48, 192−204.

Sedimentary characteristics and depositional processes of onshore 125

Bondevik, S., Sevendsen, J.I., Mangerud, J., 1997. Tsunami sedimentary facies deposited by the Stregga tsunami in shallow marine basins and coastal lakes, western Norway. Sedimentology 44, 1115–1131.

Clague, J.J., Bobrowsky, P.T., 1994. Tsunami deposits beneath tidal marshes on Vancouver Island, British Colombia. Geol. Soc. Am. Bull. 106, 1293–1303.

Clague, J.J., Hutchinson, I., Mathewes, R.W., Patterson, R.T., 1999. Evidence for late Holocene tsunamis at Catala lake, British Columbia. J. Coastal Res. 15, 45–60.

Dawson, A.G., Foster, I.D.L., Shi, S., Smith, D.E., Long, D., 1991. The identification of tsunami deposits in coastal sediment sequences. Sci. Tsunami Hazards 9, 73–82.

Dawson, A.G., Long, D., Smith, D.E., 1988. The Storegga slides: evidence from eastern Scotland for a possible tsunami. Mar. Geol. 82, 271–276.

Dawson, A.G., Shi, S., Dawson, S., Takahashi, T., Shuto, N., 1996. Coastal sedimentation associated with the June 2nd and 3rd, 1994 tsunami in Rajegwesi, Java. Quat. Sci. Rev. 15, 901–912.

Ganzawa, Y., Kito, N., Sadakata, N., 1995. Tsunami of the 1993 southwest off Hokkaido earthquake and refuge behavior of people: in the case of the Taisei town in Hokkaido, Japan. Earth Sci. 49, 379–390 (in Japanese, with English abstract).

Gelfenbaum, G., Jaffe, B., 2003. Erosion and sedimentation from the 17 July. 1998 Papua New Guinea tsunami. Pure Appl. Geophys. 160, 1969–1999.

Hutchinson, I., Clague, J.J., Mathewes, R.W., 1997. Reconstructing the tsunami record on an emerging coast: a case study of Kanim lake, Vancouver Island, British Columbia, Canada. J. Coastal Res. 13, 545–553.

Kon'no, E. (Ed.), 1961. Geological Observations of the Sanriku Coastal Region Damaged by the Tsunami Due to the Chile Earthquake in 1960, 52. Contributions from the Institute of Geology and Paleontology Tohoku University, pp. 1–40.

Lowe, D.R., 1982. Sediment gravity flows. II. Depositional models with special reference to the deposits of high-density turbidity currents. J. Sediment. Petrol. 52, 279–297.

Minoura, K., Nakata, T., 1994. Discovery of an ancient tsunami deposit in coastal sequences of southwest Japan: verification of a large historic tsunami. Isl. Arc. 3, 66–72.

Minoura, K., Nakaya, S., 1991. Traces of tsunami preserved in inter-tidal lacustrine and marsh deposits: some examples from northeast Japan. J. Geol. 99, 265–287.

Minoura, K., Gusiakov, V.G., Kurbatov, A., Takeuchi, S., Svendsen, J.I., Bondevik, S., Oda, T., 1996. Tsunami sedimentation associated with the 1923 Kamchatka earthquake. Sediment. Geol. 106, 145–154.

Minoura, K., Imamura, F., Kuran, U., Nakamura, T., Papadopoulas, G.A., Takahashi, T., Yalciner, A.C., 2000. Discovery of Minoan tsunami deposits. Geology 28, 59–62.

Minoura, K., Imamura, F., Takahashi, T., Shuto, N., 1997. Sequence of sedimentation processes caused by the 1992 Flores tsunami: evidence from Babi Island. Geology 28, 59–62.

Minoura, K., Nakaya, S., Uchida, M., 1994. Tsunami deposits in a lacustrine sequence of the Sanriku coast, northeast Japan. Sediment. Geol. 89, 25–31.

Nakanishi, I., Kodaira, S., Kobayashi, R., Kasahara, M., Kikuchi, M., 1993. The 1993 Japan Sea earthquake: Quake and tsunamis devastate small town. EOS Trans. Am. Geophys. Union 74, 377.

Nanayama, F., Shigeno, K., 1998. How to make oriented samples of loose sediments by lunch box and easy-dry bond. Chishitsu News 523, 52–56 (in Japanese).

Nanayama, F., Sigeno, K., Satake, K., Shimokawa, K., Koitabashi, S., Miyasaka, S., Ishii, M., 2000. Sedimentary differences between the 1993 Hokkaido—Nansei-oki tsunami and the 1959 Miyakojima typhoon at Taisei, southwestern Hokkaido, northern Japan. Sediment. Geol. 135, 255–264.

Nishimura, Y., Miyaji, N., 1995. Tsunami deposits from the 1993 southwest Hokkaido earthquake and the 1640 Hokkaido Komagatake eruption, northern Japan. Pure Appl. Geophys. 144, 720–733.

Peters, R., Jaffe, B., Peterson, C., Gelfenbaum, G., Kelsey, H., 2001. An overview of tsunami deposits along the Cascadia margin. ITS 2001 Proc. Session 3 (3–3), 479–490.

Pinegina, T.K., Bourgeois, J., 2001. Historical and paleo-tsunami deposits on Kamchatka, Russia: long-term chronologies and long-distance correlations. Nat. Hazards Earth Syst. Sci. 1, 177–185.

126   Chapter 7

Sagayama, T., Uchida, Y., Osawa, M., Suga, K., Hamada, S., Murayama, Y., Nishina, K., 2000. Environment of Submarine Geology in the Coastal Area of Hokkaido. 2. Southwest Hokkaido. Special Report of Geological Survey of Hokkaido, no. 29 74. (in Japanese, with English abstract).

Sato, H., Shimamoto, A., Kawamoto, E., 1995. Onshore tsunami deposits caused by 1993 southwest Hokkaido and 1983 Japan Sea earthquakes. Pure Appl. Geophys. 144, 693−717.

Shi, S., Dawson, A.G., Smith, D.E., 1995. Coastal sedimentation associated with December 12th, 1992 tsunami in Flores, Indonesia. Pure Appl. Geophys. 144, 525−536.

Shimamoto, T., Tsutsumi, A., Kawamoto, E., Miyawaki, M., Sato, H., 1995. Field survey report on tsunami disasters caused by the 1993 Southwest Hokkaido Earthquake. Pure Appl. Geophys. 144, 665−691.

Tsuji, Y., Kato, K., Satake, A., 1994. Heights and Damage of the Tsunami of the 1993 Hokkaido—Nansei-Oki Earthquake in Residential Areas on the Coast of the Hokkaido Mainland. 69, Bulletin of Earthquake Research Institute, University of Tokyo, pp. 67−106 (in Japanese, with English abstract).

Tuttle, M.P., Ruffman, A., Anderson, T., Jeter, H., 2004. Distinguishing tsunami from storm deposits in eastern North America: the 1929 Grand Banks tsunami versus the 1991 Halloween storm. Seismol. Res. Lett. 75, 117−131.

Wright, C., Mella, A., 1963. Modifications to the soil pattern of south-central Chile resulting from seismic and associated phenomena during the period May to August 1960. Bull. Seismol. Soc. Am. 53, 1367−1402.

CHAPTER 8

Distribution and significance of the 2004 Indian Ocean tsunami deposits: initial results from Thailand and Sri Lanka

K. Goto[1], F. Imamura[2], N. Keerthi[3], P. Kunthasap[4], T. Matsui[5], K. Minoura[6], A. Ruangrassamee[7], D. Sugawara[2,8], S. Supharatid[9]

[1]*Department of Earth and Planetary Science, Graduate School of Science, The University of Tokyo, Bunkyo-ku, Tokyo, Japan;* [2]*International Research Institute of Disaster Science, Tohoku University, Sendai, Miyagi, Japan;* [3]*Deepwell Drilling Technologies (PVT) Ltd., Narahenpita, Colombo, Sri Lanka;* [4]*Department of Mineral Resources, Environmental Geology Division, Ratchtawi, Bangkok, Thailand;* [5]*Department of Complexity Science and Engineering, Graduate School of Frontier Science, The University of Tokyo, Bunkyo-ku, Tokyo, Japan;* [6]*Professor Emeritus, Tohoku University, Sendai, Miyagi, Japan;* [7]*Department of Civil Engineering, Chulalongkorn University, Patumwan, Bangkok, Thailand;* [8]*Museum of Natural and Environmental History, Shizuoka, Japan;* [9]*Natural Disaster Research Center, Rangsit University, Lak-Hok, Pathumtani, Thailand*

1. Introduction

The Sumatra—Andaman Islands earthquake of December 26, 2004 (magnitude 9.0) generated one of the largest tsunamis in human history. The tsunami impacted coastal areas around the Indian Ocean. Waves reached up to 30 m above sea level on the north-western part of Sumatra, Indonesia, and about 15 m in the south-western Thailand, and on the eastern coast of Sri Lanka (Fig. 8.1A). The tsunami entrained sediment from the sea floor and beaches and transported it landward, where it was deposited.

Past tsunami deposits have been reported from outcrops and in cores from lakes and the sea (Alvarez et al., 1992; Clague and Bobrowsky, 1994; Fujiwara etv al., 2000; Kastens and Cita, 1981; Kelsey et al., 2005; Minoura and Nakaya, 1991; Minoura et al., 1987, 1997; Moore, 2000; Nanayama et al., 2003; Shiki and Yamazaki, 2000; Smit, 1999; Smit et al., 1992, 1996; Tada et al., 2002, 2003; Takashimizu and Masuda, 2000; Takayama et al., 2000).

Tsunamiites. https://doi.org/10.1016/B978-0-12-823939-1.00008-2
Copyright © 2008 Elsevier B.V. All rights reserved.

Chapter 8

Observations of thickness, grain size, and composition of tsunami deposits have been made to estimate the origin and size of past tsunamis (e.g., Bourgeois et al., 1988). Very few field surveys have been conducted immediately after the tsunami, however. Thus, the relationship between the hydraulic characteristics of tsunamis and the process of tsunami deposition is not yet well understood.

The run-up of the 2004 Indian Ocean tsunami has been well documented as a result of the efforts of international tsunami survey groups (e.g., Borrero, 2005a,b; Liu et al., 2005; http://www.drs.dpri.kyoto-u.ac.jp/sumatra/index-e.html). Thus, this tsunami provides an opportunity to increase the insight into the transport and depositional processes recorded in tsunami deposits. In this context, we investigated the stratigraphy and sedimentology of deposits left by the tsunami at sites in Thailand and Sri Lanka.

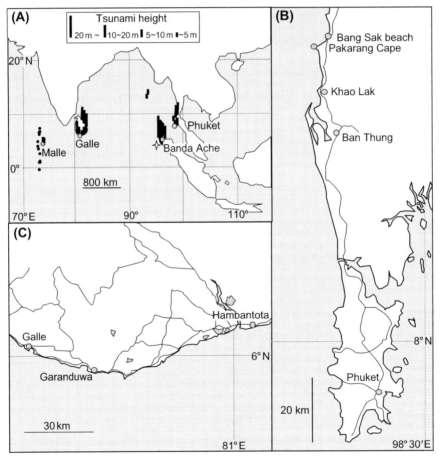

Figure 8.1
The 2004 Indian Ocean tsunami and study areas. (A) Tsunami wave heights in countries around the Indian Ocean (see http://www.drs.dpri.kyoto-u.ac.jp/sumatra/index-e.html). (B) Study area around Khao Lak in Thailand. (C) Study area around Garanduwa in Sri Lanka.

2. Localities and methods of study

A Japan—Thailand—Sri Lanka research team conducted field investigations in Thailand and Sri Lanka from March 26 to April 4, 2005. Even only 3 months after the tsunami, much of the evidence of the event had disappeared owing to recovery efforts. We, therefore, investigated undisturbed tsunami deposits at Pakarang Cape (N8°44.181′, E98°13.321′) and Bang Sak beach (N8°47.443′, E98°15.709′) near Khao Lak, Thailand (Fig. 8.1B). We also measured maximum wave heights in these areas from high watermarks (Table 8.1).

Table 8.1: Measured tsunami heights at Pakarang Cape, Bang Sak beach, and near Garanduwa.

Location	Latitude (N)	Longitude (E)	Height (m)	Distance from shoreline (m)[a]	Corrected tsunami height (m)
Thailand					
Pakarang cape					
	8°43.816′	98°13.351′	4.1[b]	150	4.5
Bang Sak beach					
	8°47.305′	98°15.733′	8.0[c]	60	7.7
	8°47.389′	98°15.736′	10.0[c]	90	10.0
	8°47.539′	98°15.722′	9.4[c]	75	9.6
	8°47.964′	98°15.583′	5.8[c]	40	6.0
	8°47.605′	98°15.680′	10.6[c]	50	11.0
Transect A					
	8°47.472′	98°15.696′	9.3[c]	25	9.6
	8°47.441′	98°15.744′	11.1[c]	110	11.3
	8°47.381′	98°15.899′	6.5[d]	300	7.0
Transect B					
	8°47.977′	98°15.662′	5.1[c]	150	4.5
	8°47.998′	98°15.811′	5.1[c]	425	4.5
Sri Lanka					
Garanduwa					
Transect C					
	5°56.318′	80°29.218′	1.0[c]	105	1.1

[a]Tsunami arrival time assumed to be 10:00 a.m., December 26, 2004, in Thailand and 9:30 a.m. in Sri Lanka (http://www.drs.dpri.kyoto-u.ac.jp/sumatra/index-e.html). Tidal corrections determined using tidal data at Ao Kaulak for Pakarang Cape and Bang Sak beach and Galle for Garanduwa (http://www.eri.u-tokyo.ac.jp/namegaya/sumatera/tide/index.htm).
[b]Tsunami height measured on a house.
[c]Tsunami height measured on a tree.
[d]Tsunami height measured on an electrical pole.

Figure 8.2
Impact of the tsunami. (A) Tsunami-transported reef blocks near Pakarang Cape, Thailand. (B) Satellite image of Pakarang Cape after the tsunami on December 29, 2004. This image is provided by Space Imaging/CRISP-Singapore. Note that there are no blocks inland of the *dotted line*. (C) Photograph showing the landward limit of reef blocks.

Pakarang Cape is 13 km north of Khao Lak (Fig. 8.1B). We did not recognize tsunami-deposited sandy sediments at Pakarang Cape, but found abundant reef blocks (Fig. 8.2A). We found sandy tsunami deposits on both natural and artificial beaches at Bang Sak beach, ∼7 km north of Pakarang Cape (Fig. 8.1B). The maximum horizontal inundation distance of the tsunami from the shoreline was 650–800 m, determined from the landward limit of dead grasses as well as comments of residents. We made a map of an ∼700 × 700 m area at Bang Sak beach (Fig. 8.3) and investigated the distribution of tsunami deposits within that area. We established transect lines perpendicular to the shoreline of an artificial beach (Figs. 8.3A and 8.4A, transect A) and a natural beach (Figs. 8.3B and 8.5A, transect B), where the landward flow had been approximately perpendicular to the shoreline. We investigated variations in thickness, grain size, and sedimentary structures of the tsunami deposits along these transect lines. We measured tsunami wave heights at three sites (Fig. 8.4A) and studied tsunami deposits at four sites

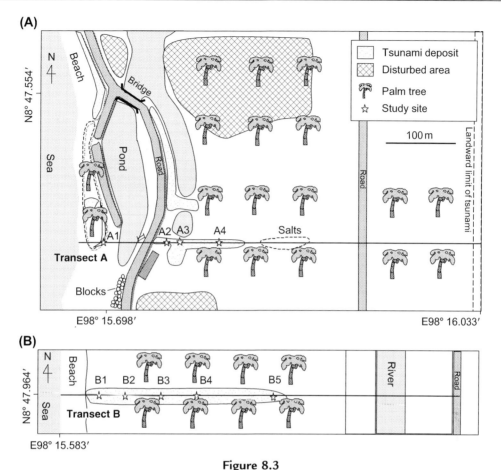

Figure 8.3
Map of the Bang Sak beach study area showing locations of (A) transect A and (B) transect B and sites of detailed observation and sampling.

along transect A (Fig. 8.4B; A1–A4). We measured tsunami wave heights at two sites (Fig. 8.5A) and studied tsunami deposits at five sites along transect B (Fig. 8.5B; B1–B5). The positions of the shorelines along transects A and B were determined at 9:43 a.m. on March 30 and at 10:06 a.m. on March 31, 2005, respectively. We reinvestigated tsunami deposits along transect B at Bang Sak beach from September 24 to October 1, 2005, 9 months after the tsunami, to investigate the condition and variation of thickness of the tsunami deposits.

Natural beaches on the west and south coasts of Sri Lanka are rare due to fishery, agriculture, and residential developments. We found a natural beach near Garanduwa (N5°56.318′, E80°29.218′), ~30 km east of Galle (Fig. 8.1C). The maximum horizontal inundation distance of the tsunami in this area was more than 1 km, according to residents. We established a transect line perpendicular to the shoreline (Fig. 8.6A, transect C),

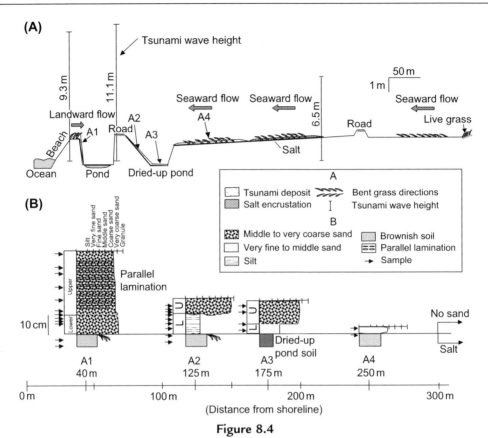

Figure 8.4

Transect A. (A) Schematic cross section along transect A at Bang Sak beach. Measured tsunami wave heights, which are not corrected for the tidal level at the time of the tsunami, are also shown. (B) Stratigraphy of sediments at study sites along transect A.

measured the tsunami wave height at one site, and studied tsunami deposits at four sites (Fig. 8.6B; C1—C4). The position of the shoreline on transect C was determined at 10:15 a.m. on April 3, 2005.

3. Distribution and significance of the tsunami deposits
3.1 Pakarang Cape, Thailand

At Pakarang Cape, the tsunami wave height ranged from 4 to 9 m (Table 8.1; Matsutomi et al., 2005). Abundant reef blocks occur on the west side of the cape (Fig. 8.2A). The blocks are fragments of corals up to 4 m in diameter. The larger blocks, which are visible as white dots on satellite images (Fig. 8.2B), extend more than 1 km along the shore. Very few blocks were found on the north and east sides of the cape (Fig. 8.2B). The landward limit of distribution of reef blocks is a line ~100 m west of the north-south road on the

Distribution and significance of the 2004 Indian Ocean tsunami 133

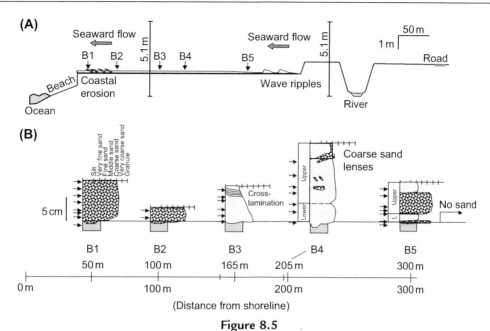

Figure 8.5

Transect B. See Fig. 8.4 for legend. (A) Schematic cross section along transect B at Bang Sak beach. (B) Stratigraphy of sediments at study sites along transect B.

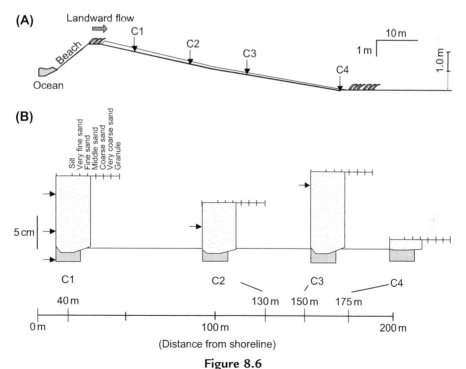

Figure 8.6

Transect C. See Fig. 8.4 for legend. (A) Schematic cross section along transect C at Garanduwa. (B) Stratigraphy of sediments at study sites along transect C.

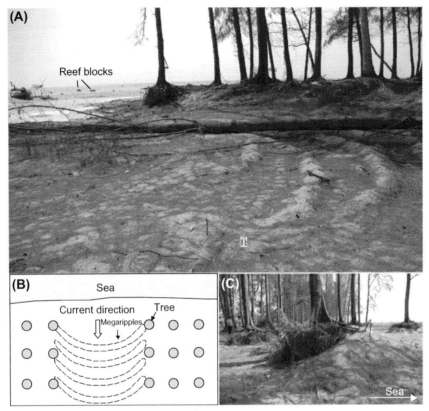

Figure 8.7
Tsunami-induced sedimentary features at Pakarang Cape. (A) Megaripples. (B) Schematic view of megaripples. (C) Localized erosion and sedimentation around trees.

cape (Fig. 8.2B and C). The distribution suggests that the blocks were transported from west to east by the tsunami wave.

We found megaripples in a small forest near the beach (Fig. 8.7A). The megaripples developed where the spacing between trees is greater than 20 m. The wavelength of the ripples was ~1 m at the shore, decreasing inland to ~20 cm in the backshore area. The height of the megaripples was less than 20 cm. The megaripples had the form of concentric semicircles open toward the sea (Fig. 8.7B), suggesting that they were formed by eastward-flowing currents.

The ground surface in a small forest near the coast was eroded up to 60 cm, and roots on the seaward side of the trees were covered by sand transported by eastward-flowing currents (Fig. 8.7C).

3.2 Bang Sak beach, Thailand

3.2.1 Transect A (artificial beach)

The horizontal inundation distance of the tsunami along transect A was ~650 m (Figs. 8.3A and 8.4A). The wave height was 9–11 m near the shoreline and 6.5 m at 400 m from the shoreline (Table 8.1 and Fig. 8.4A).

Grasses within 50 m of the shoreline along transect A were bent landward (Fig. 8.8A). The 43-cm-thick tsunami deposit at site A1 overlay landward-bent grasses (Fig. 8.8B). The lower 10 cm of the deposit consisted of light gray, poorly sorted, very coarse sand, mainly silicate minerals, shell fragments, and coral skeletal fragments (Fig. 8.9A), which had been transported from the beach or shallow sea floor by the tsunami wave. The upper 33 cm of

Figure 8.8
Features along transect A. (A) Small hill 40 m from the shore along transect A. Grasses and trees on this hill were bent landward by the incoming waves. (B) Tsunami deposit at site A1. S = pretsunami soil; T = tsunami deposit. (C) Close-up of the tsunami deposit at site A1. Note weakly developed parallel lamination.

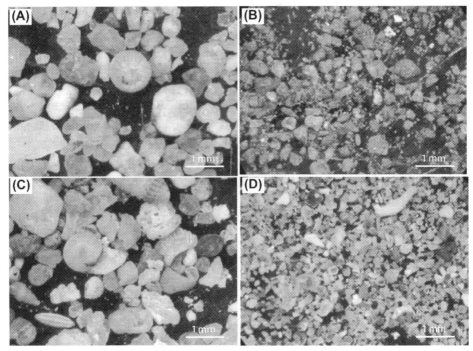

Figure 8.9
Tsunami sand becoming finer inland. The sand consists mainly of silicate minerals and fragments of shell and coral, eroded from the beach or shallow sea floor by the tsunami. (A) site A1, (B) site A4, (C) site B1, and (D) site B5.

the deposit, which gradationally overlay the lower part, consisted of white, poorly sorted, coarse to very coarse sand with vague parallel lamination (Fig. 8.8C).

The tsunami deposit at site A2 overlay dark brown soil with a sharp contact (Figs. 8.4B and 8.10A). The deposit was 18-cm thick and consisted of two units (Fig. 8.4B). The lower 10-cm unit was composed of dark gray, well-sorted silt and very fine sand. The upper 8-cm unit overlay the lower unit with a sharp, erosional contact and comprised white, poorly sorted, very coarse sand, similar to the upper sand at site A1.

The tsunami deposit at site A3 overlay the dark gray soil of a dried-up pond with a sharp contact (Figs. 8.4B and 8.10B). The deposit was 17-cm thick and consisted of two units. The lower 6-cm unit was composed of light gray, massive fine sand. The upper 11-cm unit overlay the lower unit with a conformable, gradual contact and consisted of white, coarse to very coarse sand.

The tsunami deposit at site A4 overlay a dark brown soil across a sharp contact (Figs. 8.4B and 8.10C). The deposit was 3-cm thick and was composed of white, well-sorted, medium sand, mainly silicate minerals (Fig. 8.9B), with a small amount of coral skeletal fragments.

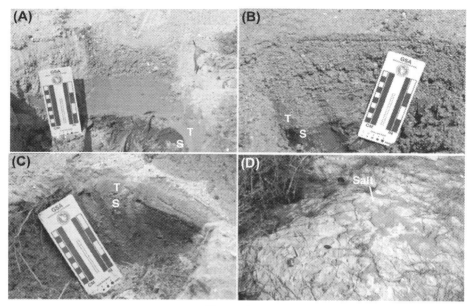

Figure 8.10
Tsunami deposits and salt encrustation. S = pretsunami soil; T = tsunami deposit. (A) Deposit at site A2, (B) deposit at site A3, (C) deposit at site A4, and (D) salt crust 300 m away from the shoreline along transect A.

The tsunami deposits did not extend more than 300 m from the shoreline along transect A (Fig. 8.4A and B). Salt incrustations, however, were observed 300–400 m from the shoreline (Fig. 8.10D). Grasses and trees were bent seaward (Fig. 8.4A) between 200 and 650 m from the shoreline due to backwash in this zone.

3.2.2 Transect B (natural beach)

Strong erosion was observed near the beach along transect B (Fig. 8.11A). The ground surface was almost flat from the beach to points 400 m inland from the shoreline (Fig. 8.5A), and grasses and trees along transect B were bent seaward due to backwash. The maximum tsunami wave height was ~5 m along this transect (Table 8.1 and Fig. 8.5A).

Tsunami deposits at sites B1–B5 overlay dark brown soil across a sharp, irregular boundary (Fig. 8.5B). The deposit at site B1 was 10-cm thick and composed of white, poorly sorted, coarse to very coarse sand (Fig. 8.11B), mainly silicate minerals, shell fragments, and coral skeletal fragments (Fig. 8.9C) similar to those observed in the tsunami deposits along transect A. The 5-cm-thick tsunami deposit at site B2, which was 5-cm thick, was composed of white, poorly sorted, coarse to very coarse sand similar to that at site B1 (Fig. 8.11C). The tsunami deposit at site B3 was ~10-cm thick (Fig. 8.11D) and fined upward from medium sand to silt. Faint cross-lamination with a seaward dip was observed in the upper 3–4 cm of the deposit, indicating influence of backwash.

Figure 8.11

Tsunami sediments and erosion. S = pretsunami soil, T = tsunami deposit. (A) Erosion near the beach along transect B. (B) Tsunami deposits at site B1. (C) Tsunami deposit at site B2. (D) Tsunami deposit at site B3. Faint seaward-dipping cross-lamination is visible in the upper 3–4 cm of the tsunami sand.

The tsunami deposit at site B4 was 17–22-cm thick (Fig. 8.5B) and consisted mainly of light gray, well-sorted, fine to medium sand with white coarse sand lenses in the upper half of the deposit (Fig. 8.12A). The coarse sand lenses extended steeply upward toward the surface and resemble water escape structure formed by dewatering during the rapid deposition of water-saturated unconsolidated sediments discussed by Lowe (1975). The tsunami deposit at location B5 was 12-cm thick (Fig. 8.5B) and comprised two upward-fining layers (Fig. 8.12B). The lower layer was 3-cm thick and graded upward from coarse to very fine sand. It was sharply overlain by a layer of 9-cm thickness that grades upward from very coarse sand to fine sand.

There were no beach sand deposits more than 300 m inland from the shoreline along transect B. As at transect A, salt encrustations were present from 300 to 400 m from the shoreline. Ripples were observed between 350 and 400 m (Fig. 8.12C), and the current direction was seaward, indicating that they were formed by backwash.

On September 2005, 9 months after the tsunami, we reinvestigated the tsunami deposits along transect B at Bang Sak beach. The tsunami deposits were observable on the ground surface at the end of March 2005 (Fig. 8.13A), but they were covered by vegetation and were disturbed by roots at the end of September 2005 (Fig. 8.13B). The deposits were

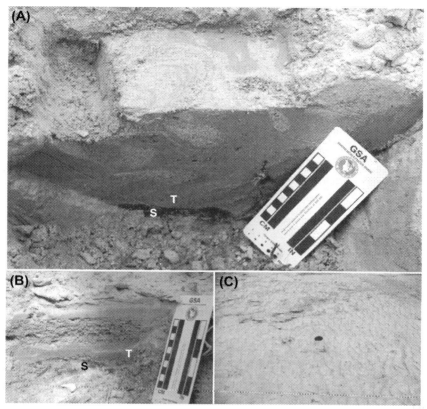

Figure 8.12

Tsunami deposits and sedimentary structures. (A) Tsunami deposit at site B4. Note coarse sand lenses in the upper half of the deposit. S = pretsunami soil, T = tsunami deposit. (B) Tsunami deposit at site B5. (C) Current ripples about 350 m from the shoreline along transect B. The ripples were produced by seaward-directed flow (backwash).

eroded by heavy squall and also by dense vegetation growth, and the thickness of the deposits had reduced ~70% of the thickness that was measured along transect B in March.

3.3 Garanduwa, Sri Lanka

3.3.1 Transect C (natural beach)

Nearshore topography at Garanduwa differs from that at Bang Sak beach (Fig. 8.6A). Along transect C, the ground rises steeply to a point ~10 m from the shore and then drops gradually farther landward (Fig. 8.6A). The maximum wave height was ~1.0 m above sea level (2.3 m above ground level) at 105 m inland from the shoreline along transect C (Table 8.1). However, the wave at the shoreline was at least 1.6 m above sea level because it overtopped the highest point near the beach (Fig. 8.6A). All grasses and trees along transect C were bent landward in the direction of the incoming tsunami.

Figure 8.13

Photographs along transect B. (A) At the end of March 2005. (B) At the end of September 2005.

Tsunami deposits at sites C1–C4 overlay dark brownish soil across a sharp contact (Fig. 8.14A–D). The thicknesses of the tsunami deposits were 10, 5, 10, and 2 cm at sites C1, C2, C3, and C4, respectively (Fig. 8.6B). The deposits were very poorly sorted, light gray, massive, fine to medium sand, mainly silicate minerals and shell fragments. No obvious grain-size variation was observed in any of the tsunami deposits along transect C. The deposits did not extend more than 200 m from the shoreline along transect C, even though the tsunami inundated the land to a distance of more than 1 km.

Figure 8.14
Tsunami deposits at sites (A) C1, (B) C2, (C) C3, and (D) C4. S = pretsunami soil; T = tsunami deposit.

4. Discussion

We recognized tsunami-transported reef blocks on the west side of Pakarang Cape. Although similar blocks have been previously reported and attributed to historic tsunamis (e.g., Kato and Kimura, 1983), no definitive evidence supporting their tsunami origin has been presented. The blocks at Pakarang Cape indicate that a large tsunami can transport blocks as large as 4 m in size inland (Goto et al., 2007). Their landward transport is supported by the orientation of megaripples and selective erosion and sedimentation around trees.

The sheet-like sand deposits at Bang Sak beach were composed mostly of silicate minerals, shell fragments, and skeletal fragments that were transported from the shallow sea floor and beach by the tsunami. The deposits generally thin and fine landward (Figs. 8.4 and 8.5), although the pattern is complicated by variations in local topography. Landward fining and thinning have been documented for several historical tsunamis (Minoura et al., 1997; Moore, 2000). The tsunami deposits along transects A and B consisted of two units (Figs. 8.4B and 8.5B), implying that at least two large waves crossed the beach. This inference is consistent with reports of local residents that large tsunami waves inundated the coast at least twice.

142 Chapter 8

Except at site A1, bent grass stems and wave ripples provide evidence for backwash along transects A and B. Grass stems at site A1 were bent landward and palm trunks had fallen in that direction (Fig. 8.8A). The lack of evidence for backwash at this site is due to the presence of topographic depressions in the backshore (Fig. 8.3). Because ponds in this area are lower than the surrounding ground (Fig. 8.4A), backwash filled these ponds and no strong currents passed site A1.

The tsunami deposits do not extend more than about 300 m inland along transects A and B, although the tsunami inundated the land to about 700 m from the shore. The absence of tsunami deposits at a distance of 300−700 m can be explained by low sediment-transport capacity of the tsunami waves. This capacity is a function of the settling velocity of the sediment particles, topography, and the forward velocity of the tsunami wave. The sediment at Bang Sak beach is composed mainly of medium to very coarse sand, with minor amount of clay and silt. We infer that the settling velocity of the sediment particles was high as compared to the velocity of the onrushing waves, so that particles settled out within ∼300 m of the shoreline. Alternatively, fine sediment particles may have been eroded and transported seaward by backwash.

No evidence was found for backwash along transect C, probably because of the topography of the transect. Because of the ground crests near the beach (Fig. 8.6A), water receded where the ground level was lower rather than along transect C. The deposits along transect C are massive, without sedimentary structures. Furthermore, no obvious grain-size variation was observed along transect C. These observations indicate that the deposits are the product of rapid sedimentation of beach sand entrained by the tsunami.

5. Conclusion

We conducted a preliminary field investigation of deposits of the 2004 Indian Ocean tsunami at two sites in Thailand and Sri Lanka. The tsunami deposits are composed of silicate minerals, shell fragments, and coral skeletal fragments, which were eroded from the beach or the shallow sea floor by the tsunami. The tsunami deposits generally thin and fine landward, although the trends are complicated by variations in local topography. The tsunami deposits are present over less than half of the inundation zone at Bang Sak beach in Thailand and near Garanduwa in Sri Lanka, probably due to the local landward decrease in the sediment transport capacity. Future detailed analyses of grain size and provenance and other quantitative studies of the deposits will provide additional information that may increase the understanding of the formation of the deposits and their relationship to the magnitude and inundation pattern of the tsunami.

Acknowledgments

The authors thank W. Kanbua, N. Chaimanee, P. Foytong, N. Saelem, Y. Chidtong, S. Nawawitphisit, and W. Buatong in Thailand and S. S. Wijayasinghe in Sri Lanka for the support they provided during our field survey. They also thank T. Shiki and J. Clague, who critically read the manuscript and gave them many valuable suggestions. The survey was supported by research funds donated to the University of Tokyo by T. Yoda, Central Japan Railway Company, Suntory Limited, WAC Inc., and by a Grant-in-Aid for Scientific Research from the Japan Society for the Promotion of Science (no. 17740331).

References

Alvarez, W., Smit, J., Lowrie, W., Asaro, F., Margolis, S.V., Claeys, P., Kastner, M., Hildebrand, A., 1992. Proximal impact deposits at the Cretaceous-Tertiary boundary in the Gulf of Mexico: a restudy of DSDP, leg 77 sites 536 and 540. Geology 20, 697–700.

Borrero, J.C., 2005a. Field data and satellite imagery of tsunami effects in Banda ache. Science 308, 1596.

Borrero, J.C., 2005b. Field survey of northern Sumatra and Banda Acheh, Indonesia after the tsunami and earthquake of 26 December 2004. Seismol. Res. Lett. 76, 309–317.

Bourgeois, J., Hansen, T.A., Wiberg, P.L., Kauffman, E.G., 1988. A tsunami deposit at the Cretaceous-Tertiary boundary in Texas. Science 241, 567–570.

Clague, J.J., Bobrowsky, P.T., 1994. Evidence for a large earthquake and tsunami 100–400 years ago on western Vancouver Island, British Columbia. Quat. Res. 41, 176–184.

Fujiwara, O., Masuda, F., Sakai, T., Irizuki, T., Fuse, K., 2000. Tsunami deposits in Holocene bay mud in southern Kanto region, Pacific coast of central Japan. Sediment. Geol. 135, 219–230.

Goto, K., Chavanich, S.A., Imamura, F., Kunthasap, P., Matsui, T., Minoura, K., Sugawara, D., Yanagisawa, H., 2007. Distribution, origin and transport process of boulders transported by the 2004 Indian Ocean tsunami at Pakarang Cape, Thailand. Sediment. Geol. 202, 821–837.

Kastens, K.A., Cita, M.B., 1981. Tsunami-induced sediment transport in the abyssal Mediterranean Sea. Geol. Soc. Am. Bull. 92, 845–857.

Kato, Y., Kimura, M., 1983. Age and origin of so-called "Tsunami-ishi," Ishigaki island, Okinawa Prefecture. J. Geol. Soc. Jpn. 89, 471–474.

Kelsey, H.M., Nelson, A.R., Hemphill-Haley, E., Witter, R.C., 2005. Tsunami history of an Oregon coastal lake reveals a 4600 yr record of great earthquakes on the Cascadia subduction zone. Geol. Soc. Am. Bull. 117, 1009–1032.

Liu, P.L.F., Lynett, P., Fernando, H., Jaffe, B.E., Fritz, H., Higman, B., Morton, R., Goff, J., Synolakis, C., 2005. Observations by the international tsunami survey team in Sri Lanka. Science 308, 1595.

Lowe, D.R., 1975. Water escape structures in coarse-grained sediments. Sedimentology 22, 157–204.

Matsutomi, H., Takahashi, T., Matsuyama, M., Harada, K., Hiraishi, T., Supartid, S., Nakusakul, S., 2005. The 2004 off Sumatra earthquake tsunami and damage at Khao Lak and Phuket island in Thailand. Ann. J. Coast. Eng. JSCE 52, 1356–1360.

Minoura, K., Nakaya, S., 1991. Traces of tsunami preserved in intertidal lacustrine and marsh deposits: Some examples from northeast Japan. J. Geol. 99, 265–287.

Minoura, K., Nakaya, S., Sato, H., 1987. Traces of tsunamis recorded in lake deposits: an example from Jusan, Shiura-mura, Aomori. J. Seismol. Soc. Jpn. 40, 183–196.

Minoura, K., Imamura, F., Takahashi, T., Shuto, N., 1997. Sequence of sedimentation processes caused by the 1992 Flores tsunami: evidence from Babi island. Geology 25, 523–526.

Moore, A.L., 2000. Landward fining in onshore gravel as evidence for a late Pleistocene tsunami on Molokai, Hawaii. Geology 28, 247–250.

Nanayama, F., Satake, K., Furukawa, R., Shimokawa, K., Atwater, B.F., Shigeno, K., Yamaki, S., 2003. Unusually large earthquakes inferred from tsunami deposits along the Kuril trench. Nature 424, 660−663.

Shiki, T., Yamazaki, T., 2000. Tsunami-induced conglomerates in Miocene upper bathyal deposits, Chita Peninsula, central Japan. Sediment. Geol. 104, 175−188.

Smit, J., 1999. The global stratigraphy of the Cretaceous-Tertiary boundary impact ejecta. Annu. Rev. Earth Planet Sci. 27, 75−113.

Smit, J., Montanari, A., Swinburne, N.H.M., Alvarez, W., Hildebrand, A.R., Margolis, S.V., Claeys, P., Lowrie, W., Asaro, F., 1992. Tektite-bearing, deep-water clastic unit at the Cretaceous-Tertiary boundary in northeastern Mexico. Geology 20, 99−103.

Smit, J., Roep, T.B., Alvarez, W., Claeys, P., Grajales-Nishimura, J.M., Bermudez, J., 1996. Coarse-grained, clastic sandstone complex at the K/T boundary around the Gulf of Mexico: deposition by tsunami waves induced by the Chicxulub impact?. In: Ryder, G., Fastovsky, D., Gartner, S. (Eds.), "Cretaceous-Tertiary Event and Other Catastrophes in Earth History" Geol. Soc. Am. Spec. Pap, vol. 307, pp. 151−182.

Tada, R., Nakano, Y., Iturralde-Vinent, M.A., Yamamoto, S., Kamata, T., Tajika, E., Toyoda, K., Kiyokawa, S., Delgado, D.G., Oji, T., Goto, K., Takayama, H., et al., 2002. Complex tsunami waves suggested by the Cretaceous-Tertiary boundary deposit at the Moncada section, western Cuba. In: Koeberl, C., Macleod, G. (Eds.), "Catastrophic Events and Mass Extinctions: Impact and Beyond" Geol. Soc. Am. Spec. Pap., vol. 356, pp. 109−123.

Tada, R., Iturralde-Vinent, M.A., Matsui, T., Tajika, E., Oji, T., Goto, K., Nakano, Y., Takayama, H., Yamamoto, S., Rojas-Consuegra, R., Kiyokawa, S., García-Delgado, D., et al., 2003. K/T boundary deposits in the proto-Caribbean basin. Mem. Am. Assoc. Petrol. Geol. 79, 582−604.

Takashimizu, Y., Masuda, F., 2000. Depositional facies and sedimentary successions of earthquake-induced tsunami deposits in Upper Pleistocene incised valley fills, central Japan. Sediment. Geol. 135, 231−239.

Takayama, H., Tada, R., Matsui, T., Iturralde-Vinent, M.A., Oji, T., Tajika, E., Kiyokawa, S., García, D., Okada, H., Hasegawa, T., Toyoda, K., 2000. Origin of the Peñalver Formation in northwestern Cuba and its relation to K/T boundary impact event. Sediment. Geol. 135, 295−320.

CHAPTER 9

Thickness and grain-size distribution of Indian Ocean tsunami deposits at Khao Lak and Phra Thong Island, South-Western Thailand

S. Fujino[1], H. Naruse[1], A. Suphawajruksakul[2], T. Jarupongsakul[2], M. Murayama[3], T. Ichihara[4]

[1]*Graduate School of Science, Kyoto University, Kyoto, Japan;* [2]*Faculty of Science, Chulalongkorn University, Bangkok, Thailand;* [3]*Center for Advanced Marine Core Research, Kochi University, Kochi, Japan;* [4]*Fukken Co., Ltd, Hiroshima, Japan*

1. Introduction

Tsunamis cause localized erosion and they transport large amounts of sediment in coastal areas. Sediments deposited by tsunamis provide a geologic record of the event. By analyzing these sediments, geologists can reconstruct ancient tsunamis and contribute to understanding the risk that they pose to nature, society, and economy, because tsunami deposits provide useful information for estimating recurrent intervals and inundation limits.

Discrimination of tsunami deposits from deposits produced by other processes, for example, storms, can be difficult. The difficulty arises, in part, from insufficient observations of modern tsunamis and their deposits. Published information is still inadequate to define diagnostic features of tsunami deposits, although key contributions have been made during the last decade (Benson et al., 1997; Gelfenbaum and Jaffe, 2003; Nanayama et al., 2000; Nishimura and Miyaji, 1995; Shi et al., 1995).

A magnitude-9.0 earthquake north-west of Sumatra in the morning of December 26, 2004 generated a large tsunami that struck coasts around the Indian Ocean, taking more than 200,000 lives. We surveyed the deposit of this tsunami in south-western Thailand in March 2005. Coastal towns in the study area were severely damaged by the event, and more than 5000 people were killed.

Tsunamiites. https://doi.org/10.1016/B978-0-12-823939-1.00009-4
Copyright © 2008 Elsevier B.V. All rights reserved.

This chapter describes terrestrial deposits at Khao Lak and on Phra Thong Island (Fig. 9.1). Specifically, it documents distributions in thickness and grain size landward from the shore and provides examples of characteristic sedimentary structures. Our objective is to add information that can be used to infer ancient tsunamis from their deposits.

2. Study areas

Phra Thong Island and Khao Lak have flat coastal plains bordered by straight beaches. Phra Thong Island (Fig. 9.2A) is about 100-km north of Phuket Island. The west side of the island is flat and mainly covered with grass, with small sandy ridges near the shoreline. The tsunami inundated this part of the island. We conducted a survey along a transect parallel to the run-up direction of the tsunami.

The Khao Lak study area extends 13 km from Khao Lak to Pakarang Cape (Fig. 9.2B). It is a 2-km-wide coastal plain, vegetated with coconut trees. The tsunami eroded the shore and inundated most of the plain. We surveyed three transects perpendicular to the shore.

3. Impact of the tsunami

The maximum heights of the tsunami on Phra Thong Island and Khao Lak were 7 and 10 m, respectively, judging from broken branches of standing trees. The run-up direction of the tsunami waves was from east to south-west, based on directions of bent grasses and palm trees (Fig. 9.3A). According to eyewitnesses, two tsunami waves occurred.

The coastlines of Phra Thong Island and the Khao Lak area were severely eroded by the tsunami, leaving 1—4-m-high scarps. Some beaches around Pakarang Cape, where the tsunami was the highest in Thailand, were eroded backward for more than 100 m. Erosion, however, was localized and decreased inland (Fig. 9.3B).

The tsunami left a sheet of sand that is massive in some places but graded in other places (Fig. 9.4A). Multiple-graded units, bounded by erosion surfaces, were also observed (Fig. 9.4B). The tsunami deposit consists of fine sand and silt with shell fragments. It is easily identified as a white to gray bed overlying a black muddy soil with rootlets.

4. Thickness and grain-size distribution

The thickness of the tsunami deposit was measured at 18 sites along the transect on Phra Thong Island and at 28 sites along the three transects at Khao Lak (Fig. 9.2). Samples for grain-size analysis were collected at 15 sites at Phra Thong Island and 24 sites at

Thickness and grain-size distribution 147

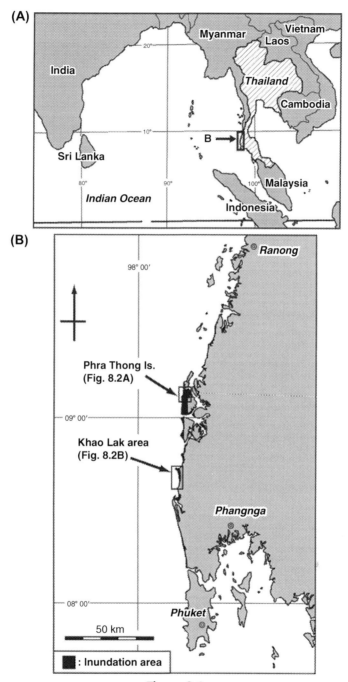

Figure 9.1
Location of the study area in south-western Thailand. *Modified from Environmental Geology Division, 2005. Andaman Sea Coastal Erosion in Changwat Ranong, Pangnga, Puket, and Krabi after Tsunami. Department of Mineral Resources, Bangkok, pp. 17.*

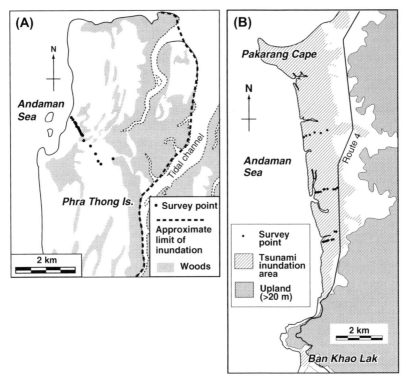

Figure 9.2
Areas inundated by the tsunami and survey points. (A) Phra Thong Island; (B) Khao Lak area. *Modified from Environmental Geology Division, 2005. Andaman Sea Coastal Erosion in Changwat Ranong, Pangnga, Puket, and Krabi after Tsunami. Department of Mineral Resources, Bangkok, pp. 17.*

Khao Lak. Grain-size analysis involved sieving of sediment coarser than 1 mm and processing of finer sediment with a laser granulometer (Mastersizer, 2000; Malvern Instruments). The precise location of the sites was established using Global Positioning System. Flooding distances along each transect were determined on the basis of a tsunami flooding direction of N80E.

The tsunami deposit extends more than 1500 m inland at Phra Thong Island and Khao Lak (Fig. 9.2). It forms a continuous sheet in both study areas, except at eroded sites. The sand sheet thins rapidly at its inland limit, although the thinning is not uniform (Figs. 9.5 and 9.6). The deposit generally is 1–10-cm thick but locally exceeds 20 cm. On Phra Thong Island, its thickness ranges considerably from the shore to the site 1200 m inland, then thins uniformly from 1200 to 2000 m along the transect (Fig. 9.5). In the Khao Lak area, the thickness of the tsunami deposit shows no relationship with the distance from the shore (Fig. 9.6).

Figure 9.3
Impact of the tsunami. (A) Bent grass, indicating flow from right to left on Phra Thong Island. (B) Shoreline erosion in the Khao Lak area. The maximum depth of the erosion is 4 m.

The tsunami deposit fines landward, with local variations. On Phra Thong Island, the deposit is fine to medium sand (1.9–2.2 φ) from the shore to the site along the transect, that is 170 m inland. It fines abruptly fine–very fine sand (2.5–3.8 φ) to 250 m from the shore (Fig. 9.5). In the Khao Lak area, the deposit is composed of fine to medium sand (1.3–2.4 φ) to 540 m from the shore and fine–very fine sand (2.5–3.8 φ) further inland (Fig. 9.6).

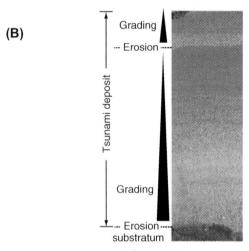

Figure 9.4
Graded tsunami deposits. (A) Vaguely graded tsunami deposit in the Khao Lak area. (B) Two graded layers of the tsunami deposit in the Khao Lak area. The base of each layer is erosional.

5. Discussion

Multiple-graded units reflect multiple waves. Unlike storm waves, tsunami waves decelerate and come to standstill prior to backwash, because tsunamis have long wavelengths (up to 200 km). As the flow wanes, entrained sediment settles from the water column, which results in graded beds (Dawson and Shi, 2000).

The widespread distribution and inland extent of the tsunami deposit in south-western Thailand reflect the large size of the tsunami. In contrast, the thickness and grain size of the tsunami deposit provide little indication of the size and destructive nature of the event. The deposit's thickness (mostly 1–10 cm) and grain size (1.3–3.8 φ) are similar to those of the 1998 Papua New Guinea tsunami (Gelfenbaum and Jaffe, 2003).

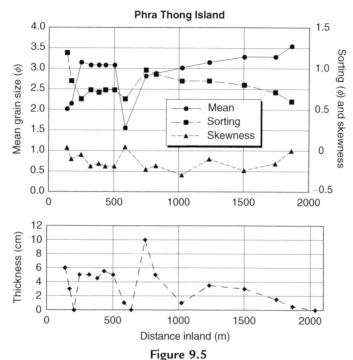

Figure 9.5
Trends in grain size, sorting, skewness, and thickness of the tsunami deposit along the transect on Phra Thong Island. The deposit thins at its landward margin.

Nanayama et al. (2000) describe thicker (up to 30 cm) and coarser (0.67 φ) deposits left by the 1993 Hokkaido-Nansei-Oki tsunami. Thickness and grain size are less influenced by the scale of the tsunami than by sediment availability and texture and by local topography.

Landward fining of the deposit and thinning at the edge of its distribution are consistent with observations of other tsunami deposits (Benson et al., 1997; Minoura et al., 2001; Nanayama and Shigeno, 2004; Nishimura and Miyaji, 1995; Sato et al., 1995; Shi et al., 1995). Deposits in coastal areas with ridges such as Khao Lak are more variable than those in flat areas such as the study sites on Phra Thong Island (Fig. 9.4), suggesting that the local topography strongly affects the characteristics of the deposits. Relatively thick sand also concentrated in depressions eroded by the tsunami waves.

Sugawara et al. (2004) suggest that the total amount of a tsunami deposit changes according to the magnitude of the tsunami. It was found, however, that the thickness of the deposit is affected by the topography, as described earlier. Thus, the wave height of a tsunami should be determined from the total volume rather than the thickness of the resulting deposit.

152 Chapter 9

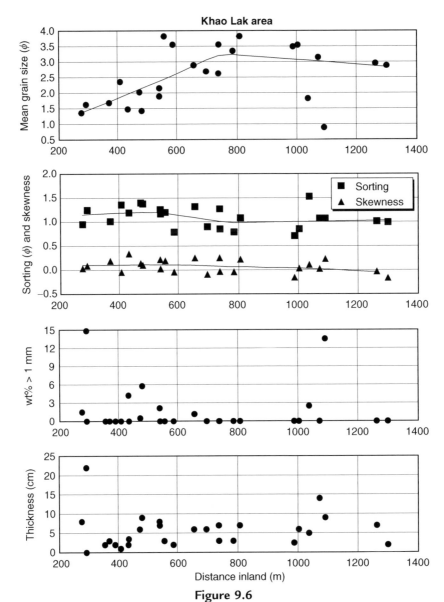

Figure 9.6
Trends in grain size, sorting, skewness, coarse grain content, and thickness of the tsunami deposit in the Khao Lak area. All graphs are compiled from the data of the three transects. The deposit fines inland. The thickness has no clear trend.

6. Conclusions

The following conclusions are drawn from the data collected along shore-normal transects on Phra Thong Island and the Khao Lak area in south-western Thailand.

1. The tsunami at both sites is a continuous sheet extending 1.5 km inland from the shore.
2. The deposit is graded or massive and locally contains multiple-graded layers resulting from multiple waves.
3. The deposit fines inland and thins near the limit of its extent. Its thickness, however, is strongly affected by the local topography.
4. The thickness and grain size of the deposit are influenced more by local factors, such as topography and source sediment, than by the size of the tsunami.

Acknowledgments

The authors are grateful to the staff and researchers of Chulalongkorn University and in particular to Professor Hashizume, for guidance throughout this study. Professor Masuda of Kyoto University is also gratefully acknowledged for helpful discussion and advice. The research was supported by KAGI21 of the 21st Century Center of Excellence program from the Japanese Ministry of Education, Culture, Sports, Science and Technology.

References

Benson, B.E., Grimm, K.A., Clague, J.J., 1997. Tsunami deposits beneath tidal marshes on Northwestern Vancouver island, British Columbia. Quat. Res. 48, 192−204.

Dawson, A.G., Shi, S., 2000. Tsunami deposits. Pure Appl. Geophys. 157, 875−897.

Environmental Geology Division, 2005. Andaman Sea Coastal Erosion in Changwat Ranong, Pangnga, Puket, and Krabi After Tsunami. Department of Mineral Resources, Bangkok, p. 17.

Gelfenbaum, G., Jaffe, B., 2003. Erosion and sedimentation from the 17 July, 1998 Papua New Guinea tsunami. Pure Appl. Geophys. 160, 1969−1999.

Minoura, K., Imamura, F., Sugawara, D., Kono, Y., Iwashita, T., 2001. The 869 Jogan tsunami deposit and recurrence interval of large-scale tsunami on the Pacific coast of northeast Japan. J. Nat. Disaster Sci. 23, 83−88.

Nanayama, F., Shigeno, K., 2004. An overview of onshore tsunami deposits in coastal lowland and our sedimentological criteria to recognize them. Mem. Geol. Soc. Jpn. 58, 19−33 (in Japanese with English abstract).

Nanayama, F., Shigeno, K., Satake, K., Shimokawa, K., Koitabashi, S., Miyasaka, S., Ishii, M., 2000. Sedimentary differences between the 1993 Hokkaido-nansei-oki tsunami and the 1959 Miyakojima typhoon at Taisei, southwestern Hokkaido, northern Japan. Sediment. Geol. 135, 255−264.

Nishimura, Y., Miyaji, N., 1995. Tsunami deposits from the 1993 Southwest Hokkaido earthquake and the 1640 Hokkaido Komagatake eruption, northern Japan. Pure Appl. Geophys. 144, 719−733.

Sato, H., Shimamoto, T., Tsutsumi, A., Kawamoto, E., 1995. Onshore tsunami deposits caused by the 1993 Southwest Hokkaido and 1983 Japan Sea earthquakes. Pure Appl. Geophys. 144, 693−717.

Shi, S., Dawson, A.G., Smith, D.E., 1995. Coastal sedimentation associated with the December 12th, 1992 tsunami in Flores, Indonesia. Pure Appl. Geophys. 144, 525−536.

Sugawara, D., Minoura, K., Imamura, F., Hirota, T., Sugawara, M., Ohkubo, S., 2004. Hydraulic experiments regarding tsunami sedimentation. Mem. Geol. Soc. Jpn. 58, 153−162 (in Japanese with English abstract).

CHAPTER 10

Lessons from the 2011 Tohoku-oki tsunami: implications for Paleotsunami research

D. Sugawara[1,2]

[1]*Museum of Natural and Environmental History, Shizuoka, Japan;* [2]*International Research Institute of Disaster Science, Tohoku University, Sendai, Miyagi, Japan*

1. Introduction

Research on tsunami deposits has clarified the role that tsunamis play as a geologic and geomorphic agent in coastal zones, as well as providing information on the recurrence interval and size of past events that have aided tsunami hazard assessment programs and disaster management activities. One of the main objectives of tsunami deposit research is to identify traces of past tsunamis within coastal sedimentary sequences and to decipher the geological record to constrain the timing and size of paleotsunami events (e.g., Bourgeois and Minoura, 1997). Since the 1980s, post-tsunami field surveys have been conducted in coastal areas across the globe to collect data on tsunami deposit sedimentological characteristics and associated tsunami hydrodynamics as modern analogs to paleotsunami events, as well as to better understand tsunami-induced erosion, transport, sediment deposition, and the resulting morphological changes.

Out of the recent disastrous tsunamis, the 2004 Indian Ocean tsunami, due to the Sumatra Earthquake, has been a landmark event for tsunami deposit research. Numerous post-tsunami surveys were conducted in the broad coastal areas of the Indian Ocean, such as Indonesia (e.g., Paris et al., 2007, 2009), Thailand (e.g., Goto et al., 2008; Umitsu et al., 2007; Matsumoto et al., 2008), and Sri Lanka (e.g., Goto et al., 2008; Morton et al., 2008; Matsumoto et al., 2010) to investigate the geological and geomorphological effects of the tsunami. These studies have investigated the nature of tsunami sedimentation, such as the deposition of tons of boulders (e.g., Goto et al., 2007; Paris et al., 2010), onshore marine sediment transport (e.g., Hawkes et al., 2007; Sawai et al., 2009), and offshore sedimentation via tsunami run-up and backwash (e.g., Sugawara et al., 2009; Sakuna et al., 2012). These studies have also raised

Tsunamiites. https://doi.org/10.1016/B978-0-12-823939-1.00010-0
Copyright © 2021 Elsevier B.V. All rights reserved.

questions associated with tsunami deposit research, such as questions regarding the unique characteristics of tsunami deposits, the quantitative relationship between tsunami hydrodynamics and deposits, and the characteristics of offshore tsunami deposits (e.g., Goto and Fujino, 2008).

The tsunami due to the 2011 off the Pacific coast of Tohoku earthquake (known as the 2011 Tohoku-oki tsunami; the Japanese term "oki" translates to "offshore" in English) was a groundbreaking event that yielded new findings on tsunami sedimentation, with the aid of an unprecedented quantity and quality of geophysical data. The geophysical data includes on-land (Ozawa et al., 2011) and seafloor (Kido et al., 2011) geodetic observations, offshore tsunami buoy records (e.g., DART and NOWPHAS), extensive inundation mapping (Haraguchi and Iwamatsu, 2011a,b), and thousands of tsunami measurements (e.g., Mori et al., 2012), as well as pre- and post-tsunami bathymetric and topographic data sets (e.g., Udo et al., 2012; Haraguchi et al., 2012). In addition, air-borne video footage taken during tsunami inundation provided opportunities to quantify spatial and temporal variations in hydrodynamic parameters, such as flow height and speed (Goto et al., 2011; Hayashi and Koshimura, 2013). The post-tsunami field observations and following studies have clarified the spatial variability of tsunami deposits (e.g., Abe et al., 2012; Richmond et al., 2012), their quantitative relationship with tsunami hydrodynamics (e.g., Goto et al., 2014a), and the characteristics of offshore tsunami sedimentation and deposits (e.g., Yokoyama et al., 2015; Yoshikawa et al., 2015). These studies have yielded significant progress in tsunami sedimentology and identified future challenges that paleotsunami studies may face.

This chapter discusses the lessons learned from the Tohoku-oki tsunami from a geological perspective with respect to tsunamis based on previously reported results throughout the last 8 years. These lessons will provide valuable implications for future advances in tsunami deposit research and contribute to the disaster management of future tsunamis across the Earth.

2. The 2011 Tohoku-oki tsunami and its precursors

The Pacific coast of northeastern Japan (Tohoku; Fig. 10.1A) faces the Japan Trench, which is a convergent boundary between the North American and Pacific Plates. According to historical records and recent seismic events, quasi-periodic earthquakes characterize this region, with moderate (M ~ 7.5) magnitudes and an average recurrence interval of 37 years (e.g., the Miyagi-oki earthquake) (e.g., Sugawara et al., 2012). Regional tsunami disaster management was predominantly based on risk assessments of the Miyagi-oki and several other tsunamigenic earthquakes, such as the 1896 Meiji-Sanriku and 1933 Showa-Sanriku earthquakes.

On March 11, 2011, the Mw = 9.0 Tohoku-oki earthquake occurred, triggering a gigantic tsunami that far exceeded the level of tsunami hazards interpreted from risk assessments of the

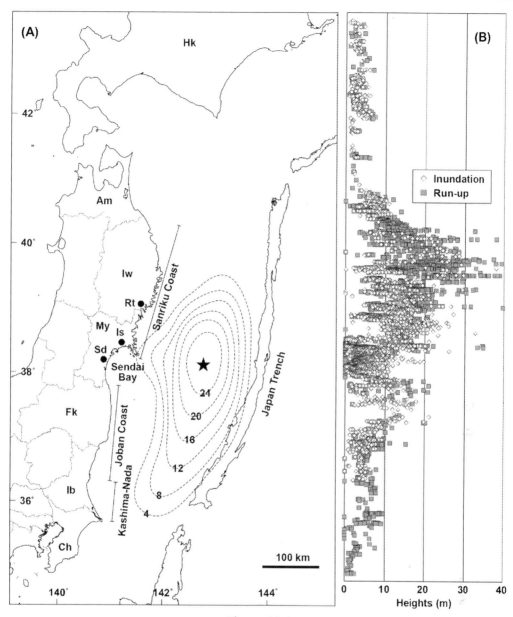

Figure 10.1
(A) Map of Tohoku showing the rupture area of the 2011 Tohoku-oki earthquake. Numbers on broken lines indicate the amount of slip estimated from geodetic observations (Ozawa et al., 2011). *Solid black circles* indicate the localities mentioned in the main text (*Is*, Ishinomaki Plain; *Rt*, Hirota Bay; *Sd*, Sendai Plain). Prefectures: *Am*, Aomori; *Ch*, Chiba; *Fk*, Fukushima; *Hk*, Hokkaido; *Iw*, Iwate; *Ib*, Ibaraki; *My*, Miyagi. (B) Inundation and run-up heights of the 2011 Tohoku-oki tsunami (Mori et al., 2012). *Modified after Sugawara, D., 2017. Tsunami sedimentation and deposits due to the 2011 Tohoku earthquake: a review of case studies from Sendai and Hirota Bays. J. Geol. Soc. Jpn. 123, 781–804. (in Japanese with English abstract).*

Miyagi-oki earthquake. Based on analyses of the geodetic data, the Tohoku-oki earthquake had a rupture area with a length and width of 450 and 150 km, respectively, which extended from the deep plate interface to the trench axis (e.g., Iinuma et al., 2012; Ozawa et al., 2011), a maximum slip on the plate interface that reached up to 27 m (Ozawa et al., 2011), and slip near the trench axis estimated at 60 m or larger (Ito et al., 2011). The rupture area on the seafloor was displaced vertically by 10 m, which generated a tsunami with a large wavelength. In addition, horizontal movement of steep slopes on the continent side of the trench axis caused additional seafloor uplift of 10—20 m, which generated an impulsive high wave (Ito et al., 2011). Sea-bottom pressure gauges and offshore GPS buoys recorded the composite waveform of the tsunami. The waveform showed an initial gradual increase, followed by a steep rise in the water level, which indicates the superimposition of the large wavelength and impulsive high wave (Takahashi et al., 2011). The mechanism that generated the impulsive high wave is still slightly controversial. Rather than a large slip near the trench axis, previous studies have proposed a submarine landslide scenario as an alternative trigger mechanism of the impulsive high wave based on pre- and post-earthquake bathymetry data and numerical modeling of tsunami generation and propagation (Tappin et al., 2014).

The tsunami arrived at the coast of Tohoku 20—70 min after the earthquake. In narrow bays with steep slopes, such as the V-shaped bays along the Sanriku Coast (Fig. 10.1A), the tsunami reached as high as 30—40 m above mean sea level (Fig. 10.1B; Mori et al., 2012). In flat, low-lying plains, such as the Sendai Plain (Fig. 10.1A), tsunami inundation proceeded as far as 4—5 km inland from the coastline (Goto et al., 2012b). The tsunami inundated 561 km^2 of land, causing the total destruction of numerous coastal communities and facilities, including the Fukushima-Daiichi nuclear power plant, as well as more than 18,000 fatalities. The erosion and deposition of coastal sediment and resulting morphological changes were commonly observed at many of the areas that were impacted by the tsunami. The tsunami transported a large amount of beach sediments and deposited them inland (e.g., Goto et al., 2011; Abe et al., 2012; Naruse et al., 2012). Large-scale sediment transport caused either total erosion or partial breaching of sandy beaches (e.g., Kato et al., 2012; Tappin et al., 2012). The quantity of onshore tsunami deposits in the main areas of tsunami devastation (Iwate, Miyagi, and Fukushima Prefectures; Fig. 10.1A) was 11 million tons, although post-tsunami recovery and reconstruction efforts, by the beginning of 2014, had already cleared away the majority of the deposits (e.g., Sugawara, 2018).

Sendai Plain and adjacent coastal lowlands are well known due to the discovery of sandy tsunami deposits that have been suggested to derive from the 869 CE Jogan earthquake and several other prehistoric events. These discoveries marked the beginning of paleotsunami research in Japan (e.g., Minoura and Nakaya, 1991; Minoura et al., 2001). The Jogan tsunami deposit was identified as a thin, medium-grained sand layer intercalated within coastal marsh deposits that extends up to 3—4 km from the present coastline (Minoura and Nakaya, 1991). The size of the Jogan earthquake was estimated at Mw = 8.3–8.4 or larger based on numerical modeling of tsunami propagation and inundation that considered the known distribution of the sandy tsunami deposits (e.g., Minoura et al., 2001; Satake et al., 2008a). These studies have discussed the implications of the hazards posed by infrequent earthquakes (i.e., those that were much larger than the M ~ 7.5 Miyagi-oki earthquake). After the 2011 disaster, various studies considered the Jogan earthquake as the precursor to the Tohoku-oki earthquake (e.g., Goto et al., 2011; Sugawara et al., 2012; Satake et al., 2013). The 1611 Keicho and 1454 Kyotoku earthquakes, both of which are known from historical and geological records of major tsunami inundation, are also possible precursors to the Tohoku-oki earthquake (e.g., Sawai et al., 2012a).

3. Lessons learned from the Tohoku-oki tsunami

3.1 Limitation of marine materials as evidence for tsunami inundation

The occurrence of marine materials within onshore event deposits provides strong evidence for seawater incursion, such as storms and tsunamis. In particular, assemblages that are characteristic of water depths below the storm wave base may support tsunami origin hypotheses based on the differences in the hydrodynamics between storms and tsunamis. Water motion due to storm waves decays with increasing depth from the water surface, that is, storm waves have a wave base. Sediment transport below the wave base is unlikely. On the other hand, tsunamis are characterized by vertically uniform horizontal motion in the water column, which may induce high flow speed near the seafloor that agitates and transports bed materials.

However, not all onshore tsunami deposits contain marine materials, as shown by previous field observations (see a brief review in Sugawara et al., 2008). Case studies of the Tohoku-oki tsunami deposits provide information on the site-specific nature of tsunami deposit provenance. In the Sendai Plain (Fig. 10.1A), the compositions of heavy minerals (Jagodziński et al., 2012), foraminifer (Pilarczyk et al., 2012), and diatom assemblages (Szczuciński et al., 2012; Takashimizu et al., 2012) are possibly indicative of onshore tsunami deposit provenance. For example, freshwater to brackish species, with the detection of no typical marine species, dominate the diatom assemblages in tsunami deposits from the central Sendai Plain (Fig. 10.2; Szczuciński et al., 2012). In the northern Sendai Plain, marine diatom species comprised ~2% of the diatom assemblage in the tsunami deposits, where the marine species occurrence frequency showed a clear trend that decreased with increasing distance inland (Takashimizu et al., 2012). Carbon-13 analyses of tsunami deposits and soil samples from the central Sendai Plain suggest the presence of a certain degree of mixing between terrestrial and marine organic matter (Chagué-Goff et al., 2012). Szczuciński et al. (2012) attempted to identify onshore

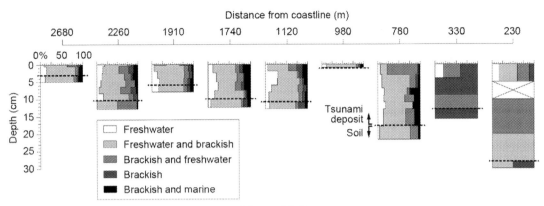

Figure 10.2

Diatom assemblages of the 2011 Tohoku-oki tsunami at sites in the central Sendai Plain (Szczuciński et al., 2012), which shows a minor contribution from marine sediments. The diatom assemblages within the tsunami deposit are dominated by freshwater and freshwater-brackish species, which is similar to the soil assemblages. *Cited from Szczuciński, W., Kokociński, M., Rzeszewski, M., Chagué-Goff, C., Cachão, M., Goto, K. and Sugawara, D., 2012, Sediment sources and sedimentation processes of 2011 Tohoku-oki tsunami deposits on Sendai Plain, Japan — Insights from diatoms, nannoliths and grain size distribution, Sediment. Geol., 282, 40–56 (Fig. 7).*

tsunami deposits based on a marine origin assumption, which can be supported by the occurrence of marine materials, such as diatoms, foraminifera, nannoliths, and sediments. As a result, they pointed out that these identifications may be of limited use, regardless of tsunami size. In fact, the tsunami that occurred on the Sendai Plain was as high as 10 m on the beach, with inundation that reached 4–5 km inland (e.g., Richmond et al., 2012). Despite the size of the tsunami, marine materials were rarely detected (Szczuciński et al., 2012).

Deposits associated with the Tohoku-oki tsunami from other sites occasionally exhibit marine signatures. In the Otomo-ura lowland in Hirota Bay (Fig. 10.1A), along the southern Sanriku Coast (inundated by 15-m-high tsunami waves), tsunami deposits include ostracod assemblages that are characteristic of inner-bay and rocky-shore environments and correspond to an inhabitation water depth of 9 m (Naruse et al., 2012; Tanaka et al., 2012). However, ostracods are practically absent from tsunami deposits in the adjacent Takata Plain at the head of Hirota Bay. Diatom assemblages in tsunami deposits from coastal lowlands along the Kashima-nada Coast, where the tsunami height ranged from 5 to 10 m, contain a mixture of assemblages from various environments that occasionally include marine species (Sawai et al., 2012b). It is clear that the nonoccurrence or nonrecognition of a marine signature within a presumed onshore tsunami deposit does not indicate that it was created by processes other than tsunami inundation.

Marine sediment dilution due to mixing with terrestrial materials (Chagué-Goff et al., 2012) can explain, to a certain extent, why specific onshore tsunami deposits have a limited marine signature. The ratio of marine sediments may decrease with increasing distance from the sea due to surficial sediment entrainment into the tsunami inundation flow, as shown by the inland decrease in marine diatoms (Takashimizu et al., 2012). For the Tohoku-oki tsunami, this process may have been controlled by land conditions in tsunami inundation areas, such as vegetation and plowing. In early spring (from March to April for Sendai Plain), rice paddies usually lack vegetation, with extensive plowing of the paddies to prepare for rice planting. These circumstances stimulate tsunami inundation, as well as the entrainment of surficial soil into the flow, which resulted in marine sediment dilution. Surficial soils were finally redeposited as a mud fraction in the tsunami deposits (Takashimizu et al., 2012).

Numerical modeling of tsunami propagation (Sugawara and Goto, 2012) and sediment transport (Sugawara et al., 2014b) provide possible hydrodynamic explanations for the lack of marine materials within the onshore tsunami deposits. The shoreface of Sendai Bay has a steep slope with water depths of up to 20 m, as well as an inner shelf that is characterized by a gentle slope (0.05%) that extends 30 km offshore with a marginal water depth of 40 m. Numerical modeling of tsunami propagation (Sugawara and Goto, 2012) has shown that the tsunami formed a bore-like wave when it propagated along the inner shelf due to shallow-water deformation. When the tsunami arrived at the coast, the energy of the front part of the tsunami bore was blocked, reflected onto the steep shoreface, and was then propagated offshore as a reflected wave. Wave reflection caused a sudden decrease in flow speed in the sea, which may have resulted in an abrupt reduction of the sediment transport capacity (Sugawara and Goto, 2012). Tsunami sediment transport modeling (Sugawara et al., 2014b) has clearly illustrated considerable differences in sediment suspension between offshore and onshore areas during tsunami run-up. The volumetric concentration of suspended sediments in the sea was, at most, 0.1% during the run-up. In contrast, the volumetric concentration of suspended sediments on the beach reached up to 2%. Although the tsunami run-up suspended a certain amount of sediments from the seafloor, decreased flow speed in the sea did not cause landward seafloor sediment transport. In addition, seafloor sediment suspension did not simultaneously occur with the passing of the wave front. As a result, suspended seafloor sediments were not transported shoreward and redeposited on the seafloor (Sugawara et al., 2014b).

These numerical results suggest that onshore marine sediment transport processes depend on the bathymetric profile of the coast. In addition, marine sediments tend to be diluted during tsunami inundation along flat coastal plains with long distances. Thorough surveys of coastal geology and geomorphology, including local bathymetry, and numerical assessments of sediment transport will help future studies use marine materials as a tsunami proxy.

162   Chapter 10

3.2 Larger extent and lower preservation potential of offshore tsunami deposits

The majority of studies on tsunami deposits associated with modern events aim to examine the distribution and sedimentological features of onshore deposits, whereas others focus on tsunami deposits from coastal ponds and lakes. Although several studies have reported the features and ages of paleotsunami deposits from offshore sedimentary environments, there are very few investigations of modern submarine tsunami deposits. Observations of erosion and deposition on the seafloor due to the Tohoku-oki tsunami have illustrated the features of offshore tsunami sedimentation, together with the potential that submarine tsunami deposits have as a geologic record of paleotsunamis.

The balance between tsunami-induced sediment erosion and deposition across the beaches of Hirota Bay (Fig. 10.1A) was quantified based on pre- and post-tsunami bathymetric data. Hirota Bay is characterized by a steep seafloor that proceeds toward the bay entrance and a narrow (i.e., a width of $\sim 2\,km$ from the coastline) coastal lowland surrounded by steep hill slopes. At the time of tsunami inundation, the steep bathymetry and narrow onshore topography of the bay stimulated a rapid transformation of the potential energy from the tsunami run-up into kinetic energy of the seawater that returned to the ocean. Thus, the backwash in Hirota Bay became energetic compared with that in flat coastal plains, such as in Sendai Bay, based on aerial video footage (Udo et al., 2015; Yamashita et al., 2016). Kato et al. (2012) estimated that the volume of beach erosion at the head of Hirota Bay was $1.86 \times 10^6\,m^3$. The sediment volume deposited on land was estimated at between 0.34 and $0.61 \times 10^6\,m^3$ based on field measurements of tsunami deposit thicknesses (Naruse et al., 2012; Udo et al., 2015). According to analyses of the pre- and post-tsunami bathymetric data, the sediment volume deposited on the seafloor near the coastline was estimated at $0.24 \times 10^6\,m^3$ (Udo et al., 2015) while the remaining sediments, with an estimated volume between 1.01 and $1.28 \times 10^6\,m^3$, may have been transported further offshore. These results suggest that only 18%−33% of the materials involved in tsunami-induced sediment transport were deposited on land, whereas the remaining materials were deposited on the seafloor. The balance between sediment erosion and deposition in Hirota Bay was further examined via the numerical modeling of sediment transport (Yamashita et al., 2016). Modeling results from Yamashita et al. (2016) suggest that the depth of beach erosion due to tsunami backwash was two times larger than erosion due to uprush. This implies that one-third of the beach sediment was deposited on land and two-thirds of the sediment was deposited offshore. The extent of the modeled onshore tsunami deposit was limited to within the narrow coastal lowland, with a maximum width of $2\,km$, while offshore deposits were transported to at least $6\,km$ from the coastline (Yamashita et al., 2016).

In contrast to Hirota Bay, the effects of tsunami backwash were limited in most areas of Sendai Bay, which is mainly due to the flat onshore topography. The backwash created

V-shaped channels, as well as breaching and scouring along the fore- and backshores of the sandy beaches (Tappin et al., 2012). Nevertheless, backwash formed embayments across the sandy beaches of the southern Sendai Plain (e.g., Yoshikawa et al., 2015). Analysis of aerial video footage (e.g., Tappin et al., 2012; Udo et al., 2013) suggests that the backwash transported beach sediments to the neighboring seafloor. Formation of erosional features on the beach and the visual observation of offshore beach sediment transport suggest that extensive seafloor erosion occurred during the backwash. High-resolution seismic bottom profiling revealed that erosional surfaces formed below the seafloor of the shoreface (Yoshikawa et al., 2015). The erosional surface extended 1.1 km from the coastline to water depths of 16.7 m. The maximum thickness of submarine tsunami deposits, inferred from the difference between the erosional surface and posttsunami water depth, was estimated at up to 1.2 m, which is consistent with deposit thicknesses estimated from the analysis of pre- and post-tsunami bathymetric data (0.6−1.1 m; Udo et al., 2013). Similar to Hirota Bay, the majority of the submarine tsunami deposits in southern Sendai Bay derived from beach erosion. Based on the pre- and post-tsunami bathymetric data, the volume percentage of sediments transported seaward by tsunami backwash was estimated at 45%−75% of the beach sediment volume above sea level (Udo et al., 2013).

These observations and analyses demonstrate that the spatial extent and thickness of submarine tsunami deposits can be much greater than that of onshore tsunami deposits. Furthermore, submarine tsunami deposits can be used as evidence of large-scale tsunami events, such as the event due to the Tohoku-oki earthquake. However, we can raise numerous questions with respect to the identification, sedimentological features, and preservation potential of submarine tsunami deposits, as well as the determination of the depositional age.

In Sendai Bay and adjacent areas, coring of bottom sediments was performed to identify submarine tsunami deposits and tsunami-related turbidites (Arai et al., 2013; Ikehara et al., 2014; Tamura et al., 2015; Yoshikawa et al., 2015). Submarine tsunami deposit identification was performed based on the lithology, extent of bioturbation, and sequential measurements of ^{134}Cs concentrations, which was released during the destruction of the Fukushima-Daiichi nuclear power plant (Ikehara et al., 2014; Tamura et al., 2015; Yoshikawa et al., 2015). Tamura et al. (2015) reported the existence of submarine tsunami deposits from the lower shoreface to the inner shelf of Sendai Bay and found that the deposits were distributed 1−3 km from the coastline (Fig. 10.3). The tsunami deposits on the lower shoreface were composed of coarse-grained sand, which suggests seaward sediment transport from the upper shoreface or backshore. On the other hand, tsunami deposits on the inner shelf were composed of silt to fine sand, which is associated with the passage of the incoming tsunami wave, rather than the backwash (Tamura et al., 2015).

164 Chapter 10

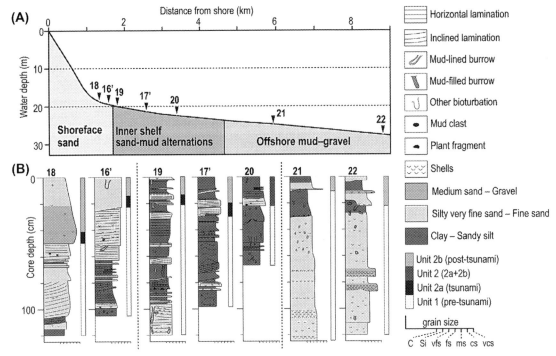

Figure 10.3
The spatial distribution and stratigraphy of the Tohoku-oki tsunami deposit in Sendai Bay (Tamura et al., 2015). The horizontal extent of the submarine tsunami deposit reached at least 3 km from the coastline. The majority of the tsunami deposits may have been reworked due to post-tsunami processes, such as wave and bioturbation activity. (A) A cross section of the bathymetric profile of the central Sendai Bay with a general classification of coastal zones (shoreface, inner shelf, and offshore). Numbers with reverse *triangles* correspond to the numbers in the geological column. (B) A geological column of the bottom sediments in Sendai Bay collected from August to September 2012. *Cited from Tamura, T., Sawai, Y., Ikehara, K., Nakashima, R., Hara, J., Kanai, Y., 2015. Shallow-marine deposits associated with the 2011 tohoku-oki tsunami in Sendai bay, Japan. J. Quat. Sci. 30, 293–297 (Fig. 2).*

Tsunami-related turbidites have been reported from areas further offshore of Sendai Bay and Sanriku Coast. Ikehara et al. (2014) found a tsunami-related turbidite from the outer margin of Sendai Bay at a water depth of 122 m. This turbidite is composed of layers of very fine sand and silt to clay, where the underlying pre-tsunami surficial deposit along the outer shelf showed evidence of seismic shaking and agitation. Arai et al. (2013) reported a tsunamigenic turbidite from deeper offshore areas in Sendai Bay and Sanriku Coast, with water depths ranging from 300 to 1400 m. The turbidite was composed of a thin sandy silt layer, where the deposit covers an extensive area of the seafloor, that is, greater than 100 km in width. Agitation of seafloor sediments by the earthquake and the subsequent resuspension due to tsunami waves have been suggested as the mechanisms that generated a turbidity current in the inner to mid-shelf of Sendai Bay (Ikehara et al., 2014).

Sediment supply through steep slopes may have triggered a self-accelerating turbidity current (Naruse, 2011), which deposited the turbidite along deeper areas of the seafloor (Arai et al., 2013). Portions of the tsunami-related turbidite can be associated with seismic shaking that triggered a turbidity current from the bathyal slope (Usami et al., 2016). Interpretations of submarine tsunami deposits and tsunami-related turbidites require careful observation and analysis.

Submarine tsunami deposits experienced post-tsunami reworking processes, such as bioturbation and storm waves (Fig. 10.3; Tamura et al., 2015). Based on ^{134}Cs measurements, previous studies have estimated that the post-tsunami reworking of tsunami deposits in Sendai Bay 15–30 months after the tsunami reached a depth of 20–30 cm (Tamura et al., 2015; Yoshikawa et al., 2015). Bioturbation plays an important role in submarine deposit preservation below the wave base and tsunami-related turbidites in deeper areas along the seafloor. Considering the strong bioturbation that characterizes hemipelagic sediments below the tsunami-related turbidites (Ikehara et al., 2014), the long-term preservation of thin fine-grained submarine tsunami deposits and turbidites is not likely. In addition, submarine tsunami deposits on the inner shelf and pre-tsunami storm deposits in Sendai Bay share similar sedimentological characteristics, which suggests that large tsunamis are unlikely to leave significant event deposits in open-sea shallow-marine settings (Tamura et al., 2015).

These results suggest that submarine tsunami deposit preservation requires a large initial thickness and high post-tsunami sedimentation rates to segregate the majority of the deposit from the effects of wave and biological activity (Tamura et al., 2015). Thus far, such an environment has only been identified in limited areas, for example, the lower shoreface of Sendai Bay (Yoshikawa et al., 2015). Further research on long-term submarine tsunami deposit preservation is required to examine the potential that deposits have to aid in the understanding of paleotsunami event recurrence.

3.3 Possible false dating of Paleotsunami events due to tsunami-induced erosion

The devastation caused by the Tohoku-oki earthquake and tsunami has raised a number of calls for the accurate estimation of the large-scale event recurrence interval. These investigations require significant work, that is, not only the thorough scans and identification of paleotsunami deposits and dating with minimal uncertainty but also spatial correlations between contemporaneous events along coastlines that face the presumed tsunami generation region. Even in Tohoku, which, over the past 30 years, has become a classic location for the study of paleotsunami deposits (e.g., Minoura and Nakaya, 1991), regional correlations and the estimation of large-scale tsunami event recurrence intervals are often problematic (e.g., Sawai et al., 2012a; Takada et al., 2016). Calendar ages estimated by radiocarbon dating usually include statistical errors that range

166 Chapter 10

from 10s to 100s of years, which is predominantly due to fluctuations in the calibration curve. Recent studies have tested and made further improvements to the resolution of tsunami deposit radiocarbon dating, as demonstrated by the sequential radiocarbon dating of bulk peat samples (Ishizawa et al., 2017). With aid from stratigraphic constraints on radiocarbon ages, sequential dating has proven that the range of statistical errors associated with the age of a tsunami event between the 16th and 17th centuries along the Pacific coast of Hokkaido, northern Japan, can be reduced by 60% compared with conventional estimation techniques (Ishizawa et al., 2017). However, tsunami-induced reworking of preexisting surficial sediments will introduce additional uncertainty to the dating and correlation of paleotsunami events.

As observed by the inclusion of rip-up clasts within sandy tsunami deposits, erosion may occur on surficial sediments during tsunami inundation. If this is the case, radiocarbon measurements of materials collected just below the tsunami deposit will yield older ages (i.e., a limiting maximum age). In addition, reworked muddy sediments, such as mud cap or drape, often cover tsunami sand layers. Avoiding the false dating of an event requires the ability to distinguish between reworked sediments, from the background layer, and sampling of materials from the appropriate horizon to obtain the limiting minimum age. Ishizawa et al. (2018) tested the sequential dating of large rip-up clasts from a paleotsunami deposit found along the Sanriku Coast to reduce the uncertainty associated with the age of the event. By setting the age of the youngest part of the rip-up clast as the maximum age of the tsunami deposit, the oldest age for the paleotsunami deposit was reduced by approximately 100 years. Ishizawa et al. (2018) pointed out that reducing the age uncertainty requires either the measurement of rip-up clasts or an appropriate choice of the sampling location, with minimal effects from tsunami-induced erosion.

These improvements to radiocarbon measurements will allow us to better constrain the age of tsunami deposits and to perform successful correlations of paleotsunami events. An example from the Tohoku-oki tsunami, however, has shown that a large-scale tsunami can occasionally disguise the age of its deposits and erase any evidence of the tsunamis that preceded it. Shinozaki et al. (2015b) examined the effects of tsunami-induced reworking of surficial sediments in a coastal lake in the Sendai Plain (Fig. 10.4). The Tohoku-oki tsunami transported sand from the beach, creating a sand layer with thicknesses ranging from 7 to 15 cm at the bottom of the lake. Prior to sand deposition, tsunami inflow into the lake caused the severe erosion of muddy bottom sediments of up to 1 m. The reworked muddy sediments settled after sand deposition. The erosion and redeposition of the mud caused the significant mixing of sediments with various ages. Radiocarbon dating of the sediments above and below the sand has shown that the age of the Tohoku-oki tsunami sand layer can be estimated at a few hundreds of years (Shinozaki et al., 2015b). In addition, a sand layer, with a possible historical tsunami event origin (i.e., either the 1454 Kyotoku or 1611 Keicho earthquake tsunami; Sawai et al., 2008a,b; Sawai et al., 2012a) has disappeared from the lake due to severe bottom sediment erosion.

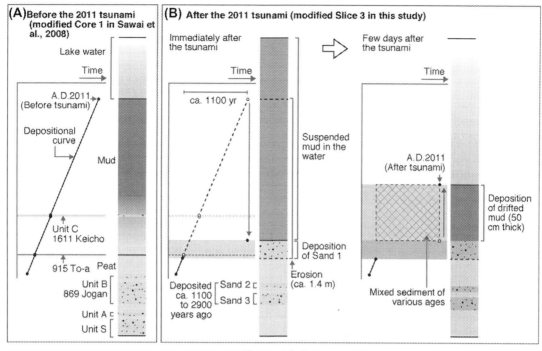

Figure 10.4
A conceptual model of tsunami-induced erosion and deposition in a coastal lake during the Tohoku-oki tsunami (Shinozaki et al., 2015b). The tsunami caused bottom sediment erosion that reached 1.4 m in depth, which included the sand layer of the 1611 Keicho tsunami and 915 CE Towada-a (To-a) volcanic tephra. The 2011 Tohoku-oki tsunami deposit was deposited onto the truncated bottom sediment, which may show an apparent older age. (A) A geological column of the pre-tsunami sediment in the lake (Sawai et al., 2008a). (B) Geological columns of the post-tsunami sediment in the lake. Linear curves plotted to the sides of each geological column show the depth-age model inferred from the radiocarbon ages of the sediments and the depositional age of the To-a tephra. *Cited from Shinozaki, T., Goto, K., Fujino, S., Sugawara, D., Chiba, T., 2015b. Erosion of a paleo-tsunami record by the 2011 Tohoku-oki tsunami along the southern Sendai Plain. Mar. Geol. 369, 127–136 (Fig. 6).*

Thus, the identification of unconformities due to tsunami-induced erosion, as well as the ability to distinguish between reworked muddy sediments, is the key to avoid false dating and perform paleotsunami event correlations. Geochemical and micropaleontological analyses can be used to distinguish between sediment types (Shinozaki et al., 2015b) while the use of sequential dating and other new dating techniques (Ishizawa et al., 2017, 2018) will be of help to reduce the uncertainty associated with the age of paleotsunami events. Although these improvements may not always be applicable to every field area, the careful investigation of sediments possibly affected by tsunami-induced reworking will yield improved age determinations of paleotsunami events.

3.4 Uncertainties in tsunami inundation distance based on deposit extent

Sandy deposits associated with the 2011 Tohoku-oki tsunami along its transect of the central Sendai Plain are distributed up to 2.8 km from the coastline; meanwhile tsunami inundation proceeded as far as 4.5 km inland (Goto et al., 2011; Richmond et al., 2012). Here, sandy tsunami deposits only covered 60% of the inundation area. In other transects in the Sendai Plain and adjacent drowned valleys, the extent of sandy deposits thicker than 0.5 cm covered nearly 100% of the inundation distance unless it exceeded a distance of 2.5 km from the coastline. For larger inundation distances, the maximum inland extent of sandy tsunami deposits was only 57%–76% of the shore-normal inundation distance (Fig. 10.5; Abe et al., 2012). In the Ishinomaki Plain (Fig. 10.1A; northern coast of Sendai Bay), the ratio of the deposit extent to inundation distance ranged from 62% to 83% (Fig. 10.5; Shishikura et al., 2012). Data compilations indicate a threshold distance of ~2 km that defines a change in the relationship between the maximum inland extent of sandy deposits and tsunami inundation distance.

The distributions of sandy tsunami deposits have been used as a primary proxy to identify the areas inundated by past tsunamis. The size of many past tsunamigenic earthquakes has been constrained based on comparisons between the simulated tsunami inundation distance and inundation distance inferred from the maximum inland extent of sandy deposits (e.g., Satake et al., 2008a,b; Ioki and Tanioka, 2016). However, as pointed out by Bourgeois and

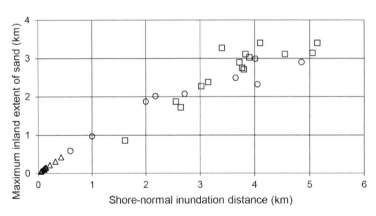

Figure 10.5

A diagram showing the relationship between the shore-normal tsunami inundation distance and maximum inland extent of the sandy tsunami deposits in the Sendai (Abe et al., 2012) and Ishinomaki (Shishikura et al., 2012) Plains, together with data from the 2006 Kuril Island tsunami (MacInnes et al., 2009). Significant differences appear between the inundation distance and sand extent at distances greater than 2 km. *Modified after Sugawara, D., 2017. Tsunami sedimentation and deposits due to the 2011 Tohoku earthquake: a review of case studies from Sendai and Hirota Bays. J. Geol. Soc. Jpn. 123, 781–804. (in Japanese with English abstract).*

Minoura (1997) and other studies, the use of paleotsunami deposits yields, at best, a minimum estimate of inundation. Post-tsunami field investigations have reported variability in the relationship between the tsunami inundation distance and maximum inland extent of the sandy deposits. For the 1992 Flores and 2004 Indian Ocean tsunamis, for example, the tsunami inundation distance was 2—4 times larger than the extent of the sandy deposits (e.g., Shi et al., 1995; Goto et al., 2008). On the other hand, the extent of sandy deposits due to the 2006 Kuril Island tsunami reached 95% of the inundation distance (Fig. 10.5; MacInnes et al., 2009).

Various factors, such as sediment source and grain size, initial tsunami height, and local bathymetry and topography, may affect the difference between the tsunami inundation distance and maximum inland extent of deposits. Numerical modeling of tsunami and sediment deposition demonstrated that initial tsunami amplitude and onshore slope largely control the difference between inundation distance and deposit extent (Cheng and Weiss, 2013). Another investigation, based on tsunami sediment transport modeling and field observations, suggested that artificial topographic features, such as coastal dikes, paved surfaces, and planted coastal forests, play an important role that may change the hydrodynamics of tsunami-induced flow and resulting transport and deposition processes for sand (e.g., Sugawara et al., 2014b). This indicates that the relationship between the tsunami inundation distances and sand deposition, based on observations from modern events, is not necessarily applicable to past events.

Geochemical approaches, such as the measurement of the water-leachable ions in seawater and the stable isotopic ratios of water and biomarkers, may improve our ability to estimate the area of tsunami inundation (e.g., Chagué-Goff et al., 2017). In the Sendai Plain, increased concentrations of water-leachable anions and cations, such as Cl, Na, and SO_4, within surficial sediments not covered by tsunami deposits provided a record of tsunami inundation. This implies that the tsunami inundation area can be determined using geochemical markers in the absence of sedimentological evidence (Chagué-Goff et al., 2012). Biomarkers with a marine origin have often been detected in tsunami deposits and adjacent soils. Shinozaki et al. (2015a, 2016) detected marine hydrocarbons, such as short-chain n-alkanes, pristane, and phytane, in soils below tsunami deposits in the Sendai Plain and the coastal lowlands of Odaka along the northern Joban Coast (Fig. 10.1A). In addition, mud caps in tsunami deposits from Odaka contain dinosterol that originates from dinoflagellates. The use of geochemical proxies is not only beneficial for the identification of paleotsunami events but also the estimation of their inundation area (e.g., Goff et al., 2012; Chagué-Goff et al., 2017).

The preservation potential of tsunami deposits causes another problem when estimating the inundation distance based on the deposit extent. Post-depositional natural and anthropogenic processes may reduce the distribution of tsunami deposits (Szczuciński, 2012).

170 Chapter 10

Vegetation and human activities may significantly disturb thin tsunami deposits near the limits of the tsunami inundation area. The anthropogenic processes that may occur after tsunami deposit formation include recovery and reconstruction efforts due to tsunami damage (e.g., Sugawara, 2018), as well as future cultivation and urbanization. The ability to obtain better constraints on the extent of the inundation due to paleotsunamis and its triggering mechanism requires the use of additional information. This additional information may include flow depths or speeds, which can be estimated from empirical quantitative relationships between flow parameters and deposit data or the inverse modeling of tsunami deposits.

3.5 Spatial variability of deposit thickness and its relation to flow depth

The estimation of flow parameters, such as flow depth and speed from paleotsunami deposits, is one of the most challenging issues in tsunami deposit research and is essential information to improve tsunami risk assessments (Goto et al., 2014b). For the first time, the Tohoku-oki tsunami offered an opportunity to establish an empirical relationship between tsunami deposit thickness and flow depth. A compilation of an extensive data set of tsunami deposit thicknesses recorded at 1300 locations in the Sendai Plain, as well as the interpolation of flow depths at corresponding points, revealed that flow depths and deposit thicknesses have positive correlations with increasing distance from the coastline, in addition to a statistical relationship between flow depth and deposit thickness (Fig. 10.6; Goto et al., 2014a). Assuming that the suspension load is dominant during tsunami sediment transport, the volumetric concentration of suspended sediment in the tsunami inundation flow can be estimated using the ratio of deposit thickness to the flow depth. In the Sendai Plain, the sediment concentration did not show a significant change with increasing distance from the coastline. Goto et al. (2014a) have assumed a geometric mean sediment concentration of 2% as a characteristic value for the Sendai Plain, which implies the saturation of suspended sediments in the tsunami inundation flow.

A hydraulic formulation for the wash load can provide a physical explanation for sediment saturation (van Rijn, 2007), which considers both the energy required to retain the sediment load in suspension and that dissipated by sediment transport. Wash load saturation occurs when both energies are balanced. The empirical relationship developed in Goto et al. (2014a) implies that the concept of saturation may be valid for tsunami-induced suspension transport of sand and mud. The formulation for suspended sediment saturation is compatible with the hydrodynamic model for turbidity current autosuspension (Naruse, 2011) and turbulence damping from flow density stratification due to the presence of suspended sediments (e.g., Winterwerp, 2001).

If the empirical relationship developed in Goto et al. (2014a) is valid for other cases, the flow depth of past tsunami inundation can be estimated based on locally averaged deposit thicknesses and predetermined (a priori, such as 2%) saturation concentration values.

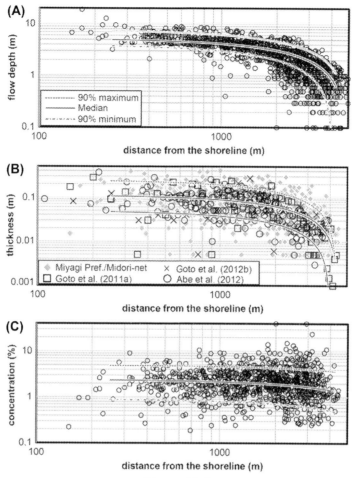

Figure 10.6

Scatter plots of (A) flow depth, (B) the thickness of the Tohoku-oki tsunami deposit, and (C) the ratio of the deposit thickness to the flow depth (concentration) (Goto et al., 2014a). Both the flow depth and deposit thickness exhibited clear maximum and minimum envelopes, as shown by the regression curves. Note that the regression curve of the median concentration values remains at ~2% and does not vary with distance from the shoreline. *Cited from Goto, K., Hashimoto, K., Sugawara, D., Yanagisawa, H., Abe, T., 2014a. Spatial thickness variability of the 2011 Tohoku-oki tsunami deposits along the coastline of Sendai Bay. Mar. Geol. 358, 38–48 (Fig. 3).*

The observed deposit thicknesses in the Sendai Plain, however, were highly variable, with estimated sediment concentrations (i.e., the ratio of the deposit thickness to flow depth) at each data point characterized by scatter from 0.1% to 10% (Fig. 10.6; Goto et al., 2014a). Although the quantitative relationship between tsunami flow depth and deposit thickness is an encouraging finding, the application of the empirical relationship is not a straightforward attempt for paleotsunami research, in which the local average thickness of the deposit is difficult to ascertain due to a limited number of data points.

3.6 Challenges to estimating earthquake size and extent from tsunami deposits

One of the important features of the 2011 Tohoku-oki earthquake was the huge slip near the trench axis, which was as large as 60 m (Ito et al., 2011), combined with the large slip (~27 m) along the deeper plate interface (Ozawa et al., 2011). The earthquake slip distribution was modeled based on an inversion analysis of the observed tsunami waveform (Fig. 10.7A; Satake et al., 2013). Tsunami hydrodynamic modeling has shown that the huge shallow slip near the trench axis generated an impulsive high wave, which caused tsunami run-up of 30–40 m along the Sanriku Coast while the large deep slip generated a tsunami with a large wavelength, which was responsible for the extensive inundation at Sendai and other coastal plains (Satake et al., 2013). Although it is not clear whether the combination of huge shallow and large deep slips is common in other subduction zones, the size and patterns of ruptures comparable with the Tohoku-oki earthquake were used for the projected scenarios of the Nankai-Trough earthquake to assess disasters from the "largest possible" earthquakes and tsunamis in Japan (e.g., Fig. 10.7B). This appears to be a practical response to the lessons learned from the Tohoku-oki earthquake and tsunami and has raised calls for the estimation of the size and extent of ruptures generated by previous earthquakes based on tsunami deposits (e.g., Goto et al., 2012a, 2014b).

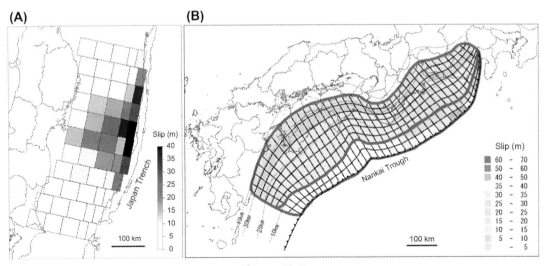

Figure 10.7

(A) The spatial distribution of fault slip that occurred in the 2011 Tohoku-oki earthquake deduced from tsunami waveform inversion (generated based on fault parameters from Satake et al., 2013). (B) The spatial distribution of fault slip from the projected scenario associated with the Nankai-Trough earthquake Case 1. *Modified after Committee evaluation of the megathrust earthquake rupture model along the Nankai Trough; http://www.bousai.go.jp/jishin/nankai/model/pdf/ kanmatsu_shiryou.pdf. Retrieved May 4, 2019.*

Paleotsunami research in Tohoku remains an important subject to test our ability to assess the size and extent of previous earthquakes because of the presence of the best-observed tsunami. The 869 CE Jogan earthquake, for example, is thought to have been accompanied by a large slip along the deeper plate interface offshore of the Sendai Plain based on a comparison between tsunami inundation numerical modeling and the spatial distribution of the sandy tsunami deposits (e.g., Satake et al., 2008a; Sugawara et al., 2011). The magnitude of the Jogan earthquake was estimated at $Mw = 8.3-8.4$ with a corresponding slip of $6-7$ m. After the Tohoku-oki tsunami, the extent and size of the rupture due to the Jogan earthquake are possibly a minimum estimate that can be revised through additional research on tsunami deposits and improved numerical modeling (Sugawara et al., 2012; Goto et al., 2014b). Considering the fact that inundation in the Sendai Plain can only be explained by deep-large slip regardless of the presence or absence of shallow-huge slip (Satake et al., 2013), estimating the presence of shallow-huge slip is an additional challenge (Goto et al., 2014b).

After the Tohoku-oki tsunami, numerous efforts have been dedicated to increasing the amount of data available on the spatial extent of the Jogan and earlier deposits to the north of the Sendai Plain (e.g., Sanriku Coast; e.g., Ishimura and Miyauchi, 2015; Takada et al., 2016; Ishimura, 2017; Inoue et al., 2017). Discoveries associated with the Jogan tsunami deposits from the northern Sanriku Coast (Inoue et al., 2017) suggest that the extent of the rupture may have been larger than the extent estimated before the 2011 Tohoku-oki tsunami. Other types of geologic records, such as from seafloor or coastal geomorphic features, can provide additional information to constrain the size and extent of ruptures generated by previous earthquakes. For example, based on the identification and dating of seismogenic turbidites in two cores recovered along the Japan Trench, Usami et al. (2018) estimated a recurrence interval of $500-800$ years for giant ($M \sim 9$) earthquakes during the last 4000 years, which includes the Jogan event. The timing of giant earthquakes was inferred from turbidite correlations between the two cores. Other geologic records of large-scale earthquakes, such as beach ridges formed by "Seismic-Driving" processes (Goff and McFadgen, 2002), can be used as evidence to assess earthquake size. Goff and Sugawara (2014) examined the links between known paleotsunami events with formation ages from beach ridges in the Sendai Plain and other coastlines in the Tohoku region, with respect to the seismic driving model. They pointed out possible synchronous beach ridge formation throughout the region, in spite of a moderately poor chronological resolution, and suggested the need for further data collection to determine the recurrence interval and magnitude of past earthquakes and tsunamis.

During the past several years, studies have been conducted to provide improved estimates of rupture size based on tsunami deposit data. For example, Namegaya and Satake (2014) proposed a method to estimate the tsunami inundation area using flow parameter constraints that can be obtained from tsunami deposits. Given that both flow depth and speed determine the amount of sediments transported by tsunami flow, there are possible flow depth and speed threshold values at a maximum inland extent of sandy tsunami deposits. In the Sendai and Ishinomaki Plains, a minimum flow depth of 1 m and flow speed of 0.6 m/s at the maximum inland extent of the sandy tsunami deposits were estimated based on the observed deposit distribution and the simulated flow depth and speed during the Tohoku-oki tsunami. The generation of the Jogan tsunami required an earthquake of Mw > 8.6 along the deep plate interface, with a corresponding slip of more than 10 m, assuming that the required flow depth at the maximum inland extent of the sandy tsunami deposit is valid for the Jogan tsunami (Namegaya and Satake, 2014).

Assumptions associated with the minimum flow depth and speed at the maximum inland extent of the sandy deposits can be considered as a type of "empirical" inverse modeling of tsunami deposits. The minimum flow depth and speed required to deposit sands may vary depending on several factors, such as grain size, bed roughness, and local topography. Another approach to estimating the flow parameters of past tsunami inundation is the numerical inverse modeling of tsunami deposits (e.g., Jaffe and Gelfenbaum, 2007; Tang and Weiss, 2015; Naruse and Abe, 2017). Data sets from the Tohoku-oki tsunami have played an important role in the development of numerical models. Jaffe et al. (2012) applied the inverse modeling of suspension grading in the sandy deposits (Jaffe and Gelfenbaum, 2007) to estimate flow speeds during Tohoku-oki tsunami inundation in the Sendai Plain. Although the choice of roughness coefficients may introduce uncertainty in the flow speeds, the estimated values of 4.2−4.3 m/s at 1149 and 1344 m from the coastline were comparable to the speed of the tsunami front measured by aerial video footage (4 m/s; Goto et al., 2011). Naruse and Abe (2017) pointed out the importance of turbulent mixing between sediments and water during tsunami inundation and nonuniform sediment transport in the flow. Their new inverse model, which incorporated effects due to turbulent mixing and nonuniform transport, was applied to the Tohoku-oki tsunami in the Sendai Plain, yielding a flow speed of 4.15 m/s and flow depth of 4.71 m. These values are comparable to the estimated values from video footage and field measurements (Naruse and Abe, 2017).

Inverse modeling of tsunami deposits uses several assumptions for flow hydrodynamics and hydraulics, such as steady uniform flow and deposition from suspension, in addition to model parameters, such as bed roughness (e.g., Jaffe and Gelfenbaum, 2007). The confident reconstruction of flow parameters highly depends on whether or not these assumptions are satisfied and if the selected parameter value is applicable. Outputs from the forward modeling of tsunami sediment transport can be used to assess the assumptions associated with inverse modeling (Jaffe et al., 2016). If the assumptions are valid, the inverse modeling output can be introduced into the forward model as boundary conditions to estimate the tsunami source parameters (hybrid modeling; Jaffe et al., 2016). However, forward modeling uses various model input data and parameters, such as topographic and bathymetric data before the tsunami, bed roughness, grain size, and initial sediment distribution, which are often poorly constrained even for modern tsunamis. The acquisition of a complete data set for model input and parameters appears to not be feasible for paleotsunamis (Sugawara et al., 2014a). The quantification of uncertainties, both in inverse and forward models, is at least necessary to assess the validity and applicability range of the models (e.g., Jaffe et al., 2016). For these reasons, the groundbreaking data set from the Tohoku-oki tsunami and its deposits will be beneficial when performing benchmark tests of the numerical models, which may reduce the uncertainties associated with tsunami sediment transport modeling.

4. Conclusions

The unexpected damage from the 2011 Tohoku-oki earthquake and tsunami reminds us of the importance of research on the geological records of past tsunamis. The discovery of the 869 CE Jogan earthquake, tsunami, and its deposits could have been used to strengthen hazard mitigation activities, such as improving tsunami hazard maps and public awareness (e.g., Goto et al., 2011, 2012a; 2014b; Sawai et al., 2012a; Sugawara et al., 2012). After the Tohoku-oki tsunami disaster, numerous studies have conducted extensive field surveys, sample analysis, and numerical modeling to improve our understanding of tsunami sedimentation and to enhance our ability to use tsunami deposits for the assessment of future tsunami hazards. The present chapter has pointed out six key findings from the literature on the Tohoku-oki tsunami deposits:

1. limitations posed when using marine materials as evidence for tsunami inundation;
2. the larger extent and lower preservation potential associated with offshore tsunami deposits;
3. possible false dating of paleotsunami events due to tsunami-induced erosion;
4. uncertainty associated with tsunami inundation distances deduced from deposit extent;
5. spatial variability in deposit thickness and its relation to flow depth; and
6. challenges associated with estimating the size and extent of earthquakes based on tsunami deposits.

These lessons are all related to the main purpose of tsunami deposit research, that is, the positive identification, dating, and quantification of a paleotsunami and its causal event. Improved understanding of the recurrence interval and size of paleotsunami events is possible by taking into account these lessons.

Since 2011, the search for paleotsunami deposits along Japanese coasts, which are faced with threats from future devastating tsunamis, has become significantly more urgent compared with the past. Findings from such tsunami deposit research are expected to improve both the hardware- and software-based countermeasures that attempt to mitigate future tsunami hazards, such as the construction of coastal dikes, the revision of

176 Chapter 10

inundation maps, the relocation of communities to inland or upland areas, and the heightening of public awareness. There are still, however, various difficulties that pose challenges to being well prepared for tsunami hazards. For public awareness, the main problem lies in the recognition of the risks posed by low-frequency, large-scale tsunamis, along with uncertainties in the recurrence interval (Goto et al., 2014b). Successful application of the results from tsunami deposit research to practical disaster management activities requires robust identification, high-confidence dating, and the quantification of the size of past tsunami events. The groundbreaking data and lessons obtained from the Tohoku-oki earthquake, tsunami, and its deposits will play a key role in future progress in tsunami deposit research.

Acknowledgments

The author thanks Prof. Tsunemasa Shiki for his encouragement and valuable comments to improve the manuscript. This contribution was supported by the grant from the Museum of Natural and Environmental History, Shizuoka.

References

Abe, T., Goto, K., Sugawara, D., 2012. Relationship between the maximum extent of tsunami sand and the inundation limit of the 2011 Tohoku-oki tsunami on Sendai Plain, Japan. Sediment. Geol. 282, 142−150.

Arai, K., Naruse, H., Miura, R., Kawamura, K., Hino, R., Ito, Y., Inazu, D., Yokokawa, M., Izumi, N., Murayama, M., Kasaya, T., 2013. Tsunami-generated turbidity current of the 2011 Tohoku-Oki earthquake. Geology 41, 1195−1198.

Bourgeois, J., Minoura, K., 1997. Paleotsunami studies − contribution to mitigation and risk assessment. In: Gusiakov, V.K. (Ed.), Tsunami Mitigation and Risk Assessment, Report of the International Workshop, Petropavlovsk-Kamchatsky, Russia, August 21−24, 1996, pp. 1−4.

Chagué-Goff, C., Andrew, A., Szczuciński, W., Goff, J., Nishimura, Y., 2012. Geochemical signatures up to the maximum inundation of the 2011 Tohoku-oki tsunami — implications for the 869 AD Jogan and other palaeotsunamis. Sediment. Geol. 282, 65−77.

Chagué-Goff, C., Szczuciński, W., Shinozaki, T., 2017. Applications of geochemistry in tsunami research: a review. Earth Sci. Rev. 165, 203−244.

Cheng, W., Weiss, R., 2013. On sediment extent and runup of tsunami waves. Earth Planet Sci. Lett. 362, 305−309.

Goff, J., Chagué-Goff, C., Nichol, S., Jaffe, B., Dominey-Howes, D., 2012. Progress in palaeotsunami research. Sediment. Geol. 243−244, 70−88.

Goff, J., McFadgen, B.G., 2002. Seismic driving of nationwide changes in geomorphology and prehistoric settlement − a 15th Century New Zealand example. Quat. Sci. Rev. 21, 2313−2320.

Goff, J., Sugawara, D., 2014. Seismic-driving of sand beach ridge formation in northern Honshu, Japan? Mar. Geol. 358, 138−149.

Goto, K., Chavanich, S.A., Imamura, F., Kunthasap, P., Matsui, T., Minoura, K., Sugawara, D., Yanagisawa, H., 2007. Distribution, origin and transport process of boulders deposited by the 2004 Indian Ocean tsunami at Pakarang Cape, Thailand. Sediment. Geol. 202, 821−837.

Goto, K., Fujino, S., 2008. Problems and perspectives of the tsunami deposits after the 2004 Indian Ocean tsunami. J. Geol. Soc. Jpn. 114, 599−617 (in Japanese with English abstract).

Goto, K., Imamura, F., Keerthi, N., Kunthasap, P., Matsui, T., Minoura, K., Ruangrassamee, A., Sugawara, D., Supharatid, S., 2008. Distribution and significance of the 2004 Indian Ocean tsunami deposits: initial results from Thailand and Sri Lanka. In: Shiki, T., Tsuji, Y., Yamazaki, T., Minoura, K. (Eds.), Tsunamiites — Features and Implications. Elsevier, Amsterdam, pp. 105—122.

Goto, K., Chagué-Goff, C., Fujino, S., Goff, J., Jaffe, B., Nishimura, Y., Richmond, B., Sugawara, D., Szczuciński, W., Tappin, D.R., Witter, R., Yulianto, E., 2011. New insights of tsunami hazard from the 2011 Tohoku-oki event. Mar. Geol. 290, 46—50.

Goto, K., Chagué-Goff, C., Goff, J., Jaffe, B., 2012a. The future of tsunami research following the 2011 Tohoku-oki event. Sediment. Geol. 282, 1—13.

Goto, K., Fujima, K., Sugawara, D., Fujino, S., Imai, K., Tsudaka, R., Abe, T., Haraguchi, T., 2012b. Field measurements and numerical modeling for the run-up heights and inundation distances of the 2011 Tohoku-oki tsunami at Sendai Plain, Japan. Earth Planets Space 64, 1247—1257.

Goto, K., Hashimoto, K., Sugawara, D., Yanagisawa, H., Abe, T., 2014a. Spatial thickness variability of the 2011 Tohoku-oki tsunami deposits along the coastline of Sendai Bay. Mar. Geol. 358, 38—48.

Goto, K., Ikehara, K., Goff, J., Chagué-Goff, C., Jaffe, B., 2014b. The 2011 Tohoku-oki tsunami — three years on. Mar. Geol. 358, 2—11.

Haraguchi, T., Iwamatsu, A., 2011a. Detailed Maps of the Impacts of the 2011 Japan Tsunami, Vol. 1, Aomori, Iwate and Miyagi Prefectures. Kokon-Shoin Publishers, Tokyo, p. 167.

Haraguchi, T., Iwamatsu, A., 2011b. Detailed Maps of the Impacts of the 2011 Japan Tsunami, Vol. 2, Fukushima, Ibaraki and Chiba Prefectures. Kokon-Shoin Publishers, Tokyo, p. 97.

Haraguchi, T., Takahashi, T., Hisamatsu, R., Morishita, Y., Sasaki, I., 2012. A field survey of geomorphic change on Kesennuma Bay caused by the 2010 Chilean tsunami and the 2011 Tohoku tsunami. J. JSCE Ser. B2 68, I_231—I_235 (in Japanese with English abstract).

Hawkes, A.D., Bird, M., Cowie, S., Grundy-Warr, C., Horton, B.P., Hwai, A.T.S., Law, L., Macgregor, C., Nott, J., Ong, J.E., Rigg, J., Robinson, R., Tan-Mullins, M., Sa, T.T., Yasin, Z., Aik, L.W., 2007. Sediments deposited by the 2004 Indian Ocean tsunami along the Malaysia—Thailand Peninsula. Mar. Geol. 242, 169—190.

Hayashi, S., Koshimura, S., 2013. The 2011 Tohoku tsunami flow velocity estimation by the aerial video analysis and numerical modeling. J. Disaster Res. 8, 561—572.

Iinuma, T., Hino, R., Kido, M., Inazu, D., Osada, Y., Ito, Y., Ohzono, M., Tsushima, H., Suzuki, S., Fujimoto, H., Miura, S., 2012. Coseismic slip distribution of the 2011 off the Pacific coast of Tohoku Earthquake (M9.0) refined by means of seafloor geodetic data. J. Geophys. Res. 117, B07409.

Ikehara, K., Irino, T., Usami, K., Jenkins, R., Omura, A., Ashi, J., 2014. Possible submarine tsunami deposits on the outer shelf of Sendai Bay, Japan resulting from the 2011 earthquake and tsunami off the Pacific coast of Tohoku. Mar. Geol. 358, 120—127.

Inoue, T., Goto, K., Nishimura, Y., Watanabe, M., Iijima, Y., Sugawara, D., 2017. Paleo-tsunami history along the northern Japan Trench: evidence from Noda Village, northern Sanriku coast, Japan. Progress in Earth and Planetary Science 4, 42.

Ioki, K., Tanioka, Y., 2016. Re-estimated fault model of the 17 th century great earthquake off Hokkaido using tsunami deposit data. Earth Planet Sci. Lett. 433, 133—138.

Ishizawa, T., Goto, K., Yokoyama, Y., Miyairi, Y., Sawada, C., Nishimura, Y., Sugawara, D., 2017. Sequential radiocarbon measurement of bulk peat for high-precision dating of tsunami deposits. Quat. Geochronol. 41, 202—210.

Ishizawa, T., Goto, K., Yokoyama, Y., Miyairi, Y., Sawada, C., Takada, K., 2018. Reducing the age range of tsunami deposits by 14C dating of rip-up clasts. Sediment. Geol. 364, 334—341.

Ishimura, D., Miyauchi, T., 2015. Historical and paleo-tsunami deposits during the last 4000 years and their correlations with historical tsunami events in Koyadori on Sanriku Coast, northeastern Japan. Prog. Earth Planet. Sci. 2, 16.

Ishimura, D., 2017. Re-examination of the age of historical and paleo-tsunami deposits at Koyadori on Sanriku Coast, Northeastern Japan. Geosci. Lett. 4, 11.

178 Chapter 10

Ito, Y., Tsuji, T., Osada, Y., Kido, M., Inazu, D., Hayashi, Y., Tsushima, H., Hino, R., Fujimoto, H., 2011. Frontal wedge deformation near the source region of the 2011 Tohoku-Oki earthquake. Geophys. Res. Lett. 38, L00G05.

Jaffe, B.E., Gelfenbaum, G., 2007. A simple model for calculating tsunami flow speed from tsunami deposits. Sediment. Geol. 200, 347–361.

Jaffe, B.E., Goto, K., Sugawara, D., Richmond, B., Fujino, S., Nishimura, Y., 2012. Flow speed estimated by inverse modeling of sandy tsunami deposits: results from the 11 March 2011 tsunami on the coastal plain near the Sendai Airport, Honshu, Japan. Sediment. Geol. 282, 90–109.

Jaffe, B., Goto, K., Sugawara, D., Gelfenbaum, G., SeanPaul, L.S., 2016. Uncertainty in tsunami sediment transport modeling. J. Disaster Res. 11, 647–661.

Jagodziński, R., Sternal, B., Szczuciński, W., Chagué-Goff, C., Sugawara, D., 2012. Heavy minerals in the 2011 Tohoku-oki tsunami deposits—insights into sediment sources and hydrodynamics. Sediment. Geol. 282, 57–64.

Kato, F., Noguchi, K., Suwa, Y., Sakagami, T., Sato, Y., 2012. Field survey on tsunami-induced topographical change. J. JSCE Ser. B3 68, I_174–I_179.

Kido, M., Osada, Y., Fujimoto, H., Hino, R., Ito, Y., 2011. Trench-normal variation in observed seafloor displacements associated with the 2011 Tohoku-Oki earthquake. Geophys. Res. Lett. 38, L24303.

MacInnes, B.T., Bourgeois, J., Pinegina, T.K., Kravchunovskaya, E.A., 2009. Tsunami geomorphology: erosion and deposition from the 15 November 2006 Kuril Island tsunami. Geology 37, 995–998.

Matsumoto, D., Naruse, H., Fujino, S., Surphawajruksakul, A., Jarupongsakul, T., Sakakura, N., Murayama, M., 2008. Truncated flame structures within a deposit of the Indian Ocean Tsunami: evidence of syn-sedimentary deformation. Sedimentology 55, 1559–1570.

Matsumoto, D., Shimamoto, T., Hirose, T., Gunatilake, J., Wickramasooriya, A., DeLile, J., Young, S., Rathnayake, C., Ranasooriya, J., Murayama, M., 2010. Thickness and grain-size distribution of the 2004 Indian Ocean tsunami deposits in Periya Kalapuwa Lagoon, eastern Sri Lanka. Sediment. Geol. 230, 96–104.

Minoura, K., Nakaya, S., 1991. Traces of tsunami preserved in inter-tidal lacustrine and marsh deposits: some examples from northeast Japan. J. Geol. 99, 265–287.

Minoura, K., Imamura, F., Sugawara, D., Kono, Y., Iwashita, T., 2001. The 869 Jogan tsunami deposit and recurrence interval of large-scale tsunami on the Pacific coast of northeast Japan. J. Nat. Disaster Sci. 23, 83–88.

Mori, N., Takahashi, T., The 2011 Tohoku earthquake tsunami Joint survey Group, 2012. Nationwide survey of the 2011 Tohoku earthquake tsunami. Coastal Eng. J. 54, 1–27.

Morton, R.A., Goff, J.R., Nichol, S.L., 2008. Hydrodynamic implications of textural trends in sand deposits of the 2004 tsunami in Sri Lanka. Sediment. Geol. 207, 56–64.

Namegaya, Y., Satake, K., 2014. Reexamination of the AD 869 Jogan earthquake size from tsunami deposit distribution, simulated flow depth, and velocity. Geophys. Res. Lett. 41, 2297–2303.

Naruse, H., 2011. A review of the autosuspension of turbidity currents: its significance and gaps in our understanding. J. Geol. Soc. Jpn. 117, 122–132 (in Japanese with English abstract).

Naruse, H., Arai, K., Matsumoto, D., Takahashi, H., Yamashita, S., Tanaka, G., Murayama, M., 2012. Sedimentary features observed in the tsunami deposits at Rikuzentakata City. Sediment. Geol. 282, 199–215.

Naruse, H., Abe, T., 2017. Inverse tsunami flow modeling including nonequilibrium sediment transport, with application to deposits from the 2011 Tohoku-Oki tsunami. J. Geophys. Res. Earth Surface 122.

Ozawa, S., Nishimura, T., Suito, H., Kobayashi, T., Tobita, M., Imakiire, T., 2011. Coseismic and postseismic slip of the 2011 magnitude-9 Tohoku-Oki earthquake. Nature 475, 373–376.

Paris, R., Lavigne, F., Wassmer, P., Sartohadi, J., 2007. Coastal sedimentation associated with the December 26, 2004 tsunami in Lhok Nga, west Banda Aceh (Sumatra, Indonesia). Mar. Geol. 238, 93–106.

Paris, R., Wassmer, P., Sartohadi, J., Lavigne, F., Barthomeuf, B., Desgages, E., Grancher, D., Baumert, P., Vautier, F., Brunstein, D., Gomez, C., 2009. Tsunamis as geomorphic crises: lessons from the December 26, 2004 tsunami in Lhok Nga, west Banda Aceh (Sumatra, Indonesia). Geomorphology 104, 59–72.

Paris, R., Fournier, J., Poizot, E., Etienne, S., Morin, J., Lavigne, F., Wassmer, P., 2010. Boulder and fine sediment transport and deposition by the 2004 tsunami in Lhok Nga (western Banda Aceh, Sumatra, Indonesia): a coupled offshore—onshore model. Mar. Geol. 268, 43—54.

Pilarczyk, J.E., Horton, B.P., Witter, R.C., Vane, C.H., Chagué-Goff, C., Goff, J., 2012. Sedimentary and foraminiferal evidence of the 2011 Tōhoku-oki tsunami on the Sendai coastal plain, Japan. Sediment. Geol. 282, 78—89.

Richmond, B., Szczuciński, W., Goto, K., Sugawara, D., Witter, R.C., Tappin, D.R., Jaffe, B., Fujino, S., Nishimura, Y., Chagué-Goff, C., Goff, J., 2012. Erosion, deposition, and landscape change on the Sendai coastal plain, Japan resulting from the March 11, 2011 Tōhoku-oki tsunami. Sediment. Geol. 282, 27—39.

Sakuna, D., Szczuciński, W., Feldens, P., Schwarzer, K., Khokiattiwong, S., 2012. Sedimentary deposits left by the 2004 Indian Ocean tsunami on the inner continental shelf offshore of Khao Lak, Andaman Sea (Thailand). Earth Planets Space 64, 931—943.

Satake, K., Namegaya, Y., Yamaki, S., 2008a. Numerical simulation of the AD 869 Jogan tsunami in Ishinomaki and Sendai plains. Ann. Rep. Active Fault Paleoearthquake Res. 8, 71—89.

Satake, K., Nanayama, F., Yamaki, S., 2008b. Fault models of unusual tsunami in the 17th century along the Kuril trench. Earth Planets Space 60, 925—935.

Satake, K., Fujii, Y., Harada, T., Namegaya, Y., 2013. Time and space distribution of coseismic slip of the 2011 Tohoku Earthquake as inferred from tsunami waveform data. Bull. Seismol. Soc. Am. 103, 1473—1492.

Sawai, Y., Fujii, Y., Fujiwara, O., Kamataki, T., Komatsubara, J., Okamura, Y., Satake, K., Shishikura, M., 2008a. Marine incursions of the past 1500 years and evidence of tsunamis at Suijin-numa, a coastal lake facing the Japan Trench. Holocene 18, 517—528.

Sawai, Y., Shishikura, M., Komatsubara, J., 2008b. A study on paleotsunami using hand corer in Sendai Plain (Sendai city, Natori city, Iwanuma city, Watari town, Yamamoto town), Miyagi, Japan. Ann. Rep. Active Fault Paleoearthquake Res 8, 17—78 (in Japanese with English abstract).

Sawai, Y., Jankaew, K., Martin, M.E., Prendergast, A., Choowong, M., Charoentitirat, T., 2009. Diatom assemblages in tsunami deposits associated with the 2004 Indian Ocean tsunami at Phra Thong Island, Thailand. Mar. Micropaleontol. 73, 70—79.

Sawai, Y., Namegaya, Y., Okamura, Y., Satake, K., Shishikura, M., 2012a. Challenges of anticipating the 2011 Tohoku earthquake and tsunami using coastal geology. Geophys. Res. Lett. 39, L21309.

Sawai, Y., Shishikura, M., Namegaya, Y., Fujii, Y., Miyashita, Y., Kagohara, K., Fujiwara, O., Tanigawa, K., 2012b. Diatom assemblages in tsunami deposits in a paddy field and on paved roads from Ibaraki and Chiba Prefectures, Japan, generated by the 2011 Tohoku tsunami. Diatom 28, 19—26.

Shi, S., Dawson, A.G., Smith, D.E., 1995. Coastal sedimentation associated with the December 12th, 1992 tsunami in Flores, Indonesia. Pure Appl. Geophys. 144, 525—536.

Shinozaki, T., Fujino, S., Ikehara, M., Sawai, Y., Tamura, T., Goto, K., Sugawara, D., Abe, T., 2015a. Marine biomarkers deposited on coastal land by the 2011 Tohoku-oki tsunami. Nat. Hazards 77, 445—460.

Shinozaki, T., Goto, K., Fujino, S., Sugawara, D., Chiba, T., 2015b. Erosion of a paleo-tsunami record by the 2011 Tohoku-oki tsunami along the southern Sendai Plain. Mar. Geol. 369, 127—136.

Shinozaki, T., Sawai, Y., Hara, J., Ikehara, M., Matsumoto, D., Tanigawa, K., 2016. Geochemical characteristics of deposits from the 2011 Tohoku-oki tsunami at Hasunuma, Kujukuri coastal plain, Japan. Isl. Arc 25, 350—368.

Shishikura, M., Fujiwara, O., Sawai, Y., Namegaya, Y., Tanigawa, K., 2012. Inland-limit of the tsunami deposit associated with the 2011 off-tohoku earthquake in the Sendai and Ishinomaki plains, northeastern Japan. Ann. Rep. Active Fault Paleoearthquake Res. 12, 45—61 (in Japanese with English Abstract).

Sugawara, D., Minoura, K., Imamura, F., 2008. Tsunamis and tsunami sedimentology. In: Shiki, T., Tsuji, Y., Yamazaki, T., Minoura, K. (Eds.), Tsunamiites — Features and Implications. Elsevier, Amsterdam, pp. 4—49.

Sugawara, D., Minoura, K., Nemoto, N., Tsukawaki, S., Goto, K., Imamura, F., 2009. Foraminiferal evidence of submarine sediment transport and deposition by backwash during the 2004 Indian Ocean tsunami. Isl. Arc 18, 513—525.

Sugawara, D., Imamura, F., Matsumoto, H., Goto, K., Minoura, K., 2011. Reconstruction of the AD 869 Jogan earthquake induced tsunami by using the geological data. J. Jpn. Soc. Nat. Disast. Sci. 29 (4), 501–516 (in Japanese, with English abstract).

Sugawara, D., Goto, K., 2012. Numerical modeling of the 2011 Tohoku-oki tsunami in the offshore and onshore of Sendai Plain, Japan. Sediment. Geol. 282, 110–123.

Sugawara, D., Goto, K., Imamura, F., Matsumoto, H., Minoura, K., 2012. Assessing the magnitude of the 869 Jogan tsunami using sedimentary deposits: prediction and consequence of the 2011 Tohoku-oki tsunami. Sediment. Geol. 282, 14–26.

Sugawara, D., Goto, K., Jaffe, B.E., 2014a. Numerical models of tsunami sediment transport – current understanding and future directions. Mar. Geol. 352, 295–320.

Sugawara, D., Takahashi, T., Imamura, F., 2014b. Sediment transport due to the 2011 Tohoku-oki tsunami at Sendai: results from numerical modeling. Mar. Geol. 358, 18–37.

Sugawara, D., 2017. Tsunami sedimentation and deposits due to the 2011 Tohoku earthquake: a review of case studies from Sendai and Hirota Bays. J. Geol. Soc. Jpn. 123, 781–804 (in Japanese with English abstract).

Sugawara, D., 2018. The geological record of tsunamis in the Anthropocene. In: Yasuda, Y., Hudson, M.J. (Eds.), Multidisciplinary Studies of the Environment and Civilization: Japanese Perspectives. Routledge, London, pp. 43–56.

Szczuciński, W., 2012. The post-depositional changes of the onshore 2004 tsunami deposits on the Andaman Sea coast of Thailand. Nat. Hazards 60, 115–133.

Szczuciński, W., Kokociński, M., Rzeszewski, M., Chagué-Goff, C., Cachão, M., Goto, K., Sugawara, D., 2012. Sediment sources and sedimentation processes of 2011 Tohoku-oki tsunami deposits on Sendai Plain, Japan — insights from diatoms, nannoliths and grain size distribution. Sediment. Geol. 282, 40–56.

Takada, K., Shishikura, M., Imai, K., Ebina, Y., Goto, K., Koshiya, S., Yamamoto, H., Igarashi, A., Ichihara, T., Kinoshita, H., Ikeda, T., River division Department of prefectural land development, Iwate prefecture Government, 2016. Distribution and ages of tsunami deposits along the Pacific coast of the Iwate prefecture. Ann. Rep. Active Fault Paleoearthquake Res. 16, 1–52 (in Japanese with English abstract).

Takahashi, S., et al., 2011. Urgent Survey for 2011 Great East Japan Earthquake and Tsunami Disaster in Ports and Coast. Technical note of the Port and Airport Research Institute 1231, p. 200.

Takashimizu, Y., Urabe, A., Suzuki, K., Sato, Y., 2012. Deposition by the 2011 Tohoku-oki tsunami on coastal lowland controlled by beach ridges near Sendai, Japan. Sediment. Geol. 282, 124–141.

Tamura, T., Sawai, Y., Ikehara, K., Nakashima, R., Hara, J., Kanai, Y., 2015. Shallow-marine deposits associated with the 2011 tohoku-oki tsunami in Sendai bay, Japan. J. Quat. Sci. 30, 293–297.

Tanaka, G., Naruse, H., Yamashita, S., Arai, K., 2012. Ostracodes reveal the sea-bed origin of tsunami deposits. Geophys. Res. Lett. 39, L05406.

Tang, H., Weiss, R., 2015. A model for tsunami flow inversion from deposits. Mar. Geol. 370, 55–62.

Tappin, D.R., Evans, H.M., Jordan, C.J., Richmond, B., Sugawara, D., Goto, K., 2012. Coastal changes in the Sendai area from the impact of the 2011 Tōhoku-oki tsunami: interpretations of time series satellite images and helicopter-borne video footage. Sediment. Geol. 282, 151–174.

Tappin, D.R., Grilli, S.T., Harris, J.C., Geller, R.J., Masterlark, T., Kirby, J.T., Shi, F., Ma, G., Thingbaijam, K.K.S., Mai, P.M., 2014. Did a submarine landslide contribute to the 2011 Tohoku tsunami? Mar. Geol. 357, 344–361.

Udo, K., Sugawara, D., Tanaka, H., Imai, K., Mano, A., 2012. Impact of the 2011 tohoku earthquake and tsunami on beach morphology along the northern Sendai coast. Coast Eng. J. 54 (15), 1250009.

Udo, K., Tanaka, H., Mano, A., Takeda, Y., 2013. Beach morphology change of southern Sendai coast due to 2011 tohoku earthquake tsunami. J. JSCE, Ser. B2 69, I_1391–I_1395 (in Japanese with English abstract).

Udo, K., Takeda, Y., Tanaka, H., Mano, A., 2015. Effect of submerged breakwaters on coastal morphology change due to tsunami. J. JSCE, Ser. B3 71, I_653–I_658 (in Japanese with English abstract).

Umitsu, M., Tanavud, C., Patanakanog, B., 2007. Effects of landforms on tsunami flow in the plains of Banda Aceh, Indonesia, and Nam Khem, Thailand. Mar. Geol. 242, 141–153.

Usami, K., Ikehara, K., Jenkins, R.G., Ashi, J., 2016. Benthic foraminiferal evidence of deep-sea sediment transport by the 2011 Tohoku-oki earthquake and tsunami. Mar. Geol. 384, 214–224.

Usami, K., Ikehara, K., Kanamatsu, T., McHugh, C.M., 2018. Supercycle in great earthquake recurrence along the Japan Trench over the last 4000 years. Geosci. Lett. 5, 11.

van Rijn, L.C., 2007. Unified view of sediment transport by currents and waves. II: suspended transport. J. Hydraul. Eng. 133, 668–689.

Winterwerp, J.C., 2001. Stratification effects by cohesive and noncohesive sediment. J. Geophys. Res. 106 (C10), 22559–22574.

Yamashita, K., Sugawara, D., Takahashi, T., Imamura, F., Saito, Y., Imato, Y., Kai, T., Uehara, H., Kato, T., Nakata, K., Saka, R., Nishikawa, A., 2016. Numerical simulations of large-scale sediment transport caused by the 2011 tohoku earthquake tsunami in Hirota bay, southern Sanriku coast. Coast Eng. J. 58, 28, 1640015.

Yokoyama, Y., Sakamoto, I., Yagi, M., Inoue, T., Nemoto, K., Fujimaki, M., Kasaya, T., Fujiwara, Y., 2015. Distribution and internal characteristics of deposits by the Tohoku-oki tsunami in Hirota Bay, Iwate prefecture. 122th Annu. Meet. Geol. Soc. Jpn. Abstr. R10-O-11.

Yoshikawa, S., Kanamatsu, T., Goto, K., Sakamoto, I., Yagi, M., Fujimaki, M., Imura, R., Nemoto, K., Sakaguchi, H., 2015. Evidence for erosion and deposition by the 2011 Tohoku-oki tsunami on the nearshore shelf of Sendai Bay, Japan. Geo Mar. Lett. 35, 315–328.

> CHAPTER 11

An overview on offshore tsunami deposits

P.J.M. Costa[1,2], L. Feist[3], A.G. Dawson[4], I. Stewart[5], K. Reicherter[3], C. Andrade[2]

[1]Department of Earth Sciences, Faculty of Science and Technology, University of Coimbra, Coimbra, Portugal; [2]Instituto D. Luiz, Faculdade de Ciências, Universidade de Lisboa, Lisboa, Portugal; [3]Neotectonics and Natural Hazards Group, RWTH Aachen University, Aachen, Germany; [4]Geography and Environmental Science, School of Social Sciences, Dundee, United Kingdom; [5]School of Geography, Earth and Environmental Sciences, Faculty of Science and Engineering, University of Plymouth, United Kingdom

1. Introduction

Sedimentological analysis of (palaeo)tsunami deposits has become a key factor for evaluating tsunami risk to coastal communities and eventually helping to mitigate their impacts. Even though the field of tsunami geoscience showed substantial progress during the last three decades, many tsunami characteristics and their related depositional mechanisms are still poorly understood, especially regarding tsunami backwash.

Many recent tsunami deposits have been well documented; however, the geological record of ancient and palaeotsunamis is still not fully understood. Erosion and postdepositional modification of these deposits play a major role in palaeotsunami research (e.g., Einsele et al., 1996; Dawson and Stewart, 2007). Imprints of tsunamis are often best preserved in affected floodplain, coastal, shallow sea, and submarine canyon environments due to their high sedimentation rates favoring rapid coverage of the tsunami deposit by other sediments. Among other factors, tsunami sediment deposition and preservation depend on sediment source from beach, nearshore, and/or offshore zones.

If an adequate sediment supply in the coastal and nearshore zone is provided, the rapid water velocities result in erosion and landward transport of sediment. This will be followed by a seaward-directed sediment-rich backwash flow moving toward the offshore and beyond. This flow can be compared to turbidity currents due to energy input from the tsunami-generating mechanism. Acting as a palimpsest, beyond storm-wave base, the offshore is an excellent location to preserve sedimentological signatures of tsunami-derived processes and of successive events. Therefore, it offers an exclusive opportunity to analyze tsunami hydro- and morphodynamic processes and its full implications allowing

Tsunamiites. https://doi.org/10.1016/B978-0-12-823939-1.00011-2
Copyright © 2021 Elsevier B.V. All rights reserved.

for higher resolution studies and widening the time window of observation with implications for the establishment of return periods.

The interpretation of tsunami deposits can be difficult due to erosion and redeposition of the transported sediment caused by following waves or backwash. Especially tsunami backwash is argued to be a relevant factor for the final tsunami deposit configuration. Furthermore, backwash flow hydraulics are believed to behave differently from inflow due to channelized concentration of the backwash turbid flows transporting sediment seaward and possibly into deeper water (Dawson and Stewart, 2007). The backwash flow velocities are influenced by local topography and bathymetry that play a crucial role in the final arrangement of the tsunami deposit. Preservation of this tsunami phase has been described in the offshore and their features are summarized in the next sections of this chapter.

2. Offshore tsunami deposits — current knowledge

On continental shelves closer to the coast, sedimentary imprints of tsunami can be identified, while it seems more unlikely that tsunami waves can have a deep-water sedimentary imprint in the far offshore areas (Dawson and Stewart, 2008). The backwash flow is channelized onshore as a high-density flow that erodes, transports, and deposits sediment seaward along the inundated onshore areas and ultimately offshore along the shelf, slope, and rise. Moreover, some authors (e.g., Le Roux and Vargas, 2005) associate tsunami backwash flows with turbidites at the shelf break and along submarine canyons. It is also important to stress the stronger preservation potential in the shallow offshore, namely beyond storm-wave base, thus making this a reliable archive of palaeotsunami evidence (Fig. 11.1).

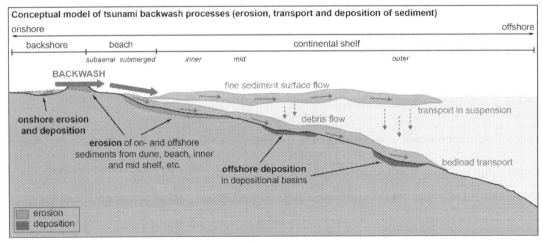

Figure 11.1
Conceptual model of tsunami backwash processes and deposition. *Adapted from Feist et al. (submitted Scientific Reports). Evidence of a 3700-year Old Offshore Deposit on the SW Iberian Continental Shelf — a Possible Lisbon AD 1755 Tsunami Predecessor? Submitted to GSA on 21st January 2020.*

Only few studies on the offshore effects of tsunami waves existed before the 2011 Tohoku-oki tsunami in Japan. Many of these older studies describe similar characteristics of onshore palaeotsunami deposits, such as extensive coarse-grained, thick-bedded event layers with specific sedimentological characteristics such as erosive basal contacts or exotic fragments from beach and nearshore zones incorporated in the deposit (e.g., Fujino et al., 2006). In the deeper offshore area, the term "deep-sea homogenite" has been used to define a massive, poorly sorted, grain-supported unit that contains large reworked shallow-marine fossils and occasional large intraclasts that have been described in association with the Bronze Age Santorini tsunami event (Cita et al., 1996). Other tsunamigenic deposits were discussed in an offshore sedimentary context and related with events such as the K/T meteoric tsunami (e.g., Smit et al., 1992), the 1755 CE tsunami (e.g., Abrantes et al., 2008), the 2003 Tokashioki earthquake (Noda et al., 2007) or to try and match earthquake-triggered turbidites with tsunamigenic events from the Saguenay (Eastern Canada) and Reloncavi (Chilean margin) (St-Onge et al., 2012). In fact, another peculiar note in terms of offshore tsunami deposits is that some have been specifically attributed to processes of tsunami backwash and the generation of gravity-driven flows of turbid water from nearshore to deep water.

After the 2011 Tohoku-oki tsunami, numerous investigations significantly improved the general understanding of tsunami deposits by analyzing sediment transport and deposition processes both onshore and offshore (e.g., Tamura et al., 2015; Yoshikawa et al., 2015). Works comparing onshore and offshore tsunami deposits have also been conducted including Smedile et al. (2012), Tipmanee et al. (2012) and Quintela et al. (2016). By applying innovative methods, Riou et al. (2019), Pilarczyk et al. (2019), and Smedile et al. (2019) were able to extend the set of features (including sedimentological, geochemical, geophysical, and paleontological) to identify offshore tsunami deposits. These and other features are described in detail in the following section.

3. Offshore tsunami deposits — their features

There are still no established criteria for the identification of offshore deposits in shallower water depths of the nearshore and inner shelf domains. Van den Bergh et al. (2003) argue that land-derived components tend to be incorporated exclusively in deposits close to the shore, while others disagree and state that no changes in seafloor morphology or sediment redistribution were left by different tsunamis in different settings of the nearshore and inner shelf (e.g., 2004 Indian Ocean tsunami: Feldens et al., 2012, Tohoku-Oki tsunami: Tamura et al., 2015). This leads to the conclusion that a strong site-specific aspect for the formation of offshore tsunami deposits exists and only a proper contextualization allows solid identification and interpretation of tsunami deposits.

186   Chapter 11

Similarly to onshore tsunami deposits, offshore deposits described so far do not allow a simplistic generalization of their characteristics; thus, their depositional processes remain to be poorly explained. Local factors such as sediment supply and bathymetry influence the character of a tsunami deposit (e.g., Le Roux and Vargas, 2005). Nevertheless, large amounts of sediment debris are eroded from beach and nearshore areas together with input of terrigenous material and are transported seaward as a density flow. These sediments are ultimately deposited in onshore and mainly in offshore areas by backwash rip currents. Such dense tsunami-driven sediment flows are moving from the coast toward offshore areas while being capable of transporting a wide range of grain sizes. Furthermore, they may allow the development of local shear (traction) carpets at their base under subcritical flow conditions while (onshore) tsunami inflow mainly presents turbulent supercritical flow conditions.

Several peculiarities of offshore tsunami deposits are presented further. This list is by no means complete but provides an accurate overview on aspects recently described in the identification of offshore tsunami deposits. Nevertheless, as stressed before, site-specific aspects control the final architecture and composition of offshore tsunami deposits.

3.1 Internal architecture

Tsunami backwash can move as a debris flow along the seafloor of the shelf, which was identified on the south-western Portuguese shelf as inverse grading at the base of the tsunami deposit followed by a massive medium sand subunit (Feist et al., submitted to Scientific Reports; Reicherter et al., 2019). Inversely graded sequences as a result of bedload transport have been previously observed in onshore tsunami deposits of the 2011 Tohoku-oki tsunami in Sendai plain (Jaffe et al., 2012). These features seem to translate the high-sediment concentration in the forefront of the seaward-directed flow and the subsequent settling of sediments transported in suspension.

In the offshore record, it is important to attempt to identify inundation and backwash flow whenever possible. This can only be achieved with very high-resolution studies. Especially, X-radiographs can reveal changes in sedimentary processes along the stratigraphic column (Noda et al., 2007; Ikehara et al., 2014). This proxy might be a good indicator to gain more information about initial erosion and resuspension of seafloor sediments produced by the friction of the propagating tsunami wave even in greater water depths. However, it is highly likely that inundation phase might not be well preserved, in particular in the deeper areas of the offshore and beyond. One can speculate that even in those areas where tsunami waves do not move sediment significantly, the difference in pressure when the tsunami waves pass is capable of disturbing the sediment—water interface, and this disturbed layer can be associated with the inundation phase, if later sealed by backwash sediments. These are somewhat speculative ideas but with further work being conducted in the near future, there are reasons to believe the offshore will prove to be an area with pristine tsunami sediment imprints that allow full reconstruction of tsunami events.

3.2 Textural and compositional aspects

Grain-size distribution in offshore tsunami layers depends greatly on the grain size of the source material and on wave energy. In general, offshore tsunami deposits are marked with an increase in grain size compared to background deposition because of the increased current velocity. Most of the tsunami layers are poorly sorted but can also be well to poorly sorted or just well-sorted. Therefore, grain-size distribution data can only be used as a supportive indicator of potential tsunami origin.

The composition of offshore tsunami deposits can include marine sediments (minerogenic and biogenic) or terrigenous to authigenic components. These components can be either autochthonous or allochthonous. In the case of possible tsunami sediments, terrigenous particles or other lithic fragments can be used as proof of a backwash process and can provide robust evidence for a tsunami-related origin for these sediments (Ikehara et al., 2014). Also, it is anticipated that differences in heavy minerals could be identified due to different modes of sediment transport and deposition. Some biogenic components (e.g., terrestrial plant fragments) can be considered as signatures of continental components and indicator for backwash processes related to tsunami events. Shell fragments can indicate allochthonous fragments from the inner shelf reworked and transported by tsunamis that created a disturbed sediment bed (e.g., Toyofuku et al., 2014). Moreover, it is also highly likely that on its seaward route, tsunami waves only displace sediments from shallower areas to slightly deeper areas. Therefore, it is possible that a tsunami deposit can present exclusively a "marine sediment source" that is simply displaced horizontally for shorter distances but still presenting a peculiar (coarser) character when compared with the under and overlying layers.

3.3 Geochemical inferences

X-ray fluorescence (XRF) scanner is used to semiquantitatively assess sediment composition by detecting minor and major element counts and thus allowing to perceive changes in the source of the deposited sediments and differentiate terrigenous material from marine sediments. In the study of offshore tsunami sediments, other XRF scanners often show well-matched results to other proxies indicating an input of terrigenous material. For example, Abrantes et al. (2008) concluded that Fe counts mimic magnetic susceptibility and Ca counts vary with the mean grain size along the core profile. In other offshore tsunami studies with samples originated from shallower areas and transported further offshore, this method also revealed consistency and promising results in the identification of allochthonous material (Sakuna-Schwartz et al., 2015; Tyuleneva et al., 2018; Riou et al., 2019). Nevertheless, the relevance of local specific context in the use of rations must be stressed. Depending on sediment availability and composition, the key ratios to identify tsunami deposits will vary if the sediment available is quartz (Si) or bioclasts (Ca); thus, identification of tsunami deposits depends on the variation and comparison with under and overlying layers and, ideally, the presence of terrigenous material. Also, a wider composition, reflecting a mixture of different environments and sediments, can be reflected in the inorganic geochemical analysis.

Pongpiachan (2014) and Bellanova et al. (2019a, b) successfully applied organic-geochemical proxies, such as n-alkanes, polycyclic aromatic hydrocarbons, steroids, fatty acids, and n-aldehydes, in onshore tsunami studies. Offshore Portugal biomarkers showed an increased input of terrestrial material in the previously identified CE 1755 tsunami backwash deposits, distinguishing them from the framing silt-dominated background sediments (Bellanova et al., 2019a; Reicherter et al., 2019).

3.4 Palaeontological features

On its route from onshore to offshore, tsunami backwash can redistribute and deposit shells and plant debris. However, if within onshore deposits the occurrence of marine species is synonym of inundation, in the offshore the redistribution of marine species is harder to assess (Riou, 2019). In several offshore tsunami deposits, allochthonous shells have been reported (Toyofuku et al., 2014; Goodman-Tchernov and Austin, 2015; Puga-Bernabéu and Aguirre, 2017; Tyuleneva et al., 2018). Only Toyofuku et al. (2014) suggest that allochthonous bivalves found in backwash deposits from the shelf after the 2011 Tohoku-Oki tsunami were dragged from the nearshore domain based on their habitat, which attests transport of the shells by backwash currents. Another argument for backwash deposits is based on the aspect and organization of these shells and sometimes corals within the deposit. They tend to show moderate fragmentation, with a mix of broken and intact shells.

There are several studies using microfossils and most of the works on offshore tsunami deposits concentrated on the study of foraminifera. Indicators for tsunami layers are the increase in coastal foraminifera abundance coupled with coarser layers (Quintela et al., 2016) or a high concentration of displaced epiphytic foraminifera coupled with grain-size changes (Smedile et al., 2011) caused by tsunami backwash. An increased ratio between agglutinated and hyaline foraminifera to total benthic foraminifera was attributed to sediment movement by tsunami wave action (Noda et al., 2007). Milker et al. (2013) developed a transfer function for water-depth reconstructions based on benthic foraminifera to reconstruct redeposition and dynamics of sediment distribution associated with the 2004 Indian Ocean tsunami. Thus, they were able to limit the maximum water depth of resuspension to 20 m offshore Khao Lak, Thailand. All the aforementioned points to the likely presence of allochthonous palaeontological content in offshore tsunami deposits. Similarly to textural, compositional, and geochemical, palaeontological characteristics must be compared with under- and overlying units to clearly discern tsunami-derived deposition from the background.

3.5 Differentiation from other high-energy events

Similarly to onshore deposits, tentative differentiation between offshore storm and tsunami deposits relies on spatial distribution of the deposits. Offshore storm deposits tend to be restrained to the shoreface and upper offshore, above the storm wave base, while offshore tsunami deposits can often be found well into the shelf domain (Weiss and Bahlburg, 2006; Goodman-Tchernov et al., 2009; Smedile et al., 2011, 2019). Despite the different physical characteristics of their generation mechanisms, offshore tsunami and storm deposits share some similarities (e.g., coarser texture). Nevertheless, some aspects can be used to differentiate them. For instance, offshore storm deposits tend to display better grain-size sorting than offshore tsunami deposits.

Comparison between tsunami and storm deposits has been attempted in the offshore domain. For example, Puga-Bernabéu and Aguirre (2017) concluded that, in the case of tsunamis, the resulting shell deposits represent all parts of the ramp while they are much more localized in the case of a storm. In addition, shells are oriented chaotically, with few very sharp fragments in tsunami deposits while they are organized horizontally, with highly abundant smooth fragments in storm deposits. Moreover, it is possible to observe a geochemical terrestrial signature in offshore tsunami deposits with, for example, higher values of Ti/Ca ratio (Sakuna et al., 2012; Veerasingam et al., 2014; Tyuleneva et al., 2018) and increases in anthropogenic material (Goodman-Tchernov et al., 2009; Goodman-Tchernov and Austin, 2015). Offshore fluvial flood deposits exhibit a similar terrestrial signature but translate a lower energetic generation mechanism (i.e., less erosional capacity and finer sediments transported). However, when compared to tsunami deposits, fluvial flood deposits tend to be composed of finer sediments (i.e., mud) and exhibit better sorting and no mud clasts (Sakuna et al., 2012; Sakuna-Schwartz et al., 2015).

Crucial in the differentiation of offshore storm and tsunami deposits is the clear establishment of the storm-wave base depth for each specific sea-level period. This facilitates the likelihood of detecting a storm event. Furthermore, anticipating two genetically different events, one can assume that in the same location, with a stable sea level, tsunami translates more energy into the environment. Therefore, it is important to identify energy-related features such as broader minerogenic assemblages, coarser grain size, more varied palaeontological content, and peculiarities in the internal architecture of the deposits.

4. Concluding remarks

Despite recent developments, studies on offshore tsunami deposits need higher resolution and site-specific multidisciplinary approaches because differences between background sediments and tsunami deposits can be small in variation and bed thickness. Nevertheless, tsunami deposits tend to exhibit peculiar traces as a result of erosion, sediment transport, and redeposition and are therefore discernible from the autochthonous sediment deposition. Hence, similar to onshore tsunami deposits, it is important to analyze

190   Chapter 11

background sedimentary conditions to differentiate them from tsunami sedimentation of allochthonous material or rather reworked material.

Little is still known about offshore tsunami deposits, although they have the potential to improve palaeotsunami reconstructions, especially in areas with limited preservation of onshore materials. As yet there are no established and widely accepted criteria for the identification of offshore tsunami deposits and a strong site-specific effect must be taken into account. The huge preservation potential (below storm-wave base) makes the investigation of these sediment bodies a major breakthrough in tsunami geoscience. So far, we were (mostly) limited to onshore imprints, and therefore, we were only able to tell the onshore half of the story of a tsunami event. If both onshore and offshore information can be coupled and merged together, then significantly more robust and accurate reconstructions of past events are possible, bringing large benefits to the accuracy of coastal hazard assessment and even the definition of return periods.

Acknowledgments

The authors acknowledge Portuguese Science Foundation project OnOff (PTDC/CTA-GEO/28941/2017). Vincent Kümmerer and Brieuc Riou are acknowledged for stimulating discussions on this topic. Professor Shiki is acknowledged for his constructive comments on an earlier version of the manuscript.

References

Abrantes, F., Alt-Epping, U., Lebreiro, S., Voelker, A., Schneider, R., 2008. Sedimentological record of tsunamis on shallow-shelf areas: the case of the 1969 AD and 1755 AD tsunamis on the Portuguese Shelf off Lisbon. Mar. Geol. 249, 283–293.

Bellanova, P., Frenken, M., Schwarzbauer, J., Deutschmann, B., Hollert, H., Costa, P.J.M., Santisteban, J.I., Vött, A., Brückner, H., Kuhlmann, J., Duarte, J.F., Schüttrumpf, H., Reicherter, K.R., Feist, L., M-152 scientific team, 2019a. Tracing the unknown — offshore tsunami deposits detected by organic geochemistry. In: AGU 2019 — Abstract. 2019 AGU Fall Meeting, San Francisco, US.

Bellanova, P., Frenken, M., Richmond, B., Schwarzbauer, J., La Selle, S., Griswold, F., Jaffe, B., Nelson, A., Reicherter, K., 2019b. Organic geochemical investigation of far-field tsunami deposits of the Kahana Valley, O'ahu, Hawai'i. Sedimentology. https://doi.org/10.1111/sed.12583.

Cita, M.B., Camerlenghi, A., Rimoldi, B., 1996. Deep-sea tsunami deposits in the eastern Mediterranean: new evidence and depositional models. Sediment. Geol. 104 (1–4), 155–173.

Dawson, A.G., Stewart, I., 2007. Tsunami deposits in the geological record. Sediment. Geol. 200, 166–183.

Dawson, A.G., Stewart, I., 2008. Offshore tractive current deposition: the Forgotten tsunami sedimentation process. In: Shiki, T., Tsuji, Y., Yamazaki, T., Minoura, K. (Eds.), Tsunamiites - Features and Implications. Elsevier, Amsterdam, pp. 153–161.

Einsele, G., Chough, S.K., Shiki, T., 1996. Depositional events and their records—an introduction. Sediment. Geol. 104, 1–9.

Feist et al. (submitted to Scientific Reports). Evidence of a 3700-year Old Offshore Deposit on the SW Iberian Continental Shelf — a Possible Lisbon AD 1755 Tsunami Predecessor? Submitted to GSA on 21st January 2020.

Feldens, P., Schwarzer, K., Sakuna, D., Szczuciński, W., Sompongchaiyakul, P., 2012. Sediment distribution on the inner continental shelf off Khao Lak (Thailand) after the 2004 Indian Ocean tsunami. Earth Planets Space 64, 875–887.

Fujino, S., Masuda, F., Tagomori, S., Matsumoto, D., 2006. Structure and depositional processes of a gravelly tsunami deposit in a shallow marine setting: lower Cretaceous Miyako Group, Japan. Sediment. Geol. 187, 127−138.

Goodman-Tchernov, B.N., Austin, J.A., 2015. Deterioration of Israel's Caesarea Maritima's ancient harbor linked to repeated tsunami events identified in geophysical mapping of offshore stratigraphy. J. Archaeol. Sci. Rep. 3, 444−454.

Goodman-Tchernov, B.N., Dey, H.W., Reinhardt, E.G., McCoy, F., Mart, Y., 2009. Tsunami waves generated by the Santorini eruption reached Eastern Mediterranean shores. Geology 37, 943−946.

Ikehara, K., Irino, T., Usami, K., Jenkins, R., Omura, A., Ashi, J., 2014. Possible submarine tsunami deposits on the outer shelf of Sendai Bay, Japan resulting from the 2011 earthquake and tsunami off the Pacific coast of Tohoku. Mar. Geol. 349, 91−98.

Jaffe, B.E., Goto, K., Sugawara, D., Richmond, B., Fujino, S., Nishimura, Y., 2012. Flow speed estimated by inverse modelling of sandy tsunami deposits: results from the 11 March 2011 tsunami on the coastal plain near the Sendai Airport, Honshu, Japan. Sediment. Geol. 282, 90−109.

Le Roux, J.P., Vargas, G., 2005. Hydraulic behavior of tsunami backflows: insights from their modern and ancient deposits. Environ. Geol. 49, 65.

Milker, Y., Wilken, M., Schumann, J., Sakuna, D., Feldens, P., Schwarzer, K., Schmiedl, G., 2013. Sediment transport on the inner shelf off Khao Lak (Andaman Sea, Thailand) during the 2004 Indian Ocean tsunami and former storm events: evidence from foraminiferal transfer functions. Nat. Hazards Earth Syst. Sci. 13, 3113−3128.

Noda, A., Katayama, H., Sagayama, T., Suga, K., Uchida, Y., Satake, K., Abe, K., Okamura, Y., 2007. Evaluation of tsunami impacts on shallow marine sediments: an example from the tsunami caused by the 2003 Tokachi-oki earthquake, northern Japan. Sediment. Geol. 200, 314−327.

Pilarczyk, J.E., Sawai, Y., Matsumoto, D., Namegaya, Y., Nishida, N., Ikehara, K., Fujiwara, O., Gouramanis, C., Dura, T., Horton, B.P., 2019. Constraining sediment provenance for tsunami deposits using distributions of grain size and foraminifera from the Kujukuri coastline and shelf, Japan. Sedimentology. https://doi.org/10.1111/sed.12591.

Pongpiachan, S., 2014. Application of binary diagnostic ratios of polycyclic aromatic hydrocarbons for identification of tsunami 2004 backwash sediments in khao lak, Thailand. Sci. World J. https://doi.org/ 10.1155/2014/485068.

Puga-Bernabéu, Á., Aguirre, J., 2017. Contrasting storm-versus tsunami-related shell beds in shallow-water ramps. Palaeogeogr. Palaeoclimatol. Palaeoecol. 471, 1−14.

Quintela, M., Costa, P.J.M., Fatela, F., Drago, T., Hoska, N., Andrade, C., Freitas, M.C., 2016. The AD 1755 tsunami deposits onshore and offshore of Algarve (south Portugal): sediment transport interpretations based on the study of Foraminifera assemblages. Quat. Int. 408, 123−138.

Reicherter, K., Vött, A., Feist, L., Costa, P.J.M., Schwarzbauer, J., Schüttrumpf, H., Holger, J., Raeke, A., Huhn-Frehers, K., 2019. Lisbon 1755, Cruise No. M152/1, 02.11.-14.11.2018, Funchal (Portugal)-Hamburg (Germany). METEOR-Berichte M152 36. https://doi.org/10.2312/cr_m152.

Riou, B., 2019. Shallow Marine Sediment Record of Tsunamis: Analysis of the Sediment-Fill of Bays of Tutuila (American Samoa) and Backwash Deposits of the 2009 South Pacific Tsunami (Ph.D. thesis). Université de La Rochelle, France, p. 189.

Riou, B., Chaumillon, E., Schneider, J.-L., Corrège, T., Chagué, C., 2019. The sediment-fill of Pago Pago Bay (Tutuila Island, American Samoa): new insights on the sediment record of past tsunamis. Sedimentology. https://doi.org/10.1111/sed.12574.

Sakuna-Schwartz, D., Feldens, P., Schwarzer, K., Khokiattiwong, S., Stattegger, K., 2015. Internal structure of event layers preserved on the Andaman Sea continental shelf, Thailand: tsunami vs. storm and flash-flood deposits. Nat. Hazards Earth Syst. Sci. 15, 1181−1199.

Sakuna, D., Szczuciński, W., Feldens, P., Schwarzer, K., Khokiattiwong, S., 2012. Sedimentary deposits left by the 2004 Indian Ocean tsunami on the inner continental shelf offshore of Khao Lak, Andaman Sea (Thailand). Earth Planets Space 64, 931−943.

Smedile, A., De Martini, P.M., Pantosti, D., 2012. Combining inland and offshore paleotsunamis evidence: the Augusta Bay (eastern Sicily, Italy) case study. Nat. Hazards Earth Syst. Sci. 12, 2557—2567.

Smedile, A., De Martini, P.M., Pantosti, D., Bellucci, L., Del Carlo, P., Gasperini, L., Pirrotta, C., Polonia, A., Boschi, E., 2011. Possible tsunami signatures from an integrated study in the Augusta Bay offshore (Eastern Sicily-Italy). Mar. Geol. 281, 1—13.

Smedile, A., Molisso, F., Chagué, C., Iorio, M., De Martini, P.M., Pinzi, S., Collins, P.E.F., Sagnotti, L., Pantosti, D., 2019. New coring study in Augusta Bay expands understanding of offshore tsunami deposits (Eastern Sicily, Italy). Sedimentology. https://doi.org/10.1111/sed.12581.

Smit, J., Montanari, A., Swinburne, N.H.M., Alvarez, W., Hilderbrand, R., Margolis, S.V., Claeys, P., Lowrie, W., Asaro, F., 1992. Tektite-bearing deep-water clastic unit at the Cretaceous-Tertiary boundary in north-eastern Mexico. Geology 20, 99—103.

St-Onge, G., Chapron, E., Salas, M., Viel, M., Mulsow, S., Debret, M., Foucher, A., Mulder, T., Desmet, M., Costa, P., Ghaleb, B., Jaouen, A., Locat, J., 2012. Comparison of earthquake-triggered turbidites from the Saguenay (Eastern Canada) and Reloncavi (Chilean margin) Fjords: implications for paleoseismicity and sedimentology. Sedimentary Geology 243—244, 89—107.

Tamura, T., Sawai, Y., Ikehara, K., Nakashima, R., Hara, J., Kanai, Y., 2015. Shallow-marine deposits associated with the 2011 Tohoku-oki tsunami in Sendai Bay, Japan. J. Quat. Sci. 30, 293—297.

Tipmanee, D., Deelaman, W., Pongpiachan, S., Schwarzer, K., Sompongchaiyakul, P., 2012. Using polycyclic aromatic hydrocarbons (PAHs) as a chemical proxy to indicate tsunami 2004 backwash in khao lak coastal area, Thailand. Nat. Hazards Earth Syst. Sci. 12, 1441—1451.

Toyofuku, T., Duros, P., Fontanier, C., Mamo, B., Bichon, S., Buscail, R., Chabaud, G., Deflandre, B., Goubet, S., Grémare, A., Menniti, C., Fujii, M., Kawamura, K., Koho, K.A., Noda, A., Namegaya, Y., Oguri, K., Radakovitch, O., Murayama, M., de Nooijer, L.J., Kurasawa, A., Ohkawara, N., Okutani, T., Sakaguchi, A., Jorissen, F., Reichart, G.-J., Kitazato, H., 2014. Unexpected biotic resilience on the Japanese seafloor caused by the 2011 Tōhoku-Oki tsunami. Sci. Rep. 4, 7517.

Tyuleneva, N., Braun, Y., Katz, T., Suchkov, I., Goodman-Tchernov, B., 2018. A new chalcolithic-era tsunami event identified in the offshore sedimentary record of Jisr al-Zarka (Israel). Mar. Geol. 396, 67—78.

Van Den Bergh, G.D., Boer, W., De Haas, H., Van Weering, T.C., Van Wijhe, R., 2003. Shallow marine tsunami deposits in Teluk Banten (NW Java, Indonesia), generated by the 1883 Krakatau eruption. Mar. Geol. 197, 13—34.

Veerasingam, S., Venkatachalapathy, R., Basavaiah, N., Ramkumar, T., Venkatramanan, S., Deenadayalan, K., 2014. Identification and characterization of tsunami deposits off-southeast coast of India from the 2004 Indian Ocean tsunami: rock magnetic and geochemical approach. J. Earth Syst. Sci. 123, 905—921.

Weiss, R., Bahlburg, H., 2006. A note on the preservation of offshore tsunami deposits. J. Sediment. Res. 76, 1267—1273.

Yoshikawa, S., Kanamatsu, T., Goto, K., Sakamoto, I., Yagi, M., Fujimaki, M., Imura, R., Nemoto, K., Sakaguchi, H., 2015. Evidence for erosion and deposition by the 2011 Tohoku-oki tsunami on the nearshore shelf of Sendai Bay, Japan. Geo Mar. Lett. 35, 315—328.

CHAPTER 12

Combined investigation of tradition archives and sedimentary relics of tsunami hazards — with reference to the Great 1700 Cascadia tsunami and other examples

Y. Tsuji

Earthquake and Tsunami Disaster Prevention Strategy Institute, Ryugasaki, Ibaraki, Japan

1. Introduction

The aim of this chapter is to discuss the possible role of tradition and archives, which can act as clues for the study of past tsunamiites and help in clarifying whole features of the specific past tsunami. We, the editors, believe that they can also help us to make an accurate estimation for future tsunami events.

In the late 19th century, many people were strongly impressed with the findings of the Trojan relic by H. Schliemann in the period 1870—1872. Since that time and that finding, many people may have come to believe that old legends and tales may sometimes reveal events that actually happened in the past, in other words, in old history.

Many people may have imagined that the old biblical story of Noah's arc and the great flood recorded in the Old Testament originated from an awful heavy flood, which occurred over the Iraq plain under the hypothermal weather during the Littorina Transgressive Period (7500—6100 BP).

Rather recently, this story has also been attributed hypothetically to a sudden pour-in of enormous amounts of sea water from the Mediterranean into the Black Sea Basin, which had been dry during the last glacial age (e.g., Ryan and Pitman, 1998). This sort of water movement may not have been classified as "tsunami" because of the unusually enormous scale of the flow. However, some parts of this water movement may have had the same kind of characteristics as a tsunami wave.

Tsunamiites. https://doi.org/10.1016/B978-0-12-823939-1.00012-4
Copyright © 2021 Elsevier B.V. All rights reserved.

Another miraculous gigantic water-affair story, which is written in the Old Testament, is the sudden disappearance and coming back of sea water that happened when the ancient Israelites led by Moses escaped from Egypt through the Suez Channel. Although some researchers may be earnest about finding some sedimentological tsunami records in these areas including the Eastern Mediterranean Sea, the present writer, however, does not have any scientific information in regard to any investigation of those particular sedimentary records.

The most well-known example of a really well investigated possible tsunamiites was discovered from tsunami traces and is named "the Bronze Age Homogenites." This has been assumed to be the deposits of a prehistoric megatsunami, which was caused by the Santorian Caldera Explosion in 1628 BCE (e.g., Cita and Aloisi, 2000; See Chapter 13). Features of these deposits and their happening mechanism are described and discussed again in the present version of this book (e.g., Cita et al., 1996; See Chapter 14).

In section 2, the author provides an example of the successful composite studies of (A) tradition and (B) material traces of a gigantic earthquake-induced tsunami, which occurred along the coastal area of North America, and (C) archives of the so-called "Orphan Tsunami," which left deposits at the opposite side of the Pacific Ocean, namely the Japanese Islands.

2. The tsunami traces of the 1700 Cascadia earthquake and corresponding archives in Japan

In recent decades, the Cascadia Subduction Zone has gained importance as a source of earthquakes of gigantic magnitude of 9 in moment magnitude scale. It is well known now that there are high potential risks by such gigantic earthquakes and accompanied tsunamis along the Cascadia Region of the Pacific Coast of North America. This matter and the geological achievements around it have been studied exhaustively by many researchers (e.g., Atwater, 1997a,b; Darienzo and Peterson, 1990).

2.1 An enormous tsunami told in stories down through tradition by native tribes of Canada

In the aforementioned frame of reference, it has been noted by researchers that a story about an enormous tsunami was handled down in the oral tradition of native tribes in the Cascadia region (Ludwin et al., 2005). The story was recorded in James Swan's diary, where he wrote about information of a tsunami that was given to him by Billy Balch, who was a leader of Makaha at Neah Bay, Washington Territory. Makaha is the name of one of the races of American Indians. He recounted that the sea flood occurred in the "not very remote" past. It began by submerging the lowland between Neah Bay and the Pacific

Ocean. Next, the water receded for 4 days. After that, sea water rose again without any swell or waves, and it submerged the whole of Cape Flattery and the whole country except the mountains back of Clyoquote (=Clayoquot at present). The tsunami dispersed the tribes, stranded canoes in trees, and caused numerous deaths. The same things happened at Quilwute (50 km south of Neah Bay).

2.2 Modern traces of sudden subsidence of Cascadia coast and an accompanying earthquake

The coastal evidence of great earthquakes, sudden subsidence of land, and following tsunami in Western Washington, USA, has been fully researched by Atwater and others (e.g., Atwater, 1997a,b, Atwater et al., 1995, Nelson et al., 1996). One of the signs of a land subsidence was "ghost forest," such as is seen still nowadays as groves of withering trunks that stand in tidal marshes of southern Washington state. Thousands of stumps of victim trees can be seen even now along banks of tidal streams there, 100 more at estuaries in Oregon and northern California (Fig. 12.1). Furthermore, a sudden subsidence

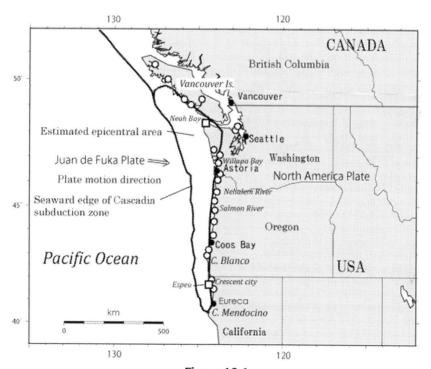

Figure 12.1
Distributions of the place where geological tsunami evidences were detected (*white circles*). *Squares* show the places where legends were handled down by American Indians. The seaward edge of Cascadia subduction zone and the epicentral area of the 1700 Cascadia Earthquake are also shown as the *fat line*.

Photo 12.1
Mud deposited layer formed by the tsunami of the Great 1700 Cascadia Earthquake at Naselle river, Willapa Bay (Atwater et al., 2005).

around these regions was proved to be the case by reading certain features of the tree rings and by the remarkable preservation of burial marsh soil with grasses.

It may do well to refer briefly to liquefaction of sediment layers, which should have accompanied the earthquake. The liquefied sand erupted through cracks onto freshly subsided land. During the present time (these days), we can easily find these conduits because water plucks sand grains up from the cracks.

2.3 Sedimentary traces of tsunami flow and their formative influence

Geologists found sedimentological evidences for tsunami (Atwater, 1997a,b) at about 20 different points along the Cascadia coast sand sheets, which were formed by a tsunami (see Fig. 12.1). Researchers could clearly judge that the sand came from the sea or not, because it tapers inland and contains the microscopic siliceous shells of marine diatoms. At most sites the sand arrived just before tidal mud became a covering of freshly subsided soil (See Fig. 22 in Atwater, 1997a,b). Neither a storm nor a tsunami of distant origin can explain this coincidence with subsidence. The simplest explanation is that a tsunami from an earthquake happened in which the crustal basement abruptly displaced the sea while lowering the adjoining coast.

Description and discussion about the Canadian Coast earthquake are not our subject in this chapter. The important points are that we could recognize the subsidence, and shaking, and that a tsunami had occurred forming a single event.

It should be noted here that an important fact had remained unsolved regarding the earlier referred researches. That was about the precise time in which the tsunami hit. In fact, what we now know are the results of radio carbon dating reported by Atwater et al., (1991), Nelson et al., 1995, combined with the information of year ring pattern matching by

Jacoby et al., 1997; Yamaguchi et al., 1997. That is to say, the designated time of the Cascadia's most recent great earthquake was most likely between 1695 and 1720 CE.

2.4 Date identification of the 1700 tsunami from documents of towns along the Japanese coast

A tsunami that occurred sometime in the past was called "Orphan tsunami" because of the lack of its known origin around the Pacific side of the Japanese Islands.

From midnight of January 27, 1700, to noon the next day, abnormal sea-level changes were recorded at several locations on the Pacific coast of Japan. The tsunamis were described in six independent documents, mostly in local government records (Fig. 12.2).

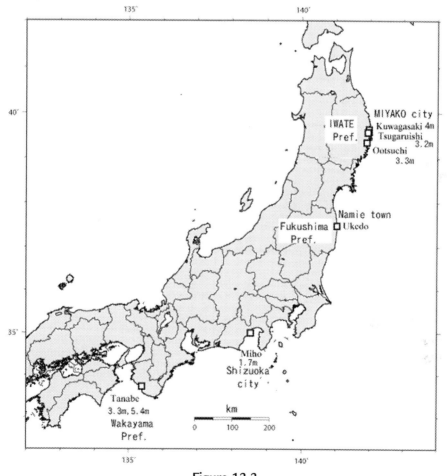

Figure 12.2
Inundation heights of the tsunami of the Cascadia Earthquake of January 16, 1700, on the coast of Japanese Island. *Squares* (□) show the places where old documents of descriptions of the tsunami exist with showing tsunami inundation heights (Tsuji et al., 1998).

The tsunamis were first noticed just before midnight of January 27 at Miyako and Otsuchi and the next morning at other locations. No one was injured or killed at towns on the tsunami-hit coast. Fig. 12.3 shows an example of the original manuscript named "Diary of Moriai Family" in Tsugaruishi village, Miyako city, Iwate prefecture. The ancestors of Moriai family were chief magistrates of Miyako city, which belonged to Nambu Territory, Iware prefecture at present. The meaning of the text is as follows: On November 8–9 of the 12th year of Genroku Era (January 27–28, 1700), high tidal waves attacked the coast,

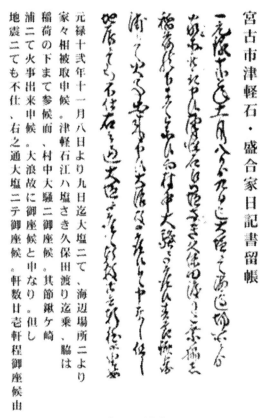

Figure 12.3

The text of "Diary of Moriai Family" in Tsugaruishi village, Miyako city, Iwate prefecture. Right figure is the original text, and left figure is the transferred text into block-type text of the present time. The meaning of the text is as follows: On November 8–9 of the 12th year of Genroku Era (January 27–28, 1700), high tidal waves attacked the coast, and houses were swept away in some villages. Sea water reached Kubota ferry point and just below Inari shrine in Tsugaruishi village (in Miyako city at present). Village people made a big disturbance. At Kuwagasaki fishery port due to the tidal wave, a big fire took place. People said, it is curious that a tsunami wave came in spite of no earthquake. In total, 21 houses burnt out there.

and houses were swept away in some villages. Sea water reached Kubota ferry point and just below Inari shrine in Tsugaruishi village (innermost coast of Miyako Bay). Village people made a big disturbance. At Kuwagasaki fishery port due to the tidal wave, a big fire took place. People said, it is curious that a tsunami wave came in spite of no earthquake. In total, 21 houses burnt out there.

This old document mentions that at Kuwagasaki, the fishery port of Miyako city, Iwate prefecture, in the north east part of Honshu Island, 20 houses were inflamed. We should notice that local people wondered that in spite of no earthquake, a large tsunami attacked. We can recognize that it was a common sense in those days that a tsunami wave is accompanied by an earthquake, and so people wondered why a tsunami came in spite of no earthquake being felt.

At Otsuchi, Iwate prefecture, in the midnight 0 h a.m., a large tide invaded into the residential area and vegetable fields behind the street were submerged. No human damage took place.

At Miho, Shizuoka city, sea water flooded in front of houses in the port area at 6 a.m. of 28th, and seven waves came till 10 a.m. on the same day. People escaped at the garden of Miho Shrine. People also wondered that tsunami waves without earthquake shaking ("Diary of Endo Family, the village headman of those days").

At Tanabe, Wakayama prefecture, from sunrise time of 9th, tsunami waves flooded in the streets and rise fields were damaged. Sea water invaded into the official store house.

Diary kept at Kyoto, Edo (Tokyo), and Morioka (capital city of Iwate prefecture) showed that the weather was clear and no stormy in the days on January 27 and 28, and so we cannot consider that the sea surface disturbance was storm surges.

2.5 Estimation of the occurrence time of the Great 1700 Cascadia earthquake

The earliest time when the initial tsunami wave was noticed was midnight 12 o'clock p.m., January 27, 1700, in Japan local time at Kuwagasaki, Miyako city, Iwate prefecture. It was 3 p.m. in universal time (London) and was 7 a.m., on 27th in Vancouver local time. We made a numerical simulation of this tsunami, and it was clarified that it takes about 10 h for the tsunami wave coming from offing of Vancouver to the coast of Miyako city. So, the occurrence time of the earthquake is estimated at 5 a.m., January 27, 1700 in universal time, that is, 9 p.m., January 26, 1700 in Vancouver local time.

The legend handled down by American Indians mentions that earthquake occurrence time was at a night in winter (Caver and Caver, 1996). The occurrence time estimated by ours is well matched with the legend handled down by American Indians.

200 Chapter 12

2.6 Estimation of magnitude of the 1700 Great Cascadia earthquake

The size of the source area of the 1700 Great Cascadia Earthquake exceeded 800 km (see Fig. 12.1). The size of the source area of the 2011 Great East Japan was 550 km in north-south direction along the trench axis, and its magnitude is estimated at M9.0. The size of the Sumatra Earthquake of December 26, 2004, was about 1200 km and its moment magnitude Mw is estimated at 9.3. The source size length of the 1960 Chilean Earthquake was about 1000 km and the magnitude is estimated at 9.5.

Considering those past gigantic earthquakes, the magnitude of the 1700 Great Cascadia Earthquake is suggested at 9.1−9.2.

2.7 Discussions on the 1700 Great Cascade earthquake

Upon overall consideration of the data from these recorded documents together with other supporting information, we have concluded that the tsunami came from far abroad across the Pacific Ocean.

We considered these data synthetically, combined with the historical and palaeontological information mentioned earlier and the known geotectonic information from around the related region. As a result we have confirmed that the source of the 1700 "Orphan tsunami" was the Cascadia Region (Satake, Shimazaki, et al.; Satake et al., 2000; Tsuji, 2004; Tsuji et al., 1998).

The examples described earlier are excellent case of successive synthetic studies of legend, archive, and sedimentary records of a tsunami affair.

It should be pointed out, however, that no sedimentary records of tsunami inundation have yet been found along the Japanese coastal areas mentioned in these examples. This is notable and will be mentioned again in one of the following section.

3. Examples from the Japanese islands

There are much amount of old documents describing past earthquakes and tsunamis in Japan. In addition to those manuscript documents, there are many stone monuments of tsunami damage found out on the coast of Sanriku district, the coast of Kii peninsula, Shikoku Island, and so on.

The oldest record of tsunami in Japan is that of the 684 Hakuho Nankai Earthquake, and we have records of large tsunamis accompanied with the series of Nankai Earthquakes. In the tsunami catalog published by Watanabe (1998), 195 near tsunamis were listed and 45 distant tsunamis are listed, most of which are trans-Pacific tsunamis.

3.1 Documented Jogan tsunami of July 13, 869 CE and its sedimentary records

As is well known, the Pacific side coasts of the Japanese Islands face toward the largest Ocean on the earth and are subject to frequent earthquakes and tsunamis. As a matter of course, these natural affairs and disasters have been studied comprehensively. General description commentary on these studies is far beyond the scope of this present chapter. Among these, the exceptional gigantic scale of inundation distance of the Jogan tsunami, which occurred in 869 CE was clearly understood in antique documents and by geosedimentological studies.

It was 1995 when Y. Iinuma, an out-of-academic circle scientist, gave caution to the possibility of hitting by a gigantic tsunami onto the Sendai Plain. This caution was based on his studies of the historical documents, which reported the 865 Jogan tsunami and other tsunamis (Iinuma, 1995, 2011). At the time,this study had not been noticed by either academic or governmental specialists.

Rather recently, before the time of the 2011 seismic tsunami, studies of the Jogan tsunami had been carried out by sedimentary geologists including Minoura et al. (2001). These studies contributed a wonderful example of successful combined researches of a tsunami event (Minoura and Nakaya, 1991; Minoura et al., 2001; Satake et al., 2008).

As a matter of fact, it was only after the 2011 Tohoku-Oki earthquake tsunami disaster that various studies considered the 864 Jogan earthquake as the precursor to the 2011 Tohoku-Oki earthquake. The recent studies concerning this situation are well recounted further by Goto et al. (2001), Sugawara et al. (2011), Satake et al. (2013), Sugawara, in this volume. In short, there have been written study reports about the gigantic scale of the 864 Jogan Tsunami and there have been warnings expressed about another possible exceptionally great tsunami event, which are based on these studies.

The actual 2011 Earthquake and Tsunami, which occurred as mentioned earlier, were out of expectation by common and special researchers both in its size and the size of the suffered areas. The tsunami left deposits in a very wide area covering the Sendai Plain, thus giving precious lessons regarding the precautions necessary in case of a tsunami event, especially the possibility of the gigantic size of it.

It is to be pointed out repeatedly that social ignorance of the warning given by a nonacademic researcher and academic geologists resulted in the misassumption of the possible size of the 2011 Tohoku Tsunami and the irrevocable accident of the Fukushima Atomic Power Station. Pertaining to these affairs, readers are encouraged to refer to the article contributed by Sugawara in this volume and other literatures. Unfortunately though, most of them are written in Japanese (Tsuji, 2012; Hatanaka, 2017; Shiki, 2018; Sugawara, this volume, and many others).

3.2 Other examples in Japan and their lessons

The narrative story of so-called "the fire of rice sheaves" is famous in Japan. It tells the story of Goryo Hamaguchi, a village headman who lived on the high ground behind the village of Hiro, Arita district, Wakayama prefecture. In the evening of December 24, 1854, a huge tsunami hit the south west coast of Kii peninsula. Just after he felt shaking of earthquake, he became aware of a large tsunami attacking the village. He set alight his newly harvested rice straws to give alert and to escape everyone urgently from the advancing tsunami. He, being the eldest in the village, could still remember the lesson of a legend about a characteristic tsunami behavior handled down from the tsunami of the 1707 Hoei earthquake. That is the tendency of initial lowering of sea water level around the area. This story has been handed down and even nowadays works as a present lesson at Arita district (70 km south of Osaka) and all over Japan.

Many documents have been written and stone monuments have been left that tell of the awful misery and the terrible lessons learned. A huge amount of studies and warnings have been carried out (e.g., Kawada, 2010, and many others). However, sedimentological investigations about the potential for another disaster still seem to be insufficient to stop endangering people (Kawada, 2010; Shiki, 2018, and many others).

Stories of sudden, awful, sea-water movement, which had caused the disappearance of coastal villages, have been told from generation to generation at some places along the Japan Sea coast. Especially, in the Wakasa Bay area, which shows rias coast, about 100 km north of Kyoto. That area is characterized by old stories of sudden subsidence of coasts or on small islands followed by gigantic tsunami hits. For example, a devastating tsunami happened at Kurumi-ura in the bay coast area on May 12, 701 CE (the first year of Taiho Era) causing damage. This story was mentioned in "Shoku Nihon-gi (The Sequel Chronology of Japan)." It tells a story of the disappearance of a village by a subsidence accompanied by a tsunami assault.

Rather recently, possible on-land tsunami deposits were found and investigated at the Takahama coast in the Wakasa Bay area (Yamamoto et al., 2015). This is another example of a scientific research aimed at checking an old archived description.

In short, archives and traditional stories of experienced tsunami hits have played a significant role in many places as lessons for the prediction of future possible tsunami disaster.

4. Lack of sedimentary relic — A supplementary discussion

As has been mentioned earlier, the documented 1700 "Orphan tsunami" left no sedimentary record on the Japanese coast. In actual fact, the depositional records of the tsunami were eagerly searched by sedimentary geologists around the districts where some tsunami effects had been recorded in antique documents. However, no correlative tsunamiites from that particular age/period have been found in those areas. This is noteworthy.

We should remember also that in the case of the 2011 Tohoku tsunami disaster, there appeared several areas of no deposition of any tsunami-related sediments. This was especially noticeable across a wide nondeposition area developed in the interior of the Sendai Plain (Minoura et al., 2001; Sugawara, this volume). As was stressed by Sugawara in his chapter in this volume, it would be safe to say that such cases of nontsunamiites deposition in historical ages/periods may have occurred in many coastal districts during a Tsunami assault. Furthermore, it should be pointed out that many tsunamis only leave records of erosion, without depositional records in some areas of invasion on land.

Therefore, there is no logical contradiction. Obviously, the sediment records should be taken seriously in the studies of past tsunami and for the precaution against future tsunami invasion. On the other hand, it is safe to say, an absence of up-rush tsunami deposits does not always mean no visitation had occurred.

At any rate, it can be said that even though there is a lack of sedimentary record of some tsunami, we may still possibly find some reasonable information for further studying the occurrence of tsunamis hits from the past and in the future.

5. Conclusive remarks

Many geologists and archeologists have studied sediments of historic and prehistoric tsunami events motivated by some traditions or archives. Such studies often clarified well the features of the concerned tsunami including divergence of wave character, invasion area, and so on. The date of a tsunami occurrence, however, cannot be determined exactly even by modern techniques. On the other hand, the absence of sedimentary record, such as in the case of the "Orphan Tsunami" along the Japanese coasts described earlier, is not an exceptional phenomenon.

In conclusion, connection of these data is very useful or might I say, it is required.

Various sedimentary records and relics, namely tsunamiites, are naturally important. Archives and even fables or passed-down legends should not be ignored. They may offer significant clues or suggestions about some geohistorical affairs such as a tsunami inundation. Only the overall study of every kind of information will effectively make it possible to obtain enough understanding about tsunami disaster prevention.

Thus, studies of old documented tsunami deposits and tsunami-related debris (namely tsunamiites) are currently being continued by investigators including Tsuji, myself. I hope my present discussion can help encourage students to join in the vitally important study of this interesting subject.

204   Chapter 12

Acknowledgments

The author wishes to express his thanks to Dr. B. F. Atwater for his permission of referring his photograph (Photo 12.1). He also thanks Prof. T. Shiki for his large support in making the text of the present paper.

References

Atwater, B.F., Yelin, T.S., Weaver, C.S., Hendley, J.W., 1995. Averting surprises in the Pacific Northwest. U.S.Geological Survey Fact Sheet 111−195, 2p. http://quake.wr.usgs.gov/prepare/factseets/PacNW/.

Atwater, B.F., 1997. Coastal evidence for great earthquakes in West Washington U.S. Geol. Surv. Prof. Pap. 1560, 77−90.

Atwater, B.F., Stuiver, M., Yamaguchi, D.K., 1991. Radiocarbon test of earthquake magnitude and the Cascadia subduction zone. Nature 353, 156−158.

Atwater, B.F., 1997. Coastal Evidence for Great Earthquakes in Western Washington. Assessing Earthquake Hazards and Reducing Risk in the Pacific Northwest, pp. 77−90. Reprinted from U. S. Geological. Survey Professional Paper 1560.

Atwater, F.B., Musumi-Rokkaku, S., Satake, k., Tsuji, Y., Ueda, K., Yamaguchi, D.K., 2005. The Orphan Tsunami of 1700; Japanese Clues to a Parent Earthquake in North America. Published jointly with University of Washington Press, DC,US, p. 133. Geological Survey Paper 1707.

Carver, G.A., Carver, J.P., 1996. Earthquake and thunder-native oral histories of paleoseismicity along the southern Cascadia subduction zone. Proc. Geol. Soc. Am. 28, 54.

Cita, M.B., Aloisi, G., 2000. Deep-sea tsunami deposits triggered by explosion of Santorini (3500y BP), eastern Mediterranean. Sediment. Geol. 135, 181−203.

Cita, M.B., Camerlenghi, aaaA., Rimold, B., 1996. Deep-sea tsunami deposits in the eastern Mediterranean. New evidence and depositional model. Sediment. Geol. 104, 135−173.

Darienzo, M.E., Peterson, C.D., 1990. Episodic tectonic subsidence of late Holocene salt marshes, northern Oregon central Cascadia margin. Tectonics 9, 1−22.

Goto, K., Chague-Goff, C., Fujino, S., Goff, J., Jaffe, B., Nishimura, Y., Richmond, B., Sugawara, D., Szczucinsky, W., Tappin, D., Witer, R., Yulianto, E., 2001. New insights of tsunami hazard from the 2011 Tohoku-oki event. Mar. Geol. 290, 46−50.

Hatanaka, A., 2017. Natural hazards and Japanese people − Folklore of earthquake, water flood and volcanic eruption. Chikuma Shinsho 251p (In Japanese).

Iinuma, Y., 1995, 2011. Tsunami in History of Sendai Plain. Honda Pub., p. 235 (in Japanese).

Jacby, G.C., Bunker, D., Benson, B., 1997. Tree-ring evidence for an A.D.1700 Cascadia earthquake in Washington and northern Oregon. Geology 25, 999−1002. https://doi.org/10.1130/0091-7613(81997) 0252.3.CO:2.

Kawada, Y., 2010. Tsunami Disaster- Need to Establish of Disaster Risk Reduction Society. Iwanami Press (in Japanese).

Ludwin, R.S., Dennis, R., Carver, D., McMillan, A.D., Losey, R., Clague, J., Jonientz-Trisier, C., Bowechop, J., Wray, H., James, K., 2005. Dating the 1700 Cascadia earthquake; great coastal earthquakes in native stories. Seismol Res. Lett. 76, 140−148.

Minoura, K., Nakaya, S., 1991. Traces of tsunami preserved in intertidal lacustrine and marsh deposits: some examples from northeast Japan. J. Geol. 99, 265−287.

Minoura, K., Imamura, F., Sugawara, D., Kono, Y., Iwashita, T., 2001. The 869 Jogan tsunami deposits and recurrent interval of large-scale tsunami of the Pacific coast of northeast Japan. J. Natl. Dis. Sci. 23, 83−88.

Nelson, A.R., Atwater, B.F., Bobrowsky, P.T., Bradley, L., Clague, J.J., Carver, G.A., Darienzo, M.F., Grant, W.C., Jrueger, H.W., Sparks, R.J., Stafford, T.W., Stuiver, M., 1995. Radiocarbon evidence for extensive plate-boundary rupture about 300 yers ago at the Cascadia subduction zone. Nature 378.

Nelson, A.R., Shennan, I., Long, A., 1996. Identifying coseismic subsidence in tidal-wetland stratigraphic sequence at the Cascadia subduction zone of western North America. J. Geophys. Res. 101, 6115−6135.

Ryan, W., Pitman, W., 1998. Noah's Flood F., Jellinek & Murray Litterary Agency. Japanese Translation by N. Toda, Shyuei Pub, Tokyo, p. 327p.

Satake, K., Wang, K., Atwater, B.F., 2000. Fault slip and seismic movement of the 1700 Cascadia earthquake inferred from Japanese tsunami descriptions. J. Geophys. Res. 108, 17. https://doi.org/10.1029/2003JB002521.

Satake, K., Namegaya, Y., Yamaki, S., 2008. Numerical simulation of the AD 869 Jogan tsunami in Ishinomaki and Sendai planes. Ann. Rep., Active Fault Paleoearthquake Res. 8, 71−89.

Satake, K., Fujii, Y., Harada, T., Namegaya, Y., 2013. Time and space distribution of coseismic slip of the 2011 Tohoku Earthquake as inferred from tsunami waveform data. Bull. Seismol. Soc. Am. 103, 1473-1192.

Shiki, T., 2018. Disaster and Disaster Prevention − Past and This Moment. Honnoizumi Pub, p. 277 (139-153), (in Japanese).

Sugawara, D., Imamura, F., Matsumoto, H., Goto, K., Minoyra, K., 2011. Reconstruction of the AD 869 Jogan earthquake induced tsunami by using the geological data. J. Jpn. Soc. Nat. Disast. Sci. 29 (4), 501−516 (in Japanese with English abstract).

Tsuji, Y., Ueda, K., Satake, K., 1998. Japanese tsunami record from the January 1700 earthquake in the Cascadia subduction zone. Zishin J. Seismol. Soc. Jpn. 51, 1−17. https://doi.org/10.4294/zisin1948.51.1_1 (in Japanese).

Tsuji, Y., 2004. Tsunami disaster mitigation on Kumano Coast, Kii peninsula, with considering the data of historical records and geologic traces, (in Japanese). Kaiyo Monthly 36 (7), 488−497.

Tsuji, Y., 2012. Millennium Earthquake Disasters-Lessons from Histories of Earthquake and Tsunamis. Diamond Press, p. pp276.

Watanabe, H., 1998. Tsunami Catalogue of Japan. Tokyo University Press, p. 236 (in Japanese).

Yamaguchi, D.K., Atwater, B.F., Bunker, D.E., Benson, B.E., Reid, M.S., 1997. Tree-ring dating the 1700.

Yamamoto, H., Urabe, A., Sasaki, N., Takashimizu, Y., Takaoka, K.S., 2015. Trench and Coring Survey of the Coastal Area along the Wakasa Bay, Takahama, Fukui Prefecture. 2015. Report of Japan Geoscience Union.

CHAPTER 13

Deep-sea homogenites: sedimentary expression of a prehistoric megatsunami in the Eastern Mediterranean

M.B. Cita

Dipartimento di Scienze, della Terra 'Ardito Desio', Università di Milano, Milano, Italy

1. Introduction

The catastrophic megatsunami of December 26, 2004 that hit the coasts of the Indian Ocean thousands of kilometers from the earthquake's epicenter shocked the world not only for the disastrous losses of lives, villages, and constructions but also for the unprecedented real-time documentation of an exceptional natural disaster by television and satellite images. Scientists familiar with tsunami-related problems worldwide are now discussing their findings in the light of the new evidence.

Submarine earthquakes are the most common cause of megatsunamis, but not the only one. Catastrophic eruptions of volcanic islands, and bolide impacts, may also trigger tsunamis capable of crossing an ocean.

The Bronze-Age ($\sim$3500 BP) destructive eruption of the volcanic island of Santorini in the eastern Mediterranean occurred in several phases (Bond and Sparks, 1976; Friedrich and Pichler, 1976; Pichler and Friedrich, 1978): first an earthquake, followed by a Plinian phase with the deposition of meters of ash and pumice; then, a large base-surge phase following the invasion of the magma chamber by the sea water, giving rise to thick ash-flow deposits that constitute the main volume of the local pyroclastic deposits; and, finally, the collapse of the magma chamber and the creation of a large and deep caldera (see Friedrich, 1999; Heiken and McCoy, 1984).

The island of Santorini is horseshoe-shaped, and the caldera rim is interrupted to the NW and SW (Fig. 13.1). The maximum elevation of Santorini is now 570 m, and the deep conical depression inside the collapsed caldera is −390 m. Larger than the Krakatoa, which exploded in 1883 triggering a tsunami with waves 30 m high that killed 36,000 persons, Santorini is considered responsible for a tsunami wave with a supposed height of

Tsunamiites. https://doi.org/10.1016/B978-0-12-823939-1.00013-6
Copyright © 2008 Elsevier B.V. All rights reserved.

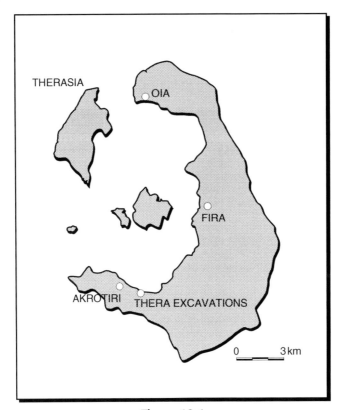

Figure 13.1
The Island of Santorini in the eastern Mediterranean.

40 m (Marinos and Melidonis, 1971; Yokoyama, 1978) or 30 m (Kastens and Cita, 1981), similar to that measured for the December 26, 2004 tsunami that hit several coastlines facing the Indian Ocean and killing some 260,000 persons.

The purpose of this invited contribution is to discuss two discrete types of deep-sea tsunamiites, thus updating the description and interpretations provided by Cita and Aloisi (2000); (see also Shiki et al., 2000).

2. Deep-sea homogenites

The deep-sea sediments under discussion were called "homogenites" since their first discovery during the "Cobblestone" scientific program, funded by the NSF and the CNR in 1978–81 (see Blechschmidt et al., 1982; Cita et al., 1982a; Kastens and Cita, 1981), and refer to the homogeneous, structureless nature of the fine-grained reworked beds that differ strongly from other nearshore tsunamiites. The original data set consisted of eight piston cores raised from the Calabrian Ridge (Area 4) and of three piston cores from the western

Mediterranean Ridge (Area 3). The present-day data set consists of 60 deep-sea cores, including a few giant cores of length 20–30 m (Fig. 13.2; see also Table 1 in Cita and Aloisi, 2000). The water depth ranges from a minimum of −2451 m to a maximum of −4071 m. Fifty cores were raised from depths between −3000 and −4000 m. All these cores have been longitudinally split, visually described, sampled, investigated microscopically after proper treatments, and correlated. Detailed analyses were carried out on a fairly large number (Cita and Camerlenghi, 1985; Cita et al., 1982a,b, 1984b; Hieke, 1984; Rimoldi, 1989; Rothwell et al., 2000; Troelstra, 1987), to which work reference is made here.

Two discrete types are presently recognized (see Cita and Aloisi, 2000; Cita and Camerlenghi, 1985; Cita and Rimoldi, 1997; Cita et al., 1984b, 1996).

2.1 Type A homogenites

2.1.1 Characteristics

Homogenites of Type A consist of thick hemipelagic, very fine-grained beds, devoid of internal structures, as also shown by X-ray analysis and computerized tomography (Cita et al., 1982a; Hieke, 1984). No Bouma intervals can be identified, nor parallel and/or convolute lamination. An increase in grain size is noticed near the base of the unit (Fig. 13.3), accompanied by a slight increase in carbonate content. The sandy base consists of tests of planktonic foraminifers and fragments of pteropods. These Type A homogenites are interpreted as pelagic turbidites.

The thickness is several meters, which is one to two orders of magnitude more than that of other turbidites in the same cores, and these units are interpreted as seismically induced (Kastens, 1984). The base of the unit is depositional, not erosional, and the stratigraphic position is very well constrained, in the middle of the Holocene, since Type A homogenites have been found exclusively above sapropel S-1, a marker bed typical of the eastern Mediterranean deep-sea record that is the sedimentary expression of basin-wide stagnation (Cita et al., 1977; Vergnaud-Grazzini et al., 1977, and later works). The termination of the anoxic episode is now dated at 6500 BP, which is consistent with the 3500 BP for the Santorini Bronze-Age catastrophic eruption (see position of the two events as recorded in the Calabrian Ridge cores illustrated in Fig. 13.4). Direct ^{14}C datings of the homogenites are meaningless due to the extremely fine-grained and mixed nature of the reworked beds. Hieke (1984) and Troelstra (1987) obtained ages from the sediments directly underlying the homogenite from the Messina abyssal plain and the Tyro Basin, respectively: they are in agreement with the chronology proposed here. Attempts to date the posthomogenite pelagic drape by means of specially tailored box cores have been unsuccessful so far (Cesare Corselli, Univ. Milano Bicocca, personal communication, 2004).

210 Chapter 13

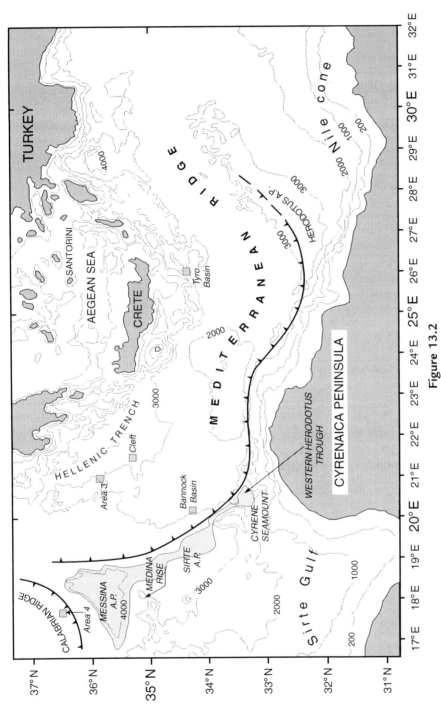

Figure 13.2

Map of the eastern Mediterranean showing the main physiographic features, the outer deformation front of the Calabrian and Mediterranean Ridges, and the location of the areas where the Holocene homogenites have been identified. Type A in cobblestone Area 4 of the southern Calabrian Ridge; Bannock Basin, cobblestone Area 3, Cleft and Tyro Basin in the Mediterranean Ridge. Type B (megabed) in the Messina abyssal plain, the Sirte abyssal plain, and the western Herodotus Trough.

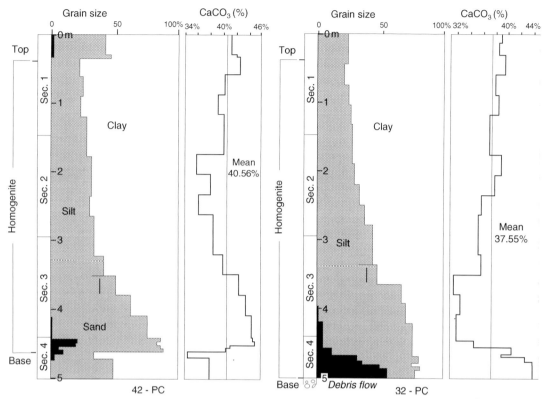

Figure 13.3
Grain-size distribution and carbonate content in two piston cores where the homogenite has comparable thicknesses, that is, Eastward Core 42 from the Calabrian Ridge and Eastward Core 32 from the western Mediterranean Ridge.

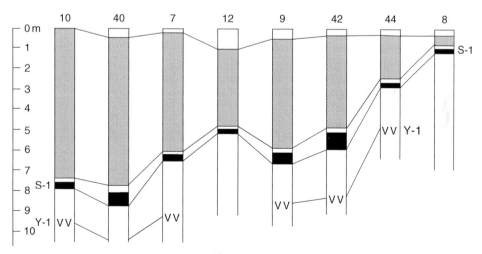

Figure 13.4
Correlation of the homogenite (grey) and of other isochronous lithologies [i.e., Sapropel S-1 (black) and Tephra Y-1] in eight piston cores raised by R/V Eastward from the southern Calabrian Ridge (cobblestone Area 4).

The source area of Type A homogenites is local, from nearby structural highs. Their physiographic setting is always at the flat bottom of small depressions and perched basins characteristic of the "cobblestone topography" (of Hersey, 1965).

The acoustic properties of Type A homogenites are well expressed in the profiles published in 1981 by Kastens and Cita, obtained by a deep-towed echo sounder in an area previously mapped in detail with a transponder-navigated survey. The thick, surficial transparent layer terminates abruptly against the walls bounding the depression, whereas the basin slopes are highly reflective. However, also with less sophisticated techniques, profiles obtained with a precision depth recorder (PDR) 3.5 kHz echo sounder located on the ship in the Cleft area of the Mediterranean Ridge (Cita and Rimoldi, 1997; Cita et al., 1982b), the surficial transparent layer recorded at the depocenter of the major depressions was calibrated by coring as a several meters thick, fine-grained Holocene homogenite, overlying sapropel S-1.

2.1.2 Interpretation

Kastens and Cita (1981) calculated the order of magnitude of the speed of the tsunami waves resulting from the caldera collapse, based on the physiography of the ocean floor (Airy law) and the path of the rays perpendicular to the wave front (see also Cita and Aloisi, 2000). Rays are evenly spaced at 5° at the source but are deflected in response to the bathymetry.

Fig. 13.5 is a conceptual model of the Type A homogenite, reproduced in several papers published by our research group in the past 20 years. The high-speed, spherical tsunami wave(s) cause(s) liquefaction of the water-saturated, unconsolidated pelagic sediments that

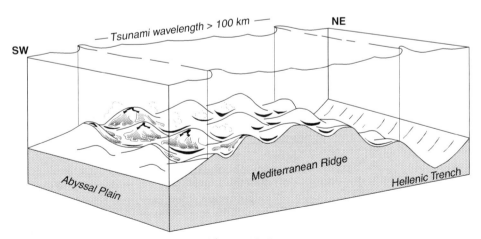

Figure 13.5
Depositional model of the Type A homogenite as recorded from the southern Calabrian Ridge and from the western Mediterranean Ridge characterized by cobblestone relief.

drape the flanks of the structural highs of the irregular "cobblestone topography." After the passage of the megatsunami, the cloud of fines settles from suspension and, if the gradient of the slope is high, the downslope flow may erode unconsolidated sediments of late or early Pleistocene or even Pliocene age (see Blechschmidt et al., 1982). The fine material accumulates at the bottom of the perched basins and forms units with thicknesses ranging from <1 m to several meters, as shown by the logs of Fig. 13.4 The grain-size distribution and the thickness of the homogenite in two cores (as depicted in Fig. 13.3) support this interpretation. The core from cobblestone Area 3, which is some 500 km away from Santorini (the source area of the tsunami), is coarser and has a debris-flow unit at the base, whereas the core from the Calabrian Ridge (~800 km from Santorini) suggests a lower energy.

It is a deep-sea context, where the unusual physiography of the sea floor plays a major role.

2.2 Type B homogenites

2.2.1 Characteristics

Type B homogenites are represented by megaturbidites with thicknesses that may be in excess of 20 m. They consist of fine-grained hemipelagic/bioclastic silts and muds that are devoid of sedimentary structures such as lamination. The thick, sandy basal part (Fig. 13.6)

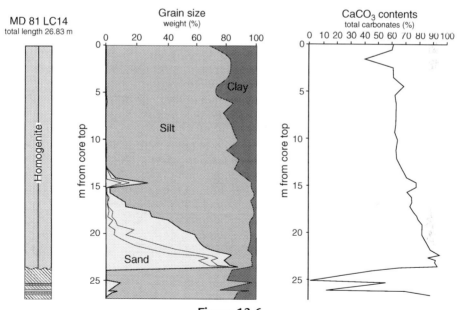

Figure 13.6
Grain-size distribution and carbonate content in the homogenite recovered in the long Core 14 from the western Herodotus Trough. See text for discussion. *Slightly modified from Cita, M.B., Aloisi, G., 2000. Deep-sea tsunami deposits triggered by the explosion of Santorini (3500 y BP), eastern mediterranean. Sediment. Geol. 135, 181–203.*

consists of bioclasts indicative of a shallow-water source area (see plate 1 in Cita and Aloisi, 2000) and includes fragments of Bryozoa, calcareous algae with numerous aragonitic needles typical of *Halimeda*, larger benthic foraminifers such as *Amphistegina, Sorites*, and *Planorbulina mediterranea*, and also mineral grains of siliciclastic composition (quartz and feldspar). The bases of these megaturbidites are erosional in the few piston cores that recovered them. Most coring ended without having reached the graded sandy base.

The stratigraphic position of Type B homogenites is mid-Holocene; since they are covered only by a thin veneer of brown pteropod ooze, which is typical of that interval in the eastern Mediterranean deep-sea record. The underlying sediments, underneath the erosional base, are of Pleistocene age.

The physiographic setting of Type B homogenites is confined to the Messina and Sirte abyssal plains, where their acoustic expression is conspicuous, with a very thick and transparent surficial layer that is recorded by all the research ships that have been operating in these areas. A few published examples are found in Della Vedova et al. (1998), Cita and Aloisi (2000), Hieke and Werner (2000), and Rebesco et al. (2000). They are also recorded in the eastward prolongation of the Sirte abyssal plain, named "western Herodotus Trough" on the UNESCO Map, which lies in the east of the Cyrene seamount (Groupe Escarmed, 1987).

The first records of Type B homogenites (Cita and Camerlenghi, 1985; Cita et al., 1984b) were from the outer deformation front of the Mediterranean Ridge, facing the Messina and Sirte abyssal plains, where these distal turbidites are indicative of a very shallow source area and with an erosional base climbing the outer slopes of the Mediterranean Ridge (see also Rimoldi, 1989). Hieke and Werner (2000), who provided an excellent record of subbottom profiles in the Ionian Basin, calculated the minimum volume of this megaturbidite (that they call "Augias turbidite") as 65 km^3. Rebesco et al. (2000) provided two new significant pieces of information. One is the record of a thick surficial transparent layer (megaturbidite) in the Matapan Trench, north of the Mediterranean Ridge. The second one is a synthetic seismogram of a compressed high-intensity radar pulse (CHIRP) line crossing the location of long Core 14, which demonstrates that only the surficial transparent layer represents the homogenite (as a descriptive term).

2.2.2 Interpretation

Fig. 13.7 depicts the conceptual model originally proposed by Cita et al. (1984b) and further elaborated by Cita and Camerlenghi (1985), Cita et al. (1996), Cita and Rimoldi (1997), and Cita and Aloisi (2000), updating and incorporating new results. The Type B homogenite is—unlike Type A—a megaturbidite of distal, African origin, triggered by the surficial tsunami wave that hit the passive margin of Africa.

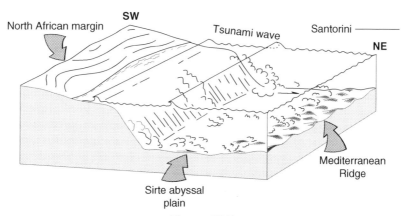

Figure 13.7
Depositional model of the Type B homogenite, as recorded in the Messina and Sirte abyssal plains, in the western Herodotus Trough and on the lowermost part of the deformation front of the Mediterranean Ridge.

The impressive images that were shown so many times on television after the catastrophic tsunami of December 26, 2004, with the third gigantic wave invading the beach and the lagoon of Phuket after a retreat of the sea, documented with unprecedented precision by satellite images, are a kind of "reality show" that strongly supports our conceptual model: it was clearly visible that the gigantic tsunami wave did not consist of just sea water but was rather a sediment-laden slurry charged with mud eroded from the sea bottom. When returning to the sea, this giant wave must have triggered a turbid current capable of reaching the bottom of the ocean.

The distance from the source of the tsunami (Santorini) to the Sirte Gulf exceeds 1000 km, which is comparable with the distance traveled by the Krakatoa-induced tsunami. Also, the distance from the Sirte Gulf to the Messina abyssal plain is of the order of 500 km and is comparable to the distance covered by gigantic megaturbidites recorded in the geological record (Mutti et al., 1984).

The occurrence of such an exceptionally large turbidite in the mid-Holocene, close to the climatic optimum and during a time of high sea-level stand, is a paradox in the context of the glacio-eustatic and palaeoclimatic history and can be accounted for only by an exceptional event, such as the megatsunami, triggered by the collapse of the Santorini caldera (see also Section 3).

3. Discussion

Four points of discussion are addressed in this section.

216 **Chapter 13**

3.1 Data

The data set includes 60 deep-sea cores raised in the years 1978–96, which contain either Type A or Type B homogenites. Reference is made to Cita and Rimoldi (1997) and Cita and Aloisi (2000) for a complete record. All are deep-sea cores raised from water depths ranging from a maximum of −4087 m (in the Messina abyssal plain) to a minimum of −2421 m (in the Katia Basin, where the collision between the Mediterranean Ridge and the North African continental margin is recorded). Fifty of the 60 cores are from water depth between −3000 and −4000 m. They are really deep-water sediment cores, whereas most tsunamiites described so far are from shallow-water, nearshore environments where the distinction between tsunamiites and tempestites seems problematic in the absence of precise historical references.

3.2 Absence of tephra Z-2 in our data set

The tephra layer deposited during the Bronze-Age eruption of Santorini is well known in the eastern Mediterranean (Keller, 1981; Keller et al., 1978; McCoy, 1978; Ninkovich and Heezen, 1965; Watkins et al., 1978) and is called Z-2. None of the cores investigated contains Z-2 tephra, however, and this is consistent with previous findings; the westerly winds during the eruption resulted in a dissemination of the airborne tephra to the east of Santorini, whereas the tsunami waves were directed preferentially to the west.

3.3 Absence of homogenites in the Herodotus abyssal plain

The total absence of Holocene homogenites on the Herodotus abyssal plain, predicted by the Kastens and Cita (1981) model, was first documented during the 1982 Bannock expedition and was reported by Cita et al. (1984b), who pointed out the strong geodynamic significance of the seismically induced versus aseismic turbidites that accumulate on the floor of the abyssal plains of the eastern Mediterranean (Cita et al., 1984a).

The transect of giant cores raised from the Mediterranean abyssal plains during the PALEOFLUX EU-funded MAST project (see Cita and Aloisi, 2000; Reeder et al., 1998; Rothwell et al., 1998, 2000) (see location in Fig. 13.8) shows that thick transparent layers, calibrated by coring, are present in all the major basins, but their ages differ considerably (Fig. 13.9). In the Balearic abyssal plain, the megaturbidite has a northern source and has been radiometrically dated as formed during the Last Glacial Maximum (16,000–18,000 BP). It has been hypothesized that it is related to gas-hydrate expulsion during a pronounced low-sea-level stand. Notice the several meters thick layered interval overlying the transparent layer, which corresponds to the late Pleistocene and Holocene. The transparent megabed recorded in the Sirte abyssal plain, as well as the same megabed visible in the subbottom profiles illustrated by Hieke (1984), Della Vedova et al. (1998),

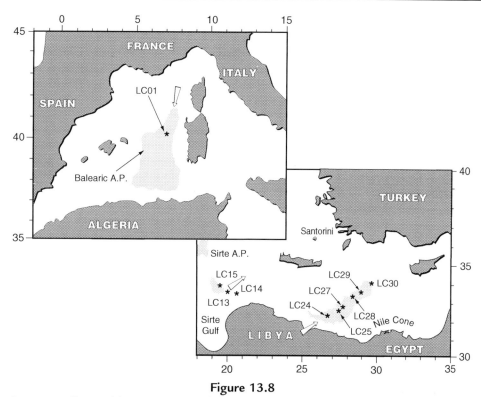

Figure 13.8
Location map of several long cores raised from the Mediterranean abyssal plains in 1995. The *open arrows* point to the direction of the high-resolution seismic profiles illustrated in Fig. 13.9 (see also Fig. 13.11). *Modified from Rothwell, R.G., Reeder, M.S., Anastasakis, G., Stow, D.A.V., Thomson, J., Kalher, G., 2000. Low-sea-level stand emplacement of megaturbidites in the Western and Eastern Mediterranean Sea. Sediment. Geol. 135, 75–88.*

Cita and Aloisi (2000), Hieke and Werner (2000), and Rebesco et al. (2000), is—instead—very close to the surface, and no layered interval exists above it. Other megaturbidites have been recorded at greater depth and are possibly correlatable with those identified in the other abyssal plains. Fig. 13.9 also shows a subbottom profile recorded across the Herodotus abyssal plain, where large turbidites are recorded in the late Pleistocene, all of them originating from the passive margin of Africa, some with a well-recognizable Nile provenance (Reeder et al., 1998; Rothwell et al., 2000).

The correlation of the long cores depicted in Fig. 13.10 is essential in understanding the paradox of Type B homogenites, which should not be there under normal circumstances. In other words, no megaturbidites of distal origin, capable of crossing a distance of hundreds of kilometers downslope of a passive margin with a vertical relief in excess of 4000 m, should be formed during the times of high sea-level stand in the middle of the Holocene. Only a quite exceptional triggering mechanism, such as a megatsunami formed

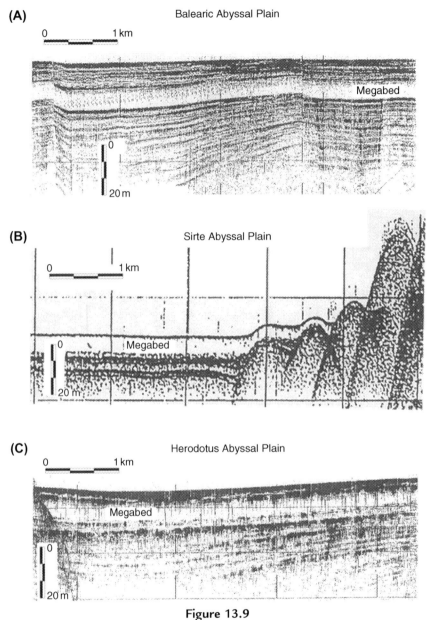

Figure 13.9
High-resolution (3.5 kHz) seismic reflection profiles recorded in the Balearic (A), Sirte (B), and Herodotus (C) abyssal plains. Each profile documents the existence of a megabed. *From Cita, M.B., Aloisi, G., 2000. Deep-sea tsunami deposits triggered by the explosion of Santorini (3500 y BP), eastern mediterranean. Sediment. Geol. 135, 181−203; Rothwell, R.G., Reeder, M.S., Anastasakis, G., Stow, D.A.V., Thomson, J., Kalher, G., 2000. Low-sea-level stand emplacement of megaturbidites in the Western and Eastern Mediterranean Sea. Sediment. Geol. 135, 75−88.*

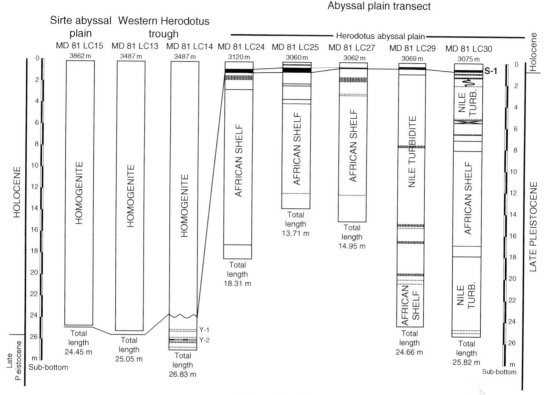

Figure 13.10
Stratigraphic correlation of eight long cores raised from the eastern Mediterranean abyssal plains. The location of the long cores is indicated in Fig. 13.8. *Slightly modified from Cita, M.B., Aloisi, G., 2000. Deep-sea tsunami deposits triggered by the explosion of Santorini (3500 y BP), Eastern mediterranean. Sediment. Geol. 135, 181–203; Rothwell, R.G., Reeder, M.S., Anastasakis, G., Stow, D.A.V., Thomson, J., Kalher, G., 2000. Low-sea-level stand emplacement of megaturbidites in the Western and Eastern Mediterranean Sea. Sediment. Geol. 135, 75–88.*

when a suprasubduction volcanic edifice collapsed after an eruption of catastrophic dimensions, could achieve this.

3.4 Comparison of type A and type B homogenites

The two types of homogenites distinguished here are similar in general (same age, same anomalous thickness, and same homogeneous character) but different in their details. The hydrodynamic situation was entirely different for the two homogenite types, which have in common only their origin by one singular prehistoric megatsunami. Type A and Type B homogenites are never found together at the same location.

3.4.1 Type A

The source area of the pelagic turbidites is close to the deposition area and is a deep-sea, pelagic setting. The base of these turbidites is depositional, never erosional, in the basin depocenters. The hydrodynamic conditions that caused the displacement of the unconsolidated pelagic sediments draping the flanks of the small-scale reliefs triggered turbid currents that moved downslope for a horizontal distance of no more than a few kilometers and vertical distance of 100−200 m.

The distribution of homogenites is strictly limited to the flat bottom, whereas the basin flanks have been eroded: core transects crossing plateaus, slopes, and basins prove that (Blechschmidt et al., 1982; Kastens and Cita, 1981). The distance of the areas investigated from the source of the megatsunami (=center of the collapsed caldera) is of the order of 500 km for cobblestone Area 3 (western Mediterranean Ridge) and 800 km for cobblestone Area 4 (Calabrian Ridge).

3.4.2 Type B

The source area of the megaturbidite is remote, shallow, and nearshore, as proven by the bioclastic composition that clearly indicates an origin from the inner shelf of North Africa. The Sirte Gulf is considered the most probable source of the megaturbidite, which expanded throughout the adjacent Messina and Sirte abyssal plains after traveling over 500 km from the shoreline to the ocean bottom, with a vertical height difference of up to 4 km. The gigantic sediment current was initiated by the megatsunami triggered by the collapse of the Santorini caldera, more than 1000 km away. The base of Type B homogenites is always erosional.

Type B homogenites have been calibrated in time by coring only close to the outer margin of the abyssal plains (Fig. 13.11), whereas their greatest thickness is located in the basin depocenter, as visualized by subbottom profiles. Coring was unsuccessful there: short cores did not reach the base, and long cores have been lost twice because of excess resistance of the coarse and thick sandy base.

4. Conclusions

Both Type A and Type B homogenites are tsunamiites, but their sedimentary context is strongly different.

Type A homogenites consist of pelagic turbidites caused by the liquefaction of hemipelagic surficial sediments. They were formed in small basinal areas of the Mediterranean and Calabrian Ridges. They are the representative of tsunamiites formed in deep-water accretionary prisms.

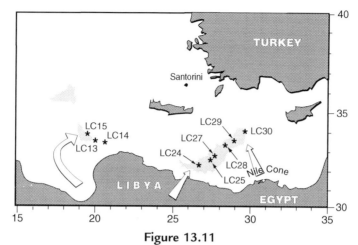

Figure 13.11
Location map of the investigated long cores that contain turbiditic megabeds from the Sirte and Herodotus abyssal plains in the Eastern Mediterranean. The *open arrows* indicate the source areas as inferred from the analytical data (Sirte Gulf for Cores LC13–15; African shelf for Cores LC24–27; Nile Cone for Cores LC29 and 30) (see also Fig. 13.8).

Type B homogenites are megaturbidites that result from an entirely different sedimentary process that does not involve liquefaction of sediments draping the ocean bottom. Instead, they were triggered by an anomalous tsunami wave hitting the shoreline of the Sirte Gulf. Type B homogenites cover the entire Ionian (or Messina) abyssal plain and its eastern prolongation (Sirte abyssal plain) as far as the collision zone. In their palaeoclimatic and glacio-eustatic context, Type B homogenites are a paradox, since in the later part of the Holocene, during a time of high and rising sea level, no megaturbidites derived from a passive continental margin should exist. Only an exceptional event, such as a megatsunami triggered by the collapse of a suprasubduction volcanic edifice, could produce such an anomalous process. Even though the Sirte Gulf is not an area of major sediment supply, the continental shelf is so large that its catastrophic erosion may well provide the material for a megabed with a volume of some 65 km^3 (Hieke and Werner, 2000).

Unlike other tsunamiites recorded in high-energy, shallow-water environments, the homogenites were deposited in the deep sea, at great distance from the source area of the tsunami wave.

5. Post scriptum

This paper was accepted for publication on March 30, 2006. Soon after three papers in a row following a marketing strategy were published on the journal *Geophysical Research Letters*, which are pertinent to the theme of prehistoric tsunami in the eastern Mediterranean.

222   Chapter 13

The first paper (Pareschi et al., 2006a; submitted February 16, 2006, revised May 2, 2006, accepted May 17, 2006; published July 1, 2006) describes submarine deposits offshore Mt. Etna in the Ionian Sea that are interpreted as debris avalanches triggered by a collapse of the eastern flank of the volcano in the early Holocene.

The second paper (Pareschi et al., 2006b; submitted June 12, 2006, revised July 19, 2006, accepted July 31, 2006; published September 21, 2006) presents a numerical modeling of the tsunami waves originated by the Santorini Late Bronze-Age volcanic eruption. The computer simulations indicate that the tsunami propagation is mainly confined to the Aegean Sea and that the tsunami impact was negligible outside it.

The third paper (Pareschi et al., 2006c; submitted August 4, 2006, revised October 6, 2006, accepted October 18, 2006; published November 28, 2006) presents computer simulations that support the occurrence of a catastrophic tsunami impacting all the eastern Mediterranean in the early Holocene. The tsunami was generated by a debris avalanche from Mt. Etna and triggered the well-known homogenites of the Ionian Sea.

The numerical modeling behind the computer simulations is certainly much more advanced than the simple calculations made by Kastens and Cita (1981), but the geological interpretation is questionable in several aspects, starting from the age of the event, which is incompatible with that of the Late Bronze-Age homogenite, calibrated by dozens of deep-sea cores. Pareschi et al.'s model is certainly interesting, but at present it has to be considered only as a working hypothesis.

Acknowledgments

The author has fond memories of the pioneering cruise with R/V Eastward in 1978. In particular, the author remembers the sedimentologist Rob Kidd and micropalaeontologist Gretchen Blechschmidt; both deceased prematurely. Vanni Aloisi, Angelo Camerlenghi, Werner Hieke, Kim Kastens, and Bianca Rimoldi have been active partners in the deep-sea homogenite research. Most sincere thanks to Isabella Premoli Silva for her invaluable help in finalizing the text and continuous support. The author thanks Tsunemasa Shiki for the invitation to write this contribution and Tsunemasa Shiki, Brian Jones, and A. J. van Loon for the careful revision that helped to improve and clarify the text.

References

Blechschmidt, G., Cita, M.B., Mazzei, R., Salvatorini, G., 1982. Stratigraphy of the western mediterranean and southern Calabrian Ridges, eastern mediterranean. Mar. Micropaleontol. 7, 101–134.

Bond, A., Sparks, R.S.J., 1976. The Minoan eruption of Santorini, Greece. J. Geol. Soc. Lond. 132, 1–16.

Cita, M.B., Aloisi, G., 2000. Deep-sea tsunami deposits triggered by the explosion of Santorini (3500 y BP), eastern mediterranean. Sediment. Geol. 135, 181–203.

Cita, M.B., Camerlenghi, A., 1985. Effetti dell'eruzione minoica di Santorino sulla sedimentazione abissale olocenica nel Mediterraneo Orientale. Rend. Sci. Accad. Lincei 8 77 (5), 177–187.

Cita, M.B., Rimoldi, B., 1997. Geological and geophysical evidence for a Holocene tsunami deposit in the eastern mediterranean deep-sea record. J. Geodyn. 24, 293–304.

Cita, M.B., Vergnaud-Grazzini, C., Robert, C., Chamley, H., Ciaranfi, N., D'Onofrio, S., 1977. Paleoclimatic record of a long deep sea core from the eastern mediterranean. Quat. Res. 8, 205−235.

Cita, M.B., Maccagni, A., Pirovano, G., 1982a. Tsunami as triggering mechanism of homogenites recorded in areas of the eastern mediterranean characterized by the "Cobblestone topography". In: Saxov, S., Niewenhuis, J.K. (Eds.), Marine Slides and Other Mass Movements. Plenum Press, London, pp. 233−261.

Cita, M.B., Bossio, A., Colombo, A., Gnaccolini, M., Salvatorini, G., Broglia, C., Camerlenghi, A., Catrullo, D., Clauzon, G., Croce, M., Giambastiani, M., Kastens, K.A., et al., 1982b. Sedimentation in the Mediterranean Ridge Cleft (DSDP site 126). Mem. Soc. Geol. Ital. 24, 427−442.

Cita, M.B., Beghi, C., Camerlenghi, A., Kastens, K.A., McCoy, F.W., Nosetto, A., Parisi, E., Scolari, F., Tomadin, L., 1984a. Turbidites and megaturbidites from the Herodotus abyssal plain (Eastern Mediterranean) unrelated to seismic events. Mar. Geol. 55, 79−101.

Cita, M.B., Camerlenghi, A., Kastens, K.A., McCoy, F.W., 1984b. New findings of Bronze age homogenites in the Ionian Sea. Geodynamic implications for the Mediterranean. Mar. Geol. 55, 47−62.

Cita, M.B., Camerlenghi, A., Rimoldi, B., 1996. Deep sea tsunami deposits in the eastern Mediterranean: new evidence and depositional models. Sediment. Geol. 104, 155−173.

Della Vedova, B., Rebesco, M., Cernobori, L., Aloisi, G., 1998. Immagini CHIRP ad alta risoluzione dell'omogenite di Santorini e delle deformazioni neotettonoche nello Jonio. GNGTS, 7° Congresso. CNR, Roma, pp. 31−32.

Escarmed Groupe, 1987. Le plateau Cyrenien: Promontoire Africain sur la marge Ionienne. Rev. Inst. Fran. Pétrole 42, 419−447.

Friedrich, W.L., 1999. Fire in the Sea. The Santorini Volcano: Natural History and the Legend of Atlantis. Cambridge University Press, Cambridge, p. 258.

Friedrich, W.L., Pichler, H., 1976. Radiocarbon dates of Santorini volcanics. Nature 262, 373−374.

Heiken, G., McCoy, F.J., 1984. Caldera development during the Minoan eruption. Thira, Cyclades, Greece. J. Geophys. Res. 89, 8441−8462.

Hersey, J.B., 1965. Sedimentary basins of the mediterranean sea. In: Whitard, W.F., Bradshaw, W. (Eds.), "Submarine Geology and Geophysics" Proceedings of the 17th Symposium. Colston Research Society, London, pp. 75−91.

Hieke, W., 1984. A thick Holocene homogenite in the Ionian abyssal plain (Eastern Mediterranean). Mar. Geol. 55, 63−78.

Hieke, W., Werner, F., 2000. The Augias megaturbidite in central Ionian Sea (Mediterranean) related to Holocene Santorini event. Sediment. Geol. 134, 205−218.

Kastens, K.A., 1984. Earthquakes as a triggering mechanism for debris flows and turbidites on the Calabrian Ridge. Mar. Geol. 55, 13−33.

Kastens, K.A., Cita, M.B., 1981. Tsunami induced sediment transport in the abyssal Mediterranean Sea. Geol. Soc. Am. Bull. 92, 591−604.

Keller, J., 1981. Quaternary tephrachronology in the Mediterranean region. In: Self, S., Sparks, R.S.J. (Eds.), Tephra Studies. Reidel, Dordrecht (The Netherlands), pp. 227−244.

Keller, J., Ryan, W.B.F., Ninkovich, D., Altherr, R., 1978. Explosive volcanic activity in the Mediterranean over the past 200,000 years as recorded in deep sea sediments. Geol. Soc. Am. Bull. 89, 591−604.

Marinos, G., Melidonis, N., 1971. On the strength of sean (tsunami) during the prehistoric eruption of Santorini. In: Int. Congr. Volcano Thera I, Proc. Acta Greek Archeol. Soc., pp. 277−382.

McCoy, F.W., 1978. The upper Thera (Minoan) ash in deep-sea sediments: distribution and comparison with other layers. In: Doumas, C. (Ed.), Thera and Aegean World, vol. II. Athens, pp. 57−78.

Mutti, E., Ricci Lucchi, F., Séguret, M., Zanzucchi, G., 1984. Seismoturbidites: a new group of resedimented deposits. Mar. Geol. 55, 103−116.

Ninkovich, D., Heezen, B.C., 1965. Santorini tephra. In: Whitard, W.F., Bradshaw, W. (Eds.), "Submarine Geology and Geophysics" Proceedings of the 17th Symposium. Colston Research Society, London, pp. 413−452.

Pareschi, M.T., Boschi, E., Mazzarini, F., Favalli, M., 2006a. Large submarine landslides offshore Mt. Etna. Geophys. Res. Lett 33 (5), L13302.

224 Chapter 13

Pareschi, M.T., Favalli, M., Boschi, E., 2006b. Impact of the Minoan tsunami of Santorini: simulated scenarios in the eastern mediterranean. Geophys. Res. Lett. **33** 6, L18607.

Pareschi, M.T., Boschi, E., Favalli, M., 2006c. Lost tsunami. Geophys. Res. Lett. **33** 6, L22608.

Pichler, H., Friedrich, W.L., 1978. Mechanism of the Minoan eruption of Santorini. In: Doumas, C. (Ed.), Thera and Aegean World, vol. II. Athens, pp. 15—31.

Rebesco, M., Della Vedova, B., Cernobori, L., Aloisi, G., 2000. Acoustic facies of Holocene megaturbidites in the eastern Mediterranean. Sediment. Geol. 134, 65—74.

Reeder, M., Rothwell, R.G., Stow, D.A.V., Kahler, G., Kenyon, N.H., 1998. Turbiditic flux, architecture and chemostratigraphy of the Herodotus basin, Levantine sea, SE mediterranean. In: Stocker, M.S., Evans, D., Cramp, A. (Eds.), Geological Processes on Continental Margins: Sedimentation Mass-Wasting and Stability, vol. 129. Geological Society, London, Special Publications, pp. 19—41.

Rimoldi, B., 1989. Upslope turbidites in the outer flank of the Mediterranean Ridge facing the Sirte abyssal plain (late Pleistocene, eastern mediterranean). Boll. Oceanol. Teor. Appl. 7, 229—249.

Rothwell, R.G., Thomson, J., Kalher, G., 1998. Low-sea-level emplacement of a very large late Pleistocene "megaturbidite" in the western Mediterranean Sea. Nature 392, 377—380.

Rothwell, R.G., Reeder, M.S., Anastasakis, G., Stow, D.A.V., Thomson, J., Kalher, G., 2000. Low-sea-level stand emplacement of megaturbidites in the western and eastern Mediterranean Sea. Sediment. Geol. 135, 75—88.

Shiki, T., Cita, M.B., Gorsline, D., 2000. Sedimentary features of seismites, seismo-turbidites and tsunamiites—an introduction. Sediment. Geol. 135, 1—5.

Troelstra, S., 1987. Late quaternary sedimentation in the Tyro and Kretheus basins, southeast of Crete. Mar. Geol. 75, 77—91.

Vergnaud-Grazzini, C., Ryan, W.F.B., Cita, M.B., 1977. Stable isotopic fractionation, climate change and episodic stagnation in the Eastern Mediterranean during the Late Quaternary. Mar. Micropaleontol. 2, 353—370.

Watkins, N.D., Sparks, R.S., Sigurdsson, H., Huang, T.C., Federman, A., Carey, S., Ninkovich, D., 1978. Volume and extent of the Minoan tephra from Santorini volcano: new evidence from deep-sea sediment cores. Nature 271, 122—126.

Yokoyama, L., 1978. The tsunami caused by the prehistoric eruption of Thera. In: Doumas, C. (Ed.), Thera and Aegean World, vol. II. Athens, pp. 277—283.

CHAPTER 14

Tsunami-related sedimentary properties of mediterranean homogenites as an example of deep-sea tsunamiite

T. Shiki[1], M.B. Cita[2]
[1]Uji, Kyoto, Japan; [2]Dipartimento di Scienze, della Terra 'Ardito Desio', Università di Milano, Milano, Italy

1. Introduction

There is only one example of a deep-sea tsunamiite known in sedimentology, as far as modern tsunamis and their sediments are concerned. That is the "homogenite" of Kastens and Cita (1981), Blechschmidt et al. (1982), Cita et al. (1984, 1996), Cita and Rimoldi (1997), and Cita and Aloisi (2000). The layer is characterized by a very thick, gray, visually structureless hemipelagic mud, with a normally graded sandy base and a sharp basal contact; it has been interpreted as the sedimentary expression of the tsunami induced by the collapse of the Santorini caldera after the gigantic eruption in the Bronze Age, about 3500 years before present.

The occurrence of homogenites from two different settings was pointed out when Cita et al. (1996) summarized the data known at the time and provided a depositional model of these layers. More recently, Cita and Aloisi (2000) and Hieke and Werner (2000) discussed these homogenites in more detail and stressed the different features of the tsunami-induced sediments from the two different settings.

It was also stated that the near-bottom oscillating currents and the pressure pulse of a tsunami were sufficient to stir up the soft, unconsolidated pelagic drape from ridges on the seafloor (Kastens and Cita, 1981). The gigantic tsunami energy can mobilize the shallow-water sediments of the distant African shelf, thus inducing a turbidity current (Cita and Aloisi, 2000; Cita et al., 1984; Hieke and Werner, 2000). On the other hand, as discussed by Minoura et al. (2000), mere shock waves by the eruption could not have affected the sediments enough to stir them up from the deep-sea floor. Anyhow, the

Tsunamiites. https://doi.org/10.1016/B978-0-12-823939-1.00014-8
Copyright © 2021 Elsevier B.V. All rights reserved.

existence of the deep-sea sediments resulting from the "Santorini tsunami" cannot be doubted, and the Mediterranean model can well be applied elsewhere.

In the present contribution, the authors will further discuss, on the basis of the currently available information, the characteristic features of the deep-sea tsunamiites, which suggest an important role of the differences in density and settling of suspended load during the transport and deposition of the sediments. They will also try to shed light on the best approach for future studies of both recent and older deep-sea tsunamiites.

2. Setting, types, and distribution of homogenites

The essential aspects regarding the sedimentary environments, settings, and distribution of homogenites are reviewed by Cita (this volume), but will be mentioned briefly here also.

Homogenites occur over large areas in the Eastern Mediterranean, though limited to locations with a flat seafloor. The most remarkable feature of homogenites is their peculiar occurrence, commonly with an exceptional thickness and with peculiar lithofacies characteristics that will be discussed further also.

Cita and Aloisi (2000) distinguished two basic types of homogenites: "type A homogenites" and "type B homogenites" (Figs. 14.1 and 14.2). The homogeneous character is less pronounced in the type B homogenite, particularly since a remarkable, fining-upward sandy interval exists at its base.

The distribution of these two types of homogenite depends on their settings, including the possible presence of topographic obstacles and their possible rebound effects. Type A homogenites are deposited in small perched basins of the Calabrian and Western Mediterranean Ridge. The presence of type B homogenites has been confirmed for the Sirte abyssal plain, the Ionian abyssal plain, and the Western Herodotus Trough. A fundamentally similar conclusion was reached by Hieke and Werner (2000). Their "isolated turbidite" coincides with the type A homogenite of Cita and Aloisi (2000), whereas their "Augias turbidite" is almost identical to the type B homogenite of Cita and Aloisi (2000).

Type A homogenites are of local origin: their source areas are "deep-water highs" (ridges). In contrast, the provenance of the type B homogenites, that is, the Augias turbidite, is a distant Northern African shallow-water shelf (Fig. 14.1). Other megaturbidites have been recorded acoustically in the Eastern Mediterranean (Rebesco et al., 2000; Rothwell et al., 2000), but have not been confirmed yet by coring.

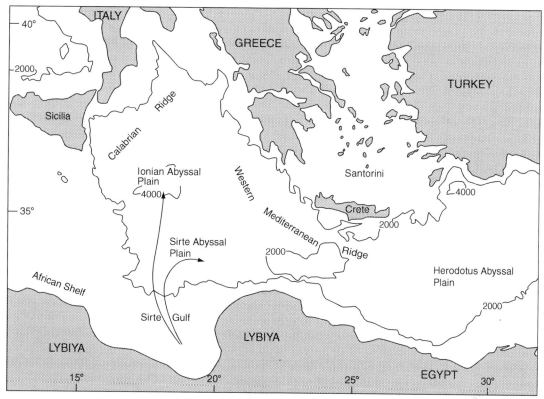

Figure 14.1
Submarine topography of the Mediterranean homogenite sedimentation area. *Arrows* show inferred paths of the sediment gravity flow that induced deposition of the type B homogenites after arrival on the abyssal plains. *Based on Cita, M.B., Aloisi, G., 2000. Deep-sea tsunami deposits triggered by the explosion of Santorini (3500yBP), Eastern Mediterranean. Sediment. Geol. 135, 181–203.*

3. Sedimentary properties

3.1 Structures and grain-size distribution

The descriptive name "homogenite" refers to the homogeneous character of the beds, especially with respect to their mud unit (division). No "Bouma sequence" or other sedimentary structures have been recognized visually, with the exception of a fining-upward sandy base of limited thickness. In fact, the grain-size distribution changes from the sandy division to the thick muddy division (Fig. 14.2). Faint vertical fluctuations in the granulometry have been found when grain-size analyses were carried out, as will be mentioned and discussed in the following. However, the fluctuations cannot be detected by the naked eye.

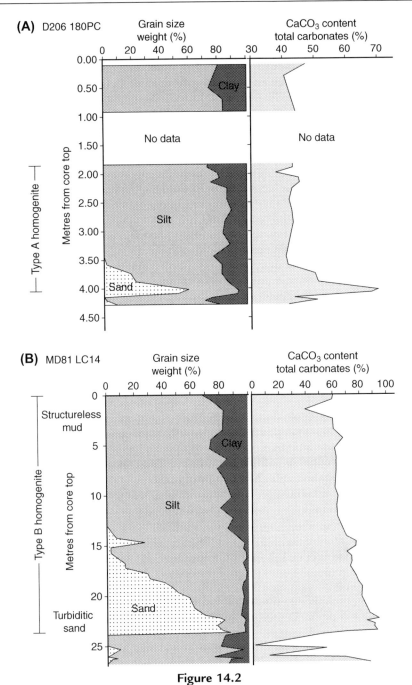

Figure 14.2
Sedimentary petrographic characteristics (sand, silt, and clay content; CaCO$_3$ content) of some homogenites. (A) Type A homogenite and (B) type B homogenite. *Based on Cita, M.B., Aloisi, G., 2000. Deep-sea tsunami deposits triggered by the explosion of Santorini (3500yBP), Eastern Mediterranean. Sediment. Geol. 135, 181–203.*

Some arguments might be put forth with respect to the exceptional thickness of the homogenites, and some may say that no sedimentological property is diagnostic for the deep-sea homogenites of the Mediterranean. In fact, numerous megaturbidites from all over the world attain thicknesses from a few to tens of meters. It must be noted, however, that the thickness of the muddy division of type B homogenites occupies as much as two-thirds of up to some tens of meters of the homogenites. In contrast, both Holocene and older megaturbidites commonly consist mainly of coarse-grained, pebbly and sandy divisions with only a thin muddy top part (d and e division of the Bouma sequence) (see Bouma, 1962; Harata et al., 1969).

Actually, the most important difference between the Mediterranean homogenites and the common thick megaturbidites is the relationship between the vertical change in granulometry in each division and the thickness of the divisions. This has not commonly been stressed sufficiently well.

Detailed grain-size analyses have not been performed on the homogenites. The ratios between the coarse, medium, and fine sands, silts, and clays, however, have been measured for tens of horizons in some homogenites (Fig. 14.2). In the type B homogenites, grading is apparent, even with the naked eye, in the basal sandy part, which is relatively thick compared to that of the type A homogenites. Fig. 14.2B also shows a decrease of the mode height of sands and grading of the mode of fine sands to coarse silts, resulting in an upward decrease of the sand/silt ratios in the lower part (lower division).

On the other hand, variations in grain-size distribution and grading are insignificant in the thick, upper, fine-grained division (the homogeneous division) of type B homogenites, showing profound mixing of a wide range of grain sizes. Faint fluctuations in the granulometry can, however, be sometimes found in this upper division. In addition, a thin sandy level (layer) occurs just above the uppermost, fine-grained part of the basal graded sandy division in core MD81LC14. This thin level has been assigned to reverse flows generated in the narrow Western Herodotus Trough (Cita and Aloisi, 2000).

More faint fluctuations in the grain-size distribution seem to occur in the upper, fine-grained mud division of both types of homogenites (Fig. 14.2). One of the present authors suggests that this fluctuation in grain-size distribution reflects the shuttle movement of the tsunami flow. This will be discussed in more detail in a following section.

As expected, the sedimentary properties at the basal contact of the Mediterranean homogenites vary in accordance with the transport mechanism. The base of type A homogenites is without obvious erosion (Cita and Aloisi, 2000). Probably, stirring up of the bottom sediments did not change the local bottom topography. On the other hand, type B homogenites (Augias turbidite) seem to have either a nonerosional or an erosional base, depending on the competence of the turbidity current.

230 Chapter 14

3.2 Constituents and their relation to other features

Careful examination of the constituent materials must help to clarify the sedimentary characteristics of tsunamiites, including the two types of homogenites, in the same way as done for other common sediments. A property that the two types of homogenites have in common is their calcareous composition. The origin and deposition of the type A homogenite are revealed by pelagic foraminifers, which make up the thin sandy base, and by the dominant silty fractions, which are composed mainly of calcareous nannofossils. On the other hand, fossils of the thick, sandy base of the type B homogenites, such as shell fragments, gastropods, bryozoans, and benthic foraminifers, indicate an origin in a shallow-water environment.

The constituents of both the coarse and fine-grained divisions of the two types of homogenites are thus predominantly calcareous fossils and their fragments. The sediments lack large (pebble-sized) calcareous fossils as well as noncalcareous (lithic) pebbles. This property is typical of the Mediterranean homogenites.

4. Discussion of sedimentological problems

4.1 Erosion of the deep-sea bottom

Tsunamis have the nature of a very shallow-water wave (long wave) in which the water velocity, together with the gigantic energy, is almost the same throughout the vertical water column, from the surface downward (see also: Shiki et al., this volume; Sugawara et al., this volume). The effect of tsunamis and tsunami-induced currents on deep-sea sediments is, however, different and the subject of sedimentological debate. The data available about the Mediterranean homogenites might therefore be most valuable, particularly for the discussion on erosion by deep-sea tsunamis.

Present-day deep ocean floors of the Pacific, Atlantic, and Indian Oceans do not show any sign of erosion or transport by tsunamis. On the other hand, transport and deposition by some tsunami-induced currents of the Miocene age were evidenced for the upper bathyal environment in central Japan (Shiki and Yamazaki, 1996; Yamazaki et al., 1989). More recently, Goto et al. (2002) and Goto et al. (this volume) estimated the depth of the ocean floor where stirring-up, erosion, and transportation of sediments occurred at the K/Pg boundary due to the impact-induced tsunami event. In this context, the statements by Kastens and Cita (1981) and Cita et al. (1996) are of interest. They stated that the pelagic sediments of the small basins between highs and ridges at a depth of 2000 m were strongly affected by the Santorini tsunami. On the other hand, however, the pelagic sediments of the 3000 m deep abyssal plain show no proof of erosion by this tsunami. Seemingly, some tsunamis can affect the seafloor at a depth of few thousands meters, but not deeper. The critical depth depends on parameters such as the tsunami's energy, the bottom topography, and so on. Although several works report the results of hydrodynamic investigations into this question, there is still little agreement as to the effect of tsunami currents on the deep-sea floor.

4.2 Genesis of the sandy division of the type B homogenite

Another, as yet unsolved, problem concerns the genesis of the sandy division of the type B homogenite. It is only logical that relatively coarse sandy material tends to concentrate in the basal part of the flow, forming a density flow that has an only slightly higher density than water at the sea bottom. The sandy material flows down the continental slope because of the coefficiency of the relative density and the ebb currents. The transport mechanism of the lower sandy division of type B homogenites thus must have been some kind of sediment gravity flow. Cita and Aloisi (2000), Hieke and Werner (2000), and especially Rebesco et al. (2000) used the term "turbidite" to describe the type of mass flow.

On the other hand, the type B homogenites are very different from common turbidites, which pile up, alternating with background, nonevent sediment beds to form a succession of commonly hundreds of meters thickness. In contrast, homogenites are rare (only one observed example) in abyssal plain successions. The sandy division of the type B homogenite is remarkably thin compared with the sandy divisions including divisions a and b (and c and d if they are sandy) of a common thick turbidite (megaturbidite). In addition, the constituent grains of the sandy part are distinctly finer than those of the common megaturbidites.

We suggest that the density of the flow was low because of dilution by a large amount of water, although it was enough to make a density-flow surge, which flowed down from the shallow shelf to the deep abyssal plain and changed its direction due to obstacles, as proposed by Cita and Aloisi (2000).

The sedimentographical differences of these two kinds of thick density-flow deposits also may or may not have been caused by different source materials. More detailed comparative investigations, especially of their constituent materials, including the grain-size distribution throughout their beds, are needed to clarify the precise nature of these flows.

4.3 High tsunami-induced suspension cloud

As stressed earlier, the thick development of the structureless fine-grained section (division) is the characteristic property that the two types of homogenites have in common. The thickness must be ascribed to the very large amounts of suspended sediment stirred up by the Santorini tsunami from a topographic high (type A homogenite) or flowed down from the shallow-water shelf (type B homogenite). That is to say, the cloud of suspended sedimentary particles possibly had a large vertical extent in the depositional area of the deep Mediterranean. This is especially true for the genesis of the type B homogenite (Fig. 14.3).

A sediment gravity flow surge accompanied by a cloud of fine-grained particles can be compared to the lower part (basal surge) of some pyroclastic flow surges and with the upper part of a glowing cloud. The cloud develops throughout the accommodation space, from just above the sandy gravity flow (density flow) to the shallow part of the water column; the vertical range decreased slowly with time.

The cloud and its materials, together with the density-flow materials, travel far across the continental slope area, until they arrive above the deep-sea floor, that is, the abyssal plain, and the materials settle there forming the homogeneous fine-grained sediments.

The amount of sediment settling from the cloud is enormous, and the depositional thickness attained is tens of meters, as mentioned earlier. Much time (order of days) is spent before all the fine-grained particles of the cloud can settle on the sea bottom and create the unusually thick sediments.

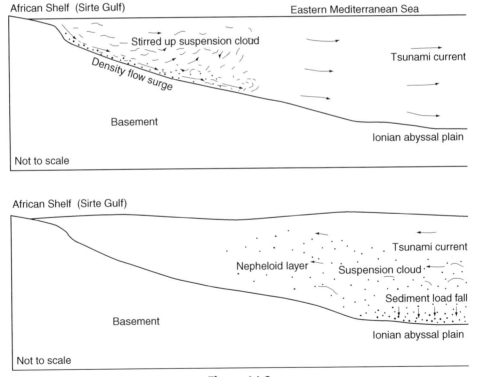

Figure 14.3
Genetic model of the type B Mediterranean homogenite. Emphasis is placed on (1) density-flow surge and suspension cloud induced by the Santorini tsunami and (2) settling from the suspension cloud and deposition of the type B homogenite on the Ionian abyssal plain.

It is logical that the silt-sized grains and clays are mixed well in the suspended cloud, even in the top part of the cloud. This is reflected by the homogeneous character of the resulting sedimentary succession, especially by the very gentle grading in the upper division of the bed (Fig. 14.2). On the other hand, sand grains fall rather rapidly and build the graded sandy basal part of the homogenites. An upward decrease of the mode height of the coarse tail (sand part) of the grain-size distribution is notable in the basal one-third part of the type B homogenite mentioned earlier.

4.4 Accumulation rate of the suspended load

Lowe (1988) has pointed out that the accumulation rate of suspended load must be included as an independent variable in the mechanical interpretation of sedimentary structures, and he has also shown a preliminary stability-field diagram for turbidity currents.

One of us (T.S.) suggests that Lowe's approach is applicable to the type B homogenates. The sandy division of the bed (the grains have a diameter of <1 mm), which shows no sign of traction-current-induced structures, implies settling from a suspension cloud with highly concentrated particles. The suspended particles originating from the African coast must sink throughout the diluted density flow, as indicated by Lowe for common turbidity currents. There is no doubt that the settling from suspension became definitively faster when the flow had arrived above the abyssal plain (Fig. 14.3). The problem is that it can still only be guessed where, when, and how the accumulation rate changed in the course of the transposition of the sandy and silty parts of the surge and cloud of the flow.

It must be kept in mind that the distribution of homogenites is limited to flat bottoms, whereas the basin flanks have been eroded as Kastens and Cita (1991) and Blechschmidt et al. (1982) have stated. Do the sediments that were deposited directly from the turbiditic, but diluted, base surge really exist somewhere? Do those sediments show some Bouma sequence? These problems have to be investigated in the future. Field work to obtain information and sediment samples for this study must be planned around the foot of the continental slope where the angle of the slope changes sometimes abruptly, sometimes gently. In reference to the earlier-proposed cloud, the classical terminology of the "nepheloid layer" concept, discussed by Ewing and Thorndike (1965), may have to be remembered.

In spite of the questions that still need to be answered, it is apparent that homogenites provide good material to evaluate the idea of Lowe (1988), which stresses the importance of suspended-load settling in the genesis of the sedimentary structures of turbidites.

4.5 Records of shuttle movement and backwash current of tsunamis

Alternating occurrences of run-up and backwash of water, the so-called "shuttle movement," are one of the most peculiar features of tsunamis. Whether the Mediterranean

234 Chapter 14

homogenites do or do not record evidence of a shuttle movement is an interesting question. The faint sawtooth-like changes in the grain-size distribution and the carbonate content in the fine-grained homogeneous upper division suggest repeated deposition with a high accumulation rate of relatively coarse calcareous material due to the shuttle movements.

On the other hand, the density flow discussed earlier also must have met the pulsed run-up currents of the tsunami wave repeatedly during its travel from the shelf to the abyssal plain, because the distance is so long that several tsunami pulses must have passed in the meantime. Obviously, the pulsed backwash currents downslope must have been more powerful than the run-up currents, and they must have played some role in bringing down the sediments and depositing the type B homogenite. Though it can hardly be deduced from the sedimentary record, the faint saw-toothed changes in grain size and the mineral composition of the sandy lower division possibly reflect this disturbance of the gravity flow. The cloud plumed up from the flow was probably also affected by the pulses of the tsunami waves and resulted in the faint changes in grain size in the fine-grained upper division.

There is no general agreement as to whether this delicate change in the grain-size distribution is only apparent, caused by measurement inaccuracy. More precise analysis is therefore needed to examine other possible causes.

5. Comparison with other deep-sea tsunamiites

As mentioned earlier, the Mediterranean homogenites are the only examples of deep-sea tsunamiites known in sedimentology, as far as modern sediments are concerned. However, some meteorite-impact-related sediments from the geological past (e.g., Alvarez et al., 1992; Bourgeois et al., 1988; Pong et al., 1992; Schlager et al., 1984; Smit et al., 1996) provide noteworthy examples of "homogenite-like deep sea beds." The K/Pq boundary deposits that crop out onshore, or that were drilled from the seafloor, around the Gulf of Mexico and the Brazilian coast exhibit the following three remarkable combinations of sedimentary features that are identical to those of the Mediterranean homogenites.

1. For example, the upper part of the K/Pq-boundary deep-sea formation in northern Cuba, which is a few meters up to 180 m thick, is characterized by a homogeneous appearance, it has coarse-tail normal grading, it comprises abundant reworked shallow marine fossils and so on (Goto et al., 2002; this volume). These sedimentary properties are considered to result from supply from a shallow sea area by some density currents. As deposition from a suspension cloud is the common mechanism forming the homogeneous upper part of the deep-sea Mediterranean tsunamiites, settling of suspended load must have played an important role also in the genesis of the homogeneous division of the probable tsunamiites at the K/Pq boundary.

2. As for the Mediterranean homogenites, the shuttle movement of water currents, which is one of the most peculiar features of tsunamis, is only faintly present in the form of a saw-toothed change in grain size and carbonate content, as described earlier. In contrast, grain-size fluctuations can be clearly observed in the K/Pq-boundary deep-sea tsunamiite. They are considered to reflect an intermittent lateral current by a series of tsunami waves (see, Goto et al., 2002, this volume; Takayama et al., 2000). This lateral current in the deep-sea environment can possibly be ascribed to the extremely large energy of the tsunami. Anyway, both the Mediterranean and the K/Pq-boundary tsunamiites have the sign of lateral currents in common. The problem concerning the deposition from the tsunami's lateral currents in the deep-sea environment is discussed by Goto et al. (this volume).

3. As described earlier, a predominance of calcareous material is a distinct feature of the Mediterranean homogenites. The K/Pq-boundary deep-sea tsunamiites are also more or less calcareous. It is obvious that the calcareous character of these deep-sea sediments reflects the constituent materials in the provenance (source) area of the sediments. One should realize, however, that this calcareous character is not a prerequisite for identifying a deep-sea tsunamiite: many noncalcareous deep-sea tsunamiites may occur in geological record.

6. Concluding remarks

The Mediterranean homogenites provide much more interesting information than has been recognized till now. They are the only known Holocene examples of abyssal tsunamiites. The different occurrences of the two types of homogenites suggest that the water depth in which tsunami waves can affect the deep-sea floor and pile up bottom sediments is limited. Sediment-graphical and sedimentological features of the type B homogenites indicate behavior as a "pyroclastic flow cloud" for the transport and depositional mechanisms of the tsunami-induced and the tsunami-affected sediments. The homogeneous characteristics in the grain size and constituents of the beds suggest the importance of the role of settling of suspended load from the cloud. The shuttle movement of tsunami currents may be reflected in the faint vertical changes in grain-size distribution within the homogenites. These special characteristics suggest a need for further detailed studies of the Mediterranean homogenites.

Appendix: reflections on terminology

Any discussion of terminology is out of scope in the present contribution. Some readers might ask, however, whether the terms "homogenite" and "tsunamiite" are, or are not, applicable to the type B homogenite. The reason behind such a question is that the

236   Chapter 14

sediment is essentially generated from some kind of "sediment gravity flow," and even the term "megaturbidite" has been given to one of them, as mentioned earlier.

In this context, it must be pointed out that the flow was induced by a specific tsunami event and was affected by its backwash current during its flow downslope, although the backwash current was not very powerful (as discussed earlier).

In this context, the original definition of the term "tsunamiite" is to be remembered. The word "tsunamiite" allows usage in a wider sense. Thus, the type B homogenite, which was induced by a tsunami and which was affected by a tsunami current, can be regarded as a kind of tsunamiite (see Shiki et al., 2008, this volume).

The term "homogenite" is a specific name first used by Kastens and Cita (1981) and redefined by Cita and Aloisi (2000) as deposits induced by the Santorini tsunami event. The name is descriptive, emphasizing the homogeneous aspect of the sediments. It does not imply any sedimentation mechanism. Therefore, this term is applicable to the sediment gravity flow deposits of the type B homogenites.

On the other hand, we do not oppose the application of the term "megaturbidite," because the flowdown mechanism of the flow must have been essentially a kind of "sediment gravity flow." Furthermore, the acoustic properties of the sediments seem to resemble those of other megaturbidites underneath (Rothwell et al., 2000).

Acknowledgments

The present contribution is based on data obtained during several research cruises in the Mediterranean. The authors thank all cruise and on-board scientists who helped them and who allowed them to use the data and related information. Thanks are due to T. Tachibana for his discussion and preparation of the text figures. The authors are also indebted to K. Suzuki, K. Goto, and other sedimentologists who kindly checked the manuscript and gave valuable suggestions.

References

Alvarez, W., Smit, J., Lowrie, W., Asano, F., Margolis, S.V., Clays, P., Kastner, M., Hildebrand, A.R., 1992. Proximal impact deposits at the Cretaceous-Tertiary boundary in the Gulf of Mexico: a restudy of DSDP, leg 77 sites 536 and 540. Geology 20, 697−700.

Blechschmidt, G., Cita, M.B., Mazzei, R., Salvatorini, G., 1982. Stratigraphy of the western Mediterranean ridge and Southern Calabrian ridge, eastern Mediterranean. Mar. Micropaleontol. 7, 101−104.

Bouma, A.H., 1962. Sedimentary of Some Flysch Deposits: A Graphic Approach to Facies Interpretation. Elsevier, Amsterdam, p. 168.

Bourgeois, J., Hansen, T.A., Wiberg, P.L., Kauffman, E.G., 1988. A tsunami deposit at the Cretaceous-Tertiary boundary in Texas. Science 241, 567−570.

Cita, M.B., Aloisi, G., 2000. Deep-sea tsunami deposits triggered by the explosion of Santorini (3500yBP), Eastern Mediterranean. Sediment. Geol. 135, 181−203.

Cita, M.B., Rimoldi, B., 1997. Geological and geophysical evidence for the Eastern Mediterranean deep-sea record. J. Geodyn. 24, 293−304.

Cita, M.B., Camerlenghi, A., Kastens, K.A., McCoy, F.W., 1984. New findings of Bronze age homogenites in the Ionian sea. Geodynamic implications for the Mediterranean. Mar. Geol. 5, 47−62.

Cita, M.B., Camerlenghi, A., Rimoldi, B., 1996. Deep sea tsunami deposits in the Eastern Mediterranean: new evidence and depositional models. Sediment. Geol. 104, 155−173.

Ewing, M., Thorndike, E.M., 1965. Suspended matter in deep ocean water. Science 147, 1291−1294.

Goto, K., Tajika, E., Iturralde-Vinent, M.A., Takayama, H., Nakano, Y., Yamamoto, S., Kiyokawa, S., Garcia-Delgado, D., Rojas-Consuegra, R., Oji, T., Matsui, T., 2002. Depositional mechanism of the Penalver Formation: the K/T boundary deep-sea tsunami deposit in North Western Cuba. In: 16th International Sedimentological Congress Abstract Volume, pp. 122−124.

Harata, T., Shiki, T., Tokuoka, T., 1969. Sedimentary structures of sandy flysch deposits at the coast of the Kirimezaki, Wakayama Prefecture (part I)—internal sedimentary structures and Bouma's model. Mem. Fac. Educ. Wakayama Univ. Nat. Sci. 19, 31−36. +Appended figure (in Japanese).

Hieke, W., Werner, F., 2000. The Augias megaturbidite in central Ionian sea (Mediterranean) related to Holocene Santorini event. Sediment. Geol. 134, 205−218.

Kastens, K.A., Cita, M.B., 1981. Tsunami induced sediment transport in the abyssal Mediterranean Sea. Geol. Soc. Am. Bull. 92, 591−604.

Lowe, D.R., 1988. Suspended-load fallout rate as an independent variable in the analysis of current structures. Sedimentology 35, 765−776.

Minoura, K., Imamura, F., Kuran, U., Nakamura, T., Papadopoulou, G.A., Takahashi, T., Yalciner, A.C., 2000. Discovery of Minoan tsunami deposits. Geology 28, 59−62.

Polonia, A., Bonatti, E., Camerienghi, A., Lucci, R.G., Panieri, G., 2013. Medierranean megaturbidite triggered by the AD365 Crete earthquak and tsunami. Sci. Rep. 3, 1285.

Polonia, A., Vaiani, S.C., Nelson, C.H., Ramano, S., Gasparotto, G., Gasperini, L., 2015. AGU 2015 Fall Meeting, Abstracts.

Pong, C.W., Powars, D.S., Poppe, L.J., Mixon, R.B., Edwards, I.E., Folger, D.W., Bruce, S., 1992. Deep Sea Drilling Project Site 642 Bolide event: new evidence of a late Miocene impact-wave deposit and a possible impact site, U.S. east. Geology 20, 771−774.

Rebesco, M., Della Vedova, B., Cernobori, L., Aloisi, G., 2000. Acoustic facies of Holocene megaturbidites in the Eastern Mediterranean. Sediment. Geol. 134, 65−74.

Rothwell, R.G., Reeder, M.S., Anastasakis, G., Stow, D.A.V., Thomson, J., Kalher, G., 2000. Low-sea-level stand emplacement of megaturbidites in the western and Eastern Mediterranean Sea. Sediment. Geol. 135, 75−88.

Schlager, W., Buffler, R.T., Angstadt, D., Phair, R., 1984. Geologic history of the Southeastern Gulf of Mexico. Init. Rep. Deep-Sea Drill. Proj. 77, 715−738.

Shiki, T., Yamazaki, T., 1996. Tsunami-induced conglomerates in Miocene upper bathyal deposits, Cita Peninsula, central Japan. Sediment. Geol. 104, 175−188.

Shiki, T., Tachibana, T., Fujiwara, O., Goto, K., Nanayama, F., Yamazaki, T., 2008. Characteristic features of tsunamiites. In: Shiki, T., Tsuji, Y., Yamazaki, T., Minoura, K. (Eds.), Tsunamiites—Features and Implications. Elsevier, Amsterdam, pp. 319−340.

Smedile, A., De Martini, P.M., Pantosti, D., Beiiucci, L., Del Carlo, P., Gasperini, L., Pirrotta, C., Polonia, A., Boschi, E., 2011. Possible tsunami signatures from an intergrated study in the augusta Bay offsore (Easern Sicily −Italy). Mar. Geol. 281, 1−13.

Smedile, A., Molisso, F., Chaque, C., Lorio, M., De martini, P.M., Collins, P.F.F., Sagnotti, L., Pantosti, D., 2019. New coring study in Augusta Bay expands understanding of offshore tsunami deposits (Eastern Siciriy, Italy). Sediomentology. https://doi.org/10.1111/sed.12581.

Smit, J., Roep, T.B., Alvarez, W., Claeys, P., Grajales-Nishimura, J.M., Bermudez, J., 1996. Coarse-grained, clastic sandstone complex at the K/T boundary around the Gulf of Mexico: deposition by tsunami waves induced by the Chicxulub impact?. In: Ryder, G., Fastovsky, D., Gartner, S. (Eds.), Cretaceous-Tertiary Event and Other Catastrophes in Earth history, vol. 307, pp. 151−182. Geol. Soc. Am. Spec. Pap.

Takayama, H., Tada, R., Mastsi, T., Iturralde-Vinent, M.A., Oji, T., Tajika, E., Kinokawa, S., Grcia, D., Okuda, H., Haseqawa, T., Toyada, K., 2000. Origin of the Penalver Formation in Northwestern Cuba and its relation to K/T boundaty impact event. Sediment. Geol. 135, 295−320.

238 Chapter 14

Yamazaki, T., Yamaoka, M., Shiki, T., 1989. Miocene offshore tractive current-worked conglomerates-Tubutegaura, Chita Peninsula, central Japan. In: Taira, A., Masuda, M. (Eds.), Sedimentary Facies and the Active Plate Margin. Tera Publ, Tokyo, pp. 438–494.

Additional comment to Chapter 14

T.Shiki

For the sake of better comprehension for readers, the writer will refer to "Himogenite" in Chapters 13 and 14 as a very thick homogeneous deep-sea sedimentary layer situated at a unique horizon developed under the Eastern Mediterranean Sea floor. It is evident now that some other thick clastic layers also develop under the Mediterranean Sea floor and this was documented by the Bologna Group of sciences and others (i.e., Polonia et al., 2013, 2015, Smedile et al., 2011, 2019, and many others cited in the reference of the Chapter 14). These articles, however, are not referred in the present two chapters of this second edition, which mainly focuses on original Homogenite.

I, Shiki, think that referring to and considering the original concept of homogenite, which was the first put forward by M.B. Cita herself, is still worthy of further discussion in the field of tsunami sedementology. This sdiment's (homogenite's) features could possibly reveal some referential information, which could shed some light on the studies of deep-sea floor "tsunamiites." Furthermore, I intended to show here, in Chapter 14, that the vertical diversities of the sedimentary features such as grain-size distribution, mineral and micro-fossil composition through each unit layer, reveal sedimentary mechanism of the tsunami-induced deposits that are called "Tsunamiites."

Any general description of the recent accumulated knowledge in concern with the tsunami-generated sediments in the Mediterranean Sea is out of the subject matter of this present volume of our book. I think it is needless to say that studies of such Homogenites in a wider science, which are commonly called "Unifites," are also significant and interesting subject for researchers and students.

Further reading

Goto, K., Tada, R., Tajika, E., Matsui, T., 2008. Deep-sea tsunami deposits in the proto-Caribbean sea at the Cretaceous/Tertiary boundary. In: Shiki, T., Tsuji, Y., Yamazaki, T., Minoura, K. (Eds.), Tsunamiites—Features and Implications. Elsevier, Amsterdam, pp. 251–275.

Sugawara, D., Minoura, K., Imamura, F., 2008. Tsunamis and tsunami sedimentology. In: Shiki, T., Tsuji, Y., Yamazaki, T., Minoura, K. (Eds.), Tsunamiites—Features and Implications. Elsevier, Amsterdam, pp. 9–49.

CHAPTER 15

Volcanism-induced tsunamis and tsunamiites

Y. Nishimura

Institute of Seismology and Volcanology, Hokkaido University, Sapporo, Japan

1. Introduction

As can be deduced from the 2004 Indian Ocean disaster, tsunamis can cause large amount of casualties. For the assessment of future risk posed by tsunamis, it is necessary to have sufficient information for understanding past tsunami events. The work on tsunami deposits has provided new information on palaeotsunami events, for instance, about their recurrence interval and the size of the tsunamis (e.g., Atwater et al., 2005; Nanayama et al., 2003). Tsunamis are caused not only by earthquakes but also by volcanic eruptions. Because volcanogenic tsunamis occur less frequently than earthquake-related tsunamis, it is even more important to find and study geological evidence for past eruption-related tsunami events. Here, I review published works on volcanogenic tsunamis and their onshore deposits. This is the first review of tsunami deposits of this type.

2. Volcanism-induced tsunamis

Simkin and Siebert (1994) compiled volcanic activity from all over the world for the last 10,000 years. They listed tsunamis of volcanic origin from 42 volcanoes and 62 eruptions, with the largest number of eruptions (5) belonging to the Stromboli in Italy and the Taal in the Philippines. After the book by Simkin and Siebert (1994) had been published, four more volcanism-induced tsunami events have been reported; these are associated with eruptions of the Rabaul in 1994 (Nishimura et al., 2005), the Karymsky in 1996 (Belousov and Belousova, 2001), the Stromboli in 2002 (Bonaccorso et al., 2003), and the Montserrat in 2003 (Pelinovsky et al., 2004). Thus, a total of 66 events from 44 volcanoes have to be considered. It should be noted that 32 of these events (48%) occurred at volcanic islands or coastal volcanoes in western Pacific countries such as Indonesia, the Philippines, and Papua New Guinea. Tsunamis are one of the major volcanism-induced disasters for these countries. Fig. 15.1 shows volcanoes that have generated tsunamis.

Tsunamiites. https://doi.org/10.1016/B978-0-12-823939-1.00015-X
Copyright © 2008 Elsevier B.V. All rights reserved.

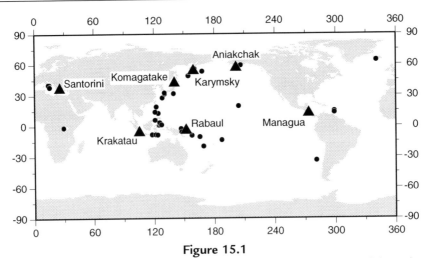

Figure 15.1
Locations of volcanoes that have generated tsunamis (*closed circles*) and of the volcanogenic tsunami deposits studied (*closed triangles*).

Volcanism-induced tsunamis are rare events, but they have caused huge damage not only around the volcanoes themselves, but also farther away. The 1883 Krakatau tsunami was one of the worst volcanic disasters in the world. More than 36,000 people were killed by the destructive tsunami in Indonesia and the surrounding countries. Among the 28 volcanic disasters after 18th century, which caused more than 1000 human deaths, 3 (11%) were caused by tsunamis: Krakatau (1883), Unzen (1741), and Oshima-Ohshima (1741). The death toll from these three tsunamis is 53,000, which is 21% of the total loss by the 28 volcanic disasters.

The genesis of a tsunami as a result of a volcanic eruption is complex (e.g., Beget, 2000; Latter, 1981). Latter (1981) compiled historical volcanogenic tsunami events and listed possible volcanic phenomena as the tsunami source, including caldera collapse, entry of pyroclastic surges and flows into the ocean, underwater volcanic explosions, and so on. He also pointed out that there are many volcanism-related tsunami events due to eruptions of which the volcanic process is not clear.

Here I describe five major historical tsunami-generating volcanic eruptions (Hokkaido-Komagatake, Oshima-Ohshima, Unzen, Krakatau, and Ritter) and one recent example (Rabaul).

2.1 The 1640 CE Hokkaido-Komagatake eruption and tsunami

The Hokkaido-Komagatake is a truncated stratovolcano (1133-m high), located in the southern part of Hokkaido, northern Japan. The most recent major eruption occurred in 1929 and was one of the largest volcanic events in Japan in the last century. On July 31,

1640, an even larger eruption took place after more than 1000 years of dormancy. During the first stage of the activity, the summit portion of the volcano collapsed by phreatic explosions and a large debris avalanche was generated. The total volume of the ejected material has been estimated to be 3.5 km^3. Part of a debris avalanche traveled down for 20 km eastward from the summit to end up in the sea. There it spread across an area as wide as 15 km and covered 126 km^2 of the seafloor. The subaqueous volume of the avalanche, estimated by extrapolating the preeruptive topography from the surrounding area, was in the range of 0.92—1.20 km^3 (Yoshimoto et al., 2003).

The sudden displacement of water due to the debris avalanche at the eastern shore of Komagatake triggered a large tsunami. Old documents indicate that more than 100 ships were destroyed by sea waves and more than 700 people drowned. Tsuji (1989) examined an old document describing the tsunami run-ups around the main building of Zenko-ji, which is one of the oldest temples in Hokkaido, located on the shore of the bay opposite to Komagatake. He estimated the maximum run-up height there to be 8.5 m above sea level. This might be the only actual observation from which the height of the 1640 tsunami can be estimated. Nishimura and Satake (1993) revealed by simple computer simulation that the tsunami reached the opposite shore across the bay in about 40 min.

2.2 The 1741 CE Oshima-Ohshima eruption and tsunami

Oshima-Ohshima is a volcanic island located about 50 km south-west of Hokkaido, northern Japan. The island is 4 km in EW and 3.5 km in NS direction, and its volcanic peak is about 700 m above sea level. The last eruptive activity occurred during 1741—90 CE. This activity is known from both geological studies and old documents (Katsui and Yamamoto, 1981). The first eruption started with the ejection of pumice and ash on August 18, 1741. The most violent eruption took place on August 29 and caused a sector collapse of the north flank. The activity continued to May 1742, when the central cone was formed. Smaller eruptions were documented for 1759, 1789, and 1790.

From the same day when the most violent eruption occurred, a destructive tsunami was documented. The tsunami swept up onto the shores of the Oshima Peninsula, strongly damaging low-lying villages and destroying many houses. Estimates indicate more than 2000 casualties on the Hokkaido coast. The maximum tsunami heights on the coast were estimated to be about 15 m (Hatori, 1984). Significant tsunami damage was also reported in central and western Japan and along the eastern coast of Korea.

As the tsunami occurred during the maximum eruptive activity of the Oshima-Ohshima volcano, it is believed that the tsunami was caused by the volcanic eruption or by a landslide associated with the eruption. Aida (1984) calculated the characteristics of the tsunami on the basis of the volume of the subaerial slide and found that the observed tsunami was much larger than the computed one. Hence, another model—in which the

242 Chapter 15

tsunami was generated by an earthquake triggered by volcanic activity—has been considered. This earthquake model was not accepted, however, because there are no reports of strong earthquakes just before the tsunami occurred. Satake and Kato (2001) reestimated the slide volume, which might have been responsible for the tsunami generation. They examined bathymetric data from around the island and found that the debris extended to ~16 km north from the volcano. The total volume, both subaerial and submarine, is 2.3 km^3. This is one of the largest debris avalanches in historic time and an order of magnitude larger than that estimated by Aida (1984). Thus, Satake and Kato (2001), taking into account the result of their numerical simulation, suggested that the tsunami was caused by this large collapse of the volcano's cone.

2.3 The 1792 CE Unzen eruption and tsunami

The Unzen is a stratovolcano consisting mainly of dacitic rocks. It is located at the western end of the Beppu—Shimabara Graben, Kyushu, southern Japan. After 198 years of dormancy, the Unzen volcano erupted on November 17, 1990. The first stage of the eruption was characterized by phreatic to phreatomagmatic activity, and this was followed by the formation of a new lava dome at the summit of Mt. Fugen (the main peak of the Unzen volcano) on May 20, 1991. Continuous growth of the lava dome and fall of lava blocks generated frequent pyroclastic flows of Merapi type. On June 3, 1991, a large pyroclastic flow came down along a preexisting channel for about 4 km. This resulted in a disaster in which 40 persons were killed and 3 were missing.

Two historical eruptions before the 1990 have been noticed, namely in 1663 and 1792. In 1792, dacite magma erupted at the north-east flank of Mt. Fugen and flowed down. A month after the lava extrusion, a strong earthquake occurred, and Mt. Mayuyama, a parasitic lava dome situated behind Shimabara city, collapsed. A large debris flow rushed into the shallow sea and resulted in the generation of a tsunami. The tsunami traveled across the Ariake Bay and hit the opposite shore. Tsunami heights of about 10 m are estimated in old documents. In total, 15,000 persons were killed by the tsunami. This is the largest volcanic disaster in Japan. The volume of the collapsed body of Mt. Mayuyama has been estimated to be about 0.48 km^3 (Ui, 1983). Aida (1975) carried out numerical computations of tsunamis assuming continuous water transport perpendicular to the shore as the tsunami source. He concluded that the observed tsunami behavior could be explained by the transport of 18,000 m^3/min for 2—4 min at the landslide area.

2.4 The 1883 CE Krakatau eruption and tsunami

Krakatau is an active volcanic island located in the Sunda Strait of Indonesia, about 40 km west of Java. The 1883 eruption, having VEI (Volcanic Explosivity Index) 6, is one of the largest explosive eruptions in historic time. Before the 1883 eruption, the volcanic complex consisted of three main islands: Krakatau, Setrung, and Panjang. Krakatau, the largest of the islands, consisted of three overlapping volcanic cones. In 1883, a large part of Krakatau collapsed and a caldera with a 7 km diameter was formed.

The 1883 eruptive activity began with an explosive eruption on May 20 (Sigurdsson et al., 1991). A paroxysmal eruption started in the afternoon of August 26, when the intensity of the explosive eruptions increased and a large ash column rose up to a height of 26 km. The evolution of the eruption could be well reconstructed from the volcanic deposits in and around Krakatau Island, such as alternating layers of pyroclastic-flow/surge deposits and pumice and ash falls.

During the paroxysmal stage of the eruption, tsunamis were generated that struck the coasts of Java and Sumatra. One of the tsunamis killed about 36,000 persons and destroyed numerous coastal villages (Simkin and Fiske, 1983). The tsunami heights were obtained through survey immediately after the eruption; the observation results were compiled by Symons (1888). The tsunami heights before run-up were reported to be 50 ft (15 m) at coastal locations along the Sunda Strait. The heights were much smaller outside the Sunda Strait: less than 10 ft (3 m). A tide gauge at Batavia (now Jakarta) recorded the tsunami. The largest tsunami wave was estimated to have been originated at 10:00 a.m. (Krakatau time) on August 27 by Nomanbhoy and Satake (1995). The tide-gauge and pressure-diagram data also suggest the generation of multiple tsunamis of a smaller scale.

The responsible mechanism for the genesis of the tsunami is not well understood. Francis (1985) mentioned the following four mechanisms as possible causes of the tsunami: (1) lateral blast (Camus and Vincent, 1983), (2) large-scale collapse of the northern part of Krakatau island (Self and Rampino, 1981; Sigurdsson et al., 1991; Verbeek, 1984), (3) pyroclastic flows (Francis, 1985; Latter, 1981; Self and Rampino, 1981; Sigurdsson et al., 1991), and (4) submarine explosion (Nomanbhoy and Satake, 1995; Yokoyama, 1981, 1987). These hypotheses have been discussed by the just mentioned authors on the basis of reexamination of the tide-gauge record and pressure-wave data, volcanological interpretation, and numerical simulation of the tsunamis, but the cause of the tsunami is not yet understood.

2.5 The 1888 CE Ritter tsunami

Ritter is a volcanic island located 23 km west of Papua New Guinea. Ritter was a 780-m-high conical volcano when first observed by Europeans in the 1700s (Cooke, 1981; Johnson, 1987). On March 13, 1888, the major part of the cone collapsed and only a part of the eastern margin of the original island remained. The present-day island is 1.9 km long, 200–300-m wide, and 140-m high. There is a small islet off the southern tip of the main arcuate island. A bathymetric survey confirmed the existence of an amphitheater structure west of the present island, with a maximum diameter of 4.4 km and a breach width of 3.5 km. The volume of the collapsed body has been estimated as large as 4–5 km^3 (Johnson, 1987).

244　Chapter 15

The collapse also generated a tsunami. Cooke (1981) and Johnson (1987) compiled documentary accounts on the tsunami damage on the western coast of New Britain, where the tsunami run-up was estimated to reach 12−15 m above sea level. Abnormal waves were also observed at Rabaul, 475 km to the east-north-east. The number of fatalities caused by the tsunami is unknown, but probably hundreds of villagers along the West New Britain coastline and elsewhere were killed. Ward and Day (2003) simulated the Ritter island landslide as constrained by a sonar survey of its debris field and compared the calculated tsunami with historical observations. Their best-fit model is that the debris dropped ~800 m vertically and moved slowly compared with a landslide.

2.6 The 1994 CE Rabaul eruption and tsunamis

The 1994 Rabaul eruption series is one of the most recent eruptive events that were accompanied by significant tsunamis. The Rabaul is a young caldera, located at the north-eastern end of the Gazelle Peninsula, Papua New Guinea. The latest caldera-forming eruption took place 1400 BP, when it erupted 11 km^3 of volcanic material. Two postcaldera volcanoes on opposite sides of the caldera, Vulcan and Tavurvur, erupted simultaneously in 1878, 1937, and 1994. At 06:15 (local time) on September 19, 1994, the Tavurvur volcano erupted, and about 1 h later the Vulcan volcano followed. The eruptions were preceded by vigorous seismicity with a maximum earthquake magnitude of 5.1 (Blong and McKee, 1995). Soon after its onset, the eruption of the Vulcan increased in strength and included a phase of strong Plinian activity. Pyroclastic flows reached as far as 2 km from the crater, and the eruption column rose up to 20 km above sea level.

Continuous sea-level changes before and during the eruption period were registered by a tide gauge at Rabaul harbor (Nishimura et al., 2005). The first tsunami was associated with one or both of the two earthquakes of magnitude 5 that occurred at 02:51 (local time) on September 18, 1994, a day before the eruption. The hypocenters of these earthquakes are estimated to have been beneath the Tavurvur and the Vulcan volcanoes, respectively (Blong and McKee, 1995). The first tsunami reached the tide gauge about 6 min after the earthquakes. The zero-peak amplitude of the tsunami is about 1 ft (30 cm), and the tsunami continued for several hours with simply decaying amplitude.

The next tsunami phase was observed from 07:43 on September 19. At that time, there were no large earthquakes, but the Vulcan volcano had started to erupt at 07:17. The tide-gauge record shows that the zero-peak amplitude of the first wave was about 0.8 ft (25 cm). Compared with the earthquake-generated tsunami on the previous day, the amplitude did not decay simply, but continued for more than 24 h, with fluctuating amplitude. It suggests that the tsunamis were excited several times or almost continuously during the eruption.

Nishimura (1997) carried out numerical experiments for the 1994 Rabaul tsunami, using simple source models and a bathymetric map of Simpson Harbour. He showed that a tsunami generated from the northern shore of the Vulcan volcano reached Rabaul ~5 min after it had been excited. The distribution pattern of tsunami heights along the coast of Simpson Harbour varied significantly with source duration and location. In most cases, however, the largest tsunami heights were observed along the western shore of Matupit Island and tsunamis at the eastern side of the island were less than half the amplitude of their western counterparts.

3. Volcanism-induced tsunamiites

Geological investigations of coastal sediments provide important information on past tsunami events (Dawson and Shi, 2000). In some cases, palaeotsunamis are represented by single sand layers. Elsewhere, tsunami deposits can consist of complex layers containing abundant stratigraphic evidence for sediment reworking and redeposition. These onshore sandy layers used as geological evidence for tsunami are called "tsunami deposits" or "tsunamiites." Tsunami deposits have been studied in recent years for the reconstruction of prehistorical tsunami recurrence and magnitude (e.g., Atwater et al., 2005; Nanayama et al., 2003; Penegina et al., 2003).

The identification process of tsunami deposits has been discussed through studying modern tsunamis of large earthquake or landslide origin and of their deposits (e.g., Dawson et al., 1996; Gelfenbaum and Jaffe, 2003; Minoura et al., 1997; Nishimura and Miyaji, 1995). Commonly observed features of tsunami deposits are the following: (1) a sheet-like distribution, (2) landward thinning, (3) landward fining, (4) upward fining (normal grading), (5) multiple units of layers (not always), (6) inclusion of beach materials, and (7) inclusion of rip-up clasts (eroded surface materials).

Case studies of tsunami deposits associated with volcanogenic tsunamis are scarce. Such deposits have been reported for seven eruptions (Fig. 15.1): (1) Managua Lake, Nicaragua, 3000–6000 BP; (2) Santorini, Greece, ~3500 BP; (3) Aniakchak, Alaska, 3500 BP; (4) Hokkaido-Komagatake, 1640 CE; (5) Krakatau, Indonesia, 1883 CE; (6) Rabaul, Papua New Guinea, 1994 CE; and (7) Karymsky, Kamchatka, 1996 CE. The three prehistorical eruptions, (1)–(3), have not been documented but tsunami occurrences are suggested by legends, volcanological aspects, and geological evidence. For the four historical events, (4)–(7), tsunamigenic volcanic activities are well documented or reported, and their deposits have been found and studied.

Four historical tsunami episodes have accounts but no geological evidence. These are (1) the 1650 Columbo (Thera Island) eruption, Greece; (2) the 1741 Oshima-Ohshima eruption, Japan; (3) the 1792 Unzen eruption; and (4) the 1883 Augustine eruption,

246   Chapter 15

Alaska. Dominey-Howes et al. (2000) examined seaside trenches in the area where the 1650 Columbo tsunami is supposed to have invaded, but they could not find any deposits of marine origin. Nishimura et al. (2000) found a sandy layer in a beach dune deposit of ~2-m thick in Kumaishi, where the 1741 Oshima-Ohshima tsunami occurred. The layer shows clear landward thinning and fining, but the age of the deposit is not well constrained. The coast around the Ariake Bay, where the 1792 Unzen tsunami is said to have traveled, has been completely developed and is densely populated so that there are no suitable places left to look for and investigate the possible 1792 tsunami deposit. Waythomas (2000) conducted a careful search for anomalously looking deposits associated with the 1883 Augustine tsunami, but could find only some possible (not unambiguous) tsunami deposits in lower Cook Inlet.

Deposition and preservation processes of tsunami deposits are complex and commonly affected by factors such as beach material and slope and shape of the coastal area. For the four aforementioned cases, the absence of tsunami deposits prevents analysis of the tsunami characteristics at these sites.

3.1 Managua tsunami deposits of 3000—6000 BP

Freundt et al. (2006) investigated the Mateare Tephra representing the fallout of a Plinian eruption from the Chiltepe Peninsula at the western shore of Lake Managua, Nicaragua, that occurred 3000—6000 BP. They found anomalous layers in the series of Mateare Tephra and related them to a tsunami associated with the lakeside explosion. The Mateare Tephra includes four units, from base to top: (A) a high-silica dacite fallout deposit generated by unsteady subplinian eruption; (B) a less silicic, massive dacite fallout deposit from the main, steady Plinian phase of the eruption; (C) an andesitic fallout deposit; and (D) an ash unit from the phreatomagmatic terminal phase of the eruption. A massive, well-sorted sand layer (Mateare Sand) replaces laterally the variable part of unit A and the lowermost part of unit B. The Mateare Sand shows a widespread continuous distribution and outcrops up to 32 m above the present lake level. It includes fallout material, and entrained fallout in the sand decreases with distance from the lake. Based on the chronological evidence of both the tephra units and the sandy layers on the one hand, and the composition on the other hand, Freundt et al. (2006) interpreted that the Mateare Sand is associated with a lake tsunami triggered by an eruption during the initial, unsteady phase of activity. The tsunami-forming activity stopped when the steady Plinian phase of the eruption was established.

The structure of the sand layers varies from site to site; probably it has been affected by local topography and/or coastal conditions. For some locations, the Mateare Sand has a wavy contact with the overlying unit B fallout pumice layer, and the sand layer partly eroded the underlying layers that are mostly composed of older erupted material.

The Mateare Sand overlies a large-scale erosional unconformity with a thickness of 2—15 cm. The thickness does not systematically vary with elevation above lake level. Tree molds are common in the Mateare Sand, but most abundant in outcrops that lie high above the present lake level. These tree molds are probably relics of driftwood washed up to near the maximum water run-up.

Freundt et al. (2006) examined the variation in chemical composition of the Mateare Sand (which includes angular to rounded, highly vesicular Mateare pumice) and discussed its depositional process. They found that the chemical composition of pumice of unit A fallout (under the Mateare Sand) differs significantly from that of the unit B fallout (above the Mateare Sand) and that the bulk composition of the pumice in the sand represents a mixture of these two kinds of pumice in different proportions. Close to the lake's shore, the sand consists almost exclusively of unit B pumice, while at the highest elevation it consists exclusively of unit A pumice. This suggests that the sand-layer forming event occurred during the unit A phase and lasted into the unit B phase and that the sizes of the early events were larger than those of the later ones. The later, smaller, events could erode the predepositional surface up to some extent inland from the shore.

The Mateare Sand is composed of reworked beach sand (mainly basaltic/andesitic lava sand) and dacitic fallout tephra. The grain-size distribution of the sand is unimodal. The mean grain size does not vary systematically with the thickness of the sand layer or with the elevation above lake level. The sandy deposit is well sorted because both sources provided well-sorted materials. Pumice in the layer is not rounded, probably because deposition was rapidly followed by entrainment. The high-density sand and the low-density pumice are neither mixed well nor well segregated, which suggests highly turbulent transportation.

Based on the aforementioned geological and volcanological features of the sandy deposit, Freundt et al. (2006) suggested that a series of tsunamis occurred during the early stage of the eruptive period of the Managua Lake volcano 3000—6000 BP. Because there is no evidence for cone collapse or for large pyroclastic flows, it is not likely that the tsunamis were generated by large-scale emplacement of water in the lake. Tsunami deposits from the Managua Lake eruption are, as described later, very different from those of the violent eruption or caldera-forming activities of the Santorini 3500 BP and of the Krakatau in 1883. A possibly similar case is the tsunami caused by Surtseyan-type eruptions of the Karymsky Lake volcano, Kamchatka, in 1996 (Belousov and Belousova, 2001).

3.2 Santorini tsunami deposits of 3500 BP

The Santorini caldera, Aegean Sea, Greece, was formed about 3500 BP. A large Plinian eruption destroyed major parts of Santorini (Thera), and a huge amount of pumice was

248 Chapter 15

ejected, coinciding with the end of the Minoan culture on Crete in the eastern Mediterranean. The eruption may have contributed to the decline of this culture. The eruption or caldera collapse probably generated a large tsunami, which destroyed the coastal settlements on Crete. The Santorini tsunami hypothesis has been put forward through archaeological and volcanological research and is also supported by onshore geological evidence (e.g., McCoy and Heiken, 2000; Minoura et al., 2000) and submarine evidence (e.g., Cita and Aloisi, 2000; Cita and Rimoldi, 1997), although Dominey-Howes (2004) criticized some of these geological findings. Dominey-Howes examined over 40 new geological sites at Crete and Kos and failed to identify any evidence for the tsunami.

Minoura et al. (2000) have reported tsunami sediments at Didim and Fethiye, western Turkey ($\sim$200 km west of Santorini), and at Gouves in Crete ($\sim$100 km south). They did trenching on the coast of Didim and Fethiye, where no human activity has taken place. In the trench wall of Didim, they found a fine sand layer composed of carbonate grains. The layer is overlain by a 10−15-cm-thick layer of fallout tephra, which is confirmed to have been ejected from the Santorini 3500 BP. The lack of an erosional contact between the tephra and the underlying sand layer implies that the deposition of the sand was followed without much interruption by the fallout. The sand layer includes offshore foraminifers; in contrast, the underlying silty mud contains intertidal foraminifers. At the Fethiye section, a 5−10-cm-thick sand layer including abundant shell fragments of shallow-marine origin is covered by the Santorini tephra and overlies nonmarine sandy silt. At Gouves, Crete, a 10−20-cm-thick sand layer has been found on a late Minoan potter's workshop at an archaeological excavation site. The site is located 30−90 m inland from the Minoan harbor and 2−3-m high above present sea level. The sand layer is covered directly by a 10−20-cm-thick pumice layer from the 3500 BP Santorini eruption. Minoura et al. concluded that the Santorini eruption and the following tsunami are well recorded in the coastal successions of western Turkey and Crete. It is important to note that the sand layers do not include any primary volcanic materials. This implies that the tsunami deposition ended before the start of tephra fall at both sites, even though the eruption triggered the tsunami.

McCoy and Heiken (1997, 2000) found a tsunami layer at a near-source site. This tsunami deposit is situated within the tephra succession at Pori, on the east coast of Santorini; the researchers suggested that it was caused by the late Bronze Age eruption of the volcano. They found a 3.5-m-thick deposit between the third and fourth phase tephra of the 3500 BP eruption. The deposit is a mixture of reworked pumice, ash, and lithics and is divided into two distinct layers. The lower layer is 1.5-m thick and extends 36 m inland to an elevation of 6 m on a gentle (2−16°) slope. It is lenticular in shape and thins inland. The basal contact is sharp with erosional scouring into the underlying tephra layer. The internal structures of the deposit include planar and cross-bedded layers, some with normal grading, clastic tails at the lee side of larger clastic particles, and layers of lithic-rich gravels, often in sharp contact with erosional features (McCoy and Heiken, 2000). The upper layer is 2-m thick and has no distinct upper boundary. It is a chaotic mixture regarding the composition of the particles, extremely badly sorted, and lacks internal structures, but it includes coarser lithic fragments than the lower layer. The contact between the upper and the lower layers is distinct, indicating erosion. McCoy and Heiken (2000) regarded the lower deposits as water-laid deposits by the oscillatory flow of the tsunami and the upper layer as a deposit reworked by the tsunami. The internal structure of the lower tsunami deposit indicates multiple inundation phases associated with pyroclastic flows entering into the sea or with seiches related to the inundation by the tsunami. They also suggested a debris flow during the fourth phase of the eruption as a possible source of the upper tsunami deposit.

3.3 Aniakchak tsunami deposits of 3500 BP

The Aniakchak is a circular caldera with a diameter of 10 km, located 670 km south-west of Anchorage on the Alaska Peninsula. The caldera was formed during a catastrophic eruption ~3500 BP. Activity continued inside the caldera, and the most recent eruption took place in 1931. The volcano faces Bristol Bay, where a pyroclastic flow from the 3500 BP caldera-forming eruption entered the sea. Waythomas and Neal (1998) hypothesized a large tsunami excitation at the time. Waythomas and Watts (2003) simulated the Aniakchak tsunami by modeling a pyroclastic flow entering into the sea and confirmed that such a pyroclastic flow can generate a significant far-field tsunami.

Waythomas and Neal (1998) found an unusual pumiceous sand sheet at 10 sites along the coastline of the northern Bristol Bay and regarded them as volcanogenic tsunami deposits caused by the 3500 BP Aniakchak eruption. The Aniakchak volcano is on the opposite side of the bay, 100—200 km south of the observation sites. They suggested that these special pumiceous sand layers have been formed by tsunamis that carried floating pumice inland.

As described by Waythomas and Neal (1998), the pumiceous sand is a discontinuous, poorly to moderately well sorted, fine to coarse sand deposit of <40-cm thick. The sand layer is situated near the top of a 1—4-m-thick Holocene peat. At some localities, the sand layer is preserved as a single, conspicuous lens or stringer of coarse sand lenses in the woody peat. At two sites, a thin, fine ash deposit is present at the base of the pumiceous sand. The basal ash represents an early phase of tephra fallout, prior to the paroxysmal caldera-forming eruption.

The pumiceous sand is a mixture of pumice and beach sand. The average composition of the pumiceous sand samples is 60% pumice, 22% glass shards and crystal fragments, and 18% lithic grains. Rip-up clasts of peat, up to 20-cm long, also occur. The unit consists of subangular to moderately well-rounded pumice. At the site where the pumiceous sand is coarsest, the maximum pumice size is about 1 cm. At some sites, the pumiceous sand contains marine diatoms. Sedimentary structures associated with bidirectional currents have not been observed, and in only a few places does the deposit exhibit faint lamination. The pumiceous sand is massive, and faint bedding and thin lenses of varying grain size are present in some places. They fill irregularities in the underlying peat.

Waythomas and Neal (1998) proposed that the pumiceous sand layer was deposited by a tsunami caused by a large pyroclastic flow during the 3500 BP eruption of the Aniakchak volcano. Entrainment of drifting pumice and erosion of local beach sand were the possible sources of the pumiceous sand. The tsunami inundated a peat-covered lowland, so that the pumiceous sand deposited on the peaty wetland was buried also by peat.

3.4 The 1640 CE Komagatake tsunami deposits

Nishimura and Miyaji (1995) and Nishimura et al. (1999) identified 1640 CE Komagatake tsunami deposits at four sites around the Uchiura Bay, in outcrops and by using a soil auger. They confirmed similar patterns of landward thinning and fining as shown by the deposits of the 1993 CE Hokkaido Nansei-Oki tsunami. The sandy sheets were deposited on a volcanic ash soil and were covered at most sites directly by the primary fallout tephra from the 1640 CE eruption of the Komagatake.

At Washinoki, located less than 5 km from the Komagatake volcano, Nishimura et al. (1999) saw an outcrop of about 4 m height (Fig. 15.2) where a sand and gravel layer of less than 20-cm thickness covers the original surface and is overlain directly by primary tephra from the volcano. The tsunami deposit is clearly covered by a thin (<1 cm) fine ash

Figure 15.2
Vertical section of the 1640 CE Komagatake outcrop at Washinoki. Gravels traced at the bottom of the thick white pumice layer (Ko-d) are tsunami deposits of the 1640 CE Komagatake tsunami.

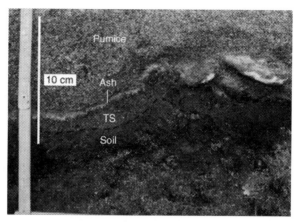

Figure 15.3
Close-up of the tsunami deposit and overlying Ko-d tephra at Washinoki. TS indicates the tsunami sand sheet. Thin white layer is the blast ash fall from the Komagatake ejected at the initial stage of the eruption.

layer that is the first blast deposit from the 1640 CE Komagatake eruption (Fig. 15.3). The ash layer is covered by pumice (1-m thick) derived from fallout during the paroxysmal stage of the eruption. It is inferred that the tsunami occurred at the first stage of the eruption associated with a sector collapse of the volcano. As there is no contamination of ash in the sandy tsunami layer, the first ash fall must have occurred after the tsunami had receded.

At a grass field at Shadai, 40 km north of the volcano, Nishimura et al. (1999) examined a vertical section perpendicular to the shoreline and performed grain-size analyses of the underlying sand deposits. They identified this sand layer as a tsunamiite that was deposited by the 1640 CE Komagatake tsunami run-up wave. The layer shows weak upward fining. The mean grain size of the deposit tends to decrease with increasing distance from the sea. The deposits contain significant amounts of pumiceous sand (a mixture of the old pumice, beach sand, and dune sand) at 60 m or more landward. These pumiceous sand layers are considered to consist of eroded material that was deposited on the original surface. The investigations also showed that the thickness of the tsunami deposits is affected by undulations that existed in the local surface at the time of the tsunami. Consequently, the deposit is relatively thin on top of dunes and thick in the depressions between them.

3.5 The 1883 CE Krakatau tsunami deposits

Carey et al. (2001) found tsunami deposits on the islands of Lagoendi, Sebuku, and Sebeshi, located about 30, 20, and 15 km north of Krakatau Island, respectively. Major deposits on the island were caused by pyroclastic flows occurring during the 1883 eruption. The thickness of the deposits is < 2 m and the eruption-triggered succession has been studied in detail by analyses of the tephra. Carey et al. (2001) found several unique sections at places less than 90 m above sea level. These sections contain two types of units, both with unusual granulometric and lithologic features that distinguish them from truly eruptive deposits. The first type consists of well-sorted deposits that range in thickness from 10 to 25 cm. These deposits resemble Plinian pumice fall deposits, with the exception that they are strongly depleted in crystals and lithic components and that the pumice grains are rounded. The second type of diverging deposits are more than 30-cm thick and consist of poorly sorted silty to sandy ash with abundant coral fragments and rounded pumice clasts.

Carey et al. (2001) regard both types as tsunami deposits associated with the 1883 CE Krakatau tsunami. For the first type of deposits, they confirmed that the pumice particles in the layer are significantly rounded, whereas the original pumice is not. They propose that the pumice in this type of deposits was originally deposited in the sea during fallout or by a pyroclastic surge traveling over the sea. Abrasion of the tephra would have occurred during floating and during tsunami run-up. Through these processes, segregation of lithics from the pumice occurred in the water and only low-density pumice was carried up and deposited inland. In contrast, the second type of tsunami deposits likely formed by the same process as tsunami deposits of earthquake origin. The tsunami eroded the original beach and deposited the "marine" pumice inland together with pumice and ash particles that had been deposited on the beach.

3.6 The 1994 CE Rabaul tsunami deposits

Nishimura et al. (2005) conducted reconnaissance geological mapping of the 1994 tsunami deposits around the caldera bay. They found characteristic sandy layers that were distinguished from primary eruptive deposits of the 1994 eruption in coastal sediment at six localities. The sandy layers are mainly composed of marine sand and volcanic pumice with flakes of coral, marine shell, wood branches, and man-made plastic scraps and differ from primary pumice and ash layers. The thickness of the layers is commonly less than 10 cm. In most cases, the researchers traced the layers up to 100 m inland and confirmed that their distribution had a sheet-like form. The thickness of the sandy layers tends to decrease with increasing distance from the sea. They regarded these sandy layers as the 1994 tsunami deposits because of the following common features: (1) the distribution of the sandy layers is limited to coastal areas; (2) the thickness of the layers tends to decrease with increasing distance from the beach; (3) the layers contain beach sand and flakes of coral and shells of marine origin; (4) the layers are intercalated between, or immediately overlain by, tephra units from the 1994 Vulcan and Tavurvur eruptions; and (5) the weather was good and the sea was not rough during the eruption, so that wind-driven waves could not have produced these deposits.

Volcanism-induced tsunamis and tsunamiites 253

Figure 15.4
Tsunami deposit intercalated between volcanic tephra at Rabaul. Distance from the beach to this pit is about 50 m. TS indicates the tsunami sand sheet. The thin-bedded alternation of sand and pumice layers probably resulted from successive deposition, reworking, and erosion processes by tsunami waves. (V) and (T) represent tephra from the Vulcan and Tavurvur volcanoes, respectively.

The tsunami deposits are continuous along the coasts, changing their lithofacies with distance from the beach and with local topography. Representative depositional facies of the tsunami sand sheets observed in pits along a 100 m profile at a site south of the old Rabaul airport are shown in Figs. 15.4 and 15.6. Fig. 15.4 shows a pumiceous sand layer with a thickness of 6 cm, at a point 50 m from the shore. The upper and lower boundaries of the layer are not clear, and a vague lamination can be seen. At the 60 m point, landward of a small local mound, a single clear sandy layer is intercalated between volcanic ashes (Fig. 15.5). There is little contamination of volcanic material in the layer. From the 70 m

Figure 15.5
Tsunami deposit intercalated between tephra at Rabaul. Distance from the beach to this pit is about 60 m. The thin sand unit is a tsunami deposit.

Figure 15.6
Tsunami deposit intercalated between tephra at Rabaul. Distance from the beach to this pit is about 100 m. A pumice-rich layer was deposited at the inland boundary of the inundated area, with a clear contrast between the sand-rich and the pumice-rich units in the tsunamiite.

point to the 90 m point, two or three sandy layers are present between volcanic tephra. At the 100 m point, the material in the uppermost sandy unit changes significantly from a sand-rich layer to a pile of pumice and wood branches (Fig. 15.6).

Thus, the thickness and lithofacies of the tsunami sand sheets change with distance from the sea and are also affected by local topography. These features might be interpreted as follows. At this site, the tsunami inundated the coast at least three times. The resulting sedimentary unit is complicated, not only by the run-up processes that may be responsible for its deposition but also by an interval of erosion (successive tsunami waves may result in the erosion of preexisting tsunami deposits). The pattern of tsunami sedimentation became, in addition, complicated by an interval of backwash currents that were more powerful and closer to the beach. These processes produced a single, complex sandy layer from the beach to the 50 m point. Separation of light and heavy material occurred as the tsunami waves traveled inland over a small mound at the 60 m point. One possible interpretation is that the pumice in the tsunami was concentrated in the upper part of the wave while it was suspended in the water and was carried inland by the run-up wave, but also was transported back to sea by the backwash currents. On the other hand, the heavy particles concentrated in the lower part of the tsunami waves; they were carried along by the run-up waves but the backwash currents were not sufficiently powerful to bring them back to sea. Thus, only relatively heavy sand particles were left onshore (Fig. 15.5).

It is known that debris composed of light materials, such as dried branches or grasses, are commonly deposited along the upper boundary of a tsunami inundation area (e.g., Nishimura and Miyaji, 1995). In the case of the Rabaul tsunami, pumice and wood branches were deposited at this boundary. It is shown in Fig. 15.6 how the components of the tsunami deposit change from sand-rich to pumice-rich. Usually such geological evidence is quickly disturbed and eventually disappears from the surface. In the Rabaul case, the cover of tephra probably played an important role in preserving the original features of the tsunami deposit.

Chronological studies indicate that the tsunami was not generated by the first small eruption of the Vulcan volcano though it occurred close to the sea. The tsunamis were excited several times or continuously by large pyroclastic flows and base surges during the later, paroxysmal stage of the eruption. The run-up heights of the tsunami waves have been estimated from the distribution of the tsunami deposits to reach about 8 m at Sulphur Creek and more than 3.5 m around the western to southern shore of Matupit Island. These values are consistent with other evidence, such as eyewitness accounts, reported damage, and numerical simulations.

3.7 The 1996 CE Karymsky tsunami deposits

Karymsky Lake is located in the uninhabited part of the eastern volcanic belt of Kamchatka. The diameter of the lake is about 4 km. The 1996 eruption started in the afternoon of January 2, 1996 and continued for 10−20 h. In total, 100−200 subaqueous explosions are, based on the average interval of the observed explosions, estimated to have occurred. The eruption formed an underwater tuff ring off the northern shore. Detailed descriptions of the eruption sequence have been given by Belousov et al. (2000) and by Belousov and Belousova (2001).

The 1996 eruption series of Karymsky Lake caused tsunamis in the lake (Belousov and Belousova, 2001; Belousov et al., 2000). The northern shore of the lake adjacent to the tuff ring was violently eroded by the tsunamis and also by explosions. The degree of shore erosion diminishes with distance from the tuff ring. Along the rest of the shoreline, more than 1.3 km from the crater, the tsunamis carved new cliffs extending 3-m high and eroded the upper soil layer, which was up to 50-cm thick. The area affected by the tsunamis is marked around the lake by strong erosion of the shore, damaged bushes, and distinctive deposits.

Belousov and Belousova (2001) divided the components of the tsunami deposit around the lake shore into two classes, nonfloating materials and floating objects. Nonfloating constituents include individual blocks that originated from newly eroded cliffs and patches of sand. The blocks are abundant on the north-eastern shore of the lake, and some of them were transported by waves up to 60 m inland. Patches of sand occur on the lakeside slope and are usually meters to tens of meters across and up to 35-cm thick. Sometimes up to four parallel layers with thicknesses of 2−6 cm can be distinguished. They are inferred to reflect the multiple tsunami run-up and backwash currents. The composition of the deposit tends to vary with distance from the source tuff ring: the deposit is composed of fresh pyroclasts in the zone near the crater and of beach sand at sites far away from the crater. Deposits of floating material are composed of old pumice, which was eroded by the tsunami, and of branches of bushes. These floating materials were piled up on the lakeside's slope.

It is inferred that the largest tsunamis were generated during the last stage of the series of eruptions because the eroded surface was not covered by tephra. Run-up heights of the tsunami waves (for the maximum events) were estimated at 24 sites around the lake on the basis of geological evidence. The highest run-up of 20–30 m occurred on the shore nearest the crater, and the run-up shows rapid attenuation with distance to 1.3 km. The run-up height decreases slowly from that point to 2–3 m at the most distant points.

4. Discussion and summary

When an eruption-triggered tsunami travels to a site where no tephra fell, its deposits there must be similar to those produced by an earthquake-induced tsunami. There are, however, some potential differences in sedimentary structures, such as the degree of upward fining because the periods of both tsunami wave types may be different. Yet, no reports or evidence for this is available at present. By contrast, in areas near the erupted volcano, a unique sediment layer composed of beach sand and reworked pumice is often produced by tsunami run-up. This characteristic "pumiceous sand layer" is one of the major evidences for volcanogenic tsunami deposits.

Pumice deposited onshore together with beach sand can have three origins. One is that pumice, which was ejected from the volcano and was washed ashore on a beach, was carried up and deposited inland with beach materials by a tsunami. As Carey et al. (2001) pointed out, layers composed of drift pumice can be distinguished from those with fresh pumice by examining the roundness of the pumice clasts themselves and the existence (or lack) of heavier lithic materials in the layer. The second possible origin is that tephra fell and were deposited in the surf zone before the tsunami arrived. These tephra could also be carried up inland easily by a tsunami. The third possible origin is that primary pumice fell on land during the inundation by a tsunami. Pumice in tsunami deposits may also be a result of a combination of two or all three of these possible origins.

As there is a significant difference in density between beach material (heavy) and volcanic pumice (light), segregation of these materials can occur in the water, both while drifting on the sea and while traveling with a tsunami.

Tsunami deposits with pumice can show lithofacies characteristics that differ from those of tsunami deposits without pumice. That is the reason why, whereas fining upward (normal grading) is one of important characteristics of tsunami deposits, it is sometimes hard to distinguish grading in pumiceous tsunami deposits because the pumice particles tend to concentrate in the upper part of the layer regardless of their size. In addition, normal tsunami deposits tend to show landward thinning and landward fining, but pumiceous tsunami deposits show the opposite. Even large pumice particles can be carried along by a tsunami much farther than other, more dense materials, which necessarily become deposited earlier. Sometimes, large piles of pumice are accumulated along the landward boundary of the inundated area. The aforementioned features are well developed in the 1994 Rabaul tsunami deposits, as described in Section 3.6.

Chronological correlations of tephra and tsunami deposits provide a rare opportunity to study the tsunami-generating process during an eruption. Fig. 15.7 shows schematic columnar sections for four types of chronological relationship between primary tephra and volcanogenic tsunami deposits. Type A shows a tsunami deposit that is covered directly by tephra (Komagatake, Santorini), Type B shows a tsunami deposit that overlies tephra (Aniakchak), Type C shows a tsunami deposit that is intercalated in a series of tephra-fall deposits (Krakatau, Rabaul), and Type D is similar to Type C but is overlain by a next unit of tephra (Managua, Santorini, Rabaul). The chronological information helps to understand the tsunamigenic volcanic process. For example, studies of tsunami deposits related to the 1640 Komagatake eruption showed that there had been no preceding eruption before the tsunamigenic cone collapse.

As has been shown in the case studies of the Managua, Aninakchak, and Santorini tsunamis, tsunami deposits of prehistorical eruption tsunamis can yield much information. There are two main reasons. One is that tsunami deposits are the only geological evidence

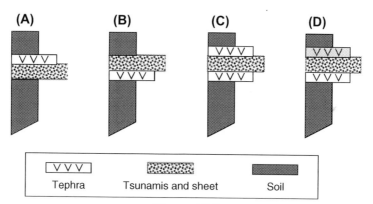

Figure 15.7
Schematic columnar sections for four types of chronological relationships between primary tephra and volcanogenic tsunami deposits. The tsunami deposit is covered directly by tephra (Type A), the tsunami deposit overlies tephra (Type B), the tsunami deposit is intercalated in a series of tephra deposits (Type C), and the tsunami deposit is similar to that of Type C but overlain by a next unit of tephra deposits (Type D).

258　**Chapter 15**

that can record the dynamic processes of a huge volcanic eruption. For modern eruption phases, seismic or infrasonic data related to the explosion and tide-gauge records concerning the tsunami can be used to construct or constrain dynamic features of the eruption. Tsunami deposits, if well preserved, are in some cases equally informative as tide-gauge records. The other reason is that tsunami deposits often still show their original features extremely well because they have been covered by the following tephra fallout soon after their formation. One good example is the 3000−6000 BP Managua eruption process, which was revealed by detailed study of the evolution of erupted material and related local tsunami deposits.

It will be important to carry out detailed fieldwork in sediments from two or more tsunamigenic episodes originating from the same volcano and to compare the processes to one another. Rabaul has experienced similar volcanic eruptions in the past. At least the previous eruption in 1937 excited some tsunamis (e.g., Johnson and Threlfall, 1985), and Arculus and Johnson (1981) translated a document that indicates the 1937 tsunami behavior. They quoted that "the water receded several times over 50 yards from the shoreline" and "the waters of the harbour at Rabaul rose and fell continuously, taking some 4−5 min between the least and greatest depth." The 1937 tsunami deposits also gave important information on the timing and scale of the volcanic eruption and related tsunami, though they were covered by a thick tephra unit. This information must be useful to evaluate future risks due to tsunamis of volcanic origin.

Acknowledgments

The author is grateful to Professor T. Shiki for the opportunity to contribute this chapter. The author also thanks Professors A. Moore and A. J. van Loon for useful comments on the paper and for polishing the English.

References

Aida, I., 1975. Numerical experiments of the tsunami associated with the collapse of Mt. Mayuyama in 1972. Bull. Seismol. Soc. Jpn. 28, 449−460 (in Japanese with English abstract).

Aida, I., 1984. An estimation of tsunamis generated by volcanic eruptions—the 1741 eruption of Oshima-Ohshima, Hokkaido. Bull. Earthquake Res. Inst. Univ. Tokyo 59, 519−531 (in Japanese with English abstract).

Arculus, A., Johnson, R.W., 1981. 1937 Rabaul eruptions, Papua New Guinea: translations of contemporary accounts by German missionaries. In: Bur. Miner. Res. Geol. Geophys. 78. Australia Rept. 229, BMR Microform MF 158.

Atwater, B.F., Musumi-Rokkaku, S., Satake, K., Tsuji, Y., Ueda, S., Yamaguchi, D.K., 2005. The orphan tsunami of 1700—Japanese clues to a parent earthquake in North America. U.S. Geol. Surv. Prof. Paper 1707 133.

Beget, J.E., 2000. Volcanic tsunamis. In: Sigurdsson, H., Houghton, B., McNut, S.R., Ryners, H., Stix, J. (Eds.), Encyclopedia of Volcanoes. Academic Press, San Diego, pp. 1005−1014.

Belousov, A., Belousova, M., 2001. Eruptive process, effects and deposits of the 1996 and the ancient basaltic phreatomagmatic eruptions in Karymskoye lake, Kamchatka, Russia. Special Publ. Int. Assoc. Sedimentol. 30, 35–60.

Belousov, A., Voight, B., Belousova, M., Muravyev, Y., 2000. Tsunami generated by subaquatic volcanic explosions: unique data from 1996 eruption in Karymskoye lake, Kamchatka, Russia. Pure Appl. Geophys. 157, 1135–1143.

Blong, R.J., McKee, C.O., 1995. The Rabaul eruption 1994: destruction of a town. In: Natural Hazards Research Center. Macquarie University, Australia, p. 52.

Bonaccorso, A., Calvari, S., Garfi, G., Lodato, L., Patane, D., 2003. Dynamics of the December 2002 flank failure and tsunami at Stromboli volcano inferred by volcanological and geophysical observations. Geophys. Res. Lett. 30. SDE 6-1–6-4.

Camus, G., Vincent, P.M., 1983. Discussion of a new hypothesis for the Krakatau volcanic eruption in 1883. J. Volcanol. Geoth. Res. 19, 167–173.

Carey, S., Morelli, D., Sigurdsson, H., Bronto, S., 2001. Tsunami deposits from major explosive eruptions: an example from the 1883 eruption of Krakatau. Geology 29, 347–350.

Cita, M.B., Aloisi, G., 2000. Deep-sea tsunami deposits triggered by the explosion of Santorini (3500 y BP), eastern Mediterranean. Sediment. Geol. 135, 181–203.

Cita, M.B., Rimoldi, B., 1997. Geological and geophysical evidence for a Holocene tsunami deposit in the eastern Mediterranean deep-sea record. J. Geodyn. 24, 293–304.

Cooke, R.J.S., 1981. Eruptive history of the volcano at Ritter island. In: Johnson, R.W. (Ed.), "Cooke-Ravian Volume of Volcanological Papers" Geological Survey of Papua New Guinea. Port Moresby, pp. 115–123.

Dawson, A.G., Shi, S., 2000. Tsunami deposits. Pure Appl. Geophys. 157, 875–897.

Dawson, A.G., Shi, S., Dawson, S., Takahashi, T., Shuto, N., 1996. Coastal sedimentation associated with the June 2nd and 3rd, 1994 tsunami in Rajegwesi, Java. Quat. Sci. Rev. 15, 901–912.

Dominey-Howes, D.T., 2004. A re-analysis of the Late Bronze age eruption and tsunami of Santorini, Greece, and the implications for the volcano-tsunami hazard. J. Volcanol. Geoth. Res. 130, 107–132.

Dominey-Howes, D.T., Papadopoulus, G.A., Dawson, A.G., 2000. Geological and historical investigation of the 1650 Mt. Volumbo (Thera island) eruption and tsunami, Aegean Sea, Greece. Nat. Hazards 21, 83–96.

Francis, P.W., 1985. The origin of the 1883 Krakatau tsunamis. J. Volcanol. Geoth. Res. 25, 349–363.

Freundt, A., Kutterolf, S., Wehrmann, H., Schmincke, H.U., Strauch, W., 2006. Eruption of the dacite to andesite zoned Mateare Tephra, and associated tsunamis in Lake Managua, Nicaragua. J. Volcanol. Geoth. Res. 149, 103–123.

Gelfenbaum, G., Jaffe, B., 2003. Erosion and sedimentation from the 17 July, 1998 Papua New Guinea tsunami. Pure Appl. Geophys. 160, 1969–1999.

Hatori, T., 1984. Reexamination of wave behavior of the Hokkaido-Oshima (the Japan Sea) tsunami in 1741—their comparison with the 1983 Nihonkai-Chubu tsunami. Bull. Earthquake Res. Inst. Univ. Tokyo 59, 115–125 in Japanese with English abstract).

Johnson, R.W., 1987. Large-scale volcanic cone collapse: the 1888 slope failure of Ritter volcano. Bull. Volcanol. 49, 669–679.

Johnson, R.W., Threlfall, N.A., 1985. Volcano Town—The 1937–43 Rabaul Eruptions. Robert Brown and Assoc, Bathurst, N.S.W, p. 151.

Katsui, Y., Yamamoto, M., 1981. The 1741–1742 activity of Oshima-oshima volcano, north Japan. J. Fac. Sci. Hokakido Univ. Ser. IV 19, 527–536.

Latter, J.H., 1981. Tsunamis of volcanic origin: summary of causes, with particular reference to Krakatau, 1883. Bull. Volcanol. 44, 467–490.

McCoy, F.W., Heiken, G., 1997. Tsunami generated by the late Bronze age volcanic eruption of Santrini, Greece. EOS Trans. Am. Geophys. Union 78, 452.

McCoy, F.W., Heiken, G., 2000. Tsunami generated by the late Bronze age eruption of Thera (Santrini), Greece. Pure Appl. Geophys. 157, 1227–1256.

Minoura, K., Imamura, F., Takahashi, T., Shuto, N., 1997. Sequence of sedimentation process caused by the 1992 Flores tsunami. Geology 25, 523—526.

Minoura, K., Imamura, F., Kuran, U., Nakamura, T., Papadopoulos, G.A., Takahashi, T., Yalciner, A.C., 2000. Discovery of Minoan tsunami deposit. Geology 28, 59—62.

Nanayama, F., Satake, K., Furukawa, R., Shimokawa, K., Atwater, B.F., Shigeno, K., Yamaki, S., 2003. Unusually large earthquakes inferred from tsunami deposits along the Kuril trench. Nature 424, 660—663.

Nishimura, Y., 1997. Evaluation of phreatomagmatic eruptions by means of tsunamis of volcanic origin. In: Proceedings of the International Seminar on Vapor Explosions and Explosive Eruptions, pp. 231—236. May 22—25, Sendai, Japan.

Nishimura, Y., Miyaji, N., 1995. Tsunami deposits from the 1993 southwest Hokkaido earthquake and the 1640 Hokkaido Komagatake eruption, northern Japan. Pure Appl. Geophys. 144, 719—733.

Nishimura, Y., Satake, K., 1993. Numerical computations of tsunamis from the past and future eruptions of Komagatake volcano, Hokkaido, Japan. In: Proceedings of the IUGG/IOC International Tsunami Symposium. Wakayama, Japan, pp. 573—583.

Nishimura, Y., Miyaji, N., Suzuki, M., 1999. Behavior of historic tsunamis of volcanic origin as revealed by onshore tsunami deposits. Phys. Chem. Earth 24, 985—989.

Nishimura, Y., Suzuki, M., Miyaji, N., Yoshida, M., Murata, T., 2000. Paleo-tsunami deposit found in Ayukawa, Kumaishi-cho, oshima Peninsula, Hokkaido. Gekkan Chikyu. Gogai 28, 147—153 in Japanese).

Nishimura, Y., Nakagawa, M., Kuduon, J., Wukawa, J., 2005. Timing and scale of tsunamis caused by the 1994 Rabaul eruption, east new Britain, Papua New Guinea. In: Satake, K. (Ed.), Tsunamis: Case Studies and Recent Developments. Springer, Berlin, pp. 43—56.

Nomanbhoy, N., Satake, K., 1995. Generation mechanism of tsunamis from the 1883 Krakatau eruption. Geophys. Res. Lett. 22, 509—512.

Pelinovsky, E., Zahibo, N., Dunkley, P., Edmonds, M., Herd, R., Talipova, T., Kozelkov, A., Nikolkina, I., 2004. Tsunami generated by the volcano eruption on July 12—13, 2003 at Montserrat, Lesser Antilles. Sci. Tsunami Hazards 22, 44—57.

Penegina, T.K., Bourgeois, J., Bazanova, L.I., Melekestev, I.V., Braitseva, O.A., 2003. A millennial-scale record of Holocene tsunamis on the Kronotskiy Bay coast, Kamchatka, Russia. Quatern. Res. 59, 36—47.

Satake, K., Kato, Y., 2001. The 1741 Oshima-Ohshima eruption: extent and volume of submarine debris avalanche. Geophys. Res. Lett. 28, 427—430.

Self, S., Rampino, M.R., 1981. The 1883 eruption of Krakatau. Nature 294, 699—704.

Sigurdsson, H.S., Carey, S., Mandeville, C., 1991. Submarine pyroclastic flows of the 1883 eruption of Krakatau volcano. Natl. Geogr. Res. Explor. 7, 310—327.

Simkin, T., Fiske, R., 1983. Krakatau, 1883 — the Volcanic Eruption and its Effect. Smithonian Institute Press, Washington, DC, p. 464.

Simkin, T., Siebert, L., 1994. Volcanoes of the World, second ed. Geoscience Press, Tuscon, AZ, p. 349.

Symons, G.J., 1888. The eruption of Krakatau and subsequent phenomenon. In: Report of the Krakatau Committee of the Royal Society, p. 494.

Tsuji, Y., 1989. Tsunami from the eruption of Hokkaido Komagatake on July 13, 1640. Abstr. Seismol. Soc. Jpn. 1, 336 (in Japanese).

Ui, T., 1983. Volcanic dry avalanche deposits—identification and comparison with nonvolcanic debris steam deposits. J. Volcanol. Geoth. Res. 18, 135—150.

Verbeek, R., 1984. The Krakatoa eruption. Nature 20, 10—15.

Ward, S., Day, S., 2003. Ritter Island volcano—lateral collapse and the tsunami of 1888. Geophys. J. Int. 154, 891—902.

Waythomas, C., 2000. Reevaluation of tsunami formation by debris avalanche at Augustine Volcano, Alaska. Pure Appl. Geophys. 157, 1145—1188.

Waythomas, C., Neal, C., 1998. Tsunami generation by pyroclastic flow during the 3500-year B. P. caldera-forming eruption of Aniakchak volcano, Alaska. Bull. Volcanol. 60, 110—124.

Waythomas, C.F., Watts, P., 2003. Numerical simulation of tsunami generation by pyroclastic flow at Aniakchak volcano Alaska. Geophys. Res. Lett. 30, 5−1:4.

Yokoyama, I., 1981. A geophysical interpretation of the 1883 Krakatau eruption. J. Volcanol. Geoth. Res. 9, 359−378.

Yokoyama, I., 1987. A scenario of the 1883 Krakatau tsunami. J. Volcanol. Geoth. Res. 34, 123−132.

Yoshimoto, M., Furukawa, R., Nanayama, F., Nishimura, Y., Nishina, K., Uchida, Y., Takarada, S., Takahashi, R., Kinoshita, H., 2003. Subaqueous distribution and volume estimation of the debris-avalanche deposit from the 1640 eruption of Hokkaido-Komagatake volcano, southwest Hokkaido, Japan. J. Geol. Soc. Jpn. 109, 595−606 in Japanese with English abstract).

CHAPTER 16

Tsunamiites—conceptual descriptions and a possible example at the Cretaceous—Paleogene boundary in the Pernambuco Basin, Northeastern Brazil

G.A. Albertão[1], P.P. Martins, Jr.[2], F. Marini[3]

[1]*PETROBRAS (UN-BC/RES/GGER), Macaé, Rio de Janeiro, Brazil;* [2]*Universidade Federal de Ouro Preto, Departamento de Geologia, Campus do Morro do Cruzeiro, Ouro Preto, Minas Gerais, Brazil;* [3]*Ecole Nationale Polytechnique de Geologie—ENSG, CNRS/CRPG Centre de Recherches Petrographyques et Geochimiques, Vandoeuvre-les-Nancy, France - In memorian*

1. Prologue

This paper is an update of a previous work with almost the same title, published in 2008 (Albertão et al., 2008). In this present version, we have added some new observations on the studied area, which are introduced in the following paragraphs.

First, there is still a very hard discussion in the scientific community about the abandonment of the use of the term Tertiary (https://ncs.naturalsciences.be/paleogene-neogene), which has been omitted from the stratigraphic charts of the IUGS/ICS (International Union of Geological Sciences/International Commission on Stratigraphy) since 1989. But, according to Edwards et al. (2013), both terms "Tertiary" and "Cretaceous-Tertiary (K/T) boundary" are still very popular and also used by many authors in the scientific community (e.g.,: the Geological Society of America — GSA — still presents the term "Tertiary" in its geological chart). However, as the ICS has not taken any initiative to reintroduce it, the term "Tertiary" has increasingly being fallen into disuse in the scientific literature and replaced by Paleogene and Neogene, officially and respectively the oldest and youngest periods of the Cenozoic Era. For this reason, in recent years, most of scientific papers relating to this subject refer the previous term K/T boundary as K/Pg (Cretaceous—Paleogene) boundary, and we follow this tendency in the present publication.

Tsunamiites. https://doi.org/10.1016/B978-0-12-823939-1.00016-1
Copyright © 2021 Elsevier B.V. All rights reserved.

Second, we briefly relate and discuss two publications that have contested some of our interpretations, exposed in our original paper of 2008. As we do not have performed new research in the area, it is not possible to discuss them in detail. However, it is important to explain that after the beginning of the year 2000, the main studied area (Poty quarry, Pernambuco Basin) underwent heavy flooding and access to most of the studied outcrops became impracticable. In this way, samples used in the study of such publications do not necessarily reproduce those discussed in the present publication, since most of our results come from samples collected in the original research (decade of 1990 and beginning of the year 2000) when the whole area of the quarry was available. Recently, in order to transform the studied area into a geosite, which was inaugurated in November 2018, a continuous exhibition outcrop was opened to the research and will allow a more adequate comparison in future studies.

2. Introduction and previous studies

Dott (1983) characterized some sedimentary catastrophic ("episodic," *sensu* Dott, 1983, or "convulsive," *sensu* Clifton, 1988) deposits that mostly concerned with tempestites, inundites, and turbidites. Since then, numerous works have focused on these types of deposit. More recently, some works have discussed tsunami processes and their by-products (Bourgeois et al., 1988; Bryant and Young, 1996; Koenig, 2001; Lawton et al., 2005; Minoura et al., 1996; Shiki et al., 2000a,b), but the tsunami processes have not yet been sufficiently explored so as to provide a general theory. For this reason, Shiki (1996) and Shiki et al. (2000a) requested theoretical papers about tsunamiites.

This proposition is still valid and may be considered as an invitation for much more geological and theoretical research; a theoretical and synthesizing view will necessarily progress step by step. The present authors intend to provide an introductory but general view with a focus on the peculiarities of two sedimentary sections in the Pernambuco Basin, northeastern Brazil [the Poty quarry section near Recife and one in the Ponta do Funil area close to the border between Pernambuco and Paraíba states (Fig. 16.1)]. These Brazilian sections contribute valuable information to an emerging general view of tsunamiites.

A scientific revolution was started by the work of Alvarez et al. (1980): they claimed that the mass extinction at the Cretaceous/Paleogene (K/Pg) boundary was triggered by the impact of an asteroid or a comet. The evidence included the anomalous enrichment of rare chemical elements [platinum group, in particular iridium (Ir)], right at the K/Pg boundary of sedimentary successions in Italy, Denmark, and New Zealand.

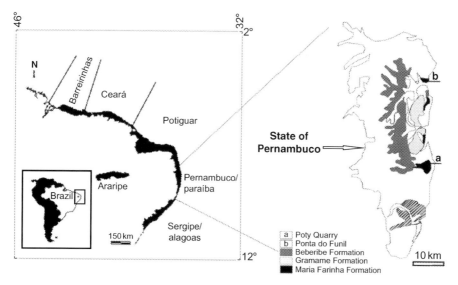

Figure 16.1
Locations of the outcrops in the Poty quarry [a: Universal Transverse Mercator (UTM) 9 152 000N/300 000E] and the Ponta do Funil area [b: UTM 9 117 000N/296 000E], Pernambuco. The two localities are about 30 km apart. Starting points for UTM coordinates: 10 Mm S of the equator and 0.5 Mm W of the meridian 39°W of Greenwich, respectively. Both localities belong to the Pernambuco/Paraíba Basin. *From Albertão, G.A., Martins Jr., P.P., 1996a. Sediment. Geol. 104, 180–201, and also from Albertão, G.A., Martins Jr., P.P., 2002, with kind permission of Springer Science and Business Media.*

Further research (Albertão and Martins, 1996a; Alvarez et al., 1992; Bourgeois et al., 1988; Hildebrand et al., 1991; Smit, 1990; among many others) pointed to peculiarities in the boundary beds at different locations all over the world. These were in particular, besides the Ir anomaly, shocked quartz, microspherules, microtektites, soot, tsunamiites, and an impact crater.

Preliminary investigations of the sedimentary section spanning the K/Pg boundary in the Pernambuco Basin provided direct evidence for a bolide-impact event (Albertão, 1993). This is a particularly interesting occurrence, because it is still the only K/Pg boundary reference section exposed in southern low-latitude regions of South America, without evidence for a significant hiatus or reworking.

Some previous detailed stratigraphic studies in the Poty quarry section (Fig. 16.1) have been carried out by Beurlen (1967a,b), Tinoco (1967, 1971, 1976), Mabesoone (1967), Mabesoone et al. (1968), and Stinnesbeck (1989) but only after the work of Albertão (1993), a series of papers (Albertão and Martins, 1996a,b, 2002; Albertão et al., 1993, 1994) described this section as one of the most complete K/Pg boundary sections in South America. They emphasized the presence of a major fossils extinction, Ir and total organic carbon (TOC) anomalies, microspherules, shocked-quartz grains, and a possible tsunamiite.

Detailed micropaleontological studies were also performed with the aim of determining the faunal extinction and new oncoming populations, based on foraminifers (Koutsoukos, 1994, 1996), nannofossils (Grassi, 2000), marine ostracods (Fauth, 2000), and palynomorphs (Sarkis, 2002). Albertão and Martins (1995, 2002) and Albertão et al. (2004) focused on the geochemical aspects to distinguish between different environments in the various layers, as well as to discuss their peculiarities.

Finally, more extensive studies were dedicated to the microspherules, under the scope of the International Geological Correlation Programme, IGCP, project number 348 (Albertão, 1997; Albertão et al., 2004; Delício et al., 2000; Marini and Albertão, 1999, 2000; Marini et al., 2000).

Some of the following descriptions of the sedimentological features of the possible tsunamiite at this Brazilian K/Pg boundary section are derived and updated from the original paper of Albertão and Martins (1996a).

Although focusing on the sedimentary section of the Pernambuco Basin, the present contribution also presents a general and comparative view on tsunamiites, allowing one to establish some methods and to determine peculiar features to distinguish impact-derived tsunamiites from event deposits formed because of other processes, such as earthquakes, volcanism, and landslides.

3. A theoretical approach toward the identification of tsunamiites

Certain key characteristics are *sine qua non* conditions for the recognition of tsunami deposits. The most important one concerns the relationship between the sedimentological products and the wave energy as estimated by length and height, according to observations by Bourgeois et al. (1988). The parameters of waves—energy, height, and length—can be reconstructed from clasts sizes, primary sedimentary structures, and other specific aspects. Simple stratigraphic separation of sandy and muddy parts may be an event-related characteristic but is not a determining one except for the situation in which the characteristics are diagnostic of tempestites. Bioturbation is a complication in every situation. Fragmented pieces of plants and animals may be an important aspect but are not decisive, for they can also be found in tempestites, although this is not the case for the dramatic mixing observed in most cases of tsunamiites.

A theoretical approach that might support a decision about whether characteristics are diagnostic of tsunami deposits needs to include the details about the following main items:

1. sedimentary deposits
2. models for the conditions of the sediments' genesis
3. disturbances that may hinder the recognition of details of the sedimentary deposits
4. comparison of different tsunami deposits in various types of coastal areas under different geographic and ecological conditions
5. fauna and flora from the local-invaded area
6. fauna and flora eventually carried by ocean waves
7. mixing of sediments and mixing of residues within the sediments by the final retreat of waves
8. season of the year, in case of a pre-existing snow or ice sheet covering the coastal area
9. degree of destruction of biotic residues, in the case of tropical areas with mangroves, dunes, and other types of ecosystems
10. possible existence of an incipient separation of sediments by size
11. possible faunal and floral extinctions in the underlying sediment
12. detailed correlation and comparison of different deposits all over the world as produced at any interval of the geological timescale

Tables 16.1 and 16.2 present a summary of characteristics that are valuable for the identification of two types of tsunamiite: (1) those resulting from impact events and (2) those resulting from any other type of event, such as earthquakes, volcanism, and landslides.

Common aspects to both, as well as specific aspects, must in any situation be distinguished from tempestite, turbidite, and inundite characteristics, although this is not a simple task, as has been extensively discussed by Coleman (1968), Dott (1983), Aigner (1985), Bourgeois et al. (1988), Clifton (1988), Einsele and Seilacher (1991), Smit et al. (1994), Bryant and Young (1996), Einsele et al. (1996), Minoura et al. (1996), Shiki (1996), Shiki et al. (2000a,b), Koenig (2001), and Dypvik and Jansa (2003), among others. In the case of a meteorite impact, an unusual concentration of siderophile elements in sediments may be an indication of the extraterrestrial contribution (Alvarez et al., 1980; Bourgeois et al., 1988). Impact-derived evidence, such as soot, high-pressure metamorphic rocks, microtektites, microspherules, and shocked quartz are part of the signature of an impact event as remaining evidence of the mechanical process of the impact itself (Albertão, 1993).

Table 16.1: Characteristics of, and theoretical considerations on, impact-derived tsunamiites (Bourgeois et al., 1988; Bryant and Young, 1996; Dypvik and Jansa, 2003; Koenig, 2001; Maurrasse and Sen, 1991; Minoura et al., 1996; Shiki et al., 2000a,b; Smit et al., 1994).

Tsunamis and tsunamiites	Mixing of sediments	Mixing of biotic material	Fossils	Layering of sediments	Direction of currents	System extent	Impact-derived material	Unusual elements
Actual-event horizon	Layer of mixing sediments	Mixing of fragments; recent (micro) fauna or fossil types	Mixing of fossil fauna and flora	Sandy sediment at the base and more silty or muddy on the top—nondecisive	Front invasion, channel retreat, turmoil, out-of-phase waves	Littoral, receding and overpassing the littoral, or far inside the continent (or island)	Spherules, soot, derived shocked quartz, etc.	Unusual concentration of rare elements in sediments (such as the Pt-family elements, derived from impacts)
Underlying horizon	Absence of mixed-type sediments		Old fauna and flora; extinction of flora and fauna				Absence of impact-derived material	Absence of unusual elements
Overlying horizon	New sediments deposited by different geological process		New fauna and flora	New layer from other type of subaerial and/or subaquatic sediments			Absence of impact-derived material	Absence of unusual elements
Geographic extension		Deep inside the land or only on littoral areas				Subaerial and/or subaquatic	Regional and/or worldwide	Occasionally present

Coastal type system(s)				Littoral and neritic zones		Occasionally present
Season of the year	Winter—soil covered with snow and/or ice, mangroves, dunes, etc.	Local fossils and/or allochthonous fossils Dominant pollen and other microfossils	Layer(s) covering previous habitats			Occasionally present

Table 16.2: Characteristics of tsunamiites induced by nonimpact processes (earthquakes, volcanism, and landslides; same references as Table 14.1).

Tsunamis and tsunamiites	Mixing of sediments	Mixing of biotic material	Fossils	Layering of sediments	Direction of currents	System extent	Sources of wave energy	Impact-derived material	Unusual elements
Actual-event horizon	Mixing of sediments and subaerial material (biotic and/or mineral)	Mixing of local flora and fauna	Fossils roughly belonging to the same time interval as under- and overlying horizons	Distinct separation of sandy and muddy layers	The same as those from many more tsunamis	Littoral, receding and overpassing the littoral, or far inside the continent (or island)		Absent	Absent
Underlying horizon	Mixing of successive tsunamis	Mixing of local flora and fauna	Fossils roughly belonging to the same time interval as under- and overlying horizons	Distinct separation of sandy and muddy layers	The same as those from many more tsunamis			Absent	Absent
Overlying horizon	Mixing of successive tsunamis	Mixing of local flora and fauna	Fossils roughly belonging to the same time interval as under- and overlying horizons	Distinct separation of sandy and muddy layers	The same as those from many more tsunamis			Absent	Absent

Geographic extension	Over the whole littoral area, occasionally far inland					Absent	Absent
Coastal type system(s)	Postevent permanence of coastal types		Plate boundaries, transcurrent faults, other fault zones, volcanoes, and landslides			Absent	Absent

272 Chapter 16

All these aspects are prerequisites for a generally applicable approach to a theory of the tsunami processes that produce tsunamiites (Bourgeois et al., 1988; Bryant and Young, 1996; Coleman, 1968; Dypvik and Jansa, 2003; Einsele et al., 1996; Koenig, 2001; Lawton et al., 2005; Martins, 2000; Martins et al., 2000; Minoura et al., 1996; Shiki, 1996; Shiki et al., 2000a,b; Smit et al., 1994).

4. Methods and data analyses

Despite the lack of specific studies aimed at the identification of a complete section with a K/Pg transition in the Brazilian sedimentary basins at the beginning of the 1990s, some areas (that yielded outcrop data) and drilling wells (that yielded subsurface data) were selected on the basis of Petróleo Brasileiro S.A. (PETROBRAS, the state oil company of Brazil) data. Albertão (1993) has more systematically studied these. Drilling wells from the Campos, Espírito Santo, and Sergipe-Alagoas Basins and some outcrops from the Pernambuco Basin, focused on the present contribution, were then selected.

For cutting samples that had been collected from those wells, instrumental neutronic activation analysis was performed by the Los Alamos National Laboratory with the objective of determining the Ir content. Although an Ir anomaly was not observed in these samples, a by-product was the determination of concentrations for 45 other elements. Accuracy obtained from these concentration analyses varied in general from 5% to 8%. Geochemical results have been presented by Albertão (1993) and have been updated and summarized by Albertão et al. (2004).

Outcrops in the Pernambuco Basin (Poty quarry and Ponta do Funil area) permitted more detailed sedimentological investigations in the field. In addition, a series of 60 samples was collected at the main stratigraphic succession, in the Poty quarry, at mutual distances of 1−2 cm across the K/Pg boundary. Most of the samples were analyzed using petrographic analysis, micropaleontological analysis (foraminifers and palynomorphs), X-ray diffractometry, stable-isotope analysis (carbon and oxygen), insoluble-residue and TOC analysis, and instrumental neutronic activation analysis, which provided concentrations for 46 chemical elements, including Ir. Details on these methods have been described by Albertão (1993) and have been updated and summarized by Albertão et al. (1994) and Albertão and Martins (1996a,b, 2002).

Spherules and quartz grains were manually separated from the rock samples and analyzed with a binocular lens following the procedures described by Albertão (1997) and Delício et al. (2000). A Philips XL30 SEM (Scanning Electron Microscope) for standardless, EDS (Energy Dispersive Spectrometer/Spectrometry) semiquantitative analysis and a Camebax SX50 EPMA (Electron Probe Micro-Analyzer) with convenient standards, for WDS (Wavelength Dispersive Spectrometer/Spectrometry), fully quantitative analysis, permitted the determination of the morphology and chemical content of the spherules (Marini and Albertão, 1999, 2000; Marini et al., 2000); a chemostratigraphic search for fluorine has been performed later (Marini et al., 2000).

5. Geological setting of the study area

The study area is part of the Pernambuco–Paraíba Basin, a passive-margin rift basin with its origin related to the opening of the South Atlantic Ocean (Chang et al., 1988; Mabesoone et al., 1968). The sedimentary succession under study was deposited during the Maastrichtian and Danian and forms a marine regressive megasequence, as defined by Chang et al. (1988).

Outcrops studied in the Pernambuco part of the basin, especially those in the Poty quarry (Fig. 16.2), show the following succession: the Paleocene Maria Farinha Formation overlies the Maastrichtian Gramame Formation (marly biomicrites of deep-neritic to upper-bathyal environment). The former consists of alternations of limestones (biomicrites, biosparites, and calcilutites) and shales deposited in a middle-deep neritic environment. There is an erosive contact between these two formations (Fig. 16.3).

The sedimentary structures present at the transition from the uppermost Gramame Formation to the basal parts of the Maria Farinha Formation (hummocky cross-stratification,

Figure 16.2
General view of the Poty quarry. The first level (1) of the quarry (white) is mainly composed of the Gramame Formation. The second and third levels (2 and 3) are composed by the basal part of the Maria Farinha Formation. The fourth and fifth levels (4 and 5) are upper and weathered parts of the Maria Farinha Formation. The total height of the quarry is about 25 m.

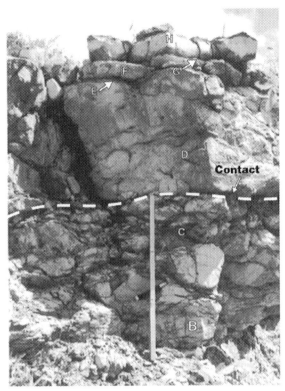

Figure 16.3

Contact (*dashed line*) between the Gramame (*beneath the line*) and Maria Farinha (*above the line*) Formations (Poty quarry). From base to top, beds B, C, D, F, and H are clearly visible. Beds E and G are thin marl horizons between beds D and F and beds F and H, respectively. Bed I overlies bed H, but is not visible in the photo. Bed A underlies bed B, but is covered by debris. From Albertão, G.A., Martins Jr., P.P., 2002, with kind permission of Springer Science and Business Media.

fining upwards and wavy bedding) characterize a carbonate ramp controlled by storms during progressive marine regression (interpretation supported by the ichnofossil, geochemical, paleontological, and mineralogical data: Albertão, 1993).

6. Characteristics of the Poty quarry K/Pg boundary

The succession directly spanning the K/Pg boundary is composed of a series of 14 beds (A–N, from base to top) that reflects a sequence of geological events. The main characteristics of this succession are summarized in Tables 16.3 and 16.4.

The base of the Maria Farinha Formation, particularly bed D (the first bed of the Maria Farinha succession; see Figs. 16.3 and 16.4), is interpreted as deposited under much higher energy conditions than the other beds. Beds D–I show features of a K/Pg boundary cocktail (Bralower et al., 1998; see also Section 8). Details on the sedimentary structures are schematically presented in Fig. 16.4.

Tsunamiites—conceptual descriptions 275

Table 16.3: Summary of field data and petrographic analyses of rocks from the neighborhood of the contact between the Gramame Formation (beds A—C) and the Maria Farinha Formation (beds D—H); see also Figs. 14.3—14.5, 14.9, 14.11, and 14.12.

Beds	Local name	Average thickness (cm)	Rock types and content
A	Marga I	40	Wackestone/packstone with planktonics/phosphatized grains/fossils/some worm tubes (*Hamulus*)/bioturbation (mainly *Thalassinoides*)/pelecypods, echinoderms, foraminifers
B	Poty I	25	Nodular limestone/wackestone—locally packstone/bioturbation/more planktonic than benthic foraminifers/*Hamulus*, echinoderms, ostracodes, calcispherulids/phosphatized fragments
C	Without local name	15	Same as B, though C is more "marly"/spherule occurrences/first observation of very rare and thinly Danian foraminifera
D	Capim	50	Erosive basal contact/packstone grading upward to wackestone-mudstone/rare bioturbation / spherule occurrences/rare, possibly shocked-quartz grains/bio- and siliciclastic coarse sands/phosphatized (more frequent) and nonphosphatized fragments (partially glauconitized—pyritized) of foraminifers, gastropods, pelecypods, worm tubes, echinoderms, shark teeth, wood (rare)
E*	Without local name	2	Continuous marl/spherule occurrences/benthic and planktonic foraminifers/echinoderms/phosphatized fragments/siliciclasts
F*	Topo do Capim	3	Similar to upper bed D/mudstone-wackestone/spherule occurrences

Continued

276 Chapter 16

Table 16.3: Summary of field data and petrographic analyses of rocks from the neighborhood of the contact between the Gramame Formation (beds A—C) and the Maria Farinha Formation (beds D—H); see also Figs. 14.3—14.5, 14.9, 14.11, and 14.12.—cont'd

Beds	Local name	Average thickness (cm)	Rock types and content
G*	without local name	2	Marl/spherule occurrences/ planktonic foraminifers/ echinoderms, fragments of worms tubes/less siliciclasts and phosphatized fragments than in bed E
H*	Batentinho	4	Recrystallized limestone/ mudstone/rare bioclasts, mainly foraminifers/spherule occurrences/ bioturbation—*Chondrites*, *Planolites*, worm tubes/E*, F*, G*, and H* are alternating mudstone/marl thin beds with complex interference ripple structures (symmetrical wave-rippled beds) throughout the quarry

From Albertão, G.A., Martins, Jr., P.P., 2002. With kind permission of Springer and Business Media.

Table 16.4: Summary of field data and petrographic analyses of the Maria Farinha Formation (beds I—N).

Beds	Local names	Average thickness (cm)	Rock types and content
I	Without local name	2	"Marly" mudstone/most probable K-Pg boundary/ spherule occurrences/rare, possibly shocked-quartz grains/ globigeriniform planktonic foraminifers, benthic ones/few siliciclasts and phosphatized fragments/Ir, fluorine, and TOC anomalies
J	Vidro	58	Apparently recrystallized micritic mudstone/tiny planktonic and benthic foraminifers/echinoderms, rare calcispherulids/slightly bioturbated
K	Without local name	20	Similar to J/bioturbation/ bioclasts with glauconite grains and rare phosphatized grains

Table 16.4: Summary of field data and petrographic analyses of the Maria Farinha Formation (beds L—N).—cont'd

Beds	Local names	Average thickness (cm)	Rock types and content
L	Topo do Vidro	23	Intensively bioturbated/almost brecciated/fragments of gastropods/foraminifers/ phosphatised and glauconitized grains
M	Enfornação do Vidro	35	With some elements of **L**/wackestone-packstone/fining-upward/abundant bioclasts/ some phosphatized fragments/large (up to 7 mm) gastropods fragments/worm tubes (serpulids), arthropods, mainly benthic foraminifers, rare bryozoans/phosphatized pellets possibly from the arthropod *Calianassa*
N	Batente	28	Similar to **M**/coarser grain size at the base of **N** than at the top of **M**

Notes: See also Figs. 16.4, 16.5 and 16.12.
From Albertão, G.A., Martins Jr., P.P., 2002, with kind permission of Springer and Business Media.

The K/Pg boundary was first defined (Albertão, 1993; Albertão and Martins, 1992; Albertão et al., 1993) in a thin (2 cm), continuous marly/shaly bed, which is found only in the Poty quarry (bed I; Figs. 16.4 and 16.5). The boundary has been defined through micropaleontological analysis (palynomorphs and foraminifers) based on major extinctions. Palynomorphs characterize the end of the Cretaceous with the extinction of many species of *Dinogymnium*, *Cricotriporites almadaensis*, and *Ariadnaesporites* sp., while the beginning of the Paleogene shows the appearance of many pollens/ spores, such as *Echitriporites trianguliformis*, *Schizeoisporites eocenicus*, *Proxapertites cursus*, and palms. Foraminifers characterize the end of the Cretaceous with the disappearance of species such as *Rugoglobigerina* ex gr. *rugosa*, *Contusotruncana contusa*, and *Pseudoguembelina costulata*, whereas the beginning of the Paleogene shows a continuity of *Guembelitria cretacea* as well as the appearance of *Eoglobigerina eobulloides* and *E. edita* and *Parvularugoglobigerina eugubina*, among other species (Fig. 16.5).

At the same level where a biotic crisis is recorded, the geochemical analysis shows significant Ir and TOC anomalies. The Ir anomaly (0.69 ppb) is identified in bed I (Figs. 16.5 and 16.6), under conditions that are very similar to those in other sections

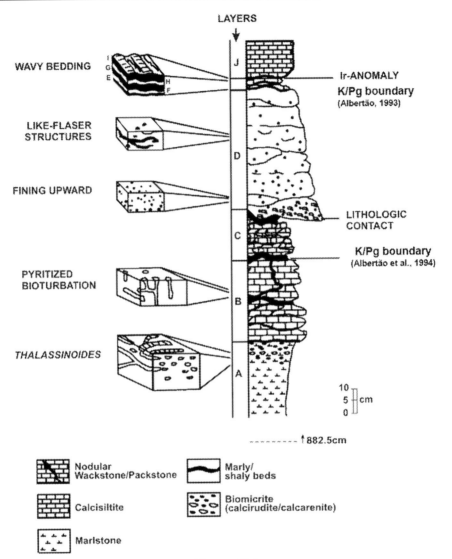

Figure 16.4
Schematic section from the Poty quarry, presenting beds A–J with their main sedimentary structures. Two possible levels for the K/Pg boundary can be considered: the upper one coincides with the Ir anomaly and the lower one with the first occurrence of rare Danian foraminifer species. The lithological contact defines the boundary between the Gramame Formation (lower succession, mostly Maastrichtian in age) and the Maria Farinha Formation (upper succession, mostly Danian in age). *From Albertão, G.A., Martins Jr., P.P., 2002, with kind permission of Springer Science and Business Media.*

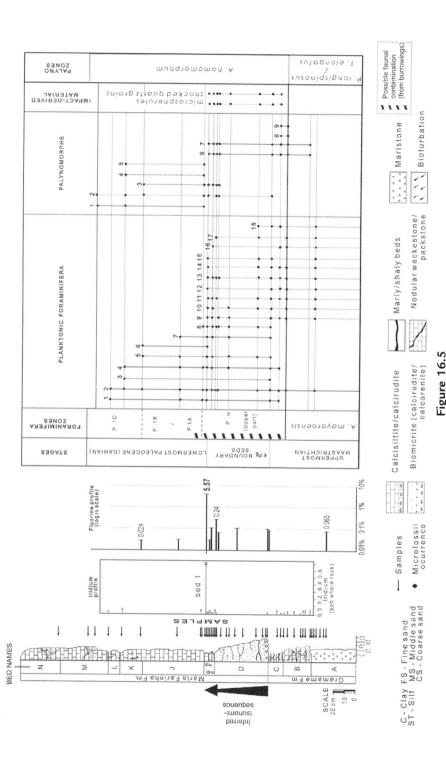

Figure 16.5

Litho- and chemostratigraphy [iridium (Ir) and fluorine] and distribution of microfossils across the K/Pg boundary section in the Poty quarry. Note the Ir and fluorine anomalies in a 2–3-cm-thick claystone (bed I, in planktonic foraminifer zone upper P.α). Original foraminifer zones from Koutsoukos (1996) and Albertão et al. (1994) are taken into account, although beds D–I are considered as K/Pg boundary beds. Possible faunal contamination under bed I is also taken into account. Planktonic Foraminifera (analysis by Dr. Eduardo A. M. Koutsoukos): (1) *Woodringina hornerstownensis*, (2) *Guembelitria cretacea*, (3) *Parasubbotina pseudobulloides*, (4) *W. claytonensis*, (5) *P.* aff. *pseudobulloides*, (6) *Eoglobigerina eobulloides*, (7) *Praemurica taurica*, (8) *Parvulorugoglobigerina eugubina*, (9) *Pseudotextularia nuttalli*, (10) *Pseudoguembelina costulata*, (11) *P. palpebra*, (12) *Rugoglobigerina* ex gr. *rugosa*, (13) *R. scotti*, (14) *Contusotruncana contusa*, (15) *R. reicheli*, (16) *Globotruncana aegyptiaca*, (17) *Racemiguembelina fructicosa*, and (18) *Globotruncana falsocalcarata*. Palynomorphs (analysis by Dr. Marília S. P. Regali): (1) *Proxapertites cursus*, (2) *Pterospermopsis* sp., (3) *Veryhachium reductum*, (4) *Schizeosporites eocenicus*, (5) *Echitriporites trianguliformis*, (6) *Ariadnaesporites* sp., (7) *Dinogymnium* spp., (8) *Cricotriporites almadaensis*, and (9) *Crassitriapertites vanderhammeni*. From Albertão, G.A., Martins Jr., P.P., 1996a. Sediment. Geol. 104, 189–201.

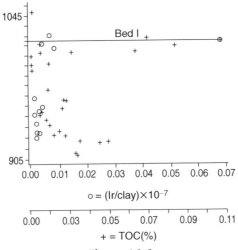

Figure 16.6

Ir/clay and total organic carbon (TOC) anomalies confirm the independence of the Ir anomaly in relation to the clay content in bed I (interval of 1021–1023 cm). Bed I also exhibits a TOC (in wt.%) anomaly. The analyzed samples belong to the base of the Maria Farinha Formation (earliest Paleocene), in the Poty quarry. The vertical scale indicates the thickness above the floor of the quarry (in cm).

around the world with an approximately identical paleoenvironment (Albertão, 1993; Baadsgaard et al., 1988; Bourgeois et al., 1988; Donovan et al., 1988; Johnson et al., 1989; Schmitz, 1992). The anomalous Ir concentration is almost 26 times higher than the mean value found in the other samples collected from the study area. High Ir concentrations may be artificially induced by an increase of the clay content in the deposit, but the Ir/Fe, Ir/Al (geochemical data), and Ir/clay (clay content obtained by X-ray diffractometry) ratios confirm the anomaly (Fig. 16.6). Bed I also records an anomalous value of fluorine (Fig. 16.5); the possible implications of this anomaly have been discussed by Marini and Albertão (1999, 2000) and Marini et al. (2000). A TOC anomaly is also observed in bed I, and this may reflect a high carbon concentration related to the presence of soot. The TOC anomaly is supported by its independence of the insoluble residue content in the analyzed samples.

Impact-derived material, such as microspherules and shattered fragments of shock-metamorphosed quartz grains, is scattered and is relatively frequent in the residues from the K/Pg boundary beds (D–I), but it does not appear or is extremely rare in the subsequent lower boundary beds (A–C) (Koutsoukos, 1996, Fig. 16.5).

The microspherules randomly scattered in beds C–I (Figs. 16.5 and 16.7A) were initially interpreted as impact-derived melt droplets, analogous to microtektites (Albertão, 1997;

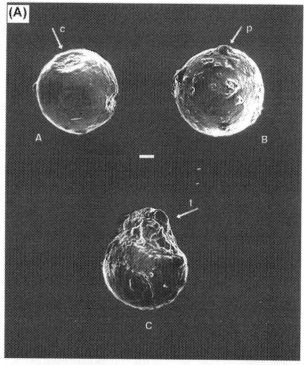
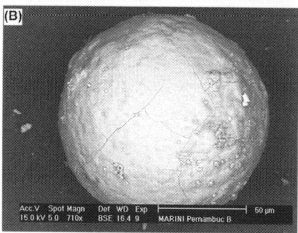

Figure 16.7
Photomicrographs of microspherules. (A) Scanning electron photomicrographs exhibiting the external aspects of three microspherules from the Poty quarry (bed I). Scale bar = 20 mm. Microspherule (A) presents conspicuous "crater-like" pits (c), circular to irregularly shaped, and a very smooth surface. Microspherule (B) exhibits a more corroded surface with protruding mounds or protrusions (p). Microspherule (C) exhibits an upper part resembling a tail (t). (B) SEM photo (BSE mode) presenting a typical "fragile" (brittle)/round spherule (bed I). The surface is covered by a thin phosphate coating. Several cracks are present in the phosphate cover. *(A) Photographs courtesy by Dr. Eduardo A.M. Koutsoukos. (B) Photograph by François Marini.*

Albertão et al., 1994). Additional, more accurate scanning electron microscopy and electron probe microanalysis performed on these spherules (Marini and Albertão, 1999; Marini et al., 2000) have revealed new information.

The majority of the sampled spherules from the Pernambuco Basin are composed mainly of F-rich (or eventually Cl-rich) apatites (Fig. 16.7B; Table 16.5) and differ strongly from the Al-rich and/or Fe-rich phosphates described from altered K/Pg boundary tektites elsewhere (Bohor, 1990; Izett, 1991; Marini et al., 2000). It was nevertheless possible, at least in one of the rarer resistant spherules, to observe a lamellar arrangement and a relict of K-feldspar resembling sanidine, and this is the only indication for an impact origin. Up until now, detailed spherule analyses have been performed on just three selected beds, because of their proximity to the K/Pg boundary: beds E, G, and I (Figs. 16.3 and 16.4). This type of analysis has not yet been performed for the spherule content of bed D, which is a much better candidate for the content of tektite remnants.

Some fragments of shocked-quartz grains with multiple sets of deformation lamellae also occur in beds C (extremely rare) to I (Fig. 16.5). The intersecting sets of straight planar lamellae (Fig. 16.8A and B) were most probably formed by shock-metamorphic processes and are similar to those of shocked-mineral grains found in rocks from known impact structures (Jansa, 1993; Robertson and Grieve, 1977) and from other K/Pg boundary sites (Bohor, 1990; Izett, 1991).

7. The controversy about the position of the K/Pg boundary

Albertão (1993) and Albertão et al. (1993) defined clay bed I (Fig. 16.4) as the K/Pg boundary, the extinction datum of nearly all the latest Maastrichtian planktonic foraminifers and of many groups of dinoflagellates, pollens, and spores. The fact that the stratigraphic record of the TOC and Ir anomalies, common shattered fragments of shock-metamorphosed quartz grains, and microspherules is coeval with the mass extinction datum in bed I gives strength to the emplacement of the K/Pg boundary in the Poty quarry section. Furthermore, these features provide direct evidence for an extraterrestrial bolide-impact origin for the terminal Cretaceous event and the related boundary beds.

Somewhat later, Koutsoukos (1994) and Albertão et al. (1994) nevertheless found the first Danian planktonic foraminifers, as well as microspherules and shocked quartz already in bed C (Figs. 16.4 and 16.5). The K/Pg boundary was consequently then considered between beds B and C, instead of bed I. Further micropaleontological studies have reinforced this point of view: marine ostracods (Fauth, 2000) and palynomorphs (Sarkis, 2002) indicate that the K/Pg boundary position is coeval with that found by foraminifer

Table 16.5: Results of electron probe microanalysis (EPMA, calibrated with Durango Fluorapatite and other classical standards, Camebax SX50 Microanalyzer).

	N	SiO_2	Al_2O_3	FeO	MnO	MgO	CaO	Na_2O	K_2O	TiO_2	P_2O5	F	Cl	SO_3	Total	Flute	Ap
E	5	0.1	0.0	0.5	0.0	0.5	54.1	1.1	0.0	0.0	30.4	4.8	0.1	2.1	93.8	4.3	71.9
H1	9	0.5	0.2	1.0	0.0	0.7	64.5	0.8	0.5	0.0	8.1	39.0	0.3	1.6	117.2	78.5	19.2
H2	5	1.1	0.7	1.1	0.0	0.7	59.2	0.7	0.4	0.1	9.0	35.1	0.2	2.6	110.9	70.5	21.2

Mean values for E, H1, and H2 samples (respectively from beds E and H). *Ap*, computed percentage of F-Apatite; *N*, number of analyses; *Flute*, computed percentage of fluorite *versus* F-Apatite. Some of the observed discrepancies in the sum of oxides result from OH and CO_3 in the phosphatic component. Other discrepancies seem to be due to disseminated, nanometric carbonate crystals. These analytical results are part of the work of Marini et al. (2000).

284   Chapter 16

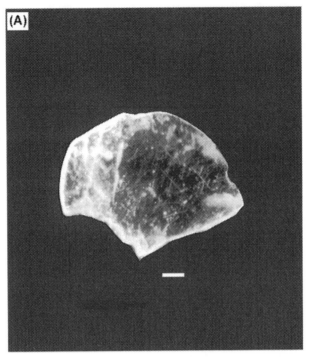

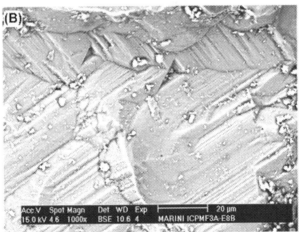

Figure 16.8
Photomicrographs of shocked quartz. (A) Grain from bed D in the Poty quarry, exhibiting intersecting sets of sharp and straight planar lamellae. Scale bar = 120 mm. (B) Photomicrograph exhibiting details on intersecting sets of sharp and straight planar lamellae, present in a grain of shocked quartz (bed D). *(A) Photograph courtesy by Dr. Eduardo A. M. Koutsoukos. (B) Photograph by François Marini.*

studies (Koutsoukos, 1994). Albertão et al. (1994) pointed to the possibility that the Ir anomaly in bed I actually represents a geochemical event related to another impact event during the Danian. The aforementioned data should also be considered in the context of the following two observations.

1. Very rare burrows can, in some parts of the Poty quarry, be found in bed D (Fig. 16.9A and B); some burrows, which have a total length of almost 60 cm, reach the top of bed C. These burrows were analyzed in some detail and reveal a preferential downward development, with evidence that they were formed after partial consolidation of the sediment. They must thus have been formed after the event responsible for deposition of bed D (Fig. 16.9A). In the burrows, reaching the top of bed C, some concentration of siliciclastics and microspherules was found, indicating some possible contamination from the upper succession. The possibility should therefore be considered that bioturbation across these beds has mixed up microfossils of different ages; specific micropaleontological and geochemical analyses of these burrows and adjacent areas are necessary to clarify this.

2. Another important aspect concerns the representativity of the sampling methods used in some of the studies. Most of the paleontological analyses by Koutsoukos (1994), Fauth (2000), and Sarkis (2002) were performed on a thin core recovered from the area during the study of Albertão (1993). It is not clear whether the sediment in this core has been affected by later bioturbation or even by possible artificial downward contamination during core drilling (the results obtained by Albertão, 1993 and Albertão et al., 1993 were based on samples collected from several parts of the Poty quarry that were analyzed through different biostratigraphic methods; all analyses indicated the major extinction at bed I).

The controversial position of the K/Pg boundary makes the following two interpretations of the events possible.

1. If the stratification was affected by bioturbation (and the K/Pg boundary is thus in bed I), two hypotheses considering the conspicuous lateral continuity of bed D and an oceanic gateway between the early Gulf of Mexico and the northern part of the South Atlantic Ocean can be considered (Koutsoukos, 1992): (i) a close coeval and genetic relationship between the Yucatán impact at the K/Pg boundary (the Chicxulub Crater; Hildebrand et al., 1991) and the impact-generated tsunami deposits found in Texas, the Caribbean, and bed D in Pernambuco (Fig. 16.10); or (ii) a relationship of bed D with other hypothetical impact(s) that may have happened somewhere in the South Atlantic, simultaneously with that of the Yucatán Peninsula.

2. If the K/Pg boundary occurs between beds B and C, the tsunami beds may be related to an early Danian bolide impact in the northern South Atlantic. This would be coherent with the hypothesis of multiple impacts across the K/Pg boundary (Keller et al., 2003a).

286 *Chapter 16*

A third point of view must, however, also be considered. Smit et al. (1994) and Stinnesbeck et al. (1994) strongly contributed to the earlier discussion, in spite of possible effects of bioturbation. Especially some observations by Smit et al. (1994) are consistent with comments of Shiki (Personal communication, 2003). Catastrophic events and their products require unconventional sedimentological and paleontological approaches. In the case of a sudden mass extinction, the disappearance of the older fauna/flora should be considered as the reference level for the boundary; the first appearances of new taxa should not, as they may be misleading. It is possible that animals and plants appear not earlier than at the end of the postevent stratigraphic interval, as a result of the dramatic environmental changes after the catastrophe. In addition, special attention has to be paid to sediment reworking, which may result in the occurrence of older fossils in younger sediments; in this case reworked/older and in situ/younger fossils would appear mixed in K/Pg beds (as it is the case of beds D–H in the Poty quarry; see Fig. 16.5). Finally, one has to consider that, in the case of catastrophic events, the timescale to be used is very different from that of "normal" sedimentation: a unique bed, of meters in thickness, can be deposited in some minutes or a few hours; this is well known from, for instance, turbidites (Dott, 1983, 1996). Based on these considerations and views, we propose (see Fig. 16.5) the following:

- the top of bed C is the final part of the upper Maastrichtian.
- beds D–I represent, indeed, the boundary beds. The base of bed D marks the beginning of the tsunami event, while beds E–I record its end (see also later part); it is impossible to establish any chronostratigraphy based on the fossil content, throughout this succession.
- the top of bed I and the base of bed J represent the beginning of Danian sedimentation.

For the time being, this third point of view seems reasonable, not only because of the various previous comments, but also because of the other hypotheses (see hypotheses 1 and 2) still requiring additional research for confirmation of fauna/flora contamination by bioturbation and of the occurrence of at least two bolide-impact events across the boundary.

8. Bed D, the possible tsunamiite and a tentative model

8.1 Characteristics of bed D

Bed D is a bioclastic/intraclastic packstone of thickness 50 cm (Figs. 16.3 and 16.11A). It resembles the "graded skeletal sheet" facies of Aigner (1985), although the coarse texture makes it difficult to distinguish fine sedimentary structures. Bed D, which is a conspicuous bed not only in the quarry but also in other areas, marks the lower boundary of the Maria Farinha Formation.

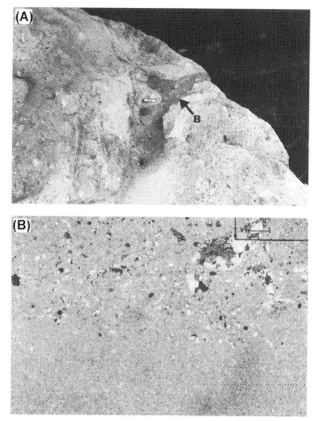

Figure 16.9
Characteristics of bed D. (A) Block from the Poty quarry with a burrow (B) from the top to the base (bottom of the photo). Marly material (dark gray) introduced from above by bioturbation. (B) Photomicrograph of a bioturbated part from the same block as in Fig. 16.9A, with coarser material introduced from above by burrowing. Scale bar = 0.4 mm.

Bed D has some peculiar stratigraphical and sedimentological characteristics, indicating its rapid deposition: (1) a sharp erosive base (Figs. 16.3 and 16.4) followed by (2) normal grading, with shells, siliciclasts, and abundant phosphatized fragments at the base (Fig. 16.11A–C); (3) a coarse grained and poorly sorted character (Fig. 16.11A and C); (4) extensive mixing and fragmentation of fossils derived from different paleodepths and reworked from older strata (Fig. 16.11A and C); (5) scattered impact-derived products (Figs. 16.7 and 16.8); (6) symmetrical wave ripples at the top of the bed (Fig. 16.12A); and (7) great lateral continuity in the basin.

The erosive nature of the base is evident (Fig. 16.3), although the bed shows no conspicuous thickness variation: the thickness varies from 50 to 90 cm. In more limited areas, for example, within the quarry limits, thickness variations are almost unnoticeable.

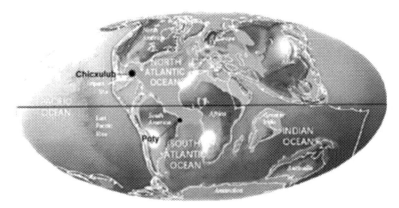

Figure 16.10
Late Cretaceous (Maastrichtian, 66 Ma) paleogeographic map, showing locations of the Chicxulub crater and the Poty quarry in the Pernambuco Basin. Diverging wave trains from the impacted area in Chicxulub may have affected the northeastern coast of South America. *Modified from Scotese, C.R., 2001. Atlas of Earth History, vol. 1. Paleogeography, PALEOMAP Project, Arlington, Texas, pp. 52.*

Normal grading (Fig. 16.11B) is sometimes subtle. Usually, the coarser material is concentrated at the base of the bed. In some parts of the outcrops, another concentration of coarse material occurs near the middle of the bed, indicating a subdivision that might be due to a second event.

Some big intraclasts and bioclasts (Fig. 16.11A and C) are also present. Their maximum diameter is around 10 cm. The sorting is poor or there is no sorting at all, and the grain-size distribution corresponds more to a coarse-tail grading than to fining upward.

Mixing of material from different sources (Fig. 16.11A and C) is one of the most important features. Continental material is present in the form of some rare wood fragments and of relatively abundant siliciclastics (mainly quartz and feldspars). Phosphatized intraclasts and pelloids indicate a contribution of shore-facies material. They usually are derived from Campanian phosphatized beds, well known in the Pernambuco Basin, which have been eroded and removed from uplifted blocks along the paleocoast. Bivalve and gastropods shells also occur and represent, together with echinoid fragments and shark teeth, the shallow-marine components. Abundant characteristically benthic foraminifers represent deeper marine environments. All these components are totally mixed up and are, depending on the size of the material, usually fragmented.

The aforementioned impact-derived products are sparsely found across bed D (Figs. 16.7 and 16.8). Microspherules are by far the most common, although their impact origin is not clear. Shocked-metamorphosed quartz fragments, although rarely, are also observed.

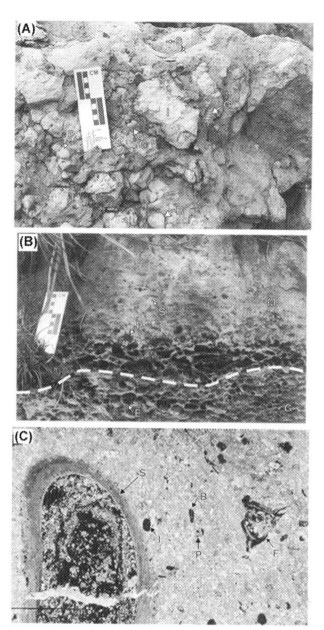

Figure 16.11

Characteristics of bed D. (A) Coarse-grained block from the Poty quarry with the characteristic heterogeneous composition of the bed: phosphatized fragments (P), fragments of gastropods (G) and bivalves (B), siliciclasts (S), and intraclasts (I). (B) Contact (*dashed line*) between the Gramame (G) and Maria Farinha (M) Formations, in the Ponta do Funil area. Note the abundant phosphatized fragments (P), siliciclasts (S), and the normal grading at the base of the Maria Farinha Formation. Some bioturbation and erosion (E) are visible at the top of the Gramame Formation. (C) Photomicrograph of a sample from the Poty quarry with phosphatized bioclasts—such as a large fragment of serpulid (S) with a phosphatized interior, planktonic (P) and benthic (B) foraminifers, and fish bones (F)—and intraclasts (I) in a micritic matrix. Crossed nicols; scale bar = 0.4 mm. *(B) From Albertão, G.A., Martins Jr., P.P., 1996. Sediment. Geol. 104, 189–201.*

Figure 16.12
Characteristics of bed I. (A) Unique continuous and nonweathered section across bed I in the Poty quarry (before the construction of the geosite) with beds D (base) to N (top) visible on the photograph. *Arrows* indicate marly beds E, G, and I (the Ir-rich bed). Bed I is highlighted with *dashed lines*. Wave ripples are present in beds E–H. Total thickness of the succession shown here is about 2.5 m. (B) Poor and rare preservation of bed I in the swales (formed by the ripples) at the top of bed H. Beds H and I are bioturbated. Bed I is usually less than 1 cm thick at these swales; it exhibits Ir, TOC, and fluorine anomalies. *(A) From Albertão, G.A., Martins Jr., P.P., 2002, with kind permission of Springer Science and Business Media.*

Bed D is overlain by wave-rippled beds of fine-grained limestones (beds F and H) and marls (beds E, G and I; see Figs. 16.3, 16.4 and 16.12A), which indicate a progressive reduction of the depositional energy. In addition, bed I (Figs. 16.5 and 16.12B) exhibits Ir and TOC anomalies.

The great lateral continuity of bed D is one more conspicuous characteristic. It makes it possible to follow this bed from the Poty quarry up to the Ponta do Funil area (Fig. 16.1). Between these areas, on the Itamaracá Island, a core was recovered from a stratigraphic well drilled by PETROBRAS in the middle of the last century; this core also contains bed D. Based on the earlier observations, it is possible to estimate a lateral continuity of at least 30 km in Pernambuco.

8.2 Semiquantitative modeling of the depositional process of bed D

The sedimentary data presented earlier, such as the mixture of fossils from different environments, and from recent and reworked strata, as well as the presence of coarse siliciclastic grains, indicate that a high-energy mechanism played a role in the deposition of bed D.

Bourgeois et al. (1988) suggest that coarse-grained sedimentary deposits at or close to the K/Pg boundary are related to huge tsunami waves generated by a meteorite impact. The occurrence of tsunamiites or deposits related to convulsive geological events has been reported from sections at Braggs and the Brazos river (Bourgeois et al., 1988), Haiti (Maurrasse and Sen, 1991), the Gulf of Mexico (Bralower et al., 1998; Lawton et al., 2005; Smit et al., 1992, 1994), Cuba (Iturralde-Vinent, 1992; Takayama et al., 2000), and at Deep Sea Drilling Project sites (Alvarez et al., 1992; Klaus et al., 2000; Olsson et al., 1997). Dypvik and Jansa (2003) presented a summarized and updated view of the sedimentary processes and sedimentary signature resulting from bolide impacts in the sea.

A semiquantitative model of the depositional processes active during the K/Pg boundary event indicates that, most likely, a major tsunami wave formed bed D. The methods described by Bourgeois et al. (1988) and the mechanism indicated by Duringer (1984; Fig. 16.13) have also been applied to bed D (Albertão and Martins, 1996a) and are summarized further. The dimension of the giant wave that was responsible for the deposition of bed D has been estimated on the basis of the assumption that the Pernambuco K/Pg boundary succession represents the same conditions that are assumed for the Brazos river K/Pg boundary succession.

The following observations were taken into account for the semiquantitative modeling (Table 16.6).

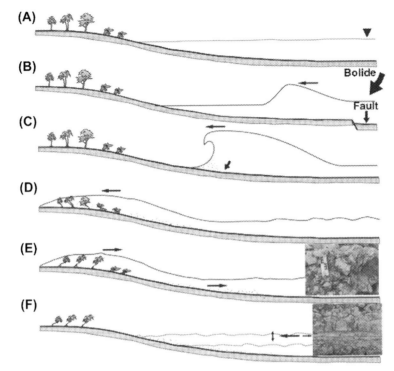

Figure 16.13

Schematic representation of a tsunami. (A) preevent stage; (B) early stage (caused by, e.g., a bolide impact or a submarine fault) generating a wave of small amplitude (2–3 m), large wave length (200 km), and high velocity (800 km/h); (C) the amplitude increases (10–30 m) and the wave velocity decreases (60 km/h) as a result of friction on the sea floor; (D) coastal flooding; (E) backwash flow (commonly responsible for the deposition of a tsunami bed); (F) subsequent oscillations of high frequency and small amplitude. Typical sedimentation in phases E and F is represented by examples from the Poty quarry (see Figs. 16.11A and 16.12A, respectively).
Adapted from Duringer, P., 1984. Tempêtes et tsunamis: Des dépôts de vagues de haute énergie intermittente dans le Muschelkalk supérieur (Trias germanique) de l'Est de la France. Bull. Soc. Géol. Fr. 26, 1177–1185. Reprinted with permission of Société Géologique de France and from Albertão, G.A., Martins Jr., P.P., 1996. Sediment. Geol. 104, 189–201.

1. Autochthonous benthic foraminifers recovered from hemipelagic clay bed I (Albertão, 1993) suggest that the deposition of bed D took place in a relatively deep neritic environment, at a water depth between 100 and 200 m (interpretation based on autochthonous benthic foraminifers).
2. The biggest clasts in bed D have diameters from 2 to 9 cm. The value of 5 cm was used in the calculations, which is the same value used by Bourgeois et al. (1988).
3. The wavy bedding structures of beds E–H (Figs. 16.4 and 16.12A) indicate wavelengths ranging from 15 to 20 cm. Bourgeois et al. (1988) used the value of 15 cm, which was also used here.

Tsunamiites—conceptual descriptions **293**

Table 16.6: Semiquantitative model based on the characteristics of bed D.

Estimated paleodepth (I) 50 −200 m (100 m)	Largest clasts (II) 2−9 (5 cm)	Wave-lengths (above layer D) (III) 15−20 cm (15 cm)
I + II U* Required 15 cm/s	I + II + U* Wave height/ velocity/length 20 m/112 km/ h/112 km	III Reduction of U*15 cm/ s ~ 1.5 cm/s

4. The exclusive action of eustatic sea-level fluctuations would not be enough to create structures such as normal grading and beds with symmetrical wave ripples.
5. Turbidity currents usually produce more asymmetrical undulations (e.g., current ripples).
6. Even the strongest storm would not have been able to generate waves to move the large clasts that occur in bed D at paleodepths of 100−200 m. Moreover, hummocky cross-stratification was not observed in the outcrops of bed D. This structure occurs below bed D in the Ponta do Funil area.

According to Bourgeois et al. (1988), shear velocities (U*) of at least 15 cm/s (but probably more than that) would be needed to erode and transport clasts of about 5 cm in diameter; U* of 15 cm/s at water depths of more than 100 m requires waves higher than 20 m (Table 16.7). Based on the wave height (Table 16.7), it is also possible to estimate (Table 16.6) the wave velocity (C = 112 km/h) and the wavelength (L = 112 km); such waves should be associated with major tsunami processes.

Table 16.7: Estimation of shear velocities versus wave heights.

T (min)	Shear velocity (cm/s) for wave heights of						
	2 m	4 m	6 m	8 m	10 m	15 m	20 m
			Water depth = 50 m				
8	–	1.4	1.9	2.4	+	+	+
12	2.1	3.8	5.4	7.0	8.5	12.1	+
16	2.7	4.9	6.9	8.9	10.9*	(15.5)	(20.0)
20	2.9	5.3	7.6	9.8	(11.9)	(17.0)	(21.9)
			Water depth = 100 m				
8	–	–	–	–	–	+	+
12	–	1.2	1.7	2.3	2.7	3.3	4.9
16	1.4	2.5	3.5	4.5	5.5*	(7.9)	(10.1)
20	1.8	3.2	4.5	5.9	(7.2)	(10.2)	(13.1)

( ), parentheses indicate extreme wave conditions; *, maximum recorded wave; –, subcritical (no transport); +, wave too steep, will collapse.
From Bourgeois, J., Hansen, T.A., Wiberg, P.L., Kauffman, E.G., 1998. Science 241, 567−570, reprinted with permission from AAAS.

294 Chapter 16

Furthermore, the wavelengths of sedimentary structures (Table 16.6) in beds E—H (15—20 cm) indicate that U* was about 1.5 cm/s (using the same principles as Bourgeois et al., 1988). Therefore, a reduction of U* from at least 15 cm/s (at the base of bed D) to 1.5 cm/s (in beds which overlie bed D) is required (Table 16.6). Such a reduction of U* is commonly attributed to tsunami processes in platform settings (Bourgeois et al., 1988). The sedimentary structures in beds E—H also corroborate the reduction of velocity toward the top of bed D (Figs. 16.4 and 16.12A). These fine-grained graded beds could have been deposited by multiple attenuated waves or through wave reflection, known from shelf settings (Bourgeois et al., 1988; Duringer, 1984). Probable impact-derived material, such as Ir, microspherules, and shattered shock-metamorphosed quartz grains are concentrated in the more slowly deposited hemipelagic bed I. Bed H and particularly bed I are bioturbated, which indicates some break in sediment accumulation.

8.3 Discussion of the tsunami process

The conclusions of Koutsoukos (1994, 1996) and Albertão et al. (1994) lead to the suggestion that beds B and C represent the tsunamiite (or another type of event bed) related to the K/Pg boundary. Nevertheless, more accurate sedimentological analyses indicate that both beds B and C were extensively bioturbated (Figs. 16.3 and 16.4) when the sediment was still unconsolidated. In addition, no diagnostic signatures were left by a tsunami-characteristic process, nor an indication of any other rapid depositional event was present.

Bed C can be regarded as a unit formed in a storm-dominated marine environment in which some kinds of animals, including burrow makers, flourished. The bed is composed of many amalgamated units (numerous erosional surfaces can be observed) and was affected and possibly disturbed by an impact-induced seismic shock as well by pressure changes because of the passage of the first tsunami wave.

The observation of possible tsunami signatures exclusively in bed D (and consequently in the fine-grained sediments represented by beds E—I) is consistent with comments by Shiki (Personal communication, 2003).

Following the earlier discussions, we consider that beds D—I form a tsunamiite that was formed by shuttle movements of tsunami waves (Fig. 16.13) induced by one single event, probably the K/Pg boundary impact. Beds D—I form couplets of alternating limestones (D, F, and H) and marls (E, G, and I), implying a different dominant origin of the sedimentary particles. Each couplet represents the sediments formed by one tsunami wave (going landward or returning back offshore). The limestones (D, F, and H) were deposited by some kind of sediment gravity flows. In contrast, the marls (E, G, and I) indicate changes in the sedimentary processes: they were deposited from suspension clouds. The wavelength of a tsunami wave is large enough to cause water movement along the sea floor, and thus the sedimentary clasts behaved as a flow; for this reason, tsunami waves may eventually show the character of a sediment gravity flow and develop features of turbidity currents (Shiki, Personal communication, 2005).

9. Updated information on the area (K/Pg boundary at Poty quarry)
9.1 Discussion on alternative interpretations

Two works (Morgan et al., 2006; Gertsch et al., 2013) have presented some additional discussion on the study area of Poty quarry. In the present paper, we do not have the intention to discuss in detail the arguments presented in those works, as we do not present a new research on the area. Nevertheless, it is important to highlight some points that are necessary to take into account when analyzing the sedimentary section of the Poty quarry. At the beginning of 2000 and up to November 2018, the limestone mining was discontinued and the area of the Poty quarry was completely flooded by pluvial, fluvial, and groundwaters. Besides that, a new mining operation started, exploring clay of the upper parts of the quarry. These facts made impossible the access to the beds of the K/Pg boundary section that were sampled and analyzed in our original work (Albertão, 1993). For these reasons, the two mentioned papers have used specific material collected from previous works or of other areas different from that of our original work.

Morgan et al. (2006) have searched for impact evidence, particularly impact quartz, in some of the most studied K/Pg boundary sections in the world; they do not relate the presence of such evidences in the analyses of the material collected from the Poty quarry. These authors quote they have analyzed the material related by Koutsoukos (1998). It is important to highlight that Dr. E. Koutsoukos is the one who performed the search for microspherules and impact quartz in our previous works, with different results from those of Morgan et al. (2006): although rare, fragments of shocked-quartz were found mainly in the sequence of beds D—I, in samples collected directly at the outcrops of the quarry before the flooding situation of the area (Albertão et al., 1994, 2008; Koutsoukos, 1994, 1996, 1998). This suggests that maybe Morgan et al. (2006) may have misunderstood the correct place to collect and identify the beds to sample the already rare materials, considering that probably the K/Pg boundary sequence would not be accessible during the flooding period, as already mentioned.

Gertsch et al. (2013), in their turn, have worked mainly in a drilling core recovered after a field campaign. We have to consider that sampled material in the sedimentary sequence recovered by drilling cores is probably not as complete as that of the outcrops. In terms of the nature of the calcium and phosphate-rich microspherules, we do agree that most of them are of diagenetic origin and not impact-related, as already discussed in item 5 of this paper. But in this same item we also comment that the pioneer work of Marini et al. (2000) suggested some of the F-rich spherules and mainly, the very rare microspherules exhibiting sanidine grains in the core (found only in the marly beds E, G, and I) as materials potentially impact-related. About the Ir anomaly, we do confirm that our geochemical analyses were performed at the most reputed laboratory and team at that time (Los Alamos National Laboratory; Albertão, 1993) and that only at bed I a relative rich concentration of that element was found (0.69 ppb). Similar values of Ir anomaly (<1 ppb) have been reported worldwide: 0.72 ppb in Marmarth (North Dakota, USA; Johnson et al., 1989), 0.57 ppb in Hell Creek (Montana, USA; Baadsgaard et al., 1988), 0.42–0.87 ppb in Scollar Canyon (Alberta, Canada; Baadsgaard et al., 1988), 0.7 ppb in Braggs (Alabama, USA; Donovan et al., 1988), 0.92 ppb in Arroyo el Mimbral (Mexico; Smit et al., 1992), 0.49 ppb in Waipara (New Zealand; Brooks et al., 1986), 0.43 ppb in Seymour Island (Antarctic; Zinsmeister et al., 1989)—just to quote some examples. Higher values of Ir anomaly are found mainly in more proximal areas of the impact crater of Chicxulub (e.g., K/Pg boundary areas in Central America; Smit et al., 1992; Iturralde-Vinent, 1992) or in deep marine (like the classical K/Pg boundary section at Gubbio, Italy, with an anomaly of 5 ppb; Alvarez et al., 1980), where sedimentary processes acting in condensed sections may increase the initial already (but not so) high Ir content. In the Poty quarry, we do not have any of these conditions. In the paper of Gertsch et al. (2013), there is no Ir analysis or samples in related stratigraphic unit as that of bed I, as we can observe from the recovered core or the exposures represented in their figures; besides that, values relatively high, around 0.5 ppb, in lower stratigraphic levels may have been influenced after bioturbation near the nonrecovered bed I. Finally, in terms of the sedimentary process Gertsch et al. (2013) interpret for the sequence across the K-Pg boundary, we can recognize that we have just the ensemble of sedimentary structures and the other described evidences to support our interpretations—but it is important to consider that models and interpretations are always based not only on the data but also on the previous concepts, experience, and background each one has to evaluate them. The research group of Gertsch et al. (2013) has extensively published about K/Pg boundary in many classical sedimentary sections worldwide, always presenting a point of view contrary to the interpretation of impact or catastrophic processes acting in this geological time; these are the cases in Brazos River, Texas, USA (Keller et al., 2007 vs. Hart et al., 2012), in many Central America sections, such as Cuba, Haiti, Mexico, and so on (e.g., in Belize: Keller et al., 2003b vs. King and Petruny, 2015), in El Kef (Keller et al., 1995 vs. Molina et al., 2006), among many others. And in the Poty quarry, this is not different. Anyway, we highlight once more the considerations presented at the end of previous item 6 "The Controversy About The Position of the K-Pg Boundary") of this work: timescale and sedimentary processes have to be reconsidered when taking into account catastrophic events, as we cannot expect the same stratigraphic results and biostratigraphic distribution as those of processes acting in "normal" sedimentation.

9.2 Present situation of the K/Pg boundary section at Poty quarry

Because of a recent campaign to transform the area as a geosite (Albertão et al., 2016), formally inaugurated in November 2018, the mining company responsible for the area opened a large front, exhibiting in three contiguous exposures a very continuous sequence of the K/Pg boundary (Fig. 16.14). The middle exposure contains the sequence of beds C–N, encompassing by this way all of the critical layers across the K/Pg boundary. This includes the complete sequence of beds D–I (Fig. 16.15), representing the interpreted tsunamiite (bed D) as well as the Ir-rich layer I, frequently bypassed by core sampling.

10. Concluding remarks

The Poty quarry outcrops present a more complete shallow-marine sedimentary succession of the K/Pg boundary than any other area in the Brazilian marginal basins. There is evidence of a terminal Cretaceous event (probably a meteorite impact):

- Ir and TOC anomalies detected for the first time at low latitudes of the southern hemisphere (particularly in South America);
- fauna and flora extinctions in underlying beds;
- particularly, a mass extinction corresponding to that known from the boundaries elsewhere between the Mesozoic and the Cenozoic eras;
- new fauna and flora in overlying strata;

Figure 16.14
Construction of a geosite in the study area at Poty quarry; it is now the geosite of the K/Pg boundary at Poty quarry (Poty Mine), Pernambuco Basin, Brazil. (A) Reopening the K/Pg boundary exposures (October 2017); (B) Inauguration of the geosite (November 2018).

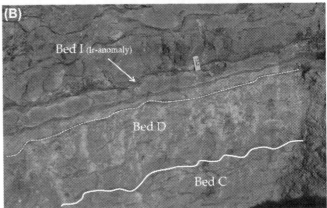

Figure 16.15
Middle (and principal) section of the geosite; the complete sequence exhibiting the K/Pg boundary. (A) General sequence of the main section: the scale bar (3.5 m) highlights the continuous sequence from bed C (base of the section) until bed N (mid-portion of the exhibited section). (B) Detail of the previous photograph [rectangle of (A)]: it is possible to observe the sequence of beds C to I; we reinforce the position of bed D (the base of the interpreted tsunamiite) and bed I (Ir-anomaly bearing bed).

- impact-derived exotic products (such as microspherules and shock-metamorphosed quartz grains);
- the presence of a possible impact-generated tsunamiite.

Semiquantitative modeling, taking into account the sedimentary characteristics of bed D, indicates that its deposition involved high-energy processes, coeval with a tsunami event. In addition, the overlying fine-grained graded beds (E—I), with out-of-phase wave bedding, indicate a rapid reduction of U^*. Microspherules and shocked quartz in these beds support an impact origin for this possible tsunamiite.

We conclude that extensive and careful sedimentological studies on tsunamiites can provide valuable information concerning the K/Pg boundary processes, which in turn would increase our understanding of tsunamiites.

Acknowledgments

We thank Prof. Tsunemasa Shiki for his interest with our technical discussions and his suggestions since his last visit to Pernambuco (2003). We have tried to take advantage of the remarks by Shiki et al. (2000a, p. viii) about "the genesis of these sedimentary features" as they indicate it as "a research area for study." We are also grateful for the technical discussions with, and the help from, other colleagues who supported us with their interest into our study of the Poty location; this holds particularly for Dr. Eduardo Apostolos Machado Koutsoukos and Dr. Marília da Silva Pares Regali (in memoriam) regarding paleontological aspects. Thanks are also due to Juan Massad Bringhenti (XEROX Co., drafting support section of PETROBRAS) for providing the final version of the figures, to David Anthony Morgan for providing an English revision, and to Profs. David T. King, Jr. and Hisashi Suzuki for their final technical review.

We thank VOTORANTIM S.A. (owner of the mining rights of the Poty Mine, where the K/Pg boundary site is located), PETROBRAS, the Department of Geology of Universidade Federal de Pernambuco, the Department of Geology of the Graduate Program of Universidade Federal de Ouro Preto and Fundação CETEC (Centro Tecnológico de Minas Gerais) for supporting our efforts from the very beginning of our studies and for permission to publish this contribution.

Last, but not least, we would like to honor our colleague, Prof. François Marini, from the University of Nancy (France), one of the coauthors of this work, who passed away in 2009, shortly after the first edition of this paper. This new publication in this way brings to our memory all the dedication that this colleague had in the field of the Geosciences and, in particular, his capital contribution to the study of the exotic and impact particles present in the K/Pg boundary of diverse localities of the world.

References

Aigner, T., 1985. Storm depositional systems: dynamic stratigraphy in modern and ancient shallow marine sequences. In: Lecture Notes in Earth Sciences, vol. 3. Springer-Verlag, Berlin, p. 174.

Albertão, G.A., 1993 (M.Sc. thesis). Abordagem Interdisciplinar e Epistemológica sobre as Evidências do Limite Cretáceo-Terciário, com Base em Leituras Efetuadas no Registro Sedimentar das Bacias da Costa Leste Brasileira, vol. 2. Federal University of Ouro Preto, p. 251.

Albertão, G.A., 1997. The Cretaceous-Tertiary boundary in Brazil—state of the art and peculiarities. Sphaerula (Int. J. IGCP 384 1, 101–114.

Albertão, G.A., Martins Jr., P.P., 1992. O limite Cretáceo-Terciário nas bacias sedimentares da Costa Leste Brasileira. In: Abstr. 35th. Congr. Geol., pp. 463–465 (Brazil).

Albertão, G.A., Martins Jr., P.P., 1995. Abordagem estatística para discriminação geoquímico-estratigráfica de depósitos da transição Meso-Cenozóica em bacias costeiras do Brasil. In: Rio Claro Bol. Res. Expandidos, VI Simp. Quantificação Geociências. UNESP-Rio Claro, São Paulo, pp. 3–7.

Albertão, G.A., Martins Jr., P.P., 1996a. A possible tsunami deposit at the Creteceous-Tertiary boundary in Pernambuco, Northeastern Brazil. Sediment. Geol. 104, 189–201.

Albertão, G.A., Martins Jr., P.P., 1996b. Stratigraphic record and geochemistry of the Cretaceous-Tertiary (K-T) boundary in Pernambuco-Paraíba, Northeastern Brazil. In: Jardiné, S., De Klasz, I., Debenay, J.-P. (Eds.), Géologie de l'Afrique et de l'Atlantique Sud, vol. 16. Elf Aquitaine Édition, pp. 403–411. Mémoire.

Albertão, G.A., Martins Jr., P.P., 2002. Petrographic and geochemical studies in the Cretaceous-Tertiary boundary, Pernambuco-Paraíba Basin, Brazil. In: Buffetaut, E., Koeberl, C. (Eds.), "Geological and Biological Effects of Impact Events". Springer-Verlag, Berlin, pp. 167–196.

Albertão, G.A., Koutsoukos, E.A.M., Regali, M.S.P., Attrep Jr., M., Martins Jr., P.P., 1993. The Cretaceous-Tertiary boundary record in eastern marginal sedimentary basins of Brazil. Acta Geol. Leopoldensia 16, 59—71.

Albertão, G.A., Koutsoukos, E.A.M., Regali, M.S.P., Attrep Jr., M., Martins Jr., P.P., 1994. The Cretaceous-Tertiary boundary in southern low-latitude regions: preliminary study in Pernambuco, Northeastern Brazil. Terra. Nova 6, 366—375.

Albertão, G.A., Grassi, A.A., Marini, F., Martins Jr., P.P., de Ros, L.F., 2004. The K-T boundary in Brazilian marginal sedimentary basins and related spherules. Geochem. J. 38, 121—128.

Albertão, G.A., Martins Jr., P.P., Marini, F., 2008. A possible tsunamiite at the Cretaceous-Tertiary boundary in Pernambuco Basin, Northeastern Brazil — reappraisal of field research and conceptual descriptions. In: Shiki, T., Minoura, K., Yamazaki, T., Tsuji, Y. (Eds.), Tsunamiites — Their Features and Implications. Developments in Sedimentology. Elsevier.

Albertão, G.A., Martins Jr., P.P., Santos, F.M.M., Barreto, A.M., Sansonowski, R., 2016. Afloramentos da Pedreira Poty (Bacia de Pernambuco), Registro Geológico do Evento Catastrófico do Limite K-Pg: um Geossítio em Implantação. XLVIII Congresso Brasileiro de Geologia. Abstracts. (Porto Alegre).

Alvarez, L.W., Alvarez, W., Asaro, F., Michel, H.V., 1980. Extraterrestrial cause for the Cretaceous-Tertiary extinction. Science 208, 1095—1108.

Alvarez, W., Smit, J., Lowrie, W., Asaro, F., Margolis, S.V., Claeys, P., Kastner, M., Hildebrand, A.R., 1992. Proximal impact deposits at the K-T boundary in the Gulf of Mexico: a restudy of DSDP leg 77 sites 536 and 540. Geology 20, 697—700.

Baadsgaard, H., Lerbekmo, J.F., MacDougall, I., 1988. A radiometric age for the Cretaceous-Tertiary boundary based upon K-Ar, Rb-Sr, and U-Pb ages of bentonites from Alberta, Saskatchewan, and Montana. Can. J. Earth Sci. 25, 1088—1097.

Beurlen, K., 1967a. Estratigrafia da faixa sedimentar costeira Recife-João Pessoa. Bol. SBG 16, 43—53.

Beurlen, K., 1967b. Paleontologia da faixa costeira Recife-João Pessoa. Bol. SBG 16, 73—79.

Bohor, B.F., 1990. Shock-induced microdeformations in quartz and other mineralogical indications of an impact event at the Cretaceous-Tertiary boundary. Tectonophysics 171, 359—372.

Bourgeois, J., Hansen, T.A., Wiberg, P.L., Kauffman, E.G., 1988. A tsunami deposit at the Cretaceous-Tertiary boundary in Texas. Science 241, 567—570.

Bralower, T.J., Paull, C.K., Leckie, R.M., 1998. The Cretaceous-Tertiary boundary cocktail: Chicxulub impact triggers collapse and extensive sediment gravity flows. Geology 26, 331—334.

Brooks, R.R., Hoek, P.L., Reeves, R.D., Strong, C.P., 1986. Geochemical delineation of the Cretaceous/Tertiary boundary in some New Zealand rock sequences. N. Z. J. Geol. Geophys. 29, 1—8.

Bryant, E.A., Young, R.W., 1996. Bedrock-sculpturing by tsunami, south coast New South Wales, Australia. J. Geol. 4, 565—582.

Chang, H.K., Kowsmann, R.O., Figueiredo, A.M.F., 1988. New concepts on the development of East Brazilian marginal basins. Episodes 11, 194—202.

Clifton, H.E. (Ed.), 1988. Sedimentologic Consequences of Convulsive Geological Events, vol. 229. Geological Society of American Special Paper, p. 157.

Coleman, P.J., 1968. Tsunamis as geological agents. J. Geol. Soc. Aust. 15, 267—273.

Delicio, M.P., Oliveira, A.D., Albertão, G.A., Martins Jr., P.P., 2000. Looking for spherules at the cretaceous Tertiary (K-T boundary in Pernambuco/Paraíba (PE/PB) basin, NE Brazil. In: Detre, C.H. (Ed.), Proceedings of the Annual Meeting of TECOS, 1998. Hungarian Academy of Science, Budapest, pp. 35—43.

Donovan, A.D., Baum, G.R., Blechschmidt, G.L., Loutit, T.S., Pflum, C.E., Vail, P.R., 1988. Sequence stratigraphic setting of the Cretaceous-Tertiary boundary in Central Alabama. In: Wilgus, C.K., Posamentier, H., Hastings, B.S., Van Wagoner, J., Ross, C.A., Kendall, C.G.S.C. (Eds.), Sea- Level Changes: An Integrated Approach, vol. 42. SEPM Spec. Publ, pp. 299—307.

Dott Jr., R.H., 1983. SEPM Presidential address: episodic sedimentation—how normal is average? How rare is rare? Does it matter? J. Sediment. Petrol. 53, 5—23.

Dott Jr., R.H., 1996. Episodic event deposits versus stratigraphic sequences—shall the twain never meet? Sediment. Geol. 104, 243—247.

Duringer, P., 1984. Tempêtes et tsunamis: Des dépôts de vagues de haute énergie intermittente dans le Muschelkalk supérieur (Trias germanique) de l'Est de la France. Bull. Soc. Geol. Fr. 26, 1177—1185.

Dypvik, H., Jansa, L.F., 2003. Sedimentary signatures and processes during marine bolide impacts: a review. Sediment. Geol. 161, 309—337.

Edwards, L.E., Orndorff, R.C., Head, M.J., Fensome, R.A., 2013. It's time to revitalize the Tertiary. In: Strati 2013, Ciências da Terra, vol. Especial VII, p. 149.

Einsele, G., Seilacher, A., 1991. Distinction of tempestites and turbidites. In: Einsele, G., Ricken, W., Seilacher, A. (Eds.), Cycles and Events in Stratigraphy. Springer, Berlin, pp. 377—382.

Einsele, G., Cough, S.K., Shiki, T., 1996. Depositional events and their records—an introduction. Sediment. Geol. 104, 1—9.

Fauth, G., 2000. The Cretaceous-Tertiary (K-T) Boundary Ostracodes from the Poty Quarry, Pernambuco-Paraíba Basin, Northeastern Brazil: Systematics, Biostratigraphy, Palaeocology, and Palaeobiogeography (Ph.D. thesis). Geologisch Paläontologisches Institut der Universität, Heidelberg, Unpubl, p. 158.

Gertsch, B., Keller, G., Adatte, T., Berner, Z., 2013. The Cretaceous—Tertiary boundary (KTB) transition in NE Brazil. J. Geol. Soc. Lond. 170, 249—262.

Grassi, A.A., 2000. O limite Cretáceo-Terciário nas Bacias de Pernambuco-Paraíba e Campos: Um estudo multidisciplinar com ênfase na bioestratigrafia de nanofósseis calcários (M.Sc. thesis). Universidade Federal do Rio Grande do Sul, Unpubl, p. 152.

Hart, M.B., Yancey, T.E., Leighton, A.D., Miller, B., Liu, C., Smart, C.W., Twitchett, R.J., 2012. The Cretaceous-Paleogene boundary on the Brazos river, Texas: new stratigraphic sections and revised interpretations. Gulf Coast Assoc. Geol. Soc. J. 1, 69—80.

Hildebrand, A.R., Penfield, G.T., Kring, D.A., Pilkington, M., Camargo, A., Jacobson, S.B., Boynton, W.V., 1991. Chicxulub crater: a possible Cretaceous/Tertiary boundary impact crater on the Yucatán Peninsula, Mexico. Geology 19, 867—871.

Iturralde-Vinent, M.A., 1992. A short note on the Cuban late Maastrichtian megaturbidite (an impact-derived deposit?). Earth Planet Sci. Lett. 109, 225—228.

Izett, G.A., 1991. Tektites in Cretaceous-Tertiary boundary rocks on Haiti and their bearing on the Alvarez impact extinction hypothesis. J. Geophys. Res. 96, 20879—20905. Internet link. https://ncs.naturalsciences. be/paleogene-neogene (visited in February 2020) – National Commission for Stratigraphy Belgium/ Paleogene-Neogene.

Jansa, L.B., 1993. Cometary impacts into ocean: their recognition and the threshold constraint for biological extinctions. Palaeogeogr. Palaeoclimatol. Palaeoecol. 104, 271—286.

Johnson, K.R., Nichols, D.J., Attrep, M., Orth Jr., C.J., 1989. High-resolution leaf-fossil record spanning the Cretaceous/Tertiary boundary. Nature 340, 708—710.

Keller, G., Li, L., MacLeod, N., 1995. The Cretaceous/Tertiary boundary stratotype section at Kl Kef Tunisia: how catastrophic was the mass extinction? Palaeogeogr. Palaeoclimatol. Palaeoecol. 119, 221—254.

Keller, G., Stinnesbeck, W., Adatte, T., Stüben, D., 2003a. Multiple impacts across the Cretaceous-Tertiary boundary. Earth Sci. Rev. 62, 327—363.

Keller, G., Stinnesbeck, W., et al., 2003b. Spherule deposits in Cretaceous—Tertiary boundary sediments in Belize and Guatemala. J. Geol. Soc. 160, 1—13.

Keller, G., Adatte, T., Berner, Z., Harting, M., Baum, G.R., Prauss, M., Tantawy, A., Stueben, D., 2007. Chicxulub impact predates K-T boundary: new evidence from Brazos, Texas. Earth Planet Sci. Lett. 269, 620—628.

King Jr., D., Petruny, L.W., 2015. Correlation of northern Belize's Cretaceous-Paleogene ('KT') boundary sections. GCAGS Trans. 65, 463—473.

Klaus, A., Norris, R.D., Kroon, D., Smit, J., 2000. Impact-induced mass wasting at the K-T boundary: Blake Nose, western North Atlantic. Geology 28, 319—322.

Koenig, R., 2001. Researchers target deadly tsunamis. Science 293, 1251—1253.

Koutsoukos, E.A.M., 1992. Late Aptian to Maastrichtian foraminiferal biogeography and palaeoceanography of the Sergipe basin, Brazil. Palaeogeogr. Palaeoclimatol. Palaeoecol. 92, 295–324.

Koutsoukos, E.A.M., 1994. The Cretaceous-Tertiary boundary in Pernambuco, Northeastern Brazil. In: Abstr. 2nd. Coloquium of Stratigraphy and Paleogeography of the South Atlantic. Angers, France, pp. 81–82.

Koutsoukos, E.A.M., 1996. The Cretaceous-Tertiary boundary at Poty, NE Brazil—event stratigraphy and palaeoenvironments. In: Jardiné, S., De Klasz, I., Debenay, J.-P. (Eds.), Géologie de l'Afrique et de l'Atlantique Sud, vol. 16. Elf Aquitaine Édition, pp. 413–431 (Mémoire).

Koutsoukos, E.A.M., 1998. An extraterrestrial impact in the early Danian: a secondary K/T boundary event. Terra. Nova 10, 68–73.

Lawton, T.F., Shipley, K.W., Aschoff, J.L., Giles, K.A., Vega, F.J., 2005. Basinward transport of Chicxulub ejecta by tsunami-induced backflow, La Popa Basin, northeastern Mexico, and its implications for distribution of impact-related deposits flanking the Gulf of Mexico. Geology 33, 81–84.

Mabesoone, J.M., 1967. Sedimentologia da faixa costeira Recife-João Pessoa. Bol. SBG 16, 57–72.

Mabesoone, J.M., Tinoco, I.M., Coutinho, P.N., 1968. The Mesozoic-Tertiary boundary in northeastern Brazil. Palaeogeogr. Palaeoclimatol. Palaeoecol. 4, 161–185.

Marini, F., Albertão, G.A., 1999. First report on KTB-related fluorine anomaly, Pernambuco area, northeastern Brazil. In: Buffetaut, E., Le Loeuff, J. (Eds.), Abstr. Workshop on Geological and Biological Evidence for Global Catastrophes—Quillan (France), Impact Programme. European Science Foundation, pp. 56–57.

Marini, F., Albertão, G.A., 2000. KT-related Fluorine anomaly: an appeal for worldwide search from recent findings in NE Brazil. In: Abstr. General Symposium. 25–5. 31st Int. Geol. Congr. Rio de Janeiro, Brazil.

Marini, F., Albertão, G.A., Oliveira, A.D., Delicio, M.P., 2000. Preliminary SEM and EPMA investigations on KTB spherules from the Pernambuco area (northeastern Brazil): diagenetic apatite and fluorite concretions, suspected fluorine anomalies, 1882. In: Detre, C.H. (Ed.), Terrestrial and Cosmic Spherule, vol. 1. Akademia Kaido (Hungary Academy of Science) Budapest, pp. 109–117. ISBN 963-05-769-37.

Martins Jr., P.P., 2000. Epistemological reappraisal of regime/rupture geology in the last 200 years. Cosmic, planetary, geophysical, atmospheric, oceanic, ecological and microscopic evidences. In: 31st Int. Geol. Congr., General Symposia. Comparative Planetology (Rio de Janeiro). Session 25–6, Impact and Extraterrestrial Spherules.

Martins Jr., P.P., Albertão, G.A., Haddad, R., 2000. The Cretaceous-Tertiary boundary in the context of impact geology and sedimentary record—an analytical review of 10 years of researches in Brazil. Brazilian Contr. 31st. Int. Geol. Congr 30, 460–465 (Revista Brasileira de Geociências).

Maurrasse, F.J.-M.R., Sen, G., 1991. Impacts, tsunamis, and the Haitian Cretaceous-Tertiary boundary bed. Science 252, 1690–1693.

Minoura, K., Gusiakov, V.G., Kurbatov, A., Takeuti, S., Svendson, J.I., Bondevik, S., Oda, T., 1996. Tsunami sedimentation associated with the 1923 Kamchatka earthquake. Sediment. Geol. 106, 145–154.

Molina, E., Alegret, L., et al., 2006. The global boundary stratotype section and point for the base of the Danian stage (Paleocene, Paleogene, "Tertiary", Cenozoic) at el Kef, Tunisia — original definition and revision. Episodes 29, 263–273.

Morgan, J., Lana, C., et al., 2006. Analyses of shocked quartz at the global K-P boundary indicate an origin from a single, high-angle, oblique impact at Chicxulub. Earth Planet Sci. Lett. 251, 264–279.

Olsson, R.K., Miller, K.G., Browning, J.V., Habib, D., Sugarman, P.J., 1997. Ejecta layer at the Cretaceous-Tertiary boundary, Bass river, New Jersey (ocean drilling Program leg 174AX). Geology 25, 759–762.

Robertson, P.B., Grieve, R.A.F., 1977. Shock attenuation at terrestrial impact structures. In: Roddy, D.J., Pepin, R.O., Merril, R.B. (Eds.), Impact and Explosion Cratering. Pergamon Press, New York, pp. 687–702.

Sarkis, M.F., 2002. Caracterização Palinoestratigráfica e Paleoecológica do Limite Cretáceo-Terciário na Seção Poty, Bacia de Pernambuco/Paraíba, nordeste do Brasil (Ph.D. thesis). Universidade Federal do Rio de Janeiro, Unpubl., p. 120

Schmitz, B., 1992. Calcophile elements and Ir in continental Cretaceous-Tertiary boundary clays from the western interior of the USA. Geochim. Cosmochim. Acta 56, 1695–1703.

Scotese, C.R., 2001. Atlas of Earth History, vol. 1. Paleogeography, PALEOMAP Project, Arlington, Texas, p. 52.

Shiki, T., 1996. Reading on the trigger records of sedimentary events—a problem for future studies. Sediment. Geol. 104, 249–255.

Shiki, T., Cita, M.B., Gorsline, D.S., 2000a. Sedimentary features of seismites, seismo-turbidites and tsunamiites—an introduction. Sediment. Geol. 135, vii–ix.

Shiki, T., Kumon, F., Inouch, Y., Kontani, Y., Sakamoto, T., Tateishi, M., Matsubara, H., Fukuyama, K., 2000b. Sedimentary features of the seismo-turbites, Lake Biwa, Japan. Sediment. Geol. 135, 37–50.

Shiki, T., 2003 and 2005. Personal Communications to G. Albertão and P.P. Martins Jr.

Smit, J., 1990. Meteorite impact, extinctions and the Cretaceous-Tertiary boundary. Geol. Mijnbouw 69, 187–204.

Smit, J., Montanari, A., Swinburne, N.H.M., Alvarez, W., Hildebrand, A.R., Margolis, S.V., Claeys, P., Lowrie, W., Asaro, F., 1992. Tektite-bearing, deep-water clastic unit at the Cretaceous-Tertiary boundary in northeastern Mexico. Geology 20, 99–103.

Smit, J., Roep, T.B., Alvarez, W., Claeys, P., Montanari, A., 1994. Deposition of channel deposits near the Cretaceous-Tertiary boundary in northeastern Mexico: catastrophic or "normal" sedimentary deposits?: Comment. Geology 22, 953–954.

Stinnesbeck, W., 1989. Fauna y Microflora en el Límite Cretácico-Terciario en el Estado de Pernambuco, Noreste de Brasil. Contribuciones de los Simposios Sobre Cretácico de America Latina, Parte A: Eventos y Registro Sedimentario, pp. 215–230.

Stinnesbeck, W., Keller, G., Adatte, T., MacLeod, N., 1994. Deposition of channel deposits near the Cretaceous-Tertiary boundary in northeastern Mexico: catastrophic or "normal" sedimentary deposits?: Reply. Geology 22, 955–956.

Takayama, H., Tada, R., Matsui, T., Iturralde-Vinent, M.A., Oji, T., Takija, E., Kiyokawa, S., Garcia, D., Okada, H., Hasegawa, T., Toyoda, K., 2000. Origin of the Peñalver Formation in northwestern Cuba and its relation to K/T boundary impact event. Sediment. Geol. 135, 295–320.

Tinoco, I.M., 1967. Micropaleontologia da faixa sedimentar costeira Recife-João Pessoa. Bol. SBG 16, 81–85.

Tinoco, I.M., 1971. Foraminíferos e a Passagem Entre o Cretáceo e o Terciário em Pernambuco. University of São Paulo, São Paulo, p. 132 (Unpubl. Ph.D. thesis).

Tinoco, I.M., 1976. Foraminíferos planctônicos e a passagem entre o Cretáceo e o Terciário, em Pernambuco, Nordeste do Brasil. In: 29th: Congr, Geol. Brazil. Anais. (SBG), vol. 2, pp. 17–35.

Zinsmeister, W.J., Feldmann, R.M., Woodburne, M.O., Elliot, D.H., 1989. Latest cretaceous/earliest Tertiary transition on Seymour Island, Antarctica. J. Paleontol. 63, 731–738.

CHAPTER 17

Deep-sea tsunami deposits in the proto-Caribbean sea at the Cretaceous/Tertiary boundary

K. Goto[1], R. Tada[1], E. Tajika[1], T. Matsui[2]
[1]*Department of Earth and Planetary Science, Graduate School of Science, The University of Tokyo, Bunkyo-ku, Tokyo, Japan;* [2]*Department of Complexity Science and Engineering, Graduate School of Frontier Science, The University of Tokyo, Bunkyo-ku, Tokyo, Japan*

1. Introduction

Alvarez et al. (1980) proposed that the Cretaceous/Tertiary (K/T)-boundary mass extinction, 65 million years ago, resulted from the impact of a large (>10 km in diameter) asteroid or comet. Their hypothesis was based on the discovery of high iridium (Ir) concentrations in the K/T-boundary clay layer at Gubbio, Italy. This idea has been widely accepted after the discovery of a circular subsurface structure of ~180 km in diameter (the Chicxulub crater) in the north-western part of the Yucatán peninsula (Fig. 17.1; Hildebrand et al., 1991). Since the discovery of the Chicxulub crater, the focus of the K/T-boundary studies shifted toward the evaluation of the magnitude and mode of the impact and its environmental consequences. Large tsunamis belonged to the major predicted consequences of the K/T-boundary impact. In fact, several probable K/T-boundary tsunami deposits have since been reported around the Gulf of Mexico (e.g., Bourgeois et al., 1988; King and Petruny, 2004; Lawton et al., 2005; Smit, 1999; Smit et al., 1992, 1996), around the proto-Caribbean Sea (Alvarez et al., 1992; Goto et al., 2002, 2008; Maurrasse and Sen, 1991; Tada et al., 2002, 2003; Takayama et al., 2000), and in the Chicxulub crater (Goto et al., 2004). Thus, the K/T-boundary impact event provides a rare opportunity to study a tsunami induced by an oceanic impact.

The ultimate cause of the tsunami waves at the K/T boundary remains controversial and their magnitude and extent are poorly defined. Especially, the influence of the tsunami on the deep-sea bottom around the impact site has not been explored in detail. To evaluate the influence and magnitude of the impact-generated tsunami around the impact site, identification of K/T-boundary tsunami deposits and clarification of their sedimentary

Tsunamiites. https://doi.org/10.1016/B978-0-12-823939-1.00017-3
Copyright © 2008 Elsevier B.V. All rights reserved.

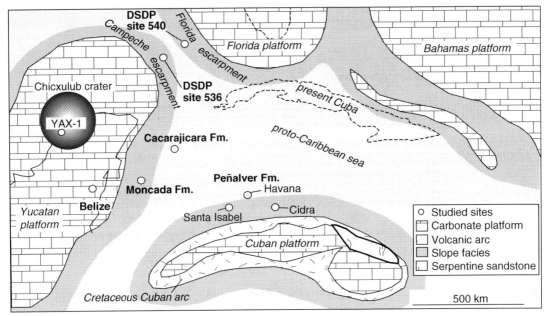

Figure 17.1
Paleogeotectonic setting of the Peñalver Formation in northwestern Cuba and other Cretaceous/Tertiary (K/T) boundary deposits at the time of the K/T-boundary impact. *Modified from Goto, K., Tada, R., Tajika, E., Iturralde-Vinent, M.A., Matsui, T., Yamamoto, S., Nakano, Y., Oji, T., Kiyokawa, S., Garcia-Delgado, D.E., Diaz-Otero, C., Rojas-Consuegra, R., 2008. Lateral lithological and compositional variations of the Cretaceous/Tertiary deep-sea tsunami deposits in northwestern Cuba. Cretac. Res. 29, 217–236; Tada, R., Iturralde-Vinent, M.A., Matsui, T., Tajika, E., Oji, T., Goto, K., Nakano, Y., Takayama, H., Yamamoto, S., Rojas-Consuegra, R., Kiyokawa, S., García-Delgado, D., et al., 2003. K/T boundary deposits in the proto-Caribbean basin. AAPG Mem. 79, 582–604.*

mechanisms are important. The mechanism responsible for the genesis of the tsunami at the K/T boundary has not been specified, despite its importance for the understanding of this impact-induced tsunami. For these reasons, the present study is focused on the clarification of the genesis of the impact-generated tsunami waves at the K/T boundary and their influence on the deep-sea bottom around the impact site.

We will first review the lateral and vertical lithological variations of the Peñalver Formation in north-western Cuba, based on the study of Goto et al. (2002, 2008), in order to clarify the influence of the tsunami on the deep-sea bottom around the impact site. The Peñalver Formation is a K/T-boundary deep-sea tsunami deposit with a thickness of ~180 m (Goto et al., 2008; Iturralde-Vinent, 1992; Pszczolkowski, 1986; Tada et al., 2003; Takayama et al., 2000). The Peñalver Formation provides a rare opportunity to evaluate the effect of the tsunami to the deep-sea bottom, because it was originally deposited on the deep-sea floor of the proto-Caribbean Sea (Brönnimann and Rigassi, 1963) and later uplifted and exposed.

We also summarize the lithology and the sedimentary mechanisms of other K/T-boundary deposits in the proto-Caribbean Sea, including the Cacarajícara Formation and the Moncada Formation in western Cuba, and at Deep Sea Drilling Project (DSDP) sites 536 and 540, in order to compare their thickness and sedimentary structures with those of the Peñalver Formation, and also with the objective to investigate the differences in influence of the tsunami regarding the depositional depth and setting around the proto-Caribbean Sea. Furthermore, the studies of the Peñalver Formation and other K/T-boundary deposits in the proto-Caribbean Sea and in the Chicxulub crater provide general geological constraints on the genesis of tsunamis. We also discuss the genesis of the tsunami at the K/T boundary.

2. Paleogeography of the proto-Caribbean sea and geological setting of the study sites

The proto-Caribbean basin was limited in the north by the Florida platform, in the northeast by the Bahamas carbonate platform, in the west by the Yucatán platform, and in the south by the Cuban carbonate platform. It continued south-eastward to the Atlantic Ocean with a north-south width of over 500 km (Fig. 17.1; see also Tada et al., 2003). In between these shallow submarine platforms, continental slopes and a deep-sea basin developed on the oceanic crust (proto-Caribbean basin) (Fig. 17.1; Tada et al., 2003).

The Peñalver Formation was deposited in a basin located to the north-northwest of the Cuban carbonate platform (Fig. 17.1; Iturralde-Vinent, 1992, 1994, 1998). The Cuban carbonate platform overlies the Cretaceous Cuban arc, which is composed of a deformed and partially metamorphosed arc complex of Aptian to Campanian age (Iturralde-Vinent, 1998). The Cretaceous Cuban arc was located 500−700 km to the south-southeast of its present position during the late Cretaceous (Fig. 17.1; Pindell, 1994; Rosencrantz, 1990). The arc complex was covered by the Cuban carbonate platform during the late Maastrichtian, and several siliciclastic sedimentary basins developed both above the arc and on the northern and southern slopes of the carbonate platform (Iturralde-Vinent, 1998).

The Cacarajícara Formation deposited in a lower slope to the basinal setting on the eastern flank of the Yucatán Platform (Fig. 17.1; Iturralde-Vinent, 1994, 1998; Kiyokawa et al., 2002; Pszczolkowski, 1987, 1999; Rosencrantz, 1990). The Moncada Formation has been interpreted by Tada et al. (2002) as deposited on the continental slope near the gateway between the Yucatán platform and the Cretaceous Cuban arc (Fig. 17.1). The depositional depth of the Moncada Formation is poorly constrained, but Tada et al. (2002) estimated that it was deposited below 200 m and above carbonate compensation depth at a depth of around 2000 m and suggested that it was deposited on an upper slope that was shallower than the depositional site of the Cacarajícara Formation.

Other examples of K/T-boundary proximal deep-sea deposits have been reported from DSDP sites 536 and 540 located ~420 and 530 km to the north-east of the rim of the Chicxulub crater (Fig. 17.1; Alvarez et al., 1992; Bralower et al., 1998), but the sedimentary mechanisms of them are still controversial. DSDP site 536 is located on a submarine ridge along the base of the Campeche Escarpment, whereas DSDP site 540 is located on the flank of an erosional valley on the Florida Escarpment of the Florida platform (Fig. 17.1; Buffler and Schlager, 1984). A northwest-southeast trending submarine valley exists between the Campeche and Florida Escarpments (Fig. 17.1; Buffler and Schlager, 1984). This submarine valley was probably located above the flat-lying abyssal Gulf plain to the north and the deep-sea proto-Caribbean basin to the south-east during the late Cretaceous (Buffler and Schlager, 1984; Schlager et al., 1984).

Under this paleogeographic configuration, the Chicxulub impact took place and the K/T-boundary deposits accumulated in the western proto-Caribbean basin (Tada et al., 2003).

3. Sedimentary processes of the Peñalver Formation

3.1 Stratigraphic setting and studied localities

The Peñalver Formation overlies, with an erosional contact (Brönnimann and Rigassi, 1963; Takayama et al., 2000), the latest Campanian to the latest Maastrichtian Vía Blanca Formation, the depositional depth of which is estimated as 600−2000 m, based on the ratio between planktonic and benthic foraminifers (Brönnimann and Rigassi, 1963). The Peñalver Formation is a calcareous clastic unit, which is fining upward gradually from calcirudite to calcilutite and which lacks any evidence of bioturbation (Goto et al., 2008; Takayama et al., 2000). Takayama et al. (2000) divided the Peñalver Formation into five members (Fig. 17.2): the Basal, Lower, Middle, Upper, and Uppermost Members in ascending order, based on variations in lithology. Tada et al. (2003) grouped the Basal plus part of the Lower Members into the Lower Unit and the part of Lower to Uppermost members as the Upper Unit, because the depositional mechanism and the origin of the Lower Unit is different from the Upper Unit, as deduced from the sedimentary structures, composition, and the grain-size distribution. Tada et al. (2003) further divided the Upper Unit into Subunits A and B, which correspond to the upper part of the Lower Member to the Upper Member and to the Uppermost Member of Takayama et al. (2000), respectively. The Peñalver Formation was deposited contemporaneously with the K/T-boundary bolide-impact event, based on biostratigraphical constraints and the occurrence of impact-induced material (Tada et al., 2003; Takayama et al., 2000). The Peñalver Formation is disconformably overlain by the Paleocene Apolo Formation.

The distribution of the Peñalver Formation is extending over 150 km in an east-west direction (Takayama et al., 2000). Goto et al. (2008) conducted a field survey of the Peñalver Formation at the type locality near Havana, at Cidra (~90 km to the east of

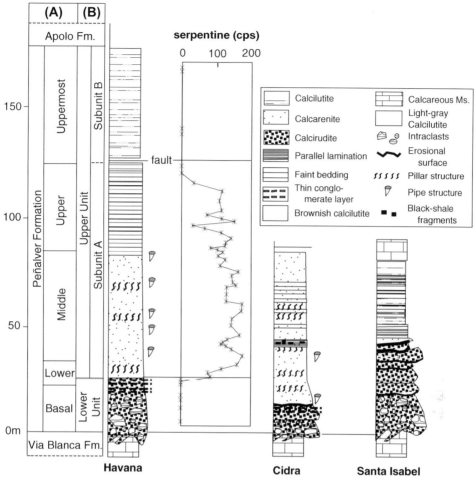

Figure 17.2
Sections of the Peñalver Formation among the type locality, Cidra and Santa Isabel. Definitions of members and units of the Peñalver Formation by (a) Takayama et al. (2000) and (b) Tada et al. (2003) are also shown. The vertical variation of the serpentine content in the insoluble residues (cps) at the type locality is based on X-ray diffraction (XRD) analysis. *Based on Goto, K., Tada, R., Tajika, E., Iturralde-Vinent, M.A., Matsui, T., Yamamoto, S., Nakano, Y., Oji, T., Kiyokawa, S., Garcia-Delgado, D.E., Diaz-Otero, C., Rojas-Consuegra, R., 2008. Lateral lithological and compositional variations of the Cretaceous/Tertiary deep-sea tsunami deposits in northwestern Cuba. Cretac. Res. 29, 217–236.*

Havana) and at Santa Isabel (~60 km to the west of Havana). The Havana area is located in the western end of the east-west-orientated Havana–Matanzas anticlinorium. Goto et al. (2008) suggested that Cidra and Santa Isabel areas are situated ~15 and 20 km to the south of the type locality, respectively, if the east-west folds are restored. The restored positions of the localities suggest that the Cidra and Santa Isabel areas were located closer to the Cretaceous Cuban arc, and thus these depositional sites were probably shallower than the type locality (Goto et al., 2008).

3.2 Lithology and petrography at the type locality near Havana

Here, we review the detailed description of the Peñalver Formation at the type locality by Goto et al. (2008) and Takayama et al. (2000). The succession from the upper part of the Vía Blanca Formation to the upper part of the Subunit A is continuously exposed at this locality (Fig. 17.3A). The Subunit B is in fault contact with the Subunit A (Fig. 17.2). The Lower Unit of the Peñalver Formation overlies the Vía Blanca Formation with an erosional surface. The Lower Unit is ~30 m thick, and its lower 25 m consists of light to medium gray, poorly sorted granule to pebble calcirudite. Large intraclast, probably derived from the underlying Vía Blanca Formation (Takayama et al., 2000), is common in

Figure 17.3
The Peñalver Formation. (A) Overview at the type locality near Havana. (B) Large intraclast in the basal part of the Lower Unit in the Victoria quarry. (C) Massive calcarenite in Subunit A at the type locality. No sedimentary structures indicative of currents are visible. (D) Faint parallel bedding in the upper part of Subunit A at the type locality. (E) Massive calcilutite in Subunit B at the type locality. No sedimentary structures indicative of currents are visible.

this part. Goto et al. (2008) reported the largest intraclast, of ~10 m, in the Lower Unit of the Peñalver Formation in the Victoria quarry (Fig. 17.3B), near the type locality. The calcirudite of the Lower Unit is composed mainly of poorly sorted, angular to subangular biomicrite fragments and shallow-water bioclasts such as fragments of rudists, mollusks, and large calcareous benthic foraminifers (Fig. 17.4A). The upper 5 m of the Lower Unit consists of light to medium gray, poorly sorted coarse to medium calcarenite. This part is characterized by intercalations of conglomerate layers that are a few centimeters thick (Takayama et al., 2000) and pillar structures, a type of the water-escape structures (Lowe, 1975).

The Upper Unit overlies the Lower Unit with a conformable and gradual contact. The Upper Unit of the Peñalver Formation, which is more than 150 m thick and subdivided into Subunits A and B (Fig. 17.2), is composed of homogeneous calcarenite and calcilutite. Subunit A is ~95 m thick and its lower 55 m consists of light to medium gray, massive, well-sorted, fine to medium calcarenite (Fig. 17.3C). Pillar structures are present throughout this part, whereas pipe structures occur at only five horizons (Fig. 17.2). The upper 40 m of the Subunit A consists of light to medium gray, massive, well-sorted, fine to very fine calcarenite. Water-escape structures have not been recognized in this part, whereas faint planar bedding of several centimeters to several meters thick becomes visible (Fig. 17.3D).

We found that the calcarenite of Subunit A is mainly composed of well-sorted and well-rounded micritic limestone fragments, crystalline carbonate fragments, and skeletons of planktonic and benthic foraminifers (Fig. 17.4B). The content of hemipelagic planktonic and benthic foraminifers increases markedly upward in the basal part of Subunit A (Goto et al., 2008). Noncarbonate grains in the calcarenite are composed mainly of well-sorted, well-rounded serpentine fragments, monocrystalline detrital quartz, and andesite

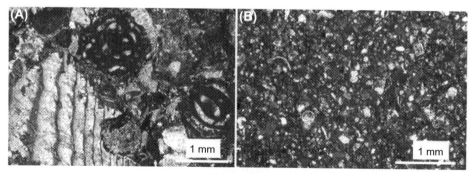

Figure 17.4

Thin sections (open nicols). (A) Calcirudite of the Lower Unit at the type locality. Abundant bioclasts are present. (B) Calcarenite of Subunit A at the type locality. Abundant micritic limestone clasts and planktonic foraminifers are present.

fragments. The serpentine fragments appear at the base of Subunit A and rapidly increase upward in abundance in the lower to the middle part of Subunit A (Fig. 17.2). In the upper part of Subunit A, the abundance of serpentine fragments gradually decreases upward, with cycles that are repeated more than six times (Fig. 17.2; Goto et al., 2008). The serpentine content is negatively correlated with the content of micritic limestone and andesite fragments. Moreover, Goto et al. (2008) found that the intervals with the higher micritic limestone and andesite lithic content coincide with intervals with a larger maximum size of the micritic limestone fragments, suggesting that the oscillations in serpentine content reflect variations in the mixing ratio between two end members: one that is characterized by serpentine fragments with smaller grain sizes, and another that is characterized by micritic limestone plus andesite fragments with larger grain sizes.

Subunit B has a tectonic (fault) contact with Subunit A at the type locality (Fig. 17.2). The estimated minimum thickness of Subunit B is 55 m. Subunit B consists of gray to light gray, massive calcilutite (Fig. 17.3E) and is mainly composed of clay-sized grains such as coccoliths, micrite, and brownish clay minerals (Goto et al., 2008). The overlying Apolo Formation is not exposed at the type locality.

3.3 Lateral and vertical variations in lithology, composition and grain size

Goto et al. (2008) investigated the thickness and the sedimentary structures of the Peñalver Formation and found significant lateral variation. The total thickness of the Peñalver Formation is more than 180 m at the type locality, more than 85 m at Cidra, and ~80 m at Santa Isabel (Fig. 17.2 and Table 17.1). The Lower Unit of the Peñalver Formation at Cidra and Santa Isabel is composed of several distinct calcirudite beds: two at Cidra and five at Santa Isabel (Fig. 17.2). These calcirudite beds, in ascending order, are called the first and second calcirudite beds at Cidra and the first to fifth calcirudite beds at Santa Isabel, respectively (Goto et al., 2008). The calcirudites in the Lower Unit at Santa Isabel and Cidra are coarse-grained than those at the type locality and contain more biomicrite grains and shallow-marine fossil fragments (Table 17.1; Goto et al., 2008). These characteristics suggest that the Peñalver Formation at Cidra and Santa Isabel was deposited in a shallower environment and closer to the source area than at the type locality (Goto et al., 2008). This is consistent with the fact that the type locality was located northward and farther away from the Cuban carbonate platform when restored to the original position (Goto et al., 2008).

We found that the Lower Unit contains abundant well-rounded macrofossil fragments at all localities, whereas well-rounded oolite and black shale particles are found only in the first calcirudite bed at Cidra; chlorite-replaced colored grains are observed only in the fifth calcirudite bed at Santa Isabel. These minor differences in composition probably reflect the heterogeneity of local source areas for the Lower Unit (Goto et al., 2008).

Table 17.1: Variations in thickness and lithology of the Cretaceous/Tertiary (K/T)-boundary deposits around the proto-Caribbean Sea.

	Cacarajícara Formation	Peñalver Formation type locality	Peñalver Formation Cidra	Peñalver Formation Santa Isabel	DSDP sites 536 and 540	Moncada Formation
Total thickness (m)	>700	>180	>85	80	>3	2
Presumed depositional environment	Deep-sea basin	Deep-sea basin	Upper to middle slope	Upper to middle slope	Escarpment of submarine valley	Upper to middle slope
Presumed depositional depth (m)	?	600~2000	?	?	?	?
Thickness of gravity-flow deposit (m)	>250	30	18	45	>2	0
Thickness of tsunami deposit (m)	>450	>150	>67	35	>1	2
Basal contact of tsunami deposit	Gradual	Gradual	Weakly eroded	Strongly eroded (grain-size gap)	? (grain-size gap)	Strongly eroded
Sedimentary structures indicative of current action	No	No	Faint parallel lamination	Parallel lamination	Bidirectional cross-lamination	Bidirectional cross-lamination
Number of cycles	?	>6	>7	>10	?	>6
Relative depositional depth	Deeper	>	>	>	?	Shallower
Presumed influence of tsunami	Weaker	<	<	<	<	Stronger

Based on Goto, K., Tada, R., Tajika, E., Iturralde-Vinent, M.A., Matsui, T., Yamamoto, S., Nakano, Y., Oji, T., Kiyokawa, S., Garcia-Delgado, D.E., Diaz-Otero, C., Rojas-Consuegra, R., 2008. Lateral lithological and compositional variations of the Cretaceous/Tertiary deep-sea tsunami deposits in northwestern Cuba. Cretac. Res. 29, 217–236, and Kiyokawa, S., Tada, R., Iturralde-Vinent, M.A., Tajika, E., Yamamoto, S., Oji, T., Nakano, Y., Goto, K., Takayama, H., Delgado, D.G., Otero, C.D., Rojas-Consuegra, R., et al., 2002. Cretaceous-Tertiary boundary sequence in the Cacarajícara Formation, western Cuba: an impact-related, high-energy, gravity-flow deposit. In: Koeberl, C., Macleod, G., (Eds.), Geol. Soc. Amer. Spec. Pap. 356, 125–144. Catastrophic Events and Mass Extinctions: Impact and Beyond, Tada, R., Iturralde-Vinent, M.A., Matsui, T., Tajika, E., Oji, T., Goto, K., Nakano, Y., Takayama, H., Yamamoto, S., Rojas-Consuegra, R., Kiyokawa, S., García-Delgado, D., et al., 2003. K/T boundary deposits in the proto-Caribbean basin. AAPG Mem. 79, 582–604, Tada, R., Nakano, Y., Iturralde-Vinent, M.A., Yamamoto, S., Kamata, T., Tajika, E., Toyoda, K., Kiyokawa, S., Delgado, D.G., Oji, T., Goto, K., Takayama, H., et al., 2002. Complex tsunami waves suggested by the Cretaceous-Tertiary boundary deposit at the Moncada section, western Cuba. In: Koeberl, C., Macleod, G., (Eds.), Geol. Soc. Amer. Spec. Pap. 356, 109–123. Catastrophic Events and Mass Extinctions: Impact and Beyond.

314 Chapter 17

An erosional surface has been recognized at the base of the Upper Unit at Cidra and Santa Isabel, and parallel lamination occurs in the middle part of the Upper Unit. A second channel-type erosional surface is observed at 2 m above the base of Subunit A at Santa Isabel; the axis has a direction of N77°W and a plunge angle of 10°W (Goto et al., 2008). This direction is different from channel directions measured in the basal part of the second, fourth and fifth calcirudite beds of the Lower Unit, which are north-south orientated and plunged 3°N, suggesting that the erosional surface in Subunit A was created by another process than those in the Lower Unit (Goto et al., 2008).

We found that the calcarenites and calcilutites in the Upper Unit at all localities consist of mainly hemipelagic grains such as micritic limestone fragments and planktonic and benthic foraminifers, suggesting a hemipelagic to pelagic origin, which is distinctly different from the Lower Unit (Goto et al., 2008; Takayama et al., 2000). The grain composition, mineral composition, and maximum sizes of the micritic limestone fragments and detrital quartz grains in the calcarenite and calcilutite of the Upper Unit are similar at all localities, suggesting regional homogeneity of the source material (Goto et al., 2008).

At Santa Isabel, Subunit B is ∼27 m thick and is characterized by 10 alternations of dark gray argillaceous calcilutite beds of 1−4 m thickness and light-gray calcilutite beds of several tens of centimeters thickness (Goto et al., 2008). The carbonate content of the light-gray calcilutite beds is 3%−8% higher than that in the adjacent dark-gray argillaceous calcilutite beds (Goto et al., 2008).

3.4 Origin and sedimentary mechanism

Takayama et al. (2000) found that the Lower Unit of the Peñalver Formation overlies the Vía Blanca Formation with an irregular erosional contact and entrains huge mudstone blocks as intraclasts. Furthermore, the calcirudite in the Lower Unit shows poor sorting and has a grain-supported fabric; it is composed of abundant well-rounded macrofossil fragments of shallow-marine origin (Goto et al., 2002; Takayama et al., 2000). Based on these features, Goto et al. (2008) suggested that the Lower Unit is a debris-flow deposit of shallow-marine origin, possibly triggered by the impact seismic wave. Differences in grain and mineral compositions and in the maximum sizes of the carbonate and silicate grains in the calcirudites of the Lower Unit among the studied localities suggest local sources for the debris flows (Goto et al., 2008).

By contrast, we found that micritic limestone fragments and planktonic and benthic foraminifer skeletons of hemipelagic to pelagic origin are the major components of the calcarenites in the Upper Unit, suggesting a source that is distinctly different from that for the Lower Unit (Tada et al., 2003). The major source of the Upper Unit was most likely the unconsolidated sediment of the Vía Blanca Formation, because similar micritic limestone fragments and planktonic and benthic foraminifer skeletons are abundant in turbidites of the Vía Blanca Formation (Takayama et al., 2000). However, Goto et al. (2008) pointed out that the presence of serpentine fragments in the Upper Unit is difficult to explain simply by resuspension of sediments from the Vía Blanca Formation, because the serpentine fragments are not present in the Vía Blanca Formation in the study area.

Goto et al. (2008) investigated grain-size distribution and roundness of serpentinite and serpentine grains in north-western and north-eastern Cuba. They concluded that the serpentinite present beneath the Vía Blanca Formation in the Havana area has grain size that is very different from serpentine fragments in the Upper Unit and suggested that the serpentine sandstones in central to north-eastern Cuba, which have grain-size distribution and roundness similar to the serpentine fragments in the Upper Unit of the Peñalver Formation, could be the source of the serpentine fragments in the Upper Unit at the type locality and at the Cidra.

The Upper Unit of the Peñalver Formation is well sorted, macroscopically fining upward, and has abundant water-escape structures, suggesting rapid sedimentation from a dense suspension cloud. Characteristics such as thicker beds and finer grain sizes than the calcirudite of the Lower Unit and a homogeneous appearance without signs of bioturbation in the Upper Unit at the type locality are similar to those of a homogenite (Tada et al., 2003; Takayama et al., 2000), which is a deep-sea deposit found in the Mediterranean, with conclusive evidence of its tsunami origin (Kastens and Cita, 1981). Such similarities imply that the dense suspension cloud could have been formed by a giant tsunami (Tada et al., 2003; Takayama et al., 2000). Furthermore, Goto et al. (2008) suggested that grain and mineral compositions and the maximum sizes of carbonate and silicate grains in the Upper Unit are similar among the localities, indicating that the Upper Unit that was deposited by settling of particles from the homogeneous, dense suspension cloud extended for at least 150 km in an east-west direction and for 20 km in the north-south direction. This also supports the hypothesis that the dense suspension cloud was generated by the tsunami because a tsunami can mix sediment particles over a wide area (e.g., Kastens and Cita, 1981). The Upper Unit contains serpentine fragments that seem to have been derived from central to north-eastern Cuba, and there is no erosional surface at the base of the Upper Unit at the type locality. Based on these observations, Goto et al. (2008) suggested that the serpentine particles were transported in suspension by a westward-moving water mass.

Examination of the lateral lithological variation of the Upper Unit by Goto et al. (2008) revealed that it is thinner at Santa Isabel and Cidra (Fig. 17.2 and Table 17.1). At these two localities, the erosional surface was recognized at the base of the Upper Unit, and parallel lamination was found in the middle part of the Upper Unit (Table 17.1). These observations suggest that the influence of the tsunami was larger at Cidra and Santa Isabel than at the type locality. According to Eq. (18.2) in Goto (2008), the near-bottom current velocity of the tsunami becomes larger with the decreasing depositional depths.

316 Chapter 17

Considering that the depositional sites at the Cidra and the Santa Isabel areas were probably shallower than those at the type locality, the presence of erosional surfaces and parallel lamination in these areas probably reflect the larger near-bottom current velocities of the tsunami due to the shallower conditions (Goto et al., 2008).

Based on the analysis of the grain and mineral composition of the calcarenite in Subunit A at the type locality, Goto et al. (2008) revealed compositional cycles that are repeated more than six times (Fig. 17.2). They mentioned that the compositional cyclicity probably reflects variations in the mixing ratio between the two end members: one characterized by serpentine fragments and the other by micritic limestone and andesite particles. The serpentine fragments were mixed with the particles of the Vía Blanca Formation that had probably become resuspended by the first tsunami wave, thus forming a suspension cloud, because they appear at the base of Subunit A, and their concentration drastically increases upward. Within Subunit A, the amount of micritic limestone and andesite fragments, as well as the maximum size of the micritic limestone fragments, increases in the intervals, whereas the serpentine particles decrease in number (Goto et al., 2008). This can be explained if the sediment particles represented by the coarser micritic limestone grains and andesite fragments were repeatedly injected into the suspension cloud that was initially characterized by finer serpentine fragments (Goto et al., 2008). Because there are no signs of erosion at the bottom of the intervals where the amount of micritic limestone and andesite fragments increases, Goto et al. (2008) suggested that the lateral supply of sedimentary particles to the suspension cloud could have been in the form of lateral injection of the suspended particles into the cloud by a northward movement of water masses such as repeated tsunami backwash currents.

4. Comparison of K/T-boundary deep-sea tsunami deposits in the proto-Caribbean sea

It has been demonstrated that the Upper Unit of the Peñalver Formation is a deep-sea deposit, which is formed by gravitational settling from a suspension cloud possibly generated by the tsunami associated with the K/T-boundary impact (Goto et al., 2008; Tada et al., 2003; Takayama et al., 2000). Additional K/T-boundary deep-sea deposits that accumulated in the proto-Caribbean Sea are the Cacarajícara and Moncada Formations from western Cuba (Fig. 17.1; Iturralde-Vinent, 1992; Kiyokawa et al., 2002; Tada et al., 2002, 2003) and from DSDP sites 536 and 540 (Alvarez et al., 1992; Bralower et al., 1998). In order to investigate the difference in influence of the tsunami with different depositional depths and settings around the proto-Caribbean Sea, we have summarized the lithologies and the sedimentation mechanisms of the Cacarajícara Formation, the Moncada Formation, and the K/T-boundary deposits at DSDP sites 536 and 540. We will also compare their thicknesses and sedimentary structures with those of the Peñalver Formation.

4.1 Cacarajícara Formation

The Cacarajícara Formation is a calcareous clastic deposit with a thickness of ∼700 m (Fig. 17.5; Kiyokawa et al., 2002). It disconformably overlies the well-bedded limestone and chert of the Cenomanian—Turonian Carmita Formation (Tada et al., 2003). The Cacarajícara Formation is subdivided, from bottom to top, into the Lower Breccia, Middle

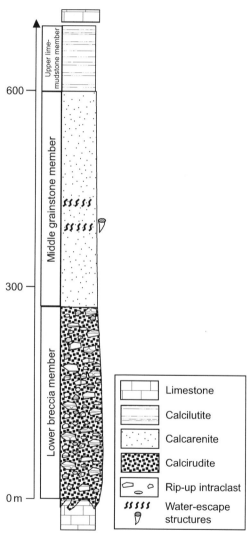

Figure 17.5
Sedimentary log of the Cacarajícara Formation. *Based on Tada, R., Iturralde-Vinent, M.A., Matsui, T., Tajika, E., Oji, T., Goto, K., Nakano, Y., Takayama, H., Yamamoto, S., Rojas-Consuegra, R., Kiyokawa, S., García-Delgado, D., et al., 2003. K/T boundary deposits in the proto-Caribbean basin. AAPG Mem. 79, 582—604.*

318   Chapter 17

Grainstone, and Upper Lime Mudstone Members (Fig. 17.5; Kiyokawa et al., 2002). The boundaries between the members are gradual (Kiyokawa et al., 2002). The Lower Breccia Member is more than 250 m thick and is composed of cobble- to pebble-sized clasts of shallow- and deep-water limestone, black chert, and reddish bedded chert, with small amounts of greenish shale and altered volcanic rocks (Kiyokawa et al., 2002). The Middle Grainstone Member, together with the Upper Lime Mudstone Member, is more than 450 m thick. These members are composed of massive calcarenites to calcilutites of pelagic origin without any sedimentary structures indicative of current action (Kiyokawa et al., 2002).

The Lower Breccia Member is interpreted as being formed by laminar flow with high-speed dilatant condition derived from the Yucatán Platform triggered by the earthquake because of the cratering event (Kiyokawa et al., 2002). These possible interpretations of the genesis are based on (1) the grain-supported fabric with rare matrix, suggesting a high-dispersive pressure; (2) the reverse grading and imbricated position of the boulders; and (3) the presence of hydrofractured clasts, suggesting high pore-pressure conditions (Kiyokawa et al., 2002). These authors also interpreted the Middle Grainstone and the Upper Lime Mudstone Members as having been formed by a high-concentration turbidity current and a low-density turbidity current, respectively, that were associated with the high-concentration laminar-flow deposit of the Lower Breccia Member (Kiyokawa et al., 2002).

Tada et al. (2003) further reported the difference in composition between the Lower Breccia Member and the Middle Grainstone plus Upper Lime Mudstone Members. Tada et al. (2003) suggested an explanation that the Middle Grainstone Member plus Upper Lime Mudstone Member represent a homogenite formed by the deep-sea tsunami associated with the K/T-boundary impact. This interpretation is more likely because of (1) the pelagic origin of the Middle Grainstone Member plus Upper Lime Mudstone Member, which is similar to the Upper Unit of the Peñalver Formation; (2) the absence of shallow-marine limestone and megafossil fragments and chert fragments, suggesting a composition that is clearly different from that of the Basal Breccia Member; and (3) the homogeneous appearance with single upward fining (Tada et al., 2003).

4.2 Moncada Formation

The Moncada Formation is a calcareous clastic deposit. It differs from the Peñalver and the Cacarajícara Formations in that it is thinner (~2 m) (Fig. 17.6; Tada et al., 2002). The Moncada Formation disconformably overlies the grayish-black, bedded micritic limestones of the Albian to Cenomanian Pons Formation (Díaz Otero et al., 2000) and is conformably overlain by marly limestones of the early Paleocene to earliest Eocene Ancon Formation (Bralower and Iturralde-Vinent, 1997). The Moncada Formation has ripple

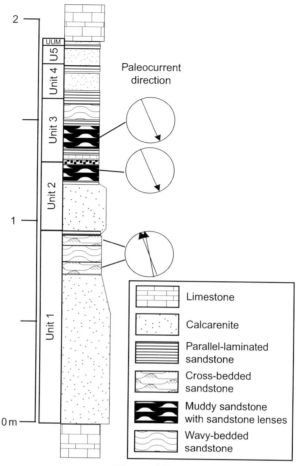

Figure 17.6
Sedimentary log of the Moncada Formations. *Based on Tada, R., Nakano, Y., Iturralde-Vinent, M.A., Yamamoto, S., Kamata, T., Tajika, E., Toyoda, K., Kiyokawa, S., Delgado, D.G., Oji, T., Goto, K., Takayama, H., et al., 2002. Complex tsunami waves suggested by the Cretaceous-Tertiary boundary deposit at the Moncada section, western Cuba. In: Koeberl, C., Macleod, G., (Eds.), Geol. Soc. Amer. Spec. Pap. 356, 109–123. Catastrophic Events and Mass Extinctions: Impact and Beyond.*

cross-laminations at several horizons that indicate north-to-south-directed paleocurrents but opposite directions occur as well (Tada et al., 2002). Changes in detrital provenance as suggested by the changes in paleocurrent directions have also been recognized by Tada et al. (2002), who proposed that the Moncada Formation was formed by repeated tsunami currents, based on (1) a repetition of upward fining units, (2) opposite current directions in the successive units, (3) systematic upward decreases in the thicknesses of the successive units, and (4) the absence of a basal erosional surface in each unit.

320 Chapter 17

4.3 DSDP sites 536 and 540

4.3.1 Stratigraphical setting

On the basis of lithology, Alvarez et al. (1992) subdivided the sedimentary succession at DSDP sites 536 and 540 into five units: 1–5 from bottom to top. Unit 1, which is composed of autochthonous limestones of early Cenomanian age, has been found only at site 540 (Alvarez et al., 1992; Buffler and Schlager, 1984). Unit 2 is a 45-m thick pebbly mudstone of probably late Cenomanian age and has been found only at site 540 (Alvarez et al., 1992; Bralower et al., 1998; Keller et al., 1993). Although this unit is suspected of having a genetic relationship with the K/T-boundary impact event (Alvarez et al., 1992; Bralower et al., 1998), no conclusive evidence supporting this relationship has been presented.

Deposits with distinct evidence of the K/T-boundary impact event such as spherules and shocked quartz grains were found in units 3 and 4 (Alvarez et al., 1992; Bralower et al., 1998). Unit 3 is composed of calcirudites and calcarenites, and unit 4 is composed of calcilutites. Unit 5 conformably overlies unit 4 and is composed of planktonic foraminiferal ooze of Danian age; it has been found only at site 536 (Alvarez et al., 1992). We reexamined the lithology and petrography of units 3 and 4 because their depositional mechanism is controversial (Alvarez et al., 1992; Bralower et al., 1998).

4.3.2 Lithology of units 3 and 4

At site 540, the interval from units 1 to 3 was only partly recovered. The contact between units 2 and 3 was not recovered (Buffler and Schlager, 1984). The thickness of the recovered part of unit 3 at site 540 is 3 m (Fig. 17.7). The lower 2 m of unit 3 consists of dark brown, poorly sorted granule-grained calcirudite to coarse-grained calcarenite. The calcirudites in unit 3 rarely contain a calcareous matrix ($<2\%$) and show a grain-supported fabric. Large bioclastic limestone fragments of up to 20 cm are present in the basal part of this unit. The upper 1 m of unit 3 at site 540 is composed of two calcarenite beds of 50 cm thickness with a conformable basal contact. Each calcarenite bed is well sorted and shows upward fining from a greenish medium-grained part to a whitish very fine-grained part. Both bidirectional (opposite directions) and monodirectional cross-laminations occur in the lower greenish medium-grained part of each bed, whereas the upper whitish very fine-grained part is massive, as described by Alvarez et al. (1992). Units 4 and 5 have not been recovered at this site.

At site 536, the sedimentary succession between unit 1 and the calcirudite part of unit 3 was not recovered, but the interval from the calcarenite part of units 3–5 has been recovered (Fig. 17.7). The calcarenite part of unit 3 overlies Albian to Aptian white limestones (Fig. 17.7; Alvarez et al., 1992). The calcarenite part of unit 3 at site 536 is also composed of hemipelagic to pelagic very fine- to fine-grained calcarenites. The

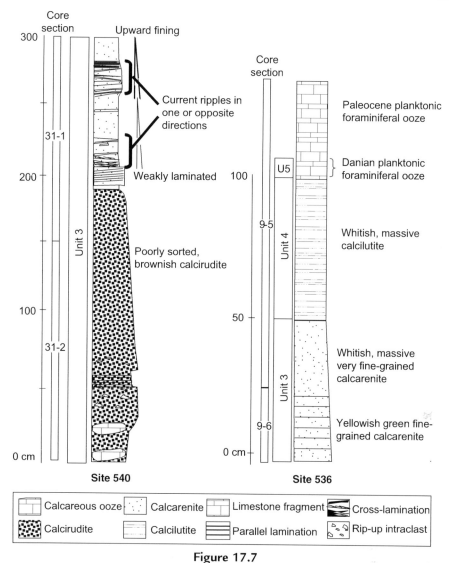

Figure 17.7
Sedimentary logs of the Cretaceous/Tertiary (K/T) boundary layers at Deep-Sea Drilling Project (DSDP) sites 536 and 540.

calcarenite shows upward fining from a yellowish green fine-grained base with parallel beds of several centimeters thickness to a whitish, massive, and very fine-grained top. The basal contact of unit 4 is gradational and conformable. The thickness of unit 4 is 50 cm. Unit 4 is composed of a well-sorted whitish massive calcilutite. The contact between the K/T-boundary deposit and the overlying Danian (P1 zone) planktonic foraminiferal ooze of unit 5 is defined by the first occurrence of Paleocene planktonic foraminifers; the conformable and gradual contact occurs at core 9–5 of site 536 (Alvarez et al., 1992).

322 Chapter 17

4.3.3 Sedimentary mechanism of units 3 and 4

Alvarez et al. (1992) have pointed out the possibility that units 3 and 4 were formed by an impact-induced gigantic tsunami. On the other hand, Bralower et al. (1998) interpreted that units 3 and 4 at these sites were formed as a result of large sediment gravity flows triggered by the K/T impact, based on their upward fining and their depositional depths well below the ordinary tsunami wave base. However, it is important to note that the upward fining can also be due to settling from a cloud of resuspended particles; in addition, not only gravity-flow deposits but also tsunami deposits show upward fining (e.g., Smit et al., 1996; Tada et al., 2002; Takayama et al., 2000). Therefore, further careful examination is necessary to clarify the depositional mechanisms of units 3 and 4.

The calcirudite part of unit 3 shows poor sorting and a grain-supported fabric. The calcirudite entrains calcareous rock fragments as large as 20 cm in its basal part that are too large to be transported in suspension by a wave-induced current in the deep sea, but they could have been transported by gravity flows. The poor sorting and the grain-supported fabric are also consistent with a gravity-flow origin (Middleton and Hampton, 1976).

On the other hand, the calcarenite parts of units 3 and 4 are well sorted, fining upward, and composed mostly of hemipelagic to pelagic grains. The upward fining in the calcarenite part of units 3 and unit 4 suggests that these units were deposited from resuspended hemipelagic to pelagic particles. Furthermore, there are cross-laminations indicating opposite current directions in the calcarenite part of unit 3 at site 540, as described by Alvarez et al. (1992). Cross-lamination with opposite directions can, in principle, be formed by the wave action (e.g., Smit et al., 1996) because water movement in opposite directions can result from waves that make water particles follow an elliptic orbit (e.g., Bryant, 2001). Most of the criticisms against the occurrence of K/T-boundary tsunami deposits around the proximal deep-sea area are, however, based on the idea that the influence of an impact-generated tsunami cannot reach the deep-sea bottom (e.g., Bohor, 1996; Bralower et al., 1998). As the current velocity of tsunamis does not diminish downward through the water column—as indicated by Eq. (18.2) in Goto (2008)—an impact-generated tsunami can, however, affect the deep-sea bottom depending on the wave height.

The calcilutite of unit 4 is homogeneous, showing upward fining but no sedimentary structures, which is consistent with settling from resuspended sediments after the influence of the tsunami had ceased (Alvarez et al., 1992).

4.4 Comparison of the thickness and sedimentary structures of the K/T-boundary deep-sea tsunami deposits in the proto-Caribbean sea

As mentioned earlier, the Middle Grainstone plus Upper Lime Mudstone Members of the Cacarajícara Formation, the entire Moncada Formation, and the calcarenite part of units 3 and 4 at DSDP sites 536 and 540 have been interpreted as having been formed by an impact-generated tsunami.

Table 17.1 shows the thicknesses, the interpreted depositional environments, the presumed depositional depths, and the sedimentary structures of the deep-sea tsunami deposits around the proto-Caribbean Sea. As shown in Table 17.1, the K/T-boundary deep-sea tsunami deposits in the proto-Caribbean Sea tend to decrease in thickness with decreasing depositional depth. Although the depositional depths of the K/T-boundary deposits at DSDP sites 536 and 540 may be similar to that of the Peñalver Formation at the type locality, the thicknesses of the K/T-boundary deep-sea tsunami deposits in the calcarenite part of units 3 and 4 at the DSDP sites and of the Upper Unit of the Peñalver Formation at the type locality differ considerably (Table 17.1). This difference is probably due to different depositional settings. Because the K/T-boundary deep-sea tsunami deposits in the calcarenite part of units 3 and 4 at the DSDP sites were made on escarpments of the Yucatán and Florida Platforms, suspended sediments formed by the tsunami could have flowed down to the deep-sea bottom of the proto-Caribbean Sea and the proto-Gulf of Mexico.

The basal contact is gradual in the Middle Grainstone plus Upper Lime Mudstone Members of the Cacarajícara Formation and the Upper Unit of the Peñalver Formation at the type locality, whereas the basal contact is an erosional surface at the other shallower sites, including the Upper Unit of the Peñalver Formation at Cidra and the Santa Isabel and the Moncada Formation, and the calcarenite parts of units 3 and 4 at DSDP sites 536 and 540 (Table 17.1). Sedimentary structures formed under the influence of current, such as parallel lamination and cross-laminations in opposite directions, are also common in the K/T-boundary deposits at these localities, whereas there is no sedimentary structure indicative of currents in the Cacarajícara Formation and the Peñalver Formation at the type locality (Table 17.1).

According to Eq. (18.2) in Goto (2008), the current velocity, U_{max}, increases with an increase in wave height and a decrease in depositional depth. Because the wave height increases with a decrease in depositional depth (Bryant, 2001), U_{max} increases with a decrease in depositional depth. Consequently, the erosional nature of the basal contact and the sedimentary structures indicative of current action in the K/T-boundary deep-sea

324 Chapter 17

tsunami deposits at the shallower sites in the proto-Caribbean Sea can be explained by the greater influence of the tsunami with decreasing depositional depth.

4.5 Compositional variations of the K/T-boundary deep-sea tsunami deposits in the proto-Caribbean sea

The K/T-boundary deep-sea tsunami deposits in the Upper Unit of the Peñalver Formation and the Middle Grainstone plus Upper Lime Mudstone Members of the Cacarajícara Formation have similar grain and mineral compositions (Kiyokawa et al., 2002; Tada et al., 2003), although the depositional areas of both formations are ~400–500 km apart. This suggests that a suspension cloud of more or less the same composition was formed and spread throughout the basin of the proto-Caribbean Sea and resulted in the deposition of the thick, homogeneous, sandy to silty graded sediments throughout the deeper part of the proto-Caribbean basin (Tada et al., 2003).

On the other hand, the K/T-boundary deep-sea tsunami deposits in the calcarenite part of units 3 and 4 at DSDP sites 536 and 540 have grain and mineral compositions that are different from those of their equivalents in the Upper Unit of the Peñalver Formation and the Middle Grainstone plus Upper Lime Mudstone Members of the Cacarajícara Formation. The difference in composition is characterized by an extremely small amount of detrital silicate grains (<2% of calcarenite by volume) at the DSDP sites compared with ~10% by volume in the Peñalver Formation (Goto et al., 2008) and by the absence of serpentine and andesite particles in the calcarenite and calcilutite at the DSDP sites. The difference in composition between the Upper Unit of the Peñalver Formation and the calcarenite parts of units 3 and 4 at the DSDP sites suggests that the suspension cloud was heterogeneous if considered on a large scale.

5. Implications for the genesis and number of tsunami currents at the K/T boundary

It has been speculated (Goto et al., 2008) that serpentine clasts in Subunit A of the Peñalver Formation were transported westward by the first tsunami wave from central to north-eastern Cuba, which is ~200–500 km to the east of the depositional area of the Peñalver Formation. Furthermore, Tada et al. (2002) reported that the first tsunami wave at the Moncada section was directed toward the north-northwest (= crater-ward), based on paleocurrent analysis of the Moncada Formation (Fig. 17.6). These data imply that the first tsunami wave was directed crater-ward in the proto-Caribbean Sea region.

Two different mechanisms have been proposed for the genesis of the giant tsunami at the K/T boundary (Matsui et al., 2002). One is a tsunami generated by landslides along the slopes of the Yucatán Platform triggered by the shock wave from the impact point (landslide-generated tsunami), and the other is a tsunami generated by ocean-water invasion into the crater and subsequent overflow of water from the crater (crater-generated tsunami).

Large-scale erosion of the Cretaceous sediments caused by the K/T-boundary impact (Bralower et al., 1998; Tada et al., 2002) and K/T-boundary deposits of over 250 m thick of probable gravity-flow origin (Grajales-Nishimura et al., 2000; Kiyokawa et al., 2002) have been documented on the continental slope around the Yucatán Peninsula. These large-scale gravity flows could have caused a landslide-generated tsunami (Bralower et al., 1998; Kiyokawa et al., 2002; Matsui et al., 2002), although the influence and magnitude of such a landslide-generated tsunami have not been well constrained yet. The flow directions of the first tsunami wave reconstructed from the Peñalver and the Moncada Formations are crater-ward, which is inconsistent with the eastward first wave interpreted for a landslide-generated tsunami in the proto-Caribbean Sea, suggesting that a landslide-generated tsunami may not have played a major role in forming the deep-sea tsunami deposits around the proto-Caribbean Sea.

By contrast, Goto et al. (2004) reported that ocean water invaded the crater immediately after the impact, based on the analysis of core YAXCOPOIL-1 (YAX-1), recovered from the Chicxulub crater. Ocean-water invasion into the crater could have induced a large north-westward water-mass movement in the proto-Caribbean Sea (Matsui et al., 2002). Such a movement would be consistent with the paleocurrent directions found in the Peñalver and the Moncada Formations. Furthermore, at least eight sedimentary pulses have been interpreted from the strata in the upper part (units 1 and 2) of the YAX-1 core (Fig. 17.8) and may imply repeated water-mass movements within the crater that were powerful enough to carry millimeter-size grains in suspension (Goto et al., 2004). Such water movements could have been caused by repeated ocean-water invasion and may have caused a crater-generated tsunami (Goto et al., 2004). It is, therefore, possible that a crater-generated tsunami occurred at the K/T boundary (Goto et al., 2004).

Goto et al. (2004) proposed, based on the study of composition and grain size variation in the suevite of the YAX-1 core (Fig. 17.8), that at least eight pulses of ocean-water invasion and subsequent overflow of water from the crater occurred after the impact and that the tsunami caused by this process propagated to the proto-Caribbean Sea and the proto-Gulf of Mexico. Cyclical changes in compositional and grain size that reflect the influence of repeated tsunami waves are sixfold present in the Upper Unit of the Peñalver Formation at the type locality (Fig. 17.2). The number of cycles in Subunit A of the Peñalver Formation at the type locality is smaller than that in YAX-1. Considering the greater depositional depth at the type locality, this smaller number of repetitions of tsunami could be explained by the decreased energy of the last tsunamis, which were no longer able to affect the sedimentation of the Peñalver Formation in the deeper part of the basin.

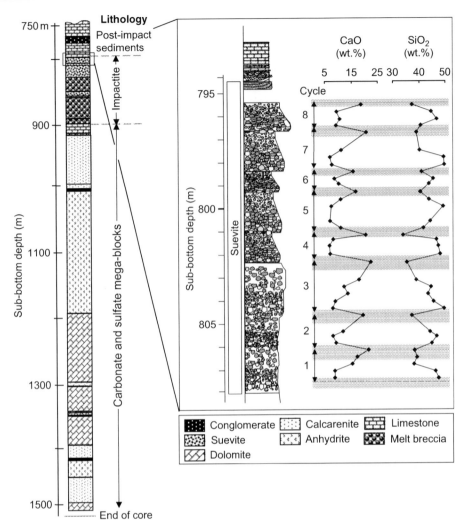

Figure 17.8

Sedimentary logs of the Cretaceous/Tertiary (K/T) boundary layer in the YAX-1 core. Vertical variations (wt.%) in the chemical composition of the suevite are also shown. *Based on Goto, K., Tada, R., Tajika, E., Bralower, T.J., Hasegawa, T., Matsui, T., 2004. Evidence for ocean water invasion into the Chicxulub crater at the Cretaceous/Tertiary boundary. Meteoritics Planet Sci. 39, 1233–1247.*

Crater-generated tsunami waves may also have propagated to the proto-Gulf of Mexico. Repeated tsunami currents are inferred for the K/T-boundary deposits, as they are characterized by bidirectional cross-lamination indicating opposite currents that repeated more than nine times at La Lajilla in the coastal region of the Gulf of Mexico (Smit et al., 1996). The number of tsunami units at La Lajilla is similar to that in YAX-1. Smit et al. (1996) have reported, however, that the first tsunami wave was directed north-westward in the K/T-boundary deposit at La Lajilla, as indicated by paleocurrent analysis of cross-lamination in what they identified as an antidune. This direction is opposite to the crater-ward direction expected in the case of a crater-generated tsunami. Identification of an antidune simply by field observations is, however, difficult (Yagishita, 1994). Therefore, further careful research is necessary to identify the paleocurrent direction of the first tsunami wave around the Gulf of Mexico.

6. Conclusions

The present study compares K/T-boundary deep-sea tsunami deposits in the proto-Caribbean Sea and reconstructs the responsible sedimentary processes in order to evaluate the influence and magnitude of the impact-induced tsunami. The genesis of tsunamis is discussed in order to understand the nature of the impact-induced tsunami at the K/T boundary. The K/T-boundary deep-sea tsunami deposits in the proto-Caribbean Sea tend to decrease in thickness with decreasing depositional depth. The erosional nature of the basal contact and the sedimentary structures indicative of currents in the K/T-boundary deep-sea tsunami deposits formed at the shallower sites in the proto-Caribbean Sea can be explained by a stronger influence of the tsunami than at deeper sites. Landslide-generated tsunami currents have not played a major role in the sedimentation of the deep-sea tsunami deposits at the K/T boundary around the proto-Caribbean Sea, whereas crater-generated tsunami currents may have been responsible for the sedimentation of deep-sea tsunami deposits at the K/T boundary around the proto-Caribbean Sea. Evidence of ocean-water invasion into the Chicxulub crater also supports this idea.

Acknowledgments

This paper is a reprint of the first version published in 2008. Original terminology (e.g., K/T-boundary) is used as it is. This research was supported by research funds donated to the University of Tokyo by NEC Corp., I. Ohkawa, T. Yoda, K. Ihara, and M. Iizuka, and by a Grant-in-Aid from the Japan Society for the Promotion of Science (JSPS) provided to T.M. (no. 17403005). This study also used samples and data provided by the Ocean Drilling Project (ODP). The authors wish to thank the important support provided with respect to the storage of samples in the East Coast Repository by G. Esmay. They also thank T. Shiki and K. Giles, who critically read the manuscript and gave them many valuable suggestions.

References

Alvarez, L.W., Alvarez, W., Asaro, F., Michel, H.V., 1980. Extraterrestrial cause for the Cretaceous-tertiary extinction. Science 208, 1095–1108.

Alvarez, W., Smit, J., Lowrie, W., Asaro, F., Margolis, S.V., Claeys, P., Kastner, M., Hildebrand, A.R., 1992. Proximal impact deposits at the Cretaceous-tertiary boundary in the gulf of Mexico: a restudy of DSDP leg 77 sites 536 and 540. Geology 20, 697–700.

Bohor, B.F., 1996. A sedimentary gravity flow hypothesis for siliciclastic units at the K/T boundary, northeastern Mexico. In: Ryder, G., Fastovsky, D., Gartner, S. (Eds.), Cretaceous-Tertiary Event and Other Catastrophes in Earth History, vol. 307. Geol. Soc. Am. Spec. Pap, pp. 183–195.

Bourgeois, J., Hansen, T.A., Wiberg, P.L., Kauffman, E.G., 1988. A tsunami deposit at the Cretaceous-Tertiary boundary in Texas. Science 241, 567–570.

Bralower, T.J., Iturralde-Vinent, M.A., 1997. Micropaleontological dating of the collision between the north American plate and the greater antilles arc in western Cuba. Palaios 12, 133–150.

328 Chapter 17

Bralower, T.J., Paul, C.K., Leckie, R.M., 1998. The Cretaceous/Tertiary boundary cocktail: Chicxulub impact triggers margin collapse and extensive sedimentary gravity flows. Geology 26, 331–334.

Brönnimann, P., Rigassi, D., 1963. Contribution to geology and paleontology of area of the city of La Habana, Cuba, and its surroundings. Eclogae Geol. Helv. 56, 193–480.

Bryant, E., 2001. Tsunami. The Underrated Hazard Cambridge University Press, Cambridge, p. 320.

Buffler, R.T., Schlager, W., 1984. Initial Reports, Deep-Sea Drilling Project, vol. 77. US Government Printing Office, Washington, p. 747.

Díaz Otero, C., Iturralde-Vinent, M., García-Delgado, D.E., 2000. The Cretaceous-tertiary boundary "cocktail" in western Cuba, greater Antilles. LPI Contrib. 1053, 76–77.

Goto, K., 2008. The genesis of oceanic impact craters and impact-generated tsunami deposits. In: Shiki, T., Tsuji, Y., Yamazaki, T., Minoura, K. (Eds.), Tsunamiites—Features and Implications. Elsevier, Amsterdam, pp. 277–297.

Goto, K., Tajika, E., Tada, R., Iturralde-Vinent, M.A., Takayama, H., Nakano, Y., Yamamoto, S., Kiyokawa, S., Garcia-Delgado, D., Rojas-Consuegra, R., Oji, T., Matsui, T., 2002. Depositional mechanism of the Penalver Formation: the K/T boundary deep-sea tsunami deposit in north western Cuba. [Abstract]. In: 16th International Sedimentological Congress. Rand Afrikaans University, Johannesburg, South Africa, pp. 122–123.

Goto, K., Tada, R., Tajika, E., Bralower, T.J., Hasegawa, T., Matsui, T., 2004. Evidence for ocean water invasion into the Chicxulub crater at the Cretaceous/Tertiary boundary. Meteoritics Planet Sci. 39, 1233–1247.

Goto, K., Tada, R., Tajika, E., Iturralde-Vinent, M.A., Matsui, T., Yamamoto, S., Nakano, Y., Oji, T., Kiyokawa, S., Garcia-Delgado, D.E., Diaz-Otero, C., Rojas-Consuegra, R., 2008. Lateral lithological and compositional variations of the Cretaceous/Tertiary deep-sea tsunami deposits in northwestern Cuba. Cretac. Res. 29, 217–236.

Grajales-Nishimura, J.M., Cedillo-Pardo, E., Rosales-Dominguez, C., Moran-Zenteno, D.J., Alvarez, W., Claeys, P., Ruiz-Morales, J., Garcia-Hernandez, J., Padilla-Avila, P., Sanchez-Rios, A., 2000. Chicxulub impact: the origin of reservoir and seal facies in the southeastern Mexico oil fields. Geology 28, 307–310.

Hildebrand, A.R., Penfield, G.T., Kring, D.A., Pilkington, M., Antonio, C.Z., Jacobsen, S.B., Boynton, W.V., 1991. Chicxulub crater: a possible Cretaceous/Tertiary boundary impact crater on the Yucatán Peninsula, Mexico. Geology 19, 867–871.

Iturralde-Vinent, M.A., 1992. A short note on the Cuban late Maastrichtian megaturbidite (an impact-derived deposit?). Earth Planet Sci. Lett. 109, 225–228.

Iturralde-Vinent, M.A., 1994. Cuban geology: a new plate-tectonic synthesis. J. Petrol. Geol. 17, 39–70.

Iturralde-Vinent, M.A., 1998. Sinopsis de la constitución geológica de Cuba. In: Melgarejo, J.C., Proenza, J.A. (Eds.), Acta Geol. Hispánica 33, Geología y Metalogénia de Cuba: Una Introducción, pp. 9–56.

Kastens, K.A., Cita, M.B., 1981. Tsunami-induced sediment transport in the abyssal Mediterranean Sea. Geol. Soc. Am. Bull. 92, 845–857.

Keller, G., MacLeod, N., Lyons, J.B., Officer, C.B., 1993. Is there evidence for Cretaceous-Tertiary boundary-age deep-water deposits in the Caribbean and Gulf of Mexico? Geology 21, 776–780.

King Jr., D.T., Petruny, L.W., 2004. Cretaceous-Tertiary boundary microtektite-bearing sands and tsunami beds, Alabama gulf coastal plain. Lunar Planet. Sci. 35, 1804.

Kiyokawa, S., Tada, R., Iturralde-Vinent, M.A., Tajika, E., Yamamoto, S., Oji, T., Nakano, Y., Goto, K., Takayama, H., Delgado, D.G., Otero, C.D., Rojas-Consuegra, R., et al., 2002. Cretaceous-Tertiary boundary sequence in the Cacarajícara Formation, western Cuba: an impact-related, high-energy, gravity-flow deposit. In: Koeberl, C., Macleod, G. (Eds.), Geol. Soc. Amer. Spec. Pap. 356, Catastrophic Events and Mass Extinctions: Impact and Beyond, pp. 125–144.

Lawton, T., Shipley, K.W., Aschoff, J.L., Giles, K.A., Vega, F.J., 2005. Basinward transport of Chicxulub ejecta by tsunami-induced backflow, La Popa basin, northeastern Mexico, and its implications for distribution of impact-related deposits flanking the Gulf of Mexico. Geology 33, 81–84.

Lowe, D.R., 1975. Water escape structures in coarse-grained sediments. Sedimentology 22, 157–204.

Matsui, T., Imamura, F., Tajika, E., Nakano, Y., Fujisawa, Y., 2002. Generation and propagation of a tsunami from the Cretaceous/Tertiary impact event. In: Koeberl, C., Macleod, G. (Eds.), Catastrophic Events and Mass Extinctions: Impact and Beyond. Geol. Soc. Amer. Spec, pp. 69–77. Pap. 356.

Maurrasse, F.J.M.R., Sen, G., 1991. Impacts, tsunamis, and the Haitian Cretaceous-Tertiary boundary layer. Science 252, 1690–1693.

Middleton, G.V., Hampton, M.A., 1976. Subaqueous sediment transport and deposition by sediment gravity flows. In: Stanley, D.J., Swift, D.J.P. (Eds.), Marine Sediment Transport and Environmental Management. Wiley, New York, pp. 197–218.

Pindell, J., 1994. Evolution of the gulf of Mexico and the Caribbean. In: Donovan, S.K., Jackson, T.A. (Eds.), Caribbean Geology: An Introduction. The University of West Indians Publishers Ass, Kingston, pp. 13–40.

Pszczolkowski, A., 1986. Megacapas del Maastrichtiano en Cuba Occidental y central. Bull. Pol. Acad. Sci., Earth 34, 81–94.

Pszczolkowski, A., 1987. Paleogeography and tectonic evolution of Cuba and adjoining areas during the Jurassic-Early Cretaceous. Ann. Soc. Geol. Pol. 57, 127–142.

Pszczolkowski, A., 1999. The exposed passive margin of North America in western Cuba. In: Mann, P. (Ed.), Caribbean Basins: Sedimentary Basin of the World, vol. 4. Elsevier, Amsterdam, pp. 93–121.

Rosencrantz, E., 1990. Structure and tectonics of the Yucatán basin, Caribbean Sea, as determined from seismic reflection studies. Tectonics 9, 1037–1059.

Schlager, W., Buffler, R.T., Angstadt, D., Phair, R., 1984. Geologic History of the Southeastern Gulf of Mexico, Initial Reports, Deep-Sea Drilling Project, vol. 77. US Government Printing Office, Washington, pp. 715–738.

Smit, J., 1999. The global stratigraphy of the Cretaceous-Tertiary boundary impact ejecta. Annu. Rev. Earth Planet Sci. 27, 75–113.

Smit, J., Montanari, A., Swinburne, N.H.M., Alvarez, W., Hildebrand, A.R., Margolis, S.V., Claeys, P., Lowrie, W., Asaro, F., 1992. Tektite-bearing, deep-water clastic unit at the Cretaceous-Tertiary boundary in northeastern Mexico. Geology 20, 99–103.

Smit, J., Roep, T.B., Alvarez, W., Claeys, P., Grajales-Nishimura, J.M., Bermudez, J., 1996. Coarse-grained, clastic sandstone complex at the K/T boundary around the Gulf of Mexico: deposition by tsunami waves induced by the Chicxulub impact? In: Ryder, G., Fastovsky, D., Gartner, S. (Eds.), Geol. Soc. Am. Spec. Pap. 307, Cretaceous-Tertiary Event and Other Catastrophes in Earth History, pp. 151–182.

Tada, R., Nakano, Y., Iturralde-Vinent, M.A., Yamamoto, S., Kamata, T., Tajika, E., Toyoda, K., Kiyokawa, S., Delgado, D.G., Oji, T., Goto, K., Takayama, H., et al., 2002. Complex tsunami waves suggested by the Cretaceous-Tertiary boundary deposit at the Moncada section, western Cuba. In: Koeberl, C., Macleod, G. (Eds.), Geol. Soc. Amer. Spec. Pap. 356, Catastrophic Events and Mass Extinctions: Impact and Beyond, pp. 109–123.

Tada, R., Iturralde-Vinent, M.A., Matsui, T., Tajika, E., Oji, T., Goto, K., Nakano, Y., Takayama, H., Yamamoto, S., Rojas-Consuegra, R., Kiyokawa, S., García-Delgado, D., et al., 2003. K/T boundary deposits in the proto-Caribbean basin. AAPG Mem 79, 582–604.

Takayama, H., Tada, R., Matsui, T., Iturralde-Vinent, M.A., Oji, T., Tajika, E., Kiyokawa, S., García, D., Okada, H., Hasegawa, T., Toyoda, K., 2000. Origin of the Peñalver Formation in northwestern Cuba and its relation to K/T boundary impact event. Sediment. Geol. 135, 295–320.

Yagishita, K., 1994. Antidunes and traction-carpet deposits in deep-water channel sandstones, Cretaceous, British Columbia, Canada. J. Sediment. Res. 64A, 34–41.

CHAPTER 18

The genesis of oceanic impact craters and impact-generated tsunami deposits

K. Goto

Department of Earth and Planetary Science, Graduate School of Science, The University of Tokyo, Bunkyo-ku, Tokyo, Japan

1. Introduction

Impacts of extraterrestrial objects on earth have occurred frequently in the geological past. A large number of impacts occurred in the oceans because these cover 70% of the surface of the earth. Such events are called "oceanic impact events" (e.g., Gersonde et al., 2002; Glikson, 1999; Jansa, 1993), "marine-target impact events" (Ormö and Lindström, 2000), or "submarine impact events" (Dypvik and Jansa, 2003). Oceanic impact processes and consequent environmental perturbations are important for the understanding of the evolution of earth and life. It is, however, difficult to identify oceanic impact craters because most of them are covered with ocean water or have vanished as a result of subduction (Gersonde et al., 2002). To date, ~ 170 impact craters have been found on the earth's surface, but only 20 of them are oceanic impact craters (Fig. 18.1; Dypvik and Jansa, 2003; Gersonde et al., 2002; Ormö and Lindström, 2000). Although new discoveries of oceanic impact craters are rare, the slowly increasing numbers of oceanic craters that are discovered facilitate the study of the geological consequences of oceanic impacts (Ormö and Lindström, 2000). Ormö and Lindström (2000) summarized the genetic processes of small (<14 km) oceanic impact craters, and Dypvik and Jansa (2003) reviewed oceanic craters of medium ($15-40$ km) and large ($>90-100$ km) size.

Large oceanic impacts generate gigantic tsunamis (impact-generated tsunamis) that can strongly affect coastlines and cause the backwash into sea of terrestrial material (Gersonde, 2000). Large impact-generated tsunamis have potential to form impact-generated tsunami deposits on both the sea floor and the land. Possible impact-generated tsunami deposits have been found, indeed, around oceanic craters (e.g., Smit, 1999; Tada et al., 2003). They are different from turbidites and gravity-flow deposits triggered by earthquakes, because they have characteristic sedimentary properties (as detailed further) and contain abundant impact-generated material such as spherules, shocked quartz, Ni-rich

Tsunamiites. https://doi.org/10.1016/B978-0-12-823939-1.00018-5
Copyright © 2008 Elsevier B.V. All rights reserved.

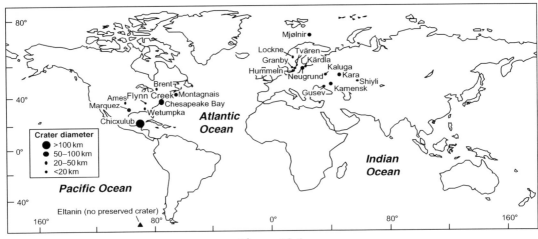

Figure 18.1

Distribution of oceanic impact craters on earth. *Based on Dypvik, H., Jansa, L., 2003. Sedimentary signatures and processes during marine bolide impacts: a review. Sediment. Geol. 161, 309—337; Gersonde, R., 2000. New field of impact research looks to the oceans. EOS Trans. 81, 21—24; Ormö, J., Lindström, M., 2000. When a cosmic impact strikes the sea bed. Geol. Mag. 137, 67—80.*

spinel, and iridium (Ir)-bearing dust (Smit, 1999). Identification of impact-generated tsunami deposits thus provides important clues for the reconstruction of oceanic impact events in the geological past (e.g., Gersonde et al., 1997; Hassler and Simonson, 2001).

In the present contribution, the properties of oceanic craters, the generation of impact tsunamis, and the genesis of impact-generated tsunami deposits are reviewed, based on studies of well-known oceanic impact events. The distribution and significance of the Cretaceous/Tertiary (K/T)-boundary tsunami deposits, which were formed by gigantic tsunamis generated by the oceanic impact (Chicxulub impact) on the Yucatán Peninsula, Mexico, 65 million years ago, are also reviewed as an example.

2. Morphology of oceanic impact craters

Impact craters on earth are roughly divided into terrestrial and oceanic impact craters (Jansa, 1993). The terrestrial craters have been well studied; they are commonly subdivided into simple and complex craters (Melosh, 1989). Simple craters tend to show a bowl shape without a central flat floor, whereas the complex craters are characterized by the presence of wall terraces, central peaks, and flat floors (Melosh, 1989). Complex craters usually have a more collapsed character than simple craters (Melosh, 1989). Terrestrial impact craters of both types are filled with a mixture of fall-back ejecta and material slumped in from the walls and crater rim during the early stage of crater formation (Montanari and Koeberl, 2000). A breccia lens underlies the apparent crater bottom that grades from locally derived monomict material at the bottom to polymict material derived from the collapsed crater wall (Ormö and Lindström, 2000). The breccia lens is overlain by thin ejecta (Ormö and Lindström, 2000).

Oceanic impact craters can also be subdivided into simple and complex craters (Ormö and Lindström, 2000), although no simple oceanic craters have been found so far (Dypvik and Jansa, 2003). Depending on the water depth, Dypvik and Jansa (2003) further subdivided the oceanic impact craters into shallow-water and deep-water craters. Only the Eltanin impact is known to have occurred in the deep-sea region (Gersonde et al., 1997); all other impacts occurred on continental shelves or slopes (Fig. 18.1; Gersonde, 2000).

The point of explosion is higher for the oceanic impacts than for a terrestrial impact because of the water column above the sea floor (Ormö and Lindström, 2000). Consequently, the size of the resulting oceanic impact craters is usually smaller than that of terrestrial craters (Ormö and Lindström, 2000). Following the formation of the crater, ocean water flows back into the crater immediately after the impact (Fig. 18.2; e.g., Ormö and Lindström, 2000). The height of the crater rim is an important parameter to control the inflow, especially for shallow oceanic impact craters, because ocean water cannot reenter into the crater immediately after the impact when the crater rim is sufficiently high above the ocean surface (Ormö et al., 2002). As a result of ocean water flowing into the crater, oceanic impact craters usually retain little or no evidence of a high crater rim (see, for instance, the Chesapeake Bay, Kärdla, Mayølr, and Montagnais craters) (Dypvik and Jansa, 2003; Jansa, 1993; Poag et al., 2002; Puura and Suuroja, 1992).

3. The generation of tsunamis by oceanic impacts

Generation of a tsunami is one of the important characteristics of a large oceanic impact, but catastrophic tsunamis may not be generated by small oceanic impacts, as reported by Melosh (2003). Four types of mechanisms that generate tsunamis can be considered (Matsui et al., 2002; Weiss et al., 2006): (1) shock-wave-induced tsunamis result from a high air pressure and wind generated by the passage of a meteorite through the atmosphere, (2) the rim wave formed at the front of the ejecta curtain (Fig. 18.3; see also Gault and Sonett, 1982; Ward and Asphaug, 2002), (3) crater-generated tsunamis (Fig. 18.2) are caused by the movement of water flowing in and out of the crater cavity after crater formation, and (4) landslide-generated tsunamis (Fig. 18.4) are caused by landslides at the margin of a continental slope that are induced by an impact-generated seismic wave. In case of the rim wave, the effect of nonlinearity is dominant on the platform, whereas the dispersion effect becomes significant in deep water (Matsui et al., 2002). The third and fourth mechanisms will be dealt with here because these mechanisms have the potential to generate gigantic tsunamis (Matsui et al., 2002).

334　*Chapter 18*

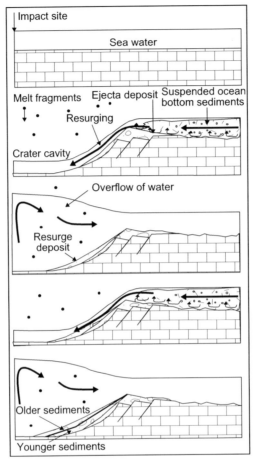

Figure 18.2
Resurge process of ocean water into an oceanic impact crater and generation of the crater-generated tsunami. *Based on Matsui, T., Imamura, F., Tajika, E., Nakano, Y., Fujisawa, Y., 2002. Generation and propagation of a tsunami from the Cretaceous/Tertiary impact event. In: Koeberl, C., Macleod, G., (Eds.), Catastrophic Events and Mass Extinctions: Impact and beyond. Geol. Soc. Amer. Spec. Pap. vol. 356, pp. 69—77.*

3.1 Crater-generated tsunamis

If the height of an oceanic impact crater rim is lower than the water depth, the surrounding ocean water will flow back into the crater (Fig. 18.2; Matsui et al., 2002). Even when the height of the crater rim is somewhat higher than the water depth, this could also occur, namely if the rim entirely or partly collapses immediately after the crater formation (Goto et al., 2004). The crater will then become overfilled with water, and a central water column will form above the crater. This water column will eventually collapse and

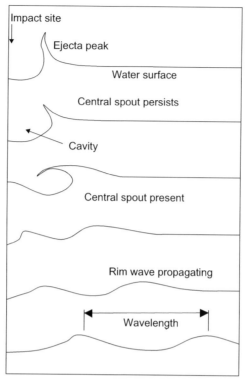

Figure 18.3

Generation of the rim wave. *Modified from Gault, D.E., Sonett, C.P., 1982. Laboratory simulation of pelagic asteroidal impact: atmospheric injection, benthic topography, and the surface wave radiation field. In: Silver, L.T., Schultz, P.H., (Eds.), Geological Implications of Impacts of Large Asteroids and Comets on the Earth. Geol. Soc. Amer. Spec. Pap. vol. 190, 69—92; Matsui, T., Imamura, F., Tajika, E., Nakano, Y., Fujisawa, Y., 2002. Generation and propagation of a tsunami from the Cretaceous/Tertiary impact event. In: Koeberl, C., Macleod, G., (Eds.), Catastrophic Events and Mass Extinctions: Impact and beyond. Geol. Soc. Amer. Spec. Pap. vol. 356, pp. 69—77.*

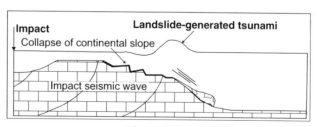

Figure 18.4

Generation of a landslide-generated tsunami. *Based on Matsui, T., Imamura, F., Tajika, E., Nakano, Y., Fujisawa, Y., 2002. Generation and propagation of a tsunami from the Cretaceous/Tertiary impact event. In: Koeberl, C., Macleod, G., (Eds.), Catastrophic Events and Mass Extinctions: Impact and beyond. Geol. Soc. Amer. Spec. Pap. vol. 356, pp. 69—77.*

336   Chapter 18

propagate as waves outward from the crater. According to a numerical simulation by Matsui et al. (2002), the amplitude and period of the waves from the crater are controlled by the depth of the water surrounding the crater, because the flow rate of water into the crater from the surrounding ocean is controlled by the water depth. The lower the inward flow velocity, the lower the wave height of the rushing wave and the longer the period. The first wave of a crater-generated tsunami is a receding wave toward the crater, and this is a significant property of a tsunami generated by this mechanism (Matsui et al., 2002).

3.2 Landslide-generated tsunamis

Shock waves and seismic waves generated by a large oceanic impact can trigger large-scale landslides at the margin of the continental slope, which may cause a landslide-generated tsunami (Fig. 18.4; Matsui et al., 2002). The magnitude of a landslide-generated tsunami is highly dependent on the volume of the landslide (Matsui et al., 2002). Information about the surface area and the thickness of the landslide, based on a seismic survey around the crater, is therefore important for a reliable estimate of the magnitude of the responsible landslide-generated tsunami.

4. Impact-generated tsunami deposits inside and outside the oceanic impact craters

Impact-generated tsunamis affect the surrounding ocean bottom and coastal areas and form impact-generated tsunami deposits inside and outside the crater. In this section, the characteristics of such deposits are summarized.

4.1 Deposits formed by water flowing into an oceanic impact crater

The crater rim of oceanic impact craters tends to collapse not only by gravity but also by the inflow of ocean water into the crater (Ormö and Lindström, 2000). The inflowing water forms gullies in the crater rim (e.g., the Gusev, Kamensk, Kara, Kärdla, Lockne, Mayølr, and Montagnais craters: Dalwigk and Ormö, 2001; Dypvik and Jansa, 2003; Ormö and Lindström, 2000; Sturkell, 1998; Sturkell and Ormö, 1997). The deposits inside the oceanic crater, which are reworked by the inflowing ocean water, are called "resurge deposits" (Ormö and Lindström, 2000). They are formed both in the gullies and inside the crater (Dalwigk and Ormö, 2001; Dypvik and Jansa, 2003; Goto et al., 2004; King et al., 2002; Lindström et al., 1994; Ormö and Lindström, 2000; Poag et al., 2002).

Resurge deposits in the gullies have been well studied for the Lockne crater in Sweden (Fig. 18.1), which is ~13.5 km in diameter and which was formed on the ocean floor at a depth of over 200 m (Ormö and Lindström, 2000). According to Ormö and Lindström (2000) and Dalwigk and Ormö (2001), the lower part of the resurge deposits in the gullies of the Lockne crater is matrix-supported and is similar to debris-flow deposits.

The upper part is clast-supported and has a relatively high proportion of crystalline ejecta (Ormö and Lindström, 2000). The uppermost fine-grained part of the resurge deposit displays current lineation, cross-bedding, and water-escape structures (Dalwigk and Ormö, 2001).

The resurge deposits inside oceanic impact craters have been well studied for several craters such as the Chesapeake Bay, Chicxulub, Mayølr, Montagnais, Tvären, and Wetumpka craters (Fig. 18.1; Dalwigk and Ormö, 2001; Dypvik and Jansa, 2003; Goto et al., 2004; Jansa, 1993; King et al., 2002; Lindström et al., 1994; Ormö, 1994; Ormö and Lindström, 2000; Poag et al., 2002). Unusually thick sedimentary breccias are reported from these craters; they are much thicker than those in terrestrial craters of the same size (Jansa, 1993; Poag et al., 2002). The presence of the thick sedimentary breccias inside the oceanic crater is probably due to inflow of ocean water: strong currents associated with the inflow must have eroded the oceanic bottom sediments and ejecta around the crater and transported them into the crater (Poag et al., 2002).

When an oceanic impact crater becomes overfilled with water, a tsunami propagates outside the crater (Fig. 18.2). The volume of the water inside the crater then becomes too small and ocean water flows back into the crater again. This can thus result in repeated inflow of water into the crater. Consequent repetitions of upward fining beds are observed in the Chesapeake Bay impact crater (Poag et al., 2002) and the Chicxulub crater (Goto et al., 2004).

Resurge deposits in the Tvären and the Chicxulub craters tend to show a reversed succession (Goto et al., 2004; Ormö, 1994; Ormö and Lindström, 2000): the lower part of the deposit consists of particles derived from surficial sediments outside the crater and the upper part consists of particles from older sediments (Ormö, 1994; Ormö and Lindström, 2000). This is because bottom sediments outside the crater were eroded by currents and transported into the crater, beginning with the uppermost (younger) sediments followed by progressively older material (Fig. 18.2; Goto et al., 2004; Ormö and Lindström, 2000).

4.2 Impact-generated deposits outside an oceanic impact crater

Around an oceanic impact site, impact-generated deposits are formed by various processes such as the fall of ejecta from the air, seismic waves, and tsunamis. Approximately 90% of the material ejected from the crater is deposited relatively nearby (<5 crater radii), and these particles are called "proximal ejecta," whereas the rest of ejecta is deposited >5 crater radii from the crater rim and are called "distal ejecta" (Melosh, 1989; Montanari and Koeberl, 2000). Distal ejecta usually consist of fine-grained clasts, minerals, and glassy melt fragments (Montanari and Koeberl, 2000). Impact ejecta are ejected into the atmosphere by the impact and then settle through the atmosphere and the ocean water (Alvarez et al., 1995). Due to the difference in grain size and density, impact ejecta tend to be sorted during their transport in the air and water column (Takayama et al., 2000). In general, coarse spherules (millimeter-sized) are present in the basal part of the deposit, micrometer-sized shocked-quartz grains and microspherules in the middle to upper part, and nanometer-sized Ir-bearing dust at the top (Smit, 1999).

Impact-generated deposits can also be influenced by seismic waves and tsunamis in the case of an oceanic impact. One of the earliest impact-related phenomena is the seismic wave (Smit, 1999; Smit et al., 1996). This seismic wave may be large enough to cause slope failure (Boslough et al., 1996). In fact, several probable gravity-flow deposits have been found around oceanic impact craters (Bralower et al., 1998; Gersonde et al., 1997; Grajales-Nishimura et al., 2000; Kiyokawa et al., 2002; Klaus et al., 2000; Smit, 1999; Smit et al., 1996; Tada et al., 2003; Takayama et al., 2000).

The influence of the tsunami probably starts after the generation of gravity flows triggered by the seismic wave because the wave velocity of a tsunami is usually lower than that of the seismic wave (Smit et al., 1996). A tsunami is a "very shallow-water wave (long wave)" characterized by a large wavelength relative to the water depth, and this is distinctly different from other large waves such as storm waves, which have a small wave length relative to the water depth (Fig. 18.5A). According to the theory for small-amplitude water waves, the maximum orbital velocity, U_{max} (m/s), of the shallow-water wave at a water depth of z (m) is:

$$U_{max} = \frac{H}{2} \frac{gT}{L} \frac{\cosh[\{2\pi(h+z)/L\}]}{\cosh\{(2\pi h)/L\}} \quad (18.1)$$

where H is the wave height (m), g is the gravitational acceleration (m/s^2), T is the wave period (s), L is the wavelength, and h is the water depth (m). According to Eq. (18.1),

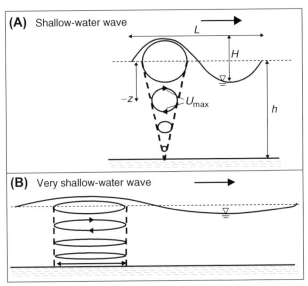

Figure 18.5

Difference between (A) a shallow-water wave and (B) a very shallow-water wave (tsunami).

U_{max} becomes 0 at the ocean bottom ($z = -h$). Thus, a shallow-water wave does not influence the sea floor. In contrast, when h/L becomes small (i.e., a very shallow-water wave), each term of cosh in Eq. (18.1) approaches to 1. Because $L = \sqrt{gh} \cdot T$, U_{max} becomes

$$U_{max} = \frac{H}{2}\sqrt{\frac{g}{h}} \tag{18.2}$$

U_{max} in Eq. (18.2) is independent of the water depth, z, indicating that the maximum orbital velocities at the top and the bottom of the water column are the same (Fig. 18.5B). This suggests that sediment particles on the ocean bottom have the potential to become suspended by passing tsunami waves, depending on their wave height. Therefore, impact-generated tsunami deposits can be formed on both shallow-sea and deep-sea bottoms in the case of a large tsunami. Some probable deep-sea tsunami deposits have been reported from around the Gulf of Mexico and the proto-Caribbean Sea, as a result of the K/T-boundary oceanic impact, as mentioned further (see also Alvarez et al., 1992; Goto et al., 2008; Maurrasse and Sen, 1991; Tada et al., 2002, 2003; Takayama et al., 2000).

Because of the time lag between the fall of impact ejecta from the air, seismic waves, and tsunami waves, impact-generated deposits around the impact site tend to be composed of a lower part consisting of gravity-flow deposits generated by the impact seismic wave, a middle part consisting of deposits generated by the impact-generated tsunami, and an upper part consisting of fine-grained ejecta (Smit, 1999).

The influence of an impact seismic wave and of a tsunami decreases with distance from the crater, and most of the ejecta are deposited near the crater as proximal ejecta. Thus, the thickness of the impact-generated deposits tends to decrease with distance from the crater. Because the arrival time of impact ejecta depends on the depositional depth as well as on the distance from the impact site, the spatial distribution of impact ejecta in impact-generated deposits can vary from site to site. In the next section, the distribution and significance of the K/T-boundary deposits will be reviewed, exemplifying the distribution and significance of impact-generated deposits around an oceanic impact crater.

5. Distribution and significance of the K/T-boundary tsunami deposits around the Chicxulub crater

5.1 The K/T-boundary impact event

The Chicxulub crater on the Yucatán Peninsula, Mexico (Fig. 18.1; Hildebrand et al., 1991), is now well accepted as an impact crater formed at the K/T boundary. The crater has a diameter of ~ 180 km and is one of the largest known impact craters on earth. It is believed that the Chicxulub impact was the major cause of the K/T-boundary mass extinction. Because the impact was in a shallow ocean (Sharpton et al., 1996), a large tsunami was one

340 Chapter 18

of the major consequences. As mentioned by Goto et al. (2008), both tsunamis triggered by landslides on the slopes of the Yucatán platform and a crater-generated tsunami could have occurred at the K/T boundary. Several probable tsunami deposits have been reported from around the Gulf of Mexico (Alvarez et al., 1992; Bourgeois et al., 1988; Lawton et al., 2005; Smit, 1999; Smit et al., 1992 1996), from the proto-Caribbean Sea (Goto et al., 2008; Maurrasse and Sen, 1991; Tada et al., 2002, 2003; Takayama et al., 2000), and from inside the crater (Goto et al., 2004), although some criticisms about their origin (e.g., Keller et al., 2003) and sedimentary process (e.g., Bohor, 1996; Gale, 2006) still remain. The K/T-boundary impact has provided a rare opportunity for the investigation of a tsunami induced by an oceanic impact.

The thickness and sedimentary structures of the K/T-boundary deposits tend to change systematically with increasing distance from the Chicxulub crater. Tsunami and/or gravity-flow deposits of several meters to several hundreds of meters thick, which are composed of a mixture of local clastics and impact ejecta, have been reported from within 1500 km from the center of the crater. In contrast, most of the K/T-boundary deposits present at >1500 km from the crater's center have a thickness of the order of millimeters, and they are mainly composed of impact ejecta. This difference probably indicates that proximal sites have been strongly influenced by the impact-generated seismic and tsunami waves, but that the influence of these waves was not truly significant at >1500 km away from the crater.

Goto et al. (2008) have summarized the lithology and sedimentary processes of the K/T-boundary deep-sea tsunami deposits in the proto-Caribbean Sea region, including the Peñalver, Cacarajícara, and Moncada Formations in Cuba and at Deep Sea Drilling Project (DSDP) sites 536 and 540. In this section, the focus is on the change of the K/T-boundary deposits with increasing distance from the crater.

5.2 On the edge of the Yucatán platform

Grajales-Nishimura et al. (2000) have reported about a thick sedimentary breccia at Campeche, in the western part of the Yucatán platform, ~350 km south-west from the center of the Chicxulub crater (Fig. 18.6). They mention that the deposit has a thickness of >300 m and is composed of an ~300 m thick lower part of brecciated dolomitized limestone and an upper bentonitic part of ~30 m thickness with reworked ejecta and breccia lenses (Fig. 18.7). The thickness of the breccia decreases to 40−60 m and that of the bentonitic bed to 5−11 m at ~600 km south-west from the center of the crater. Shocked-quartz grains were found in the upper part of the breccia and throughout the bentonite; the latter also contains greenish altered-glass fragments. The upper part of the bentonite contains thin lenses of conglomerate with current lineation. Grajales-Nishimura et al. (2000) interpreted the sedimentary process of this K/T-boundary deposit as follows: (1) the carbonate platform collapsed due to the impact, resulting in deposition of the

The genesis of oceanic impact craters 341

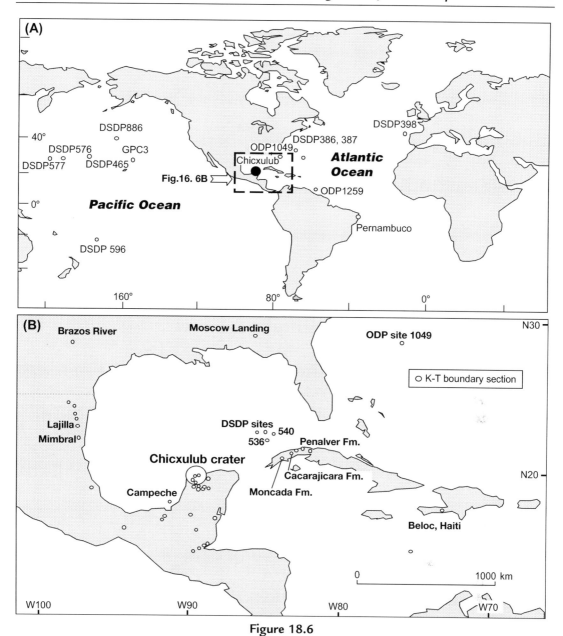

Figure 18.6
Distribution of K/T-boundary deposits in (A) the Atlantic and Pacific Oceans and (B) the Gulf of Mexico and the Caribbean Sea. Note that this figure does not include the K/T-boundary deposits in other regions such as Europe and North America. *(A) Based on Albertão and Martins (1996), Bostwick and Kyte (1996), Kyte et al. (1996), Norris and Firth (2002). (B) After Tada et al. (2003).*

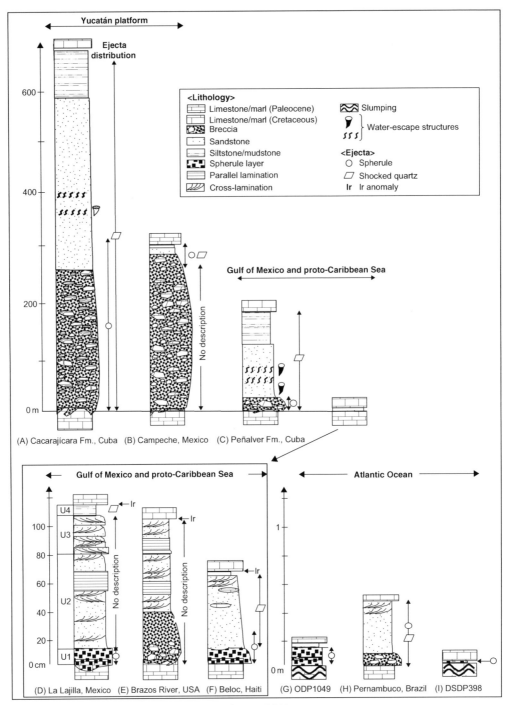

Figure 18.7
Sedimentary logs of the K/T-boundary deposits in (A) the Cacarajícara Formation, Cuba; (B) Campeche, Mexico; (C) the Peñalver Formation, Cuba; (D) La Lajilla, Mexico; (E) Brazos River, United States; (F) Beloc, Haiti; (G) Ocean Drilling Program (ODP) site 1049; (H) Pernambuco, Brazil; and (I) Deep Sea Drilling Project (DSDP) site 398. Vertical distributions of impact ejecta are also shown. *Based on Albertão and Martins (1996), Bourgeois et al. (1988), Grajales-Nishimura et al. (2000), Kiyokawa et al. (2002), Leroux et al. (1995), Maurrasse and Sen (1991), Norris and Firth (2002), Smit et al. (1996), Takayama et al. (2000).*

The genesis of oceanic impact craters   343

breccia, (2) impact ejecta fell back, and (3) reworking of the ejecta layer with coarser material took place by one or more passages of the impact-generated tsunami waves.

Kiyokawa et al. (2002) have reported a similar thick breccia with overlying thick sandstone and siltstone beds (>700 m thick) in north-western Cuba (Fig. 18.7: the Cacarajícara Formation), which was originally located in south-eastern edge of the Yucatán platform ~500 km east of the center of the crater at the time of the K/T impact (see Goto et al., 2008).

5.3 The Gulf of Mexico region

5.3.1 Brazos River section

The K/T-boundary tsunami deposit was discovered by Bourgeois et al. (1988) along the Brazos River, Texas (Fig. 18.6). Although the origin (e.g., Montgomery et al., 1992) and sedimentary process (e.g., Gale, 2006) of the K/T-boundary deposits at the Brazos River section have been still suspected, the interpretation by Bourgeois et al. (1988) is generally accepted (Schulte et al., 2006). According to Bourgeois et al. (1988), the K/T-boundary deposit overlies late Cretaceous shelf mudstones without distinct sandstone layers; they estimated the depositional depth of the mudstone as 75−200 m, based on the assemblage of benthic foraminifers. The deposit has an erosional basis, and the basal coarse-grained sandstone contains shell fragments, fish teeth, wood debris, and angular clasts of mudstone that are commonly >5 cm. This basal layer grades upward through parallel- and wave-ripple-laminated very fine-grained sandstones to siltstones and mudstones. The sandstone bed comprises two or more amalgamated graded beds; their total thickness ranges from 30 to 130 cm.

Based on the observation of sedimentary features, Bourgeois et al. (1988) considered that the deposition of the sandstone bed must have required (1) initial shear velocities of 15−100 cm/s to erode and transport the mudstone clasts and (2) a rapid drop in the shear velocities to <1 cm/s, under oscillatory flow conditions to form wave ripples in the very fine sand. Based on these considerations, they concluded that only a tsunami is likely to have produced these conditions on the continental shelf. They proposed that tsunami waves with a height of >50 m caused major erosion at the shelf break and on the shelf and that they transported and redeposited shelf sediments (Bourgeois et al., 1988).

5.3.2 North-Eastern Mexico (the K/T-boundary sandstone complex)

After the discovery of the Chicxulub crater, numerous K/T-boundary tsunami deposits have been reported from coastal areas of Mexico and United States (Smit, 1999; Smit et al., 1992, 1996). The deposits that were formed on the ocean bottom around the Gulf of Mexico are characterized by a series of unusual sandstone and siltstone beds, which are called a "K/T-boundary sandstone complex" (Smit et al., 1996). A typical K/T-boundary

344   Chapter 18

sandstone complex is present in La Lajilla, Mexico (Fig. 18.7; Smit, 1999; Smit et al., 1996). It is usually several meters thick, and it differs distinctly from the K/T-boundary deposits around the edge of the Yucatán platform by the absence of a thick breccia (Fig. 18.7). According to Smit et al. (1996), the K/T-boundary sandstone complex that is present at neritic to upper-bathyal depths around the Gulf of Mexico can be subdivided on the basis of its lithology into three to four units, which are named units 1–4, from bottom to top (Fig. 18.7).

According to Smit (1999) and Smit et al. (1996), unit 1 overlies latest Maastrichtian marls with an irregular erosional surface and is characterized by poorly sorted, coarse-grained pebbly sandstones, which are mainly composed of millimeter-sized altered spherules. Unit 2 consists of a succession of several lenticular sandstone layers made up of a mixture of foraminifers, bioclasts, plant remains, and terrestrial matter. The sandstones of unit 2 are graded and show parallel or current-ripple laminae. Unit 2 usually also contains a set of well-sorted lenticular sandstones, with a wide variety of sedimentary structures indicating paleocurrent directions in two opposite directions. Unit 3 consists of strings of fine sand ripples alternating with thin silt layers, and unit 4 is a graded siltstone to mudstone. Units 3 and 4 contain anomalous Ir concentrations and Ni-rich spinels (Smit et al., 1996).

Some criticisms about the origin (e.g., Keller et al., 2003) and sedimentary process (e.g., Bohor, 1996) for the K/T-boundary sandstone complex in Mexico still remain. Smit et al. (1996) and Smit (1999) argued against a turbidite origin of the K/T-boundary sandstone complex (Bohor, 1996) based on the following considerations: (1) A relatively thin and lenticular sandstone complex over an area that continues for at least 2000 km is present. Over this entire length, the thickness of the unit varies only between 1 and 3 m. (2) Reversals of currents, as indicated by the opposite orientations of ripple cross-bedding, are common. Such current reversals can hardly be explained in turbidites. (3) There are no turbidites known to occur within the succession of Cretaceous and Palaeocene marls, and the K/T-boundary sandstone complex is unique. Based on these data, Smit et al. (1996) and Smit (1999) interpreted that redeposition by a series of large, waning tsunami waves is the most likely explanation for the properties of the K/T-boundary sandstone complex. Similar K/T-boundary sandstone complexes have been reported from north-eastern Mexico and from Texas and Alabama, United States (King and Petruny, 2004; Lawton et al., 2005; Smit, 1999; Smit et al., 1992, 1996).

5.4 The proto-Caribbean sea region (DSDP sites, Haiti and Cuba)

K/T-boundary tsunami deposits similar to the K/T-boundary sandstone complex in north-eastern Mexico have also been found in western Cuba (the Moncada Formation) (Fig. 18.6; Tada et al., 2002, 2003; see also Goto et al., 2008), indicating that the impact-generated tsunami also propagated to the eastern side of the crater. As discussed in Goto et al. (2008), the K/T-boundary deposits at DSDP sites 536 and 540, which are located ~500 and 600 km to the north-east of the center of the Chicxulub crater and which were formed in the deep sea (Fig. 18.6), also have a thickness and other properties that are similar to those of the K/T-boundary sandstone complex in the Gulf of Mexico (Fig. 18.7). They have been interpreted as deep-sea tsunami deposits (Alvarez et al., 1992; Goto et al., 2008).

K/T-boundary tsunami deposits have also been reported from Beloc, Haiti, where the water depth was about 2000 m at the time of the impact (Fig. 18.6; Maurrasse and Sen, 1991). The thickness of the deposits varies from 40 to 70 cm and their lower part consists of impact ejecta (Fig. 18.7). The bed grades upward into fine sand and silt, with low-angle cross-bedding; thin Ir-rich layers occur at the top of the succession (Maurrasse and Sen, 1991). Maurrasse and Sen (1991) interpreted the K/T-boundary deposits in Beloc as tsunami deposits because (1) the thickness of the deposits varies in a lateral direction but its petrographical composition remains constant and (2) there is no laterally changing grain size as seen in turbidites.

The Peñalver Formation in north-western Cuba is unusually thick (>180 m) compared to the other K/T-boundary deposits within the Gulf of Mexico and the proto-Caribbean Sea regions (Fig. 18.7; Goto et al., 2008; Tada et al., 2003; Takayama et al., 2000). The lower unit of the Peñalver Formation is 30 m thick and is composed of calcirudites with grains that were derived from a shallow platform of the Cuban volcanic arc. It is considered to have been formed by debris flows triggered by the impact-related seismic wave (Goto et al., 2008; Tada et al., 2003; Takayama et al., 2000). In contrast, the upper unit of >150 m is composed of calcarenites to calcilutites with grains that were derived from hemipelagic to pelagic sediments, suggesting a distinctly different source compared to the lower part. The upper part is considered as having been formed under the influence of a tsunami, based on upward fining, the regional homogeneity of the source material of pelagic to hemipelagic origin, and the occurrence of allochthonous material (Goto et al., 2008; Tada et al., 2003; Takayama et al., 2000).

As discussed by Goto et al. (2008), the unusual thickness of the Peñalver Formation probably reflects a unique depositional setting. Because the formation occurred on the northern frank of the Cretaceous Cuban Arc, which was covered by the Cuban carbonate platform, an impact-induced seismic wave could have generated debris flows from this platform. The sand- to silt-sized particles from the upper slope were probably supplied by repeated backwash currents of the tsunami (Tada et al., 2003). These additional supplies of arc and platform sediments were probably responsible for the unusually thick K/T-boundary deposits.

The occurrence of K/T-boundary tsunami deposits in the deep sea is still criticized with the main argument that impact-generated tsunamis cannot affect the deep-sea bottom, due to the diminishing power of currents with increasing water depth (e.g., Bohor, 1996).

346   Chapter 18

Eq. (18.2) proves, however, that the influence of a tsunami does not diminish downward. Considering that the sedimentary characteristics of the K/T-boundary deposits in Mexico, United States, and Cuba, at DSDP sites 536 and 540, and on Haiti are distinctly different from those of turbidites (Alvarez et al., 1992; Goto et al., 2008; Maurrasse and Sen, 1991; Smit, 1999; Smit et al., 1996; Takayama et al., 2000; Tada et al., 2002, 2003), these deposits likely have been formed by an impact-generated tsunami.

5.5 The Atlantic Ocean

Norris and Firth (2002) summarized the sedimentary characteristics of the K/T-boundary deep-sea deposits in the Atlantic Ocean. At Ocean Drilling Program (ODP) site 1049C on the Blake Nose, east of Florida (Fig. 18.6), a ~17-cm-thick bed of greenish spherules and gray silty ooze with a high Ir content are present (Fig. 18.7; Klaus et al., 2000; Norris et al., 1999). The spherule layer overlies a deformed latest Maastrichtian ooze of several meters to several tens of meters thick. The deformation is interpreted as a result of mass wasting induced by the impact seismic wave (Klaus et al., 2000). The lower part of the bed is faintly laminated and dominated by green spherules of 1−3 mm, accompanied by large Cretaceous planktonic foraminifers, shocked quartz, and clasts of limestone and dolomite (Klaus et al., 2000). The presence of faint lamination indicates that minor reworking occurred during and after the sedimentation of spherules (Klaus et al., 2000). The K/T-boundary deposits in this region are distinctly different from the equivalent deposits around the Gulf of Mexico and proto-Caribbean Sea by their composition: they consist mainly of distal ejecta, and there are no thick sandstone and siltstone beds indicative of reworking by the tsunami. At DSDP site 398D, on the Iberian abyssal plain on the eastern North Atlantic margin (Fig. 18.6), the thickness of the K/T-boundary deposits becomes ~1 mm, and these deposits are mainly composed of greenish spherules (Fig. 18.7; Norris and Firth, 2002), implying that the influence of the tsunami was not strong enough to bring the deep-sea sediments in the Atlantic Ocean into suspension.

In contrast, a ~50-cm-thick K/T-boundary deposit is reported from Pernambuco, north-eastern Brazil, ~7000 km away from the center of the Chicxulub crater (Figs. 18.6 and 18.7). According to Albertão and Martins (1996), the K/T-boundary deposit at this site occurred at a depth of 100−200 m (based on the benthic foraminifer assemblage). This deposit overlies a latest Maastrichtian limestone bed with a sharp and erosional contact and is composed of bioclastic and intraclastic packstone with spherules and shocked-quartz grains. Current ripples are present at the top of the deposit. Albertão and Martins (1996) interpreted this deposit as a tsunami deposit based on (1) its sharp, erosive base, (2) fining-upward units, (3) extensive mixing and fragmentation of fossils derived from different paleodepths and reworked from older strata, (4) interference ripples at the top of the bed, and (5) the lateral continuity of the deposits. The deposit is unusually thick in comparison to the deep-sea deposits in the DSDP and ODP cores from a similar distance to the crater and rather similar to the K/T-boundary sandstone complex in the Gulf of Mexico. The difference may be attributed to the different depositional depths of the sites. Although the influence of the tsunami may not have been large enough to bring sandy to silty particles into suspension from the deep-sea bottom at this distance, its influence may still have been large at a depth of ~ 200 m. On the other hand, the origin of the Chicxulub impact for the deposits in Pernambuco has been questioned (e.g., Koutsoukos, 1998; Morgan et al., 2006). Therefore, further careful research is required for the deposits.

5.6 The Pacific Ocean

Bostwick and Kyte (1996) and Kyte et al. (1996) have investigated K/T-boundary deposits in the Pacific Ocean (Fig. 18.6). These sites are located $\sim 5800-11,000$ km west from the Chicxulub crater. Bostwick and Kyte (1996) found pyrite-rich clay layers of ~ 3 mm thickness, with minute shocked-quartz grains at these sites and an Ir anomaly in the top part of the deposits. Neither thick sandstone or siltstone beds (Bostwick and Kyte, 1996; Kyte et al., 1996) nor evidence of intensive reworking by a tsunami was found in these deposits, implying that the influence of the tsunami was not strong enough to whirl up the deep-sea sediments in the Pacific Ocean.

6. Significance and distribution of the K/T-boundary tsunami deposits

Although K/T-boundary tsunami deposits have different thicknesses and sedimentary characteristics, depending on the depositional setting and the distance from the crater, they seem to have the following common properties: (1) a homogeneous thickness, grain-size distribution, and grain composition over a large area (Albertão and Martins, 1996; Goto et al., 2008; Maurrasse and Sen, 1991; Smit et al., 1996); (2) opposite current directions in successive units (Alvarez et al., 1992; Goto et al., 2008; Smit, 1999; Smit et al., 1996; Tada et al., 2002, 2003) or repeated compositional cycles that probably reflect the repeated passage of tsunamis (Goto et al., 2008); (3) the presence of terrestrial matter such as plant and wood fragments (Bourgeois et al., 1988; Smit et al., 1996); and (4) the absence of the Bouma sequence that is typical for turbidites (Bourgeois et al., 1988; Maurrasse and Sen, 1991; Smit et al., 1996; Tada et al., 2002, 2003).

As mentioned earlier, the thickness of the K/T-boundary deposits tends to decrease with distance from the Chicxulub crater. Thick breccias, which were probably formed by the impact seismic wave, exist around the edge of the Yucatán platform. Most of the K/T-boundary tsunami deposits have been reported from the Gulf of Mexico and the proto-Caribbean Sea, suggesting that the influence of the tsunami was strong in these regions.

348 Chapter 18

At the time of the K/T-boundary impact, the impact site was surrounded by land in the north, south, and west, and shallow carbonate platforms existed in the south and east (Tada et al., 2003). The energy of the tsunami was therefore trapped in the Gulf of Mexico and the proto-Caribbean regions (Dypvik and Jansa, 2003). In the Atlantic and Pacific Oceans, only millimeter-thick to centimeter-thick distal ejecta layers have been found, whereas tsunami deposits of ~ 50 cm thickness have been found in the shallow sediments of Pernambuco, Brazil, suggesting that the influence of the tsunami in these oceans might have been restricted only in shallow waters (less than ~ 200 m deep).

The K/T-boundary tsunami deposits have thus several properties in common, and systematic variations occur that depend on both the depositional depth and the distance from the impact site. These characteristics are similar to those of impact-generated tsunami deposits in other oceanic impacts (e.g., Gersonde et al., 1997), so that they are useful for identification. Future comparative studies between the K/T-boundary tsunami deposits and other oceanic impact-generated tsunami deposits will provide useful information that will help to understand their precise genesis and to reconstruct the effects and magnitude of impact-generated tsunamis.

7. Summary

1. Impact craters on earth can be divided into terrestrial and oceanic ones. Oceanic impact events can, in turn, be subdivided into shallow and deep oceanic ones. In the case of an oceanic impact event, ocean water flows back into the crater immediately after the impact, eroding the crater rim. Consequently, oceanic impact craters usually show little or no evidence of a high crater rim.
2. Impact-generated tsunamis are one of the consequences of oceanic impact events and affect the surrounding ocean bottom and coastal areas; they form impact-generated tsunami deposits inside and outside the crater. Resurge deposits are formed in the gullies through which the water flowed into the crater and inside the crater.
3. K/T-boundary tsunami deposits are the best-studied impact-generated tsunami deposits. The K/T-boundary tsunami deposits have several sedimentary characteristics in common, and the thickness of the deposits decreases with distance from the Chicxulub crater.

Acknowledgments

This paper is a reprint of the first version published in 2008. Original terminology (e.g., K/T-boundary) is used as it is. The author wishes to thank T. Shiki, D. King, S. Kiyokawa, R. Tada, and E. Tajika for their critically reading of the manuscript and their many valuable suggestions. This research was partly supported by a Grant-in-Aid for Scientific Research from the Japan Society for the Promotion of Science (no. 17403005).

References

Albertão, G.A., Martins, P.P., 1996. A possible tsunami deposit at the Cretaceous-Tertiary boundary in Pernambuco, northeastern Brazil. Sediment. Geol. 104, 189–201.

Alvarez, W., Smit, J., Lowrie, W., Asaro, F., Margolis, S.V., Claeys, P., Kastner, M., Hildebrand, A., 1992. Proximal impact deposits at the Cretaceous-Tertiary boundary in the Gulf of Mexico: a restudy of DSDP, leg 77 sites 536 and 540. Geology 20, 697–700.

Alvarez, W., Claeys, P., Kieffer, S.W., 1995. Emplacement of Cretaceous-Tertiary boundary shocked quartz from Chicxulub crater. Science 269, 930–935.

Bohor, B.F., 1996. A sedimentary gravity flow hypothesis for siliciclastic units at the K/T boundary, northeastern Mexico. In: Ryder, G., Fastovsky, D., Gartner, S. (Eds.), Cretaceous-Tertiary Event and Other Catastrophes in Earth History, vol. 307. Geol. Soc. Amer. Spec. Pap, pp. 183–195.

Boslough, M.B., Chael, E.P., Trucano, T.G., Crawford, D.A., Campbell, D.L., 1996. Axial focusing of impact energy in the Earth's interior: a possible link to flood basalts and hotspots. In: Ryder, G., Fastovsky, D., Gartner, S. (Eds.), Cretaceous-Tertiary Event and Other Catastrophes in Earth History, vol. 307. Geol. Soc. Amer. Spec. Pap, pp. 541–550.

Bostwick, J.A., Kyte, F.T., 1996. The size and abundance of shocked quartz in Cretaceous-Tertiary boundary sediments from the Pacific basin. In: Ryder, G., Fastovsky, D., Gartner, S. (Eds.), Cretaceous-Tertiary Event and Other Catastrophes in Earth History, vol. 307. Geol. Soc. Amer. Spec. Pap, pp. 403–415.

Bourgeois, J., Hansen, T.A., Wiberg, P.L., Kauffman, E.G., 1988. A tsunami deposit at the Cretaceous-Tertiary boundary in Texas. Science 241, 567–570.

Bralower, T.J., Paul, C.K., Leckie, R.M., 1998. The Cretaceous/Tertiary boundary cocktail: Chicxulub impact triggers margin collapse and extensive sedimentary gravity flows. Geology 26, 331–334.

Dalwigk, I., Ormö, J., 2001. Formation of resurge gullies at impacts at sea: the Lockne crater, Sweden. Meteoritics Planet Sci. 36, 359–369.

Dypvik, H., Jansa, L., 2003. Sedimentary signatures and processes during marine bolide impacts: a review. Sediment. Geol. 161, 309–337.

Gale, A.S., 2006. The Cretaceous-Paleogene boundary on the Brazos River, Falls county, Texas: is there evidence for impact-induced tsunami sedimentation? Proc. Geol. Assoc. 117, 173–185.

Gault, D.E., Sonett, C.P., 1982. Laboratory simulation of pelagic asteroidal impact: atmospheric injection, benthic topography, and the surface wave radiation field. In: Silver, L.T., Schultz, P.H. (Eds.), Geological Implications of Impacts of Large Asteroids and Comets on the Earth, vol. 190. Geol. Soc. Amer. Spec. Pap, pp. 69–92.

Gersonde, R., 2000. New field of impact research looks to the oceans. EOS Trans 81, 21–24.

Gersonde, R., Kyte, F.T., Bleil, U., Diekmann, B., Flores, J.A., Gohl, K., Grahl, G., Hagen, R., Kuhn, G., Sierro, F.J., Volker, D., Abelmann, A., et al., 1997. Geological record and reconstruction of the late Pliocene impact of the Eltanin asteroid in the Southern Ocean. Nature 390, 357–363.

Gersonde, R., Deutsch, A., Ivanov, B.A., Kyte, F.T., 2002. Oceanic impacts—a growing field of fundamental geoscience. Deep-Sea Res. II 49, 951–957.

Glikson, A.Y., 1999. Oceanic mega-impacts and crustal evolution. Geology 27, 387–390.

Goto, K., Tada, R., Tajika, E., Bralower, T.J., Hasegawa, T., Matsui, T., 2004. Evidence for ocean water invasion into the Chicxulub crater at the Cretaceous/Tertiary boundary. Meteoritics Planet Sci. 39, 1233–1247.

Goto, K., Tada, R., Tajika, E., Matsui, T., 2008. Deep-sea tsunami deposits in the proto-Caribbean sea at the Cretaceous-Tertiary boundary. In: Shiki, T., Tsuji, Y., Yamazaki, T., Minoura, K. (Eds.), Tsunamiites—Features and Implications. Elsevier, Amsterdam, pp. 251–257.

Grajales-Nishimura, J.M., Cedillo-Pardo, E., Rosales-Dominguez, C., Moran-Zenteno, D.J., Alvarez, W., Claeys, P., Ruiz-Morales, J., Garcia-Hernandez, J., Padilla-Avila, P., Sanchez-Rios, A., 2000. Chicxulub impact: the origin of reservoir and seal facies in the southeastern Mexico oil fields. Geology 28, 307–310.

Hassler, S.W., Simonson, B.M., 2001. The sedimentary record of extraterrestrial impacts in deep-shelf environments: evidence from the early Precambrian. J. Geol. 109, 1–19.

Hildebrand, A.R., Penfield, G.T., Icring, D.A., Pilkington, M., Camargo, N., Jacobsen, S.B., Boynton, W.V., 1991. Chicxulub crater: a possible Cretaceous-Tertiary boundary impact crater on the Yucatán Peninsula, Mexico. Geology 19, 867–871.

Jansa, L.F., 1993. Cometary impacts into ocean—their recognition and the threshold constraint for biological extinctions. Palaeogeogr. Palaeoclimatol. Palaeoecol. 104, 271–286.

Keller, G., Stinnesbeck, W., Adatte, T., Stueben, D., 2003. Multiple impacts across the Cretaceous-Tertiary boundary. Earth Sci. Rev. 62, 327–363.

King Jr., D.T., Petruny, L.W., 2004. Cretaceous-Tertiary boundary microtektite-bearing sands and tsunami beds, Alabama gulf coastal plain. Lunar Planet. Sci. 35, 1804.

King Jr., D.T., Neathery, T.L., Petruny, L.W., Koeberl, C., Hames, W.E., 2002. Shallow-marine impact origin of the Wetumpka structure (Alabama, USA). Earth Planet Sci. Lett. 202, 541–549.

Kiyokawa, S., Tada, R., Iturralde-Vinent, M.A., Tajika, E., Yamamoto, S., Oji, T., Nakano, Y., Goto, K., Takayama, H., Delgado, D.G., Otero, C.D., Rojas-Consuegra, R., et al., 2002. Cretaceous-Tertiary boundary sequence in the Cacarajícara Formation, western Cuba: an impact-related, high-energy, gravity-flow deposit. In: Koeberl, C., Macleod, G. (Eds.), Catastrophic Events and Mass Extinctions: Impact and Beyond, vol. 356. Geol. Soc. Amer. Spec. Pap, pp. 125–144.

Klaus, A., Norris, R.D., Kroon, D., Smit, J., 2000. Impact-induced mass wasting at the K-T boundary: Blake Nose, western North Atlantic. Geology 28, 319–322.

Koutsoukos, E.A.M., 1998. An extraterrestrial impact in the early Danian: a secondary K/T boundary event? Terra. Nova 10, 68–73.

Kyte, F.T., Bostwick, J.A., Zhou, L., 1996. The Cretaceous-Tertiary boundary on the Pacific plate: composition and distribution of impact debris. In: Ryder, G., Fastovsky, D., Gartner, S. (Eds.), Cretaceous-Tertiary Event and Other Catastrophes in Earth History, vol. 307. Geol. Soc. Amer. Spec. Pap, pp. 389–401.

Lawton, T.F., Shipley, K.W., Aschoff, J.L., Giles, K.A., Vega, F.J., 2005. Basinward transport of Chicxulub ejecta by tsunami-induced backflow, La Popa basin, northeastern Mexico, and its implications for distribution of impact-related deposits flanking the Gulf of Mexico. Geology 33, 81–84.

Leroux, H., Rocchia, R., Froget, L., Orue-Etxebarria, X., Doukhan, J.C., Robin, E., 1995. The K/T boundary at Beloc (Haiti): compared stratigraphic distributions of the boundary markers. Earth Planet Sci. Lett. 131, 255–268.

Lindström, M., Floden, T., Grahn, Y., Kathol, B., 1994. Post-impact deposits in Tvären, a marine middle ordovician crater south of Stockholm, Sweden. Geol. Mag. 131, 91–103.

Matsui, T., Imamura, F., Tajika, E., Nakano, Y., Fujisawa, Y., 2002. Generation and propagation of a tsunami from the Cretaceous/Tertiary impact event. In: Koeberl, C., Macleod, G. (Eds.), Catastrophic Events and Mass Extinctions: Impact and Beyond, vol. 356. Geol. Soc. Amer. Spec. Pap, pp. 69–77.

Maurrasse, F.J.M.R., Sen, G., 1991. Impacts, tsunamis, and the Haitian Cretaceous-Tertiary boundary layer. Science 252, 1690–1693.

Melosh, H.J., 1989. Impact Cratering: A Geologic Process. Oxford University Press, New York, p. 245.

Melosh, H.J., 2003. Impact-generated tsunamis: an over-rated hazard. Lunar Planet. Sci. 34, 2013.

Montanari, A., Koeberl, C., 2000. Impact Stratigraphy: The Italian Record. Lecture Notes in Earth Sciences 93. Springer-Verlag, Heidelberg, p. 364.

Montgomery, H., Pessagno, E., Soegaard, K., 1992. Misconceptions concerning the Cretaceous Tertiary boundary at the Brazos River, falls county, Texas. Earth Planet Sci. Lett. 109, 593–600.

Morgan, J., Lana, C., Kearsley, A., Coles, B., Belcher, C., Montanari, S., Diaz-Martines, E., Barbosa, A., Neumann, V., 2006. Analyses of shocked quartz at the global K-P boundary indicate an origin from a single, high-angle, oblique impact at Chcixulub. Earth Planet Sci. Lett. 251, 264–279.

Norris, R.D., Firth, J.V., 2002. Mass wasting of Atlantic continental margins following the Chicxulub impact event. In: Koeberl, C., Macleod, G. (Eds.), Catastrophic Events and Mass Extinctions: Impact and Beyond, vol. 356. Geol. Soc. Amer. Spec. Pap, pp. 79–95.

Norris, R.D., Huber, B.T., Self-Trail, J., 1999. Synchroneity of the K-T oceanic mass extinction and meteorite impact: Blake Nose, western North Atlantic. Geology 27, 419–422.

Ormö, J., 1994. The pre-impact Ordovician stratigraphy of the Tvären Bay impact structure, SE Sweden. GFF (Geol. Foren. Stockh. Forh.) 116, 139–144.

Ormö, J., Lindström, M., 2000. When a cosmic impact strikes the sea bed. Geol. Mag. 137, 67–80.

Ormö, J., Shuvalov, V.V., Lindström, M., 2002. Numerical modeling for target water depth estimation of marine-target impact craters. J. Geophys. Res. 107 (E12), 10. https://doi.org/10.1029/2002JE001865.

Poag, C.W., Plescia, J.B., Molzer, P.C., 2002. Ancient impact structures on modern continental shelves: the Chesapeake Bay, Montagnais, and Toms Canyon craters, Atlantic margin of north America. Deep-Sea Res. II 49, 1081–1102.

Puura, V., Suuroja, K., 1992. Ordovician impact crater at Kärdla, Hiiumaa, Estonia. Tectonophysics 216, 143–156.

Schulte, P., Speijer, R., Mai, H., Kontny, A., 2006. The Cretaceous-Paleogene (K-P) boundary at Brazos, Texas: sequence stratigraphy, depositional events and the Chicxulub impact. Sedimet. Geol. 184, 77–109.

Sharpton, V., Marin, L.E., Carney, J.L., Lee, S., Ryder, G., Schuraytz, B.C., Sikora, P., Spudis, P.D., 1996. A model of the Chicxulub impact basin based on evolution of geophysical data, well logs, and frill core samples. In: Ryder, G., Fastovsky, D., Gartner, S. (Eds.), Cretaceous-Tertiary Event and Other Catastrophes in Earth History, vol. 307. Geol. Soc. Amer. Spec. Pap, pp. 55–74.

Smit, J., 1999. The global stratigraphy of the Cretaceous-Tertiary boundary impact ejecta. Annu. Rev. Earth Planet Sci. 27, 75–113.

Smit, J., Montanari, A., Swinburne, N.H.M., Alvarez, W., Hildebrand, A.R., Margolis, S.V., Claeys, P., Lowrie, W., Asaro, F., 1992. Tektite-bearing, deep-water clastic unit at the Cretaceous-Tertiary boundary in northeastern Mexico. Geology 20, 99–103.

Smit, J., Roep, T.B., Alvarez, W., Claeys, P., Grajales-Nishimura, J.M., Bermudez, J., 1996. Coarse-grained, clastic sandstone complex at the K/T boundary around the Gulf of Mexico: deposition by tsunami waves induced by the Chicxulub impact?. In: Ryder, G., Fastovsky, D., Gartner, S. (Eds.), Cretaceous-Tertiary Event and Other Catastrophes in Earth History, vol. 307. Geol. Soc. Amer. Spec. Pap, pp. 151–182.

Sturkell, E.F.F., 1998. Resurge morphology of the marine Lockne impact crater, Jämtland, central Sweden. Geol. Mag. 135, 121–127.

Sturkell, E.F.F., Ormö, J., 1997. Impact-related clastic injections in the marine Ordovician Lockne impact structure, Central Sweden. Sedimentology 44, 793–804.

Tada, R., Nakano, Y., Iturralde-Vinent, M.A., Yamamoto, S., Kamata, T., Tajika, E., Toyoda, K., Kiyokawa, S., Delgado, D.G., Oji, T., Goto, K., Takayama, H., et al., 2002. Complex tsunami waves suggested by the Cretaceous-Tertiary boundary deposit at the Moncada section, western Cuba. In: Koeberl, C., Macleod, G. (Eds.), Catastrophic Events and Mass Extinctions: Impact and Beyond, vol. 356. Geol. Soc. Amer. Spec. Pap, pp. 109–123.

Tada, R., Iturralde-Vinent, M.A., Matsui, T., Tajika, E., Oji, T., Goto, K., Nakano, Y., Takayama, H., Yamamoto, S., Rojas-Consuegra, R., Kiyokawa, S., García-Delgado, 2003. K/T boundary deposits in the proto-Caribbean basin. Mem. Am. Assoc. Petrol. Geol. 79, 582–604.

Takayama, H., Tada, R., Matsui, T., Iturralde-Vinent, M.A., Oji, T., Tajika, E., Kiyokawa, S., García, D., Okada, H., Hasegawa, T., Toyoda, K., 2000. Origin of the Peñalver Formation in northwestern Cuba and its relation to K/T boundary impact event. Sediment. Geol. 135, 295–320.

Ward, S.N., Asphaug, E., 2002. Impact tsunami—Eltanin. Deep-Sea Res. II 49, 1073–1079.

Weiss, R., Wünnemann, K., Bahlburg, H., 2006. Numerical modeling of generation, propagation and run-up of tsunamis caused by oceanic impacts: model strategy and technical solutions. Geophys. J. Int. 167, 77–88.

CHAPTER 19

Tsunami boulder deposits — a strongly debated topic in paleo-tsunami research

A. Scheffers

Southern Cross University, Southern Cross GeoScience, School of Environment, Science and Engineering, Lismore, Australia

1. Introduction

1.1 A short glance on tsunami and paleotsunami research

Large rocks and boulders transmit a certain magic power and have been important to humans since prehistoric times. Megalithic cultures worshiped large boulders, in particular as part of astronomical field signs and burial grounds. This kind of relationship with boulders, whether they be single boulders, boulder clusters, or distinct patterns of boulders, normally excludes those visibly fallen from steep relief like cliff breakdown or talus in mountainous regions, as the origin and the reason for the existence of these boulders were considered obvious from the human experience, as they were deposited due to gravity. However, some blocks and boulders lacked a nearby source and were distributed across the landscape and were first related to a deluge. Later these "erratics" were explained through the natural sciences, with an accepted theory that most of them were transported and deposited by glaciers, and Ice Age Science was born.

Along coastlines, boulders at a cliff foot are not considered exceptional or a matter for debate. Boulders found in these conditions require further investigation, particularly those that have been transported inland and in most cases against gravity, to understand the mechanism by which they have been transported. Waves are the primary mode of transport for coastal boulders, and extreme waves such as those that occur in extreme storms may therefore transport boulders of larger size and mass.

While some examples of boulders of extraordinary size exist, such as that of a 300 m^3 (about 500 ton) coral block at the west Java coast dislocated by the 1883 Krakatoa volcanic explosion tsunami in the Sunda Strait (photographed in 1885 CE, compare Lau et al., 2014, Fig. 2.3), research into boulder tsunamites has progressed slowly over the last 40 years and is still in its infancy (Scheffers, 2015). The main reason for this limited

---

Tsunamiites. https://doi.org/10.1016/B978-0-12-823939-1.00019-7
Copyright © 2021 Elsevier B.V. All rights reserved.

understanding is that the debate is concentrated on the question of "storm waves versus tsunami flow" as potential transport agents and that tsunamis as a transport mechanism are usually excluded. However, a number of publications investigate tsunami signatures in past geological periods as far back to the Precambrian. The majority of research has focused on the Chicxulub event at the Cretaceous/Tertiary boundary with >110 articles published on the subject (and >100 articles in the year 2000 alone), such as those by Shiki et al. (2008), and Tappin (2017, Figs. 19.2 and 19.3), and on fine sediments and their structures, such as those by Dunbar and McCullough (2012) as well as De Martini et al. (2016, 2017). Listed 25 tsunami events up to 0 BCE, 147 events for the period CE 0–1500, 50 for the 16th century, 122 for the 17th century, 210 for the 18th century, 600 for the 19th century, and 926 for the 20th century. Arcos et al. (2017) identified 32 large events over the last 30 years, including 22 along coasts of the Pacific Ocean. Catalogs for specific oceans and regions have also been produced by Altinok and Ersoy (2000), Ambrasey and Synolakis (2010), Baptista and Miranda (2009), Brauner et al. (2005), Bryant (2001, 2008), Goff (2008), Lander et al. (2002), National Geophysical Data Center NGDC/World Data Service

Figure 19.1
Very large individual boulders (>50 tons) considered to have been deposited by tsunami events. (A and B): from a 5–6 m high reef rock platform at the east coast of Bonaire (Netherland's Antilles, southern Caribbean), (C): from Quobba, western Australia, at +15 m above sea level (a.s.l.; sandstone, some with bivalve borings), (D): on a 10-m high cliff top in NW Morocco south of Rabat, Pleistocene dune sandstone, from the 1755 CE Lisbon Tsunami, partly overturned (bioerosive rock pools at their base).

Figure 19.2
Sediment sequences in tsunami deposits. (A): bimodal deposit with coarse sand, shell hash, and angular boulders (Inisheer, Aran islands, western Ireland); (B): coarse sand and angular small gravel on fine sand and silt with some stones (Galway Bay, western Ireland); (C): Chaotic mixture of sand and well-rounded pebbles and small boulders (Galway Bay, western Ireland); (D): well-rounded basalt boulders (up to 2 m across) in volcanic ashes from Sta. Lucia, southern Caribbean, at 50 m a.s.l (Pleistocene).

(2013), O'Loughlin and Lander (2003), Papadopoulos and Chalkis (1984), Papadopoulos and Fokaefs (2005), Papadopoulos et al. (2014), Soloviev et al. (2000), and Tinti and Maramai (1996).

From paleotsunami research and multiple local and regional studies on single events, it can be concluded that for the time span of the late Holocene (i.e., the past 6000–7000 years) tsunami events cannot be ruled out for any of the world's coastlines, including large inland lakes with steep banks. However, the characteristics of a tsunami, in particular run-up height and inundation distance, differ significantly between very exposed deep water coasts, shallow foreshore regions, and sheltered bays and also depend on tidal range and the time of tsunami wave arrival.

Another consideration is the type of deposition, whether it be sheets of fine sediment, single boulders (Fig. 19.1), boulder fields and clusters, or bimodal or chaotic mixtures of different grain sizes that include boulders (compare Scheffers, 2008, Fig. 17.1). Boulder deposits may simply be the result of boulder transport, or boulder transport may also be enabled and supported by their inclusion in a suspension-loaded mass during inundation,

Figure 19.3
Imbrication of boulder deposits. (A): Banded Iron rock from the Dampier Peninsula of western Australia, with individual boulders exhibiting a mass of >50 tons. (B): Imbrication inclined in a landward direction, western Ireland. (C and D): Steep imbrication, which would be disturbed if storm waves were able to reach the location. (C) is from Inishmore (Galway, Ireland), (D) from Mallorca (Spain).

which consequently results in bimodal/unsorted deposits (Felton et al., 2006; Fujino et al., 2006; Goff et al., 2010b; Kelletat and Scheffers, 2004, Figs. 19.2 and 19.4; Moore et al., 1994b, and Fig. 19.2; see also Erdmann et al., 2017). One of the differences between tsunami and storm wave deposits is that the latter tend to sort material and leave visible stratification between different events (e.g., May et al., 2015a), but usually lack backflow processes.

In general, tsunami research is undertaken using a variety of methods that differ between studies and are often poorly connected and include local field studies of deposits, laboratory tests, modeling of transport conditions (e.g., Apotsos et al., 2011; Boesl et al., 2019; Goto and Imamura, 2007; Goto et al., 2010c, 2011, 2019; Herterich et al., 2018; Hisamatsu et al., 2014; Hoffmeister et al., 2014; Imamura et al., 2008; Nandasena et al., 2011a; Nott, 2003a, 2003b; O'Boyle et al., 2017; Rovere et al., 2017b; Weiss et al., 2009), mapping of sedimentologic and morphologic evidence of tsunami across the entire impact area, and calculation of tsunami wave character and theoretical inundation behavior (e.g., Bondevik et al., 2005; Chacón-Barrantes, 2015; Harbitz, 1992; Mader, 1999; Weiss and Diplas, 2015).

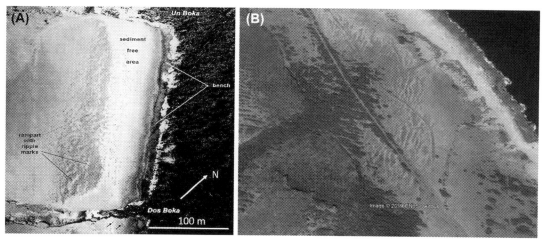

Figure 19.4
Ripple marks in old coral rubble ramparts along the east coast of Curacao, Netherlands Antilles, Caribbean. In B these ripple marks are visible for 400 m, from the low cliff in the interglacial reef rock up to about 9 m a.s.l. Easterly winds (trade wind) have cut parallel grooves in the shrubby vegetation. *Credits: (A): aerial photograph; (B): Google Earth.*

1.2 Costal boulders, tsunami boulders, and the role of their depositional environment on dislocation process and history

Ocean waves can be destructive and transport large amount of sediment onshore. However, their power is limited by:

- bathymetry,
- coastal exposure and relief,
- steepness and bed roughness of the coastal slope (or cliff),
- resistance of coastal rock (depending on geology, mineralogy, density/compactness, bedding, and joint patterns)
- intensity of weathering (also dependent on time span without disturbance, such as those studied by Kelletat et al. (2019)),
- size and total mass of sediments/clasts available for transport,
- deposition conditions in terms of sea level and tidal ranges, exact timing of an event in relation to tidal level (for short timescales) and sea-level history (for longer timescales),
- and other factors that are dependent on local and regional conditions.

People living at the coast understand energy involved in extreme conditions, but also sediment transport limitations, such as inland distance and clast size. Local experience has been considered by coastal engineers for centuries when protecting coastlines from destruction. Such projects use natural (e.g., rip-rap) or artificial (e.g., tetrapods) clasts to armor shorelines, which include a size and mass that resist dislocation by the strongest local and regional impacts under the assumption of a stable sea level. Such expertise needs to be considered when studying large contemporary and historic events, as these schemes provide a proven method for developing theoretical understanding of boulder deposition from large storm events. While storm energy levels are correctly anticipated by these schemes, energies from strong tsunamis are much more difficult to predict, either because they have not occurred in a region previously or because they have happened due to a combination of unexpected conditions.

There remain more challenges in the analysis of processes and associated characteristics and chronologies in field studies. Direct observations on geomorphologic and sedimentary deposits during an extreme event or closely following an event may be informative. However, each event is site- and energy-specific, and comparisons that can be drawn between events are therefore limited. To establish a solid fundamental basis for a new branch in coastal sciences such as boulder deposition by tsunami events, a general agreement on terminology and basic data is needed to develop sound conclusions, which includes:

- a definition of a "coastal boulder," which excludes all dislocation processes except marine ones;
- form, size, and mass of single boulders and the largest boulders in a deposit;
- the extent of horizontal and vertical movement based on place of origin (such as whether it is derived from bedrock or the tidal fringe), including information on whether the boulder has undergone multiple/repeated movement or backwash;
- age of deposition, which can be inferred from vegetation and soil cover, lichen cover, weathering characteristics and intensity, relation to nearby sediment archives, and surface age;
- coastal geology and topography, including relief and exposure to marine forces including tides;
- sea-level history over the (potential) time frame of deposition;
- the type of transport process: sliding, rotation, saltation, including whether the boulder is overturned, abraded, or with collision marks, moved individually or with other clast;
- the type of deposition: whether the boulder has been deposited individually, as a cluster, pile, ridge, strewn or dense boulder field, imbrication;
- foreshore bathymetry and recorded or potential wave heights;
- any documentation (geologic, geomorphologic, sedimentary, archaeologic, mythologic/legendary, historic) of former extreme events (storms, tsunamis).

However, this information cannot confirm the contribution of tsunami processes to boulder deposition explicitly. On the other hand, tsunami impacts should be considered in cases where the storm history of a field site is well documented and the envelope of boulder deposition exceeds the threshold of the largest storm impacts.

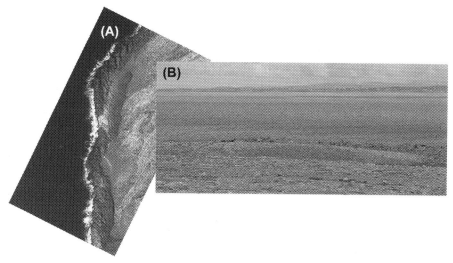

Figure 19.5
Tongues of boulders located in Galway Bay south of Black Head. The northern one (A), seen from the east (B), is over 120 m long and 30 m wide with a mean thickness in the order of 0.2—1 m. Total boulder mass is estimated to be around 15,000 tons. Boring bivalve traces found in this deposit point to an age of about 5,000 BP. The transport energy direction was at an angle of 15° from the shoreline trend (from SSW). These are two of the seven examples of boulder tongues from the Galway—Aran region of western Ireland.

Wide regions of the world's coastline are subject to similar energies and frequencies of extreme storms (tropical cyclones (TCs), hurricanes, typhoons) and tsunami events such as the case on the Hawaiian Islands, Indonesia, Japan, and Chile. Additionally, significant characteristics of boulder patterns (Fig. 19.3) such as balancing, near-vertical imbrication undisturbed by storm waves, or imbrication inclination in a landward direction, as well as geomorphological aspects that indicate extended and enduring strong flow, such as ripple marks, rhythmic ridge patterns with characteristics of megaripples and large boulder tongues, may provide strong evidence of a tsunami origin (Figs. 19.4 and 19.5).

2. Examples of tsunami boulder deposition

Onshore boulder deposition is predominantly attributed to strong storm waves, and the idea that tsunamis may have contributed to, or be solely responsible for, these deposits was not considered until the end of the 20th century in most countries. In fact, many large boulders may derive from storm impacts, although this is rarely considered typical. Examples from more recent studies are those from tropical coastlines related to cyclones, hurricanes, or typhoons; the potential for emplacement by tsunamis is mentioned infrequently (e.g., Kennedy et al., 2017; Khan et al., 2010; Lau et al., 2014; May et al.,

360 Chapter 19

2015a,b; Noormets et al., 2004; Scheffers and Scheffers, 2006; Terry et al., 2013; Zhao et al., 2009). Much of this research is concerned with extratropical winter storms (e.g., Cox et al., 2012; Etienne and Paris, 2010. The extreme cluster of six hurricane-force winter storms in the eastern Atlantic in the 2013/2014 season, which included the largest waves in living memory, gained the attention of studies in Brittany (France, see Autret et al., 2016) and along the Irish west coast (Cox et al., 2018; Erdmann et al., 2015, 2017; 2018a,b; Masselink et al., 2016). These studies are particularly valuable as a source of data on the force of storm waves and their potential for boulder transport (depending on water depth, exposure, relief type, age, sea-level history) and how existing older boulder deposits survived these impacts unaffected. These data enable a threshold between storm and tsunami power to be defined in relation to boulder movement against gravity.

2.1 Documented case studies of recent tsunami events

The most simple type of boulder dislocation by tsunamis is that triggered by rock falls from steep slopes into deep water, such as along fjord coastlines, which are frequent, regular events where glacier recession and permafrost decay are occurring. Recent examples can be found in Greenland (Dahl-Jensen et al., 2004), the Fjordlands of New Zealand's South Island (Dykstra, 2009, 2012) and Norway (Harbitz et al., 2014). A significant event was the tsunami with highest wave height ever recorded, in 1958 in Lituya Bay, SW Alaska (Fritz et al., 2009; Mader, 1999; Ward and Day, 2010), which reached 524 m above sea level. These types of tsunami events have predominantly local effects, and none of the records of these events report boulder deposition. This is thought to be due to steep slopes hindering deposition, strong backwash, or later denudation processes.

Boulder deposits have been investigated for lesser-known tsunami events, such as the 2006 event in the Kuriles (Bourgeois and McInnes, 2010), the 1993 tsunami in Japan (Nanayama et al., 2000, with a comparison of the boulder deposits left by a strong typhoon 1959 CE in the same region), and a tsunami in 2009 on the islands of Samoa and Tonga (Fritz et al., 2011). Despite a large amount of research on the 2004 Andaman—Sumatra tsunami, only a few studies document boulder deposits (e.g., Etienne et al., 2010; Kelletat et al., 2007), despite this tsunami having destructive power along thousands of kilometers of coastline in the northern Indian Ocean and large waves having reached Western Australia and the east coast of Africa. Of special interest are the studies on the Cape Pakarang boulder field in western Thailand (Goto et al., 2007, 2010c; Neugebauer et al., 2011), which highlight issues involved with paleotsunami dating (Fig. 19.6, and Table 19.1). A survey of Cape Pakarang (Thailand), just weeks after the impact of the Indian Ocean Tsunami of 2004 in Thailand, revealed a boulder field with a width of about 400 m located on an old reef platform just below mean high water level (Neugebauer et al., 2011). The field comprised about 600 boulders, with a 600 m extension in N—S direction. The boulders did not show any fresh breakage, suggesting that they had been deposited at this location a long time ago. However, a comparison of satellite images from before and after the tsunami demonstrated that the boulders were emplaced by the 2004 event. This conclusion is supported by eyewitness accounts of the tsunami, such as those by Hans Jürgen Kroll from Germany, who had lived for a number of years at the Cape and experienced the tsunami. Radiocarbon ages of nine samples of

Figure 19.6
Northern part of the Cape Pakarang (western Thailand) boulder field (A), emplaced 2004, at low water. Size distribution is irregular and not ordered by transport distance. A 16-ton overturned boulder (B) shows coral growth structures and round contours with no fresh break-off.

Table 19.1: Nine radiocarbon ages from the large boulder field emplaced by the 2004 boxing day tsunami at Cape Pakarang on Thailand's west coast.

Beta	Submitter No.	Measured age	13C/12C	Conventional age	2 Sigma calibration (click link to retrieve plot)
253089	Pak 1	4570 ± 50 BP	−7.0 o/oo	4870 ± 50 BP	Cal BC 3350 to 3070 (Cal BP 5300 to 5020)
253090	Pak 2a	4400 ± 50 BP	−3.5 o/oo	4750 ± 50 BP	Cal BC 3260 to 2900 (Cal BP 5210 to 4850)
253091	Pak 2b	4600 ± 50 BP	−3.0 o/oo	4960 ± 50 BP	Cal BC 3490 to 3270 (Cal BP 5440 to 5220
253092	Pak 3	4210 ± 50 BP	−1.0 o/oo	4600 ± 50 BP	Cal BC 2960 to 2760 (Cal BP 4910 to 4710)
253093	Pak 4	4130 ± 50 BP	−3.2 o/oo	4490 ± 50 BP	Cal BC 2960 to 2760 (Cal BP 4910 to 4710)
253095	Pak 6	4110 ± 50 BP	−3.9 o/oo	4450 ± 50 BP	Cal BC 2850 to 2540 (Cal BP 4800 to 4490)
253096	Pak 7	4100 ± 60 BP	−1.5 o/oo	4490 ± 60 BP	Cal BC 2880 to 2560 (Cal BP 4830 to 4510)
253097	Pak 8	4190 ± 50 BP	−2.7 o/oo	4560 ± 50 BP	Cal BC 2900 to 2680 (Cal BP 4850 to 4630)
253098	Pak 9	3990 ± 70 BP	−2.7 o/oo	4360 ± 70 BP	Cal BC 2760 to 2390 (Cal BP 4710 to 4340)

From Neugebauer, N.P., Brill, D., Brückner, H., Kelletat, D., Scheffers, S., Vött, A., 2011. 5000 jahre tsunami-geschichte am Kap Pakarang (Thailand). Coastline Rep. 17, 81–98.

362 Chapter 19

corals on the surfaces of boulders (Table 19.1) range from 4360 to 4960 years BP (reservoir corrected ages). These boulders are thought to have been available for transport in front of the platform edge and in water depths few meters below mean low water following an older tsunami. The age differences of the boulders are thought to be due to differential weathering over a >4000-year period, as well as the lack of a comparable protocol for sampling from boulder to boulder. This case study is a good example of possible uncertainties that arise in dating past tsunami events when no independent stratigraphic or relative environmental age control is used. At a number of other sites in Thailand, single large boulders with masses of 10 t or more were dislocated from living coral reefs and carried onshore (Kelletat et al., 2007; Scheffers, 2008). The lack of abrasion on the living outer parts of these boulders documents transport by saltation or within a hyperconcentrated flow over a distance of at least 200 m and a vertical range of about 4—5 m against gravity. Goto et al. (2010c) calculated that the strongest wave of the 2004 tsunami at Cape Pakarang may have been able to dislocate boulders weighing up to 250 t, but the largest boulder present at this site is of the order of only 12—14 m^3 and about 20 t. It is therefore thought that the older tsunami event that produced the old boulder deposit was likely of a lower magnitude, or that the reef structure from which these boulders originated did not facilitate the breaking off of larger boulders.

The large 2011 Tohoku-oki tsunami event in Japan affected numerous sites with rocky shorelines and crossed the coastal plain with inundation depths of many meters (locally >15 m) and velocities of the order of 5—6 m/s. Nandasena et al. (2013) present observations on boulders transported by the 2011 event, including their places of origin and characteristics of their transport and deposition. Most of the boulders were deposited individually with large distances between one another. Some were found to have been angular and recently dislocated from the bedrock, while others were well rounded and were carried from the surf zone. The former may have dislocated from bedrock in the immediate vicinity of their deposition location, while the latter are thought to have carried from a location hundreds of meters away. Where boulders were deposited in clusters, the boulder pattern is chaotic with poorly developed imbrication, or a lack of, which might be due to the irregular boulder forms. The maximum transport distance is 600 m, and the largest boulder has a mass of 167 t. In the Settai region, larger boulders (e.g., 285 t) date to one or more older, similar, or higher-energy events (e.g., 1611 CE tsunami). Another study of the 2011 tsunami event identified a wide range of sediment sizes (Yamada et al., 2014) including fragments from coastal protection works and large well-rounded boulders organized in three clusters and located at least 750 m to inland. These boulders are thought to have be transported together with muddy sediments by the first waves that reached 1750 m inland and had a local height of up to 28 m.

2.2 Boulders deposited by paleotsunami events

The number of studies focusing on coastal boulder deposits has increased over the past two decades, with many discussing the potential contributions of storm waves, bores, and tsunamis (e.g., Biolchi et al., 2016; Buckley et al., 2012; Causon Deguara and Gauci, 2014; Dewey and Ryan, 2017; Erdmann et al., 2015, 2017, 2018b; Furlani et al., 2011; Goto et al., 2010a,b, 2015; Hall et al., 2010; Hoffmeister et al., 2014; Kennedy et al., 2017; Lau et al., 2014; Lorang, 2011; Medina et al., 2011; Nandasena et al., 2011b; Noormets et al., 2004; Nott, 2003a; Oliveira, 2017; Regnauld et al., 2010; Richmond et al., 2011; Scheffers, 2005; Scheffers and Kelletat, 2019; Scheffers et al., 2009a,b, 2010; Weiss and Diplas, 2015; Wlliams, 2010; Williams and Hall, 2004; Yu et al., 2009). Most of the study sites are subjected to open ocean conditions with regular stormy seasons and consequential energetic wave attacks. While storm waves cannot be excluded as the transport mechanism of boulders at these sites, the contribution of tsunamis cannot be entirely eliminated; the event type is difficult to determine as single boulders do not generally carry any tsunami signatures. A tsunami event is more likely to have deposited boulders if wind waves at the site are limited by the size of the water body, water depth, climate province, and/or exposure. Additionally, documentation of historical occurrences of tsunamis may contain hints about the origin of boulder deposits (Armigliato et al., 2015a,b; Bryant, 2001, 2008; Engel and May 2012; Engel et al., 2010; Etienne et al., 2011; Felton et al., 2006; Hisamatsu et al., 2014; Kelletat and Schellmann, 2002; Kelletat et al., 2007, 2013; Maouche et al., 2009; Mastronuzzi and Pignatelli, 2012; Mhammdi et al., 2008; Mottershead et al., 2014; Scheffers, 2004; Scheffers and Kelletat, 2005, 2006b; Scheffers and Kinis, 2014; Scheffers and Scheffers, 2007; Scheffers et al., 2008; Scicchitano et al., 2007; Whelan and Kelletat, 2005).

Arguments for paleotsunamis from field evidence may be indirect or direct. Indirect evidence includes boulder deposits inland of or above known limits of storm deposits. Additionally, in remote regions without coastal settlements and infrastructure, useful evidence to inform the deposition history of boulders can be soil at their base or around a deposit or single boulder or established vegetation (Fig. 19.7). Lichen-covered boulders or boulders resting in delicately balanced positions (e.g., steep to vertical, Fig. 19.7) are unlikely to have been transported or moved due to gravity for centuries. Additional support for their antiquity is weathering mantles or ages of deposition derived from marine organisms on the boulder surface (e.g., crusts of calcareous algae, balanids, bryozoa) or in the boulder (boring bivalves). In the case of limestone boulders, especially those from older geologic units and of high density, a good method for calculating a minimum period of boulder stability is the degree of karstification, often in the form of microkarren running vertically down on the boulders. Any dislocation would have produced irregular karren on the boulder surfaces. In special cases, which occur along limestone coasts including coral

364 Chapter 19

Figure 19.7
Long-time, large stable boulders in centuries-old shrub vegetation at about 10 m asl and 80–100 m from a cliff at the southeast coast of Anguilla Island, Caribbean (A and B). Bivalve borings in some of these boulders (C) point to their origin from the intertidal or upper subtidal.

reefs, bowl-like depressions are found in the boulders that have been dislocated to a terrestrial environment. The depressions are derived from bioerosion in the higher portion of the tidal zone or lower supratidal fringes. These flat-bottom bioerosive pools may be tilted to different degrees on the boulders. Rain-water dissolution will start to form other small pools inside the older ones and in some cases dissolve a narrow channel at the lowest overflow points (Fig. 19.8). As the rate of rain-water dissolution is very slow (calculated to be on average around 0.02−0.03 mm/year in all climate zones (Häuselmann, 2008; Erdmann et al., 2018b; Kelletat et al., 2019), a few centimeters of dissolution point to centuries of stable boulder position and exclude any storm movement within this time frame.

Direct indicators include clustering of boulders, including imbrication (Fig. 19.9), landward-dipping boulders deposited by strong backflow (Scheffers and Kinis, 2014), and marks of strong and wide flow, such as ripple marks in coarse gravel or even boulder fields (Scheffers, 2006). Also telling are particular boulder accumulations that provide evidence for transport by high-energy flow and deposition of large amount of rock debris (thousands of t) in a very short time, for example, the boulder tongues documented by Erdmann et al. (2015, 2017, 2018b) on the Irish west coast (Fig. 19.5). In contrast to storm-moved boulders and boulder accumulations, these deposits are not found 90° to the coastal trend, but rather at much lower angles (15−35°), nearer to being parallel to the coastline, similar to megaripple patterns of ridge segments that are also derived from strong flows. This kind of sedimentation has not previously been documented from storms. Additional evidence for paleotsunami events can be found in nearby finer deposits and typical tsunami event

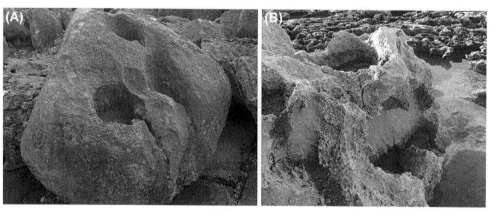

Figure 19.8
Bioerosive rock pools on limestone boulders lifted from the lower supratidal with later transformation by rain-water dissolution forming a second generation of pools (as in B) and outflow channels. The intensity of this dissolution indicates ages of immobility of these boulders of far more than 1000 years, as the terrestrial dissolution rate is only of the order of 0.02−0.03 mm/year for dense limestone. *(A) from inside Galway Bay, western Ireland, (B) from Ibiza Island, Spain.*

Figure 19.9
Imbrication of very large (10–50 tons) platy boulders results in clear pattern with high resistance, with plates dipping in direction of wave or flow attack. Example A is from the open ocean coast of Inishmore (Aran Islands) at about +5 m mhw and 60 m from the shoreline, example B at about +4 m mhw inside Galway Bay, where storm waves are limited by refraction and shallow water.

characteristics such as the chaotic/unstratified/bimodal character of clasts, inclusion of mud clasts, basal unconformities, or upward grading.

Comparing boulder deposits from recent tsunami events and those thought to have been deposited by (potential) paleotsunamis is challenging. In some cases, many similarities can be found between events, such as the 2004 CE Pakarang boulder field in Thailand and those formed by the 1771 Meiwa tsunami in southern Japan, as the two locations that include flat intertidal reef platforms that are nearly identical in form (e.g., Goto et al., 2010b). Boulder transport distances from typhoons at this latitudes are 240–350 m landward, and boulder mass is up to 50 t for Ishigaki Island (up to 216 t from the Meiwa tsunami; see Goto et al., 2010a). In contrast, boulders deposited by historical tsunamis may reach up to 1.5 km to inland (e.g., those on the southern Ryukyu Islands (see Goto et al., 2010b)) and have individual masses of between 700 t and 2500 t (e.g., the coral reef on Shimogi Island (Imamura et al., 2008),). Estimated flow velocities for the transport of these boulders is thought to be of the order of 12.5−15 m/s (Imamura et al., 2008). In the Caribbean, paleotsunami boulder fields are found on low cliff tops (5−6 m a.s.l.). Small and medium-sized boulders are spread across areas 100−300 m wide, which also include ripple marks produced by strong flow, such as those found in Curacao and Bonaire (Scheffers, 2006, see also Fig. 19.4). Fields of loosely strewn boulders of medium size with singular large ones (>50 to > 100 t, Figs. 19.1, 19.7 and 19.10) are present on Barbados, Bonaire, Curacao, Anguilla, and several Bahama Islands (Kelletat et al., 2004, 2007; Scheffers, 2004, 2005, 2008, Fig. 17.7; Scheffers and Kelletat, 2006a,b), as well as in the Mediterranean (e.g., Furlani et al., 2011; Kelletat and Schellmann, 2002; Kelletat et al., 2013; Maouche et al., 2009; Mastronuzzi and Sanso, 2000; Mottershead et al., 2014, 2017; Scheffers and Scheffers, 2007, Scheffers et al., 2008, 2014). In small embayments, old boulders can form half-circle accumulations and, if backwash is thought to have been involved in their organization, these deposits sometimes even form ridges. In central western Ireland, most boulder deposits in cliff top position are unique in their cross-sectional profile, as seaward erosion has transformed the original deposit into a concave, steep seaward slope with a sharp upper crest that transitions into the gently inclined slope of the leeward boulder field (Fig. 19.11; Scheffers et al., 2009b, 2010; Erdmann et al., 2015, 2017, 2018b).

Tsunami boulder deposits of Pleistocene to Palaeozoic age are commonly conglomeratic and covered with other rocks. Cemented or noncemented chaotic boulder deposits of Middle to Early Pleistocene age up to >300 m a.s.l. are considered to have occurred by tsunamis generated by large volcano flank collapses in the Hawaiian, Canary, and Cape Verde Islands (Carracedo, 1996; Caracedo et al., 1999; Day et al., 1997; Masson, 1996; Masson et al., 2002; Moore et al., 1989, 1994a; 1994b; Ward and Day, 2001).

The extreme variety of factors implicated in coastal boulder movement, mathematical theories, simplifications by testing in wave and flow channels, and numerical modeling may help to clarify processes involved in these events. Such parameters include boulder form, bed roughness/friction, slope angles, and boulder numbers, which influence flow thresholds and therefore the initiation of movement and continuous transport, as well as potential upper thresholds for bore and tsunami flow velocities. Film evidence at shorelines has shown that flow velocities were of the order of 3−6 m/s during the tsunami in Thailand in 2004, which is similar to fast flow of storm wave bores that occur during winter storms in western Ireland or during Hurricane Ivan in 2004 on Bonaire. Other similar footage showed that flow velocities were near 20 m/s when water was channeled between buildings and other obstacles during the 2011 Tohoku-oki tsunami in the city of

Figure 19.10

(A): A field of platy boulders (up to 90 tons) on an intertidal conglomeritic platform at Cabo de Trafalgar, south coast of Spain, likely to have been emplaced by the Great Lisbon Tsunami of 1755 CE. (B): Large boulders (the majority >50 tons) located on a 12–15 m high cliff top near Whale Point, Eleuthera, Bahamas, emplaced around 3000 BP. These boulders have been immobile for a long time, as can be seen by the significant karstification on their top.

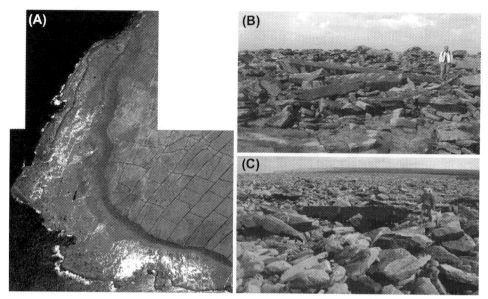

Figure 19.11
Ridge-like boulder accumulation on the SW coast of Inishmaan, Aran Islands, Galway Bay, western Ireland. (A): This 1.3 km long section of a boulder ridge has a maximum distance to the cliff of 280 m and rises from +6 m above mhw to +18 m above mhw from SE to N. (B and C): A low inclining landward facing slope containing boulders up to 6 m long leading to a field of strewn boulders. The overall ridge has formed mainly as a result of a seaward erosion by undermining the boulder deposit during very strong storms that originate in the Atlantic Ocean.

Sendai. The transport energy of these flows would also depend on boulder shapes (irregular, spherical, platy), boulder density, turbulence of flow, and suspension load in the flow. Films show a decrease in the velocity of high suspension muddy tsunami flow, which will have a higher transport energy compared to clear water conditions, on the runway of the Sendai airport.

The kind of movement (e.g., sliding, rotation, saltation, Fig. 19.12) is another important factor, as it influences the contact time and force of that contact with the surface of the boulder. Impact marks and striations are commonly left on limestone bedrock, giving direction, character, and intensity of boulder contacts during movement. The source of boulders, either loose or joint bound in bedrock, and their position related to inundation (subaerial, partly or totally submerged) are other important parameters. Freshly broken rock fragments and sedimentary rocks generally have an easily identifiable origin, and their dislocation from the intertidal zone can be determined from the attachment of fresh calcareous algae, vermetids, balanids, corals, or boring bivalves. In some cases, the origin of these types of boulders can be determined within a specific tidal range.

Figure 19.12
Platy boulders detached from old limestone rocks, lifted vertically, and dislocated without ground contact (saltation). These 0.5 m thick fragments have been emplaced by a white water bore (approaching from the upper right with max. velocity of <5 m/s) during a hurricane 4-category winter storm in early February 2014, near Doolin at the entrance of Galway Bay, western Ireland. The flat boulder has a surface of 17.3 m^2 and a mass of 21.5 t, which would require a lifting power of 1.25 t/m^2 during the dislocation process.

Three more aspects have been considered for older (i.e., paleotsunami) deposits:

(a) the preservation of a boulder, including the loss of mass by weathering and therefore its exact size, form, and mass during the transport, or multiple periods of transport if affected by more than one event;
(b) transport distance and the position of the shoreline in times when tsunamis have happened in the past, depending on the amount of coastal erosion;
(c) the sea-level position at the time of the tsunami, which is needed for any calculations or numerical modeling to determine the direction of transport horizontally and vertically.

When attempting to understand historical tsunami events, the position of the shoreline is relatively easy to determine. However, for prehistoric events, in particular those older than 2000–3000 years, the age of the deposits or individual boulders remains unknown and so relative sea level and the shoreline position are not considered. Consequently, comparisons of events from different time periods and regions are problematic. A late- to mid-Holocene deposit may have been left undisturbed in far-field regions due to glacioisostatic uplift, and as a result the deposit will be a different height above sea level to when it was deposited. Such an uplift would also limit further coastal erosion processes including extreme storm event impacts the boulder deposits would experience. This may be the case for sites of ongoing glacioisostatic uplift, such as those in western Scotland and Northern Ireland, while those on the periphery of this uplift, such as in western and southern Ireland, have experienced a significant sea level rise of about 5 m during the last 5000 years (Erdman et al., 2017, 2018b) with coastal erosion and transgression as part of their history. Without knowledge of regional sea-level history and the age of boulder deposits, conclusions remain speculative on the history such boulder deposits have experienced since deposition.

Tsunamigenic transport of the largest coastal boulders known worldwide (>500 or even >1000 t) is not disputed. Examples of events of this scale exist in the Rangiroa Atoll in French Polynesia (Bourrouilh-Le Jan and Talandier, 1985; Talandier and Bourrouilh-Le Jan 1987), which is located close to sea level. Giant blocks were deposited around 15 m a.s.l. up to 400 m from a cliff edge on Tongatapu (Frohlich et al., 2009; Hornbach et al., 2008). Other examples are at the east coast of northern Eleuthera, Bahamas, where a handful of giant boulders were deposited between sea level and a cliff top at 15 m a.s.l (Fig. 19.13; Hearty, 1997; Hearty and Tormey, 2017; Rovere et al., 2017a,b, Viret, 2008) and also a 700 t block at the same elevation on Dirk Hartog island, western Australia (Playford, 2014). According to historical sources, the largest boulder moved by a tsunami on the west coast of Shimoji island in the Miyako Island group, southwest Japan, is 1700 m^3 and around 2500 t in mass (Imamura et al., 2008). It was emplaced on land, probably by the Meiwa tsunami in 1771 CE. Except for the Eleuthera boulders, all these examples date to the second half of the Holocene when sea level was similar to modern day. Hearty and Tormey (2017, 2018) discuss whether a superstorm of unknown energy, but significantly stronger than modern-day Category 5 hurricanes, might have emplaced the Eleuthera boulders, based on their Eemian age when climate was about 2°C warmer than today and sea level was 6–9 m higher. Rovere et al. (2017a,b) modeled the position and mass of these boulders and concluded that the energy of the strongest modern hurricanes may have done the job if sea level was at least 3.5 m higher than today. It is important to recognize that the only three other sites worldwide with boulders of a similar size that have moved onshore by marine forces are Tongatapu, Rangiroa, and the Shark Bay area in western Australia, all of these regions being situated in the wave shadow of tropical cyclones approaching from the east. Thus, these western-exposed sites should not be directly compared to the windward coastline of Eleuthera. Therefore, the only other marine force able to move boulders of that size over long distances and against gravity would be tsunamis. In a recent study, Scheffers and Kelletat (2019) reconstructed the Eemian landscape on Eleuthera based on erosion rates for limestone boulder dissolution and cliff retreat. They concluded that boulder masses up to >1000 t and transport distances of the order of 500 m require an event of larger magnitude than a hurricane.

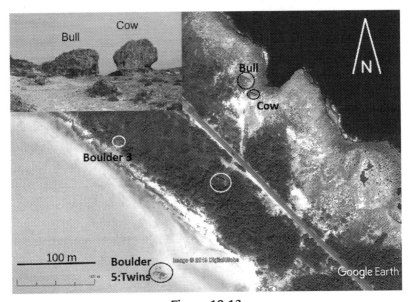

Figure 19.13
Position of giant boulders near Glass Window Bridge in the northern section of Eleuthera Island, Bahamas.

Approx. mass of individual boulders:

	Actual	In Eemian
Bull:	900 t	1.700 t
Cow:	400 t	>700 t
Twins:	600 t	1000 t

Credit: Google Earth.

Consequently, tsunamis should not be excluded as processes that transport and deposit large boulders simply based on the argument that they did not happen in the last centuries or are, so far, unknown within a given region.

3. Conclusion

Physical coastal science explores three main topics: form and shape (geomorphology), deposits (sedimentology), and marine processes (hydrodynamics). Sedimentology has become increasingly important in coastal process studies in recent decades. Paleotsunami research, however, has been predominantly based on sedimentological evidence from the beginning, and as such, research in this field has documented the kind, structure, and texture of deposits, while offering almost nothing on forms that have been made by tsunamis.

Documenting boulder transport inland against gravity by waves is firstly a question of geomorphology rather than sedimentology. It is not possible to reconstruct the dominant processes of boulder transport and deposition or the intensity and frequency of these events through investigations of dislocated coarse clasts alone. Additionally, depending on the scientific question, spatial and temporal dimensions are important for such analyses.

After decades of research on coastal coarse-clast deposits, researchers still do not employ standard and consistent methods and approaches in data acquisition. Coastal boulders may have experienced many movements at different times and by different processes. To discriminate these, we should find indicators for these potential activations, either from the organizational form of boulders or on the boulders themselves.

As in other landscapes, the importance of one extreme event depends on the energy of former events and the time span between events. If material is dislocated by an older event, the next events, which could be of stronger magnitude, maybe less efficient pending the availability of material for transport. Several potential processes may contribute to onshore boulder transport, particularly as discussion is focused on prehistoric events. No extreme event type, including frequent and exceptional rare storm waves and bores, freak waves, infragravity waves, and tsunamis should be excluded at the beginning of an investigation. They may only be eliminated in the course of research if physical factors become contradictory with other evidence.

The number of open questions in this field of research has not decreased because each observation and model force new interpretations. The lack of process observations is evident, as is the difficulty to find agreement about the most relevant physical processes in boulder movement by storm waves or tsunami flow. Numerical models can calculate natural flow phenomena and their relationship with parameters such as relief, surface roughness, sediment types, but only with very limited resolution. Consequently, obtaining key information on the intensity of tsunami events and their effects on a coastal landscape compared to "normal" and more constant wave processes remains difficult. One large tsunami event may involve thousands of times more transport energy than a set of extraordinary high storm waves, simply because of its comparable long duration and large mass of water.

Another desideratum in this field of research is extending tsunami histories to cover past periods of higher sea levels, such as those from interglacial periods or older geologic periods. Many submarine slides from volcano collapses or at the edge of continental slopes with the potential of very large tsunamis are tentatively dated but lack correlation with onshore deposits and coastal forms. In fact, only the history of younger Holocene tsunamis is partly known, which is a short period of tsunamigenic impacts along the coastlines of the world.

In nearly every case, it remains unknown whether a boulder has been dislocated from its source area during one event or during several events with shorter or longer time periods in between. Additionally, it still is not clear whether bores from storm waves are able to

374 Chapter 19

pick up a large boulder and whether the uptake of large boulders far from a cliff front is indicative of tsunami flow conditions. Onshore boulder distribution may not reflect the inundation seen during the 2004 Indian Ocean tsunami, which deposited large boulders exclusively in the surf and lowermost supratidal zones, even in areas with flow depths of 12 m in Thailand and over 30 m in Sumatra.

The most valuable and important indicators to differentiate between tsunami and storm transport are as follows:

1. Direct observation and registration of boulder transport and deposition during a historical tsunami event, and clear evidence for a deposit related to such a tsunami, such as for Krakatoa in 1883 CE, Andaman—Sumatra in 2004 CE, Tohoku-oki in 2011 CE, and one of the best historical examples, the Meiwa tsunami of 1771 in southern Japan.
2. The occurrence of geomorphological features that have so far never been observed or described for storm-induced deposits, including large boulder tongues, series of cross-way boulder ridges or megaripples in boulder fields, and long imbrication trains formed of very large boulders (>20 t).
3. Stratigraphic and temporal correlation of boulder deposits with finer sediment layers that show clear signatures of strong, wide, and long-lasting flows, such as basal unconformities, mud clasts, a mud cap, and upward grading.
4. The possibility of a tsunami should not be dismissed if coastal boulder deposits are incongruent with the maximum envelope of deposition for longer-term records of storm-induced deposition, taking into account the sea-level history at the time of transport and deposition, This can be applied to open exposures at deep-water coasts, to sheltered and shallow water sites, and to locations where strong storm waves can be excluded because of topography, the limited size of the water body, or climate.
5. The absence of tectonic, seismic, and volcanic activity in the study region is not grounds for excluding tsunamis when interpreting boulder deposits.

Future research should focus both on direct observations of boulder transport by large waves of different origins, from storms to tsunamis, and also paleodeposits with an open mind with respect to the transport process. Furthermore, hydrodynamic modeling of boulder transport scenarios remains important. Researchers must also study the sea-level history at the time the event took place. Lastly, it is important to understand the environmental history and setting at local and regional scales to provide sound conclusions that enrich the debate of storm versus tsunami events in a landscape.

References

Altinok, Y., Ersoy, S., 2000. Tsunamis observed on and near the Turkish coast. Nat. Hazards 21, 185—205.

Ambrasey, A., Synolakis, C., 2010. Tsunami catalogs for the eastern Mediterranean revisited. J. Earthq. Eng. 14, 309—330.

Apotsos, A., Buckley, M., Gelfenbaum, G., Jaffe, B., Vatvani, D., 2011. Nearshore tsunami inundation model validation: toward sediment transport applications. Pure Appl. Geophys. 168, 2097–2119.

Arcos, N., Dunbar, P., Stroker, K., Kong, L., 2017. The legacy of the 1992 Nicaragua tsunami. EOS 98. https://doi.org/10.1029/2017EO080845.

Armigliato, A., Tinti, S., Pagnoni, G., Zaniboni, F., 2015a. Is the debate on the sources of large historical tsunamigenic earthquakes along the Italian coasts closed? The tsunami research point of view. EGU Gen. Assemb. Abstr. 17, EGU2015–3859.

Armigliato, A., Tinti, S., Pagnoni, G., Zaniboni, F., 2015b. Large historical tsunamigenic earthquakes in Italy: the Neglected tsunami research point of view. In: Amer. Geophys. Union, Fall Meeting 2015. Abstract id PA42B-06.

Autret, R., Dodet, G., Fichaut, B., Suanez, S., David, L., Leckler, F., Ardhuin, F., Amann, J., Grandjean, P., Allemand, P., Filipot, J.F., 2016. A comprehensive hydro-geomorphic study of cliff-top storm deposits on Banneg Island during winter 2013–2014. Mar. Geol. 382, 37–55.

Baptista, M.A., Miranda, J.M., 2009. Revision of the Portuguese catalog of tsunamis. Nat. Hazards Earth Syst. Sci. 9, 25–42.

Biolchi, S., Furlani, S., Antonioli, F., Baldassini, N., Deguara, J., Devoto, S., Di Stefano, A., Evans, J., Gambin, T., Gauci, R., Mastronuzzi, G., Monaco, C., Scicchitano, G., 2016. Boulder accumulations related to extreme wave events on the eastern coast of Malta. Nat. Hazards Earth Syst. Sci. 16, 737–756.

Boesl, F., Engel, M., Eco, R.C., Galang, J.A., Gonzalo, L.A., Llanes, F., Quix, E., Brückner, H., 2019. Digital mapping of coastal boulders — high-resolution data acquisition to infer past and recent transport dynamics. Sedimentology. https://doi.org/10.1111/sed.12578.

Bondevik, S., Løvholt, F., Harbitz, C., Mangerud, J., Dawson, A., Svendsen, J.I., 2005. The Storegga slide tsunami — comparing field observations with numerical simulations. Mar. Petrol. Geol. 22, 195–208.

Bourrouilh-Le Jan, F.G., Talandier, J., 1985. Sédimentation et fracturation de haute énergie en milieu récifal: tsunamis, ouragans et cyclones et leurs effets sur la sédimentologie et la géomorphologie d'un atoll: motu et hoa, à Rangiroa, Tuamotu, Pacifique SE. Mar. Geol. 67, 263–333.

Bourgeois, J., MacInnes, B., 2010. Tsunami boulder transport and other dramatic effects of the 15 November 2006 central Kuril island tsunami on the island of Mantua. Z. Geomorphol. 54 (3), 175–195.

Brauner, C., Tröber, S., Bertogg, M., Zimmerli, P., 2005. Tsunami in South Asia: Building Financial Protection. Zürich, Swiss Reinsurance Company, p. 8.

Bryant, E., 2001. Tsunami — The Underrated Hazard. Cambridge, University Press, p. 320.

Bryant, E., 2008. Tsunami — The Underrated Hazard, second ed. Springer-Praxis Publishing, Chichester, p. 330.

Buckley, B., Jaffe, B.E., Wei, Y., Watt, S., 2012. Estimated velocities and inferred cause of overwash that emplaced inland fields of cobbles and boulders at Anegada, British Virgin Island. Nat. Hazards 63 (1), 133–149.

Carracedo, J.C., 1996. A simple model for the genesis of large gravitational landslide hazards in the Canary Islands. In: McGuire, W., Jones, A.P., Neuberg, J. (Eds.), Volcano Instability on the Earth and Terrestrial Planets, vol. 110. Geol. Soc. London Spec. Publ., pp. 125–135

Carracedo, J.C., Day, S.J., Guillou, H., Perez Torrado, F.J., 1999. Giant quaternary landslides in the evolution of La Palma and El Hierro, canary islands. J. Volcanol. Geoth. Res. 94, 169–190.

Causon Deguara, J., Gauci, R., 2014. Boulder and megaclast deposits on the South-east coast of Malta: signature of storm or tsunami event? In: Benincasa (Ed.), Proceedings of the Fifth International Symposium on Monitoring of Mediterranean Coastal Areas: Problems and Measurement Techniques, Livorno (Italy) 17–19 June 2014. Florence, Italy, pp. 594–603.

Chacón-Barrantes, S., 2015. Development of a Modelling Strategy for Simulation of Coastal Development Due to Tsunamis (PhD dissertation). Faculty of Mathematics and Natural Sciences, Kiel University, p. 87.

Cox, R., Zentner, D.B., Kirchner, B.J., Cook, M.S., 2012. Boulder ridges on the aran islands (Ireland): recent movements caused by storm waves, not tsunamis. J. Geol. 120, 249–272.

Cox, R., Jahn, K.L., Watkins, O.G., Cox, P., 2018. Extraordinary boulder transport by storm waves (west of Ireland, winter 2013—2014), and criteria for analysing coastal boulder deposits. Earth Sci. Rev. 177, 623—636.

Dahl-Jensen, T., Melchior, L., Dedersen, S.A., Wenig, W., 2004. Landslide and tsunami 21 November 2000 in Paatuut, west Greenland. Nat. Hazards 31 (1), 277—287.

Day, S.J., Carracedo, J.C., Guillou, H., 1997. Age and geometry of an aborted rift flank collapse: the san Andrés fault, El Hierro, canary islands. Geol. Mag. 134 (4), 523—537.

De Martini, P.M., Orefice, S., Patera, A., Paris, R., Terrinha, P., Noiva, J., Pantosti, D., 2016. The ASTARTE paleotsunami deposits data base-a web-based reference for tsunami research around Europe. EGU Gen. Assemb. Abstr. 18, EGU2016—6324.

De Martini, P.M., Patera, A., Orefice, S., Paris, R., Völker, D., Lastras, G., Terrinha, P., Noiva, J., Smedile, A., Pantosti, D., Hunt, J., Gutscher, M.-A., Migeon, S., Papadopoulos, G., Triantafyllou, I., Yalciner, A.C., 2017. The ASTARTE Paleotsunami and Mass Transport Deposits databases — web-based references for tsunami and submarine landslide research around Europe. Geophys. Res. Abstr. 19, EGU2017—15055.

Dewey, J.F., Ryan, P.D., 2017. Storm, rogue waves, or tsunami origin of megaclast deposits in western Ireland and North Island, New Zealand? Proc. Natl. Acad. Sci. U.S.A. https://doi.org/10.1073/pnas.1713233114.

Dunbar, P., McCullough, H., 2012. Global tsunami deposits database. Nat. Hazards 63 (1), 267—278.

Dykstra, J.L., 2009. Landslide-generated tsunami hazards in Fiordland, New Zealand and Norway. In: Amer. Geophys. Union, Spring Meeting 2009. Abstr. CG11A-07.

Dykstra, J.L., 2012. The Postglacial Evolution of Milford Sound, Fjordland, New Zealand (Ph.D. thesis). University of Canterbury, New Zealand, p. 311.

Engel, M., May, S.M., 2012. Bonaire's boulder fields revisited: evidence for Holocene tsunami impact on the Leeward Antilles. Quat. Sci. Rev. 54, 126—141.

Engel, M., Brückner, H., Wennrich, V., Scheffers, A., Kelletat, D., Vött, A., Schäbitz, F., Daut, G., Willershäuser, T., May, S.M., 2010. Coastal stratigraphies of eastern Bonaire (Netherlands Antilles): new insights into the palaeotsunami history of the southern Caribbean. Sediment. Geol. 231, 14—30.

Erdmann, W., Kelletat, D., Scheffers, A., Haslett, S.K., 2015. Origin and Formation of Coastal Boulder Deposits in Galway Bay and on the Aran Islands, Western Ireland. Springer Briefs in Geography, Dordrecht, p. 125.

Erdmann, W., Kelletat, D., Kuckuck, M., 2017. Boulder ridges and washover features in Galway Bay, western Ireland. J. Coast Res. 33 (5), 997—1021.

Erdmann, W., Kelletat, D., Scheffers, A., 2018a. Boulder dislocation by storms, with scenarios from the Irish west coast. Mar. Geol. 399, 1—13.

Erdmann, W., Scheffers, A.M., Kelletat, D.H., 2018b. Holocene coastal sedimentation in a rocky environment: geomorphological evidence from the Aran Islands and Galway Bay (western Ireland). J. Coast Res. 34 (4), 772—793.

Etienne, S., Paris, R., 2010. Boulder accumulations related to storms on the south coast of the Reykjanes Peninsula (Iceland). Geomorphology 114, 55—70.

Etienne, S., Buckley, M., Paris, R., Nandasena, A.K., Clark, K., Strotz, L., Chagué-Goff, C., Goff, J., Richmond, B., 2011. The use of boulders for characterising past tsunamis: lessons from the 2004 Indian Ocean and 2009 South Pacific tsunamis. Earth Sci. Rev. 107 (1—2), 76—90.

Felton, E.A., Crook, K.A.W., Keating, B.H., Kay, E.A., 2006. Sedimentology of rocky shorelines: 4. Coarse gravel lithofacies, molluscan biofacies, and the stratigraphic and eustatic records in the type area of the Pleistocene Hulopoe Gravel, Lanai, Hawaii. Sediment. Geol. 184, 1—76.

Fritz, H.M., Mohammed, F., Yoo, J., 2009. Lituya Bay landslide impact generated megatsunami 50th anniversary. Pure Appl. Geophys. 166, 153—175.

Fritz, H.M., Borrero, J., Synolakis, C.E., Okal, E.A., Weiss, R., Titov, V., Jaffe, B.E., Foteinis, S., Lynett, P.J., Chan, I.-C., Liu, P.L.F., 2011. Insights on the 2009 South Pacific Tsunami in Samoa and Tonga from field surveys and numerical simulations. Earth Sci. Rev. https://doi.org/10.1016/j.earscirev.2011.03.004.

Frohlich, C., Hornbach, M., Taylor, F., Shen, C., Moala, A., Morton, A., Kruger, J., 2009. Huge erratic boulders in Tonga deposited by a prehistoric tsunami. Geology 37, 131–134.

Fujino, S., Masuda, F., Tagamori, S., Matsumoto, D., 2006. Structure and depositional processes of a gravelly tsunami deposit in a shallow marine setting: lower Cretaceous, Miyako Group, Japan. Sediment. Geol. 187, 127–138.

Furlani, S., Biolchi, S., Devoto, S., Saliba, D., Scicchitano, G., 2011. Large boulders along the NE Maltese coast: tsunami or storm wave deposits? J. Coast Res. 61, 470. https://doi.org/10.2112/SI61-001.60.

Goff, J.R., 2008. The New Zealand Palaeotsunami Database, vol. 131. NIWA Technical Report, Christchurch, New Zealand.

Goff, J., Weiss, R., Courtney, C., Dominey-Howes, D., 2010b. Testing the hypothesis for tsunami boulder deposition from suspension. Mar. Geol. 277, 73–77.

Goto, K., Imamura, F., 2007. Numerical models for sediment transport by tsunamis. Quat. Res. 46, 463–475.

Goto, K., Chavanich, S.A., Imamura, F., Kunthasap, P., Matsui, T., Minoura, K., Sugawara, D., Yanagisawa, H., 2007. Distribution, origin and transport process of boulders deposited by the 2004 Indian Ocean tsunami at Pakarang Cape, Thailand. Sediment. Geol. 202, 821–837.

Goto, K., Miyagi, K., Kawamata, H., Imamura, F., 2010a. Discrimination of boulders deposited by tsunamis and storm waves at Ishigaki Island, Japan. Mar. Geol. 269, 34–45.

Goto, K., Kawana, T., Imamura, F., 2010b. Historical and geological evidence of boulders deposited by tsunamis, southern Ryukyu Islands, Japan. Earth Sci. Rev. 102, 77–99.

Goto, K., Okada, K., Imamura, F., 2010c. Numerical analysis of boulder transport by the 2004 Indian Ocean tsunami at Pakarang Cape, Thailand. Mar. Geol. 268, 97–105.

Goto, K., Chagué-Goff, C., Fujino, S., Goff, J., Jaffe, B., Nishimura, Y., Richmond, B., Sugawara, D., Szczucoinski, W., Tappin, D.R., Witter, R.C., Yulianto, E., 2011. New insights of tsunami hazard from the 2011 Tohoku-oki event. Mar. Geol. 290, 46–50.

Goto, T., Satake, K., Sugai, S., Ishibe, T., Harada, T., Murotani, S., 2015. Historical tsunami and storm deposits during the last five centuries on the Sanriku coast, Japan. Mar. Geol. 367, 105–117.

Goto, K., Hongo, C., Watanabe, M., Miyazawa, K., Hisamatsu, A., 2019. Large tsunamis reset growth of massive corals. Progr. Earth Planet. Sci. 6 https://doi.org/10.1186/s40645-019-0265-2.

Häuselmann, P., 2008. Surface corrosion of an Alpine karren field: recent measures at Innerbergli (Siebenhengste, Switzerland). Int. J. Speleol. (Edizione Italiana) 37 (2), 107–111.

Hall, A.M., Hansom, J.D., Williams, D.M., 2010. Wave-emplaced coarse debris and megaclasts in Ireland and Scotland: boulder transport in high-energy littoral environment: a discussion. J. Geol. 118, 699–704.

Harbitz, C.B., 1992. Model simulations of tsunamis generated by the Storegga slides. Mar. Geol. 105, 1–21.

Harbitz, C.B., Glimsdal, S., Lövholt, F., Kveldsvik, V., Pedersen, G.K., Jensen, A., 2014. Rockslide tsunamis in complex fjords: from an unstable rock slope at Åkerneset to tsunami risk in western Norway. Coast Eng. 88, 101–122.

Hearty, P.J., 1997. Boulder deposits from large waves during the last interglaciation on north Eleuthera island, Bahamas. Quat. Res. 48, 326–338.

Hearty, P.J., Tormey, B.R., 2017. Sea-level change and superstorms; geologic evidence from the last interglacial (MIS 5e) in the Bahamas and Bermuda offers ominous prospects for a warming Earth. Mar. Geol. 390, 347–365.

Hearty, P.J., Tormey, B.R., 2018. Discussion of: Mylroie, J.E. (2018): comments on Bahamian Fenestrae and boulder evidence from the last interglacial. J. Coast Res. 34 (6), 1471–1479.

Herterich, J.G., Cox, R., Dias, F., 2018. How does wave impact generate large boulders? Modelling hydraulic fracture of cliffs and shore platforms. Mar. Geol. 399, 34–46.

Hisamatsu, A., Goto, K., Imamura, F., 2014. Local paleo-tsunami size evaluation using numerical modeling for boulder transport at Ishigaki Island, Japan. Episodes 37 (4), 265—275.

Hoffmeister, D., Ntageretzis, K., Aasen, H., Curdt, C., Hadler, H., Willershäuser, T., Bareth, G., Brückner, H., Vött, A., 2014. 3D model-based estimations of volume and mass of high-energy dislocated boulders in coastal areas of Greece by terrestrial laser scanning. Z. Geomorphol. N.F. Suppl. Issue 58, 115—135.

Hornbach, M.J., Frohlich, C., Taylor, F.W., 2008. Unraveling the source of large erratic boulders on Tonga: implications for geohazards and mega-tsunamis. Geol. Soc. Amer. Abstr. Programs 40 (6), 161.

Imamura, F., Goto, K., Ohkubo, S., 2008. A numerical model for the transport of a boulder by tsunami. J. Geophys. Res. — Oceans 113, 1—12.

Kelletat, D., Scheffers, A., 2004. Bimodal tsunami deposits — a neglected feature in palaeo-tsunami research. In: Schernewski, G., Dolch, T. (Eds.), Geographie der Meere und Küsten, vol. 1. Coastline Rep., pp. 1—20

Kelletat, D., Schellmann, G., 2002. Tsunamis in Cyprus: field evidences and 14C dating results. Z. Geomorphol. 46 (1), 19—34.

Kelletat, D., Scheffers, A., Scheffers, S., 2004. Holocene tsunami deposits on the Bahaman islands of long island and Eleuthera. Z. Geomorphol. 48 (4), 519—540.

Kelletat, D., Scheffers, A., Scheffers, S., 2007. Field signatures of the SE Asian mega-tsunami along the west coast of Thailand compared to Holocene palaeo-tsunami from the Atlantic region. Pure Appl. Geophys. 164, 1—19.

Kelletat, D.H., Kinis, S., Scheffers, A.M., 2013. Recent advances in palaeotsunami research of the eastern Mediterranean. Profesör Doktor İlhan Kayan 'a Armağan. In: Öner, E. (Ed.), Ege Üniversitesi Yayınları Edebiyat Fakültesi Yayın No 181, pp. 89—116.

Kelletat, D., Engel, M., May, S.M., Erdmann, W., Scheffers, A., Brückner, H., 2019. Erosive impact of tsunami and storm waves on rocky coasts and post-depositional weathering of coarse-clast deposits. In: Engel, M., Pilarczyk, J., May, S.M., Brill, D., Garrett, E. (Eds.), Geological Records of Tsunamis and Other Extreme Waves. Elsevier (in press).

Kennedy, A.B., Mori, N., Yasuda, T., Shimizono, T., Tomiczek, T., Donahue, A., Shimura, T., Imai, Y., 2017. Extreme block and boulder transport along a cliffed coastline (Calicoan island, Philippines) during super typhoon Haiyan. Mar. Geol. 383, 65—77.

Khan, S., Robinson, E., Rowe, D.A., Coutou, R., 2010. Size and mass of shoreline boulders moved and emplaced by recent hurricanes, Jamaica. Z. Geomorphol. 54, 281—299.

Lander, J.F., Whiteside, L.S., Lockridge, P.A., 2002. A brief history of tsunamis in the Caribbean Sea. Sci. Tsunami Hazards 20, 57—94.

Lau, A.Y.A., Etienne, S., Terry, J.P., Switzer, A.D., Lee, Y.S., 2014. A preliminary study of the distribution, sizes and orientations of large reef-top coral boulders deposited by extreme waves at Makemo Atoll, French Polynesia. In: Green, A.N., Cooper, J.A.G. (Eds.), Proceedings 13th International Coastal Symposium (Durban, South Africa), J. Coast. Res., Spec. Issue, vol. 70, pp. 272—277.

Lorang, M.S., 2011. A wave-competence approach to distinguish between boulder and megaclast deposits due to storm waves versus tsunamis. Mar. Geol. 283, 90—97.

Mader, C.L., 1999. Modeling the 1958 Lituya bay mega-tsunami. Sci. Tsunami Hazards 17 (1), 57—67.

Maouche, S., Morhange, C., Meghraoui, M., 2009. Large boulder accumulation on the Algerian coast evidence for tsunami events in the western Mediterranean. Mar. Geol. 262 (1), 96—104.

Masselink, G., Castelle, B., Scott, T., Dodet, G., Suanez, S., Jackson, D., Floch, F., 2016. Extreme wave activity during 2013/14 winter and morphological impacts along the Atlantic coast of Europe. Geophys. Res. Lett. 43 (5), 2135—2143.

Masson, D.G., 1996. Catastrophic collapse of volcanic island of Hierro 15 ka ago and the history of landslides in the Canary Islands. Geology 24, 231—234.

Masson, D.G., Watts, A.B., Gee, M.J.R., Urgeles, R., Mitchell, N.C., Le Bas, T.P., Canals, M., 2002. Slope failures in the flanks of the western Canary Islands. Earth Sci. Rev. 57, 1—35.

Mastronuzzi, G., Pignatelli, C., 2012. The boulders berm of Punta Saguerra (Taranto, Italy) a morphological imprint of 4th April, 1836 Rossano calabro tsunami? Earth Planets Space 64, 829—842.

Mastronuzzi, G., Sanso, P., 2000. Boulder transport by catastrophic waves along the Ionian coast of Apulia (southern Italy). Mar. Geol. 170, 93–103.

May, S.M., Brill, D., Engel, M., Scheffers, A., Pint, A., Opitz, S., Wennrich, V., Squire, P., Kelletat, D., Brückner, H., 2015a. Traces of historical tropical cyclones and tsunamis in the Ashburton Delta (NW Australia). Sedimentology 62, 1546–1572.

May, S.M., Engel, M., Brill, D., Cuadra, C., Lagmay, A.M.F., Santiago, J., Suarez, J.K., Reyes, M., Brückner, H., 2015b. Block and boulder transport in eastern samar (Philippines) during Supertyphoon Haiyan. Earth Surf. Dynamics. 3, 543–558.

Medina, F., Mhammdi, N., Chiguer, A., Akil, M., Jaaidi, E.B., 2011. The Rabat and Larache boulder fields; new examples of high-energy deposits related to storms and tsunami waves in north-western Morocco. Nat. Hazards Medina 59, 725–747.

Mhammdi, N., Medina, F., Kelletat, D., Amahmou, M., Aloussi, L., 2008. Large boulders along the Rabat coast (Morocco); possible emplacement by the November, 1st, 1755 A.D. Lisbon tsunami. Sci. Tsunami Hazards 27 (1), 1–30.

Moore, J.G., Clague, D.A., Holcomb, R.T., Lipman, P.W., Normark, W.R., Torresan, M.E., 1989. Prodigious submarine landslides on the Hawaiian Ridge. J. Geophys. Res. 94, 17465–17484.

Moore, J.G., Normark, W.R., Holcomb, R.T., 1994a. Giant Hawaiian landslides. Annu. Rev. Earth Planet Sci. 22, 119–144.

Moore, J.G., Bryan, W.B., Ludwig, K.R., 1994b. Chaotic deposition by a giant wave, Molokai, Hawaii. Geol. Soc. Am. Bull. 106, 962–967.

Mottershead, D., Bray, M., Soar, P., Farres, P.J., 2014. Extreme waves events in the central Mediterranean: geomorphic evidence of tsunami on the Maltese Islands. Z. Geomorphol. 58 (3), 385–411. https://doi.org/10.1127/0372-8854/2014/0129.

Mottershead, D.N., Bray, M.J., Soar, P.J., 2017. Tsunami landfalls in the Maltese archipelago: reconciling the historical record with geomorphological evidence. In: Scourse, E.M., Chapman, N.A., Tappin, D.R., Wallis, S.R. (Eds.), Tsunamis: Geology, Hazards and Risks. Geological Society, London, Special Publications, p. 456. https://doi.org/10.1144/SP456.8. First published online February 23, 2017, updated 3 March 2017.

Nanayama, F., Shigeno, K., Satake, K., Shimokawa, K., Koitabashi, S., Mayasaka, S., Ishii, M., 2000. Sedimentary differences between 1993 Hokkaido-Nansei-Oki tsunami and 1959 Miyakijima typhoon at Tasai, southwestern Hokkaido, northern Japan. Sediment. Geol. 135, 255–264.

Nandasena, N.A.K., Paris, R., Tanaka, N., 2011a. Numerical assessment of boulder transport by the 2004 Indian Ocean Tsunami in Lhok Nga, west Banda Aceh (Sumatra, Indonesia). Comput. Geosci. 37, 1391–1399.

Nandasena, N.A.K., Paris, R., Tanaka, N., 2011b. Reassessment of hydrodynamic equations: minimum flow velocity to initiate boulder transport by high energy events (storms, tsunamis). Mar. Geol. 281, 70–84.

Nandasena, N.A.K., Tanaka, N., Sasaki, Y., Osada, M., 2013. Boulder transport by the 2011 Great East Japan tsunami: comprehensive field observations and whither model predictions? Mar. Geol. 346, 292–309.

National Geophysical Data Center/World Data Center (NGDC/WDC), 2013. Global Historical Tsunami Database – Tsunami Event Data. https://doi.org/10.7289/V5PN93H7. Boulder, CO, USA.

Neugebauer, N.P., Brill, D., Brückner, H., Kelletat, D., Scheffers, S., Vött, A., 2011. 5000 jahre tsunami-geschichte am Kap Pakarang (Thailand). Coastline Rep. 17, 81–98.

Noormets, R., Crook, K.A.W., Felton, E.A., 2004. Sedimentology of rocky shorelines: 3. Hydrodynamics of megaclast emplacement and transport on a shore platform, Oahu, Hawaii. Sediment. Geol. 172, 41–65.

Nott, J.F., 2003a. Tsunami or storm waves?—determining the origin of a spectacular field of wave emplaced boulders using numerical storm surge and wave models and hydrodynamic transport equations. J. Coast Res. 19, 348–356.

Nott, J., 2003b. Waves, coastal boulder deposits and the importance of the pre-transport setting. Earth Planet Sci. Lett. 210, 269–276.

O'Boyle, L., Whitakker, T., Cox, R., Elsäer, B., 2017. Experimental modelling of wave amplification over irregular bathymetry for investigations of boulder transport by extreme wave events. Vienna, Austria. European Geosciences Union. Geophys. Res. Abstr. 19, 16868.

Oliveira, M.A., 2017. Boulder Deposits Related to Extreme Marine Events in the Western Coast of Portugal (Ph.D. thesis). Faculty of Science, University of Lisbon, Portugal.

O'Loughlin, K., Lander, J.F., 2003. Caribbean Tsunamis — A 500-Year History from 1498–1998. Kluwer Academic Publishers, Dordrecht.

Papadopoulos, G.A., Chalkis, B.J., 1984. Tsunami observed in Greece and the surrounding area from antiquity up to present times. Mar. Geol. 56, 309–317.

Papadopoulos, G.A., Fokaefs, A., 2005. Strong tsunamis in the Mediterranean sea: a re-evaluation. ISET J. Earthq. Technol. 42, 159–170.

Papadopoulos, G.A., Gràcia, E., Urgeles, R., Sallares, V., De Martini, P.M., Pantosti, D., González, M., Yalciner, A.C., Mascle, J., Sakellariou, D., Salamon, A., Tinti, S., Karastathis, V., Fokaefs, A., Camerlenghi, A., Novikova, T., Papageorgiou, A., 2014. Historical and pre-historical tsunamis in the Mediterranean and its connected seas: geological signatures, generation mechanisms and coastal impacts. Mar. Geol. 354, 81–109.

Playford, P., 2014. Mega-tsunamis along the Coastlines of Western Australia. Presentation Geol. Soc. Australia, Perth, Australia.

Regnauld, H., Oszwald, J., Planchon, O., Pignatelli, C., Piscitelli, A., Mastronuzzi, G., Audevard, A., 2010. Polygenetic (tsunami and storm) deposits? A case study from Ushant Island, Western France. Z. Geomorphol. 54 (3), 197–217.

Richmond, B.M., Watt, S., Buckley, M., Jaffe, B.E., Gelfenbaum, G., Morton, R.A., 2011. Recent storm and tsunami coarse-clast deposits characteristics, southeast Hawaii. Mar. Geol. 283, 79–89.

Rovere, A., Casella, E., Harris, D., Lorscheid, T., Nandasena, N.A.K., Dyer, B., Sandstrom, M.R., Stocchi, P., D'Andrea, W.R., Raymo, M.E., 2017a. Wave models for Eleuthera, northern Bahamas. PANGAEA. Supplement to: Rovere, A., et al., 2017. Giant boulders and last interglacial storm intensity in the north Atlantic. Proc. Natl. Acad. Sci. U.S.A. 114 (46), 12144–12149. https://doi.org/10.1594/PANGAEA. 880687.

Rovere, A., Harris, D., Casella, E., Lorscheid, T., Stocchi, P., Nandasena, N., Sandstrom, M., D'Andrea, W., Dyer, B., Raymo, M., 2017b. Superstorms at the end of the Last Interglacial (MIS 5e)? Modeling paleowaves and the transport of giant boulders. Geophys. Res. Abstr. 19. EGU2017-4087, 2017 EGU General Assembly.

Scheffers, A., 2004. Tsunami imprints on the Leeward Netherlands Antilles (Aruba, Curaçao, Bonaire) and their relation to other coastal problems. Quat. Int. 120, 163–172.

Scheffers, A., 2005. Coastal response to extreme wave events — hurricanes and tsunami on Bonaire. Essener Geographische Arbeiten 37, 100.

Scheffers, A., 2006. Ripple marks in coarse tsunami deposits. Z. Geomorphol. Suppl.-Bd. 146, 221–233.

Scheffers, A., 2008. Tsunami boulder deposits. In: Shiki, T., Tsuji, Y., Yamazaki, T., Minoura, K. (Eds.), Tsunamiites — Features and Implications. Elsevier, Berlin, pp. 299–318.

Scheffers, A., 2015. Paleotsunami research — current debate and controversies. In: Shroder, J.F., Ellis, J.T., Sherman, D.J. (Eds.), Coastal and Marine Hazards, Risks, and Disasters. Elsevier, Amsterdam, pp. 59–92.

Scheffers, A., Kelletat, D., 2005. Tsunami relics on the coastal landscape west of Lisbon, Portugal. Sci. Tsunami Hazards 23, 3–16.

Tsunamis, hurricanes and neotectonics as driving mechanisms in coastal evolution. Proceedings of the Bonaire field symposium, March 2-6, 2006. In: Scheffers, A., Kelletat, D. (Eds.), Z. Geomorphol., Suppl. Bd. 146, 265.

Scheffers, A., Kelletat, D., 2006b. Recent advances in palaeo-tsunami field research in the intra-Americas-sea (Barbados, st. Martin and Anguilla). In: Proc. NSF Caribbean Tsunami Workshop. World Scientific Publishers, Puerto Rico, pp. 178–202.

Scheffers, A., Kelletat, D., 2019. Megaboulder movement by superstorms: a geomorphological debate. J. Coast Res. (in press).

Scheffers, A., Kinis, S., 2014. Stable imbrication and delicate/unstable settings in coastal boulder deposits: indicators for tsunami dislocation? Quat. Int. 332, 73—84.

Scheffers, A., Scheffers, S., 2006. Documentation of hurricane Ivan on the coastline of Bonaire. J. Coast Res. 22, 1437—1450.

Scheffers, A., Scheffers, S., 2007. First significant tsunami evidence from historical times in western Crete (Greece). Earth Planet Sci. Lett. 259 (3—4), 613—624.

Scheffers, A., Kelletat, D., Vött, A., May, S.M., Scheffers, S., 2008. Late Holocene tsunami traces on the western and southern coastlines of the Peloponnesus (Greece). Earth Planet Sci. Lett. 269, 271—279.

Scheffers, S.R., Scheffers, A., Kelletat, D., Bryant, E.A., 2009a. The Holocene palaeotsunami history of West Australia. Earth Planet Sci. Lett. 270, 137—146.

Scheffers, A., Scheffers, S., Kelletat, D., Browne, T., 2009b. Wave emplaced coarse debris and megaclasts in Ireland and Scotland: boulder transport in a high-energy littoral environment. J. Geol. 117, 553—573.

Scheffers, A., Kelletat, D., Haslett, S.K., Scheffers, S., Browne, T., 2010. Coastal boulder deposits in Galway bay and the aran islands, western Ireland. Z. Geomorphol. 54, 247—279.

Scheffers, A.M., Engel, M., May, S.M., Scheffers, S.R., Joannes-Boyau, R., Hänssler, E., Kennedy, K., Kelletat, D., Brückner, H., Vött, A., Schellmann, G., Schäbitz, F., Radtke, U., Sommer, B., Willershäuser, T., Felis, T., 2014. Potential and limits of combining studies of coarse- and fine-grained sediments for the coastal event history of a Caribbean carbonate environment. Geol. Soc. Lond., Spec. Publ. 388, 503—531. https://doi.org/10.1144/SP388.4.

Scicchitano, G., Monaco, C., Tortorici, L., 2007. Large boulder deposits by tsunami waves along the Ionian coast of south-eastern Sicily (Italy). Mar. Geol. 238, 75—91.

Shiki, T., Tsuji, Y., Minoura, K., Yamazaki, T. (Eds.), 2008. Tsunamiites — Features and Implications. Elsevier, Amsterdam, p. 411.

Soloviev, S.L., Solovieva, O.N., Go, C.N., Kim, K.S., Shchetnikov, N.A., 2000. Tsunamis in the Mediterranean Sea 2000 B.C. — 2000 A.D. Kluwer Academic Publishers, Dordrecht, p. 240.

Talandier, J., Bourrouilh-Le Jan, F., 1987. High energy sedimentation in French Polynesia: cyclone or tsunami? In: El-Sabh, M.I., Murty, T.S. (Eds.), Natural and Man-Made Hazards. Springer, Dordrecht, pp. 193—199.

Tappin, D.R., 2017. The importance of geologists and geology in tsunami science and tsunami hazard. Geol. Soc. Lond., Spec. Publ. 456, 5—38.

Terry, J.P., Lau, A.Y., Etienne, S., 2013. Reef-platform Coral Boulders. Evidence for High-Energy Marine Inundation Events on Tropical Coastlines. Springer Briefs in Earth Sciences, Dordrecht, p. 96.

Tinti, S., Maramai, A., 1996. Catalogue of tsunami generated in Italy and in Cote d'Azur, France, a step towards a unified catalogue of tsunami in Europe. Ann. Geofisc. 39, 1253—1300.

Viret, G., 2008. Mégablocs au Nord d'Eleuthera (Bahamas) Preuve de Vagues Extrêmes au Sous-stade Isotopique 5e ou Restes Erosionnels? (MS thesis) University of Geneva, Switzerland, p. 93.

Ward, S.N., Day, S.J., 2001. Cumbre Vieja volcano; potential collapse and tsunami at La Palma, canary islands. Geophys. Res. Lett. 28, 3397—3400.

Ward, S.N., Day, S., 2010. The 1958 Lituya bay landslide and tsunami — a tsunami Ball approach. J. Earthquake Tsunami 4 (4), 285—319.

Weiss, R., Diplas, P., 2015. Untangling boulder dislodgement in storms and tsunamis: is it possible with simple theories? Geochem. Geophys. Geosyst. 16 (3), 890—898.

Weiss, R., Fritz, H.M., Wünnemann, K., 2009. Hybrid modeling of the mega-tsunami runup in Lituya Bay after half a century. Geophys. Res. Lett. 36, 1—6.

Whelan, F., Kelletat, D., 2005. Boulder seposits on the southern Spanish Atlantic coast: possible evidence for the 1755 AD Lisbon tsunami. Sci. Tsunami Hazards 23 (3), 25—38.

Williams, D.M., 2010. Mechanisms of wave transport of megaclasts on elevated cliff-top platforms: examples from western Ireland relevant to the storm wave versus tsunami controversy. Ir. J. Earth Sci. 28, 13—23.

382 Chapter 19

Williams, D.M., Hall, A.M., 2004. Cliff-top megaclast deposit of Ireland, a record of extreme waves in the North Atlantic: storms or tsunamis? Mar. Geol. 206, 101−117.

Yamada, M., Fujino, S., Goto, K., 2014. Deposition of sediments of diverse sizes by the 2011 Tohoku-oki tsunami at Miyako City, Japan. Mar. Geol. 358, 67−78.

Yu, K.F., Zhao, J.X., Shi, Q., Meng, Q.S., 2009. Reconstruction of storm/tsunami records over the last 4000 years using transported coral blocks and lagoon sediments in the southern South China Sea. Quat. Int. 195, 128−137.

Zhao, J.X., Neil, D.T., Feng, Y.X., Yu, K.F., Pandolfi, J.M., 2009. High-precision U-series dating of very young cyclone-transported coral reef blocks from Heron and Wistari reefs, southern Great Barrier Reef, Australia. Quat. Int. 195 (1−2), 122−127.

| CHAPTER 20 |

Characteristic features of tsunamiites

T. Shiki[1], T. Tachibana[2], O. Fujiwara[3], K. Goto[4], F. Nanayama[3], T. Yamazaki[5]
[1]Uji, Kyoto, Japan; [2]Soil Engineering Corporation, Higashi-ku, Okayama, Japan; [3]Geological Survey of Japan, National Institute of Advanced Industrial Science and Technology (AIST), Tsukuba, Ibaraki, Japan; [4]Department of Earth and Planetary Science, Graduate School of Science, The University of Tokyo, Bunkyo-ku, Tokyo, Japan; [5]Toyonaka, Osaka, Japan

1. Introduction

It was at the beginning of 1960s that Kon'no and others carried out excellent geological observations of the Sanriku coastal region, just after the damage caused by the 1960 Chile earthquake tsunami (Kon'no et al., 1961). Tsunamiites, however, are still the most difficult deposits to identify among the many kinds of high-energy event deposits. What are the criteria for identifying tsunamiites? This is a major point of interest for all tsunamiite researchers. It must be stressed, however, that such a convenient key for identifying tsunamiites does not exist.

Features of tsunamiites of various settings reflect the characteristics of tsunami waves and currents. The features can, however, fairly be different, depending on the characteristics of tsunami waves and their related currents under different hydrographic conditions and in different sedimentary environments. Furthermore, some sedimentary features of tsunamiites can also be common in other kinds of event deposits. Diastrophic occurrence and periodicity are well-known characteristics of tsunamis and have been well studied. These characteristics are, however, not unique for tsunamis. Despite these difficulties, sedimentologists can recognize tsunamiites in many cases, if careful observations are carried out. In this section, we will discuss this problem and provide some useful suggestions for finding and identifying tsunamiites in historic and older sediments.

It is to be noted here that most of the following descriptions and explanations are focused on and safely applicable to seismic tsunamiites, but not for the case of meteorite impact-induced tsunamiite of exceptionally gigantic size, such as the K/Pq-boundary beds. Readers interested in those aspects are referred to the previous chapters (e.g., Goto, 2008, this volume).

Tsunamiites. https://doi.org/10.1016/B978-0-12-823939-1.00020-3
Copyright © 2021 Elsevier B.V. All rights reserved.

2. Characteristics of tsunamis and tsunami deposition

Generation, propagation, and quantification of tsunamis are dealt with in detail by Sugawara et al. (2008, this volume). Only some of the most important characteristic features of tsunamis will therefore be mentioned here. Fig. 20.1 shows the features schematically.

Coleman (1968) already pointed out the significance of tsunamis as a geological agent. A great number of researches have been carried out since then on the propagation, wavelength, and wave height of tsunamis; coastal disasters caused by run-up of tsunamis are better known since the studies by Miyoshi (1968), Miyazaki (1971), and Miyoshi and Makino (1972) (see also Sugawara et al., 2008, this volume). On the other hand, important hydrodynamic knowledge, which is indispensable for the understanding of the sediment-transport mechanism by tsunamis, is not yet very deep. Some aspects are known, however, about the mechanisms that are of interest for sedimentological discussions about tsunamiites (see the review by Sugawara et al., 2008, this volume). Some characteristic features of the various types of tsunamiites can therefore be dealt with here in the light of this knowledge.

2.1 Diastrophic nature of tsunamis and tsunamiite deposition

Tsunamis occur suddenly and are often diastrophic. They have a gigantic energy and propagate waves all over the globe. Thus, their deposits are distributed widely and sometimes even globally. They can be used as marker horizons across various facies in many sedimentary basins around the globe, as they are commonly exceptional in lithology

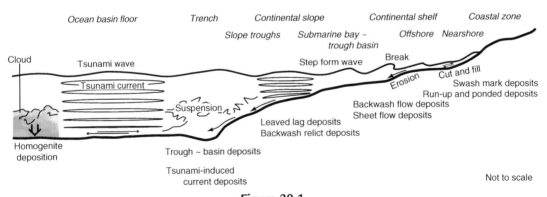

Figure 20.1
Schematic diagram showing tsunami water movement and sedimentary processes generated or affected by a tsunami from the oceanic floor to the coastal zone. Special attention is given to show the nature of the tsunami wave and current, the role of gravity in some tsunamiite deposition, and the suspension cloud.

and facies in neighboring and isolated sedimentary basins. That is, they show features of an exceptional high-energy flow regime among the more common background sediments of a lower-energy flow regime. For instance, a peculiar occurrence of exceptionally thick- and coarse-grained clastic material can be one good indication of tsunami action.

A number of tsunami-driven coastal boulder deposits are documented by Scheffers (2008, this volume). The Miocene Tsubutegaura boulder conglomerates on the Chita Peninsula, central Japan, are peculiar examples of large boulders deposited in an upper-bathyal environment (Yamazaki et al., 1989; Shiki and Yamazaki, 1996).

It must be noted that a "wide" extent does not necessarily imply a "continuous" deposit. One could easily imagine that great continuity without any gap is a characteristic of tsunami deposits, but this is a misunderstanding. The problems of the low-preservation potential, what is called "ephemeral characteristics," and of up-rush tsunamiites are discussed by Keating et al. (2008). An important point to be noted is the effect of the powerful energy of the tsunami wave (and current) itself. The gigantic energy results in strong erosion, together with the transportation and the deposition of vast amounts of material. Erosion takes place locally, especially where the topography is inclined and/or has channels, or, occasionally, where it is relatively high, as illustrated in Fig. 20.1. In some cases, the occurrence of a tsunami is documented only by large-scale scour-and-fill structures, as in the Late Albian to Cenomanian deposits of the Grajau Basin, northern Brazil, reported by Rossetti et al. (2000). Therefore, a patchy distribution is a common characteristic of tsunamiites, though the sediments can build an extensive body in some areas. On the other hand, a tsunamiite's distribution can also be very narrow, local, and sporadic.

In this context, it must be pointed out that the gigantic energy of tsunamis does not always result in thick-, coarse-grained sediments. A tsunami can transport only the material available. The grain sizes of tsunamiites depend therefore on the supplied sediments, which closely reflect the source materials.

2.2 Length of tsunami waves and sedimentary structures in tsunamiites

A well-known feature of tsunamis is the very large length of the waves. In other words, it has a "shallow-sea wave" character (see Fig. 3.3 in Sugawara et al., 2008, this volume; Fig. 16.5 in Goto, 2008, Fig. 18.5, this volume) when it propagates in the deep ocean. This means that the sea-water movement at the basal part of the wave is actually a current with a simple "shuttle" characteristic without an up-and-down element (see Fig. 16.5 in Goto, 2008, Fig. 18.5, this volume). Consequently, any kind of current-induced structure, that is, parallel lamination, current ripple marks, antidunes, imbrication, sole marks, and so on, can occur in tsunamiites.

386 Chapter 20

On the other hand, wave ripples have also been reported occasionally. Hummocky structures, which are believed to indicate wave action in combination with a current, occur rather commonly in some submarine tsunamiites (e.g., Albertao and Martins, 2002; Albertao et al., 2008, this volume; Fujiwara et al., 2000). It is still a problem why these wave-generated or wave-related structures can happen in tsunamiites. This will be discussed later.

The velocity and the length of a tsunami wave decrease (and its height increases) when the wave comes close to a shallow coastal area. However, the wavelength is still enough to inundate coastal plains very far inland and to distribute deposits there over a vast area.

The shape of a tsunami wave also changes near the coast. It forms a single step, and it may, or may not, break, depending on the specific conditions. Bottom erosion is the most violent when the wave breaks. A widespread erosion surface and gigantic scour-and-fill structures (Rossetti et al., 2000) are the sedimentological marks left by tsunami action at that zone.

It must also be noted that the velocity of water movement in long waves such as those of a tsunami is the same from top to bottom, even in very deep of sea water (Fig. 20.1; see also Fig. 3-1-1-A in Sugawara et al., 2008, this volume), but the water movement changes in the coastal area and onshore. The inner structure of tsunami waves implies that shearing does not occur in tsunami waves. However, the velocity decreases very sharply toward zero in the bottom boundary layer of the tsunami-induced bottom current just above the sea floor, and shearing is very strong there. Ripping up of the bottom material can happen very effectively because of the large lift-up effect of the vertical velocity difference in the bottom layer of a tsunami. This implies that the effect is present at a much greater depth in a tsunami than in violent wind-induced deep-water waves. After lifting-up and during transportation in the main, nonsheared layer, even very fragile materials are not broken. This hydrodynamical structure of a tsunami wave is the reason for the common occurrence of many rip-up clasts of angular shape and nonwear shells in tsunamiite beds, together with the entrainment mechanism of big boulders.

Soft-sediment rip-up clasts up to 1 m in diameter occur in the Holocene tsunami section on the southern Boso Peninsula in east Japan (Fujiwara and Kamataki, 2008, this volume; Fujiwara et al., 2000). Delicate shells commonly occur in the section (Fujiwara, 2004; Fujiwara et al., 2003a). Rip-up clasts of 1-m size are observed in a cobble conglomerate tsunamiite in the Miocene upper-bathyal Tsubutegaura tsunamiite succession (Yamazaki et al., 1989; Shiki and Yamazaki, 1996). A broken soft-sediment block, 2 m in diameter, has been found together with the rip-up clasts at the same stratigraphic level in this succession. A few fossils of starfish, which would beyond doubt have been broken if affected by shear stress, are also found in a thin tsunamiite sand bed in this succession. It thus appears that even these weak, fragile sediment blocks and fossil remains are not broken if they are lifted up into the nonshear layer of a tsunami wave current.

2.3 Shuttle movement of tsunamis and its records

Shuttle movement, that is, the repeated forward and backward oscillatory water movement resulting from successive run-up and backflow stages, is a marked feature of tsunamis and their related currents. A few to several pulses of a single tsunami event can deposit a series of units (layers) that show repeatedly opposite current directions, as will be mentioned later. A good example was provided by Nanayama et al. (2000) (Fig. 20.2). Palaeocurrents, with both landward and seaward direction, can be reconstructed on the basis of the dip of wedge-shaped cross-stratification and by the imbrication of shells and gravel, as shown by Fujiwara et al. (2000) for tsunamiite layers deposited in a small Holocene bay, as will be mentioned in some more detail further.

Another good example of shuttle movement of a tsunami regards the K/Pg-boundary tsunamiites in neritic to upper-bathyal environments. Smit et al. (1996) described a variety of sedimentary features of these deposits, containing evidence for repeated up-section changes in current direction (as interpreted from cross-lamination) and strength of the tsunami.

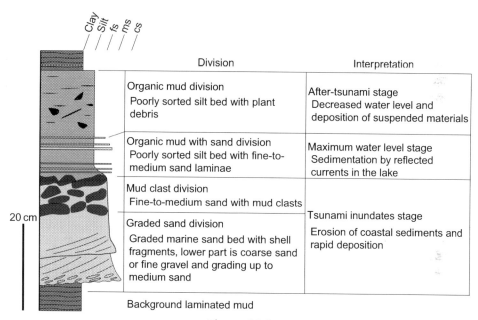

Figure 20.2

Sedimentological profile of a typical tsunamiite bed deposited in a coastal lacustrine environment. The existence of units with current structures, mud clasts, sand/silt alternation, and organic fragments in the uppermost fine-grained division are to be noted. A single bed consists of several divisions with different sedimentological features. The word "divisions" cited here does not have the same meaning as the word "units," which is used in the text. A division can involve a few units (subunits).

388 **Chapter 20**

It is well known that herringbone structures indicate shuttle movement of water and are commonly found in tidal sediments. This structure can, however, also occur in tsunamiites. Actually, the structure has been reported from late Pleistocene incised valley-fill sediments in central Japan by Takashimizu and Masuda (2000). Also this structure will be mentioned again later.

It should be mentioned here that some of the deposits of the repeated currents in opposite directions are not always preserved because of erosion by later pulses of the same tsunami event. It must also be noted that, because of some specific coastal and sea-bottom topography, a tsunami can run toward two or more directions with a high angle of divergence. Bays, capes and submarine canyons, and small basal topographic irregularities can create the conditions required for such differently directed currents.

Shuttle movement and current action into two or more directions not always result in distinct sedimentary structures. For example, such water movements have been reconstructed for the K/Pq-boundary tsunamiites in Cuba by faint fluctuations in grain size and material composition only (Tada et al., 2003; Goto et al., 2008b, this volume). Similar faint fluctuations in grain size and composition have been found in the upper, muddy division of a Mediterranean homogenite, as discussed by Shiki and Cita (2008, this volume).

Besides these sedimentary features, it is also worth noting that the shuttle movement of tsunamis is exhibited by a "saw-toothed" pattern in the vertical grain-size distribution in the successive units of a tsunamiite bed. A good example has been demonstrated by Fujiwara et al. (2003b) (see also Fujiwara and Kamataki, 2008, this volume) from the tsunami deposits in a Holocene valley mentioned earlier.

2.4 Shuttle movement versus gravity flows and tsunamiite variety

Shuttle movement is one of the most noteworthy features of tsunamis, as stressed earlier. However, nothing on Earth can escape from gravity. The kinetic energy of tsunami-induced water changes therefore during its movement in accordance with the surge direction, namely run-up and backwash, and in relation to the bottom's slope direction and its steepness, and results in lateral changes in the characteristics of its deposits.

A tsunami loses its energy when it runs up a coast and inundates the coastal area essentially because of the water movement in a direction that is "against" gravity. Friction on the bottom and percolation of water into the ground also contribute to the energy loss of a run-up tsunami. The depth (the thickness of water column) and speed of the water flow decrease until the flow stops and starts to turn over and occur in the opposite direction, toward the sea. Consequently, the run-up part of a tsunamiite thins out inland and leaves only a very thin sedimentary layer. In some cases lined lagged sediments, namely swash marks, can be left at the final place on the coastal area reached by the run-up flow.

The entrainment range of each pulse of a tsunami depends on the energy of the pulse. Consequently, the number of layers or depositional units within a tsunamiite, just like the thickness and grain size of the deposit, decreases as these sediments are positioned more inland. These features, due to gravity, are restricted to run-up inland tsunamiites and are not applicable to submarine tsunamiites.

The backwash water of a tsunami wave also must decelerate and come to a hold after the movement away from the coast. In this case, however, the current carries eroded bottom sediments from the various environments around the coastline. If the nature of the current changes into that of some sediment gravity flow (Middleton and Hampton, 1976), water must flow downslope with these particles. This is the reason why the basal division of many submarine tsunamiite beds seems to resemble turbidites with a graded (or massive) division (see the pictures of tsunamiites in Albertao et al., 2008, this volume). This mechanism, caused by gravity, cannot act above a horizontal flat floor but works to a high degree on long inclined slopes such as a continental slope. Gravity plays a very important role not only in transporting sediments from shallow environments but also in creating unique features of deep-sea tsunamiites such as in the B-type homogenite in the Mediterranean Sea (Cita, 2008; Shiki and Cita, 2008, this volume) and in some oceanic-impact tsunamiites deposited around craters (e.g., Goto, 2008, this volume; Tada et al., 2003; Takayama et al., 2000). Turbiditic features like these deposits do not happen in the run-up tsunamiites.

In short, a tsunami current moves as a shuttle. However, its sediments do not always follow and exhibit the shuttle movement. Records of the repeated shuttle flows are one of the remarkable features of tsunamiites but are restricted to depositional sites that have sufficient and continuous sediment supply in combination with a high preservation potential of the deposits, the conditions of which are related with the sedimentary conditions in and around the depositional area.

2.5 Tsunamiites as marker horizons

In spite of the only sporadic occurrences, mentioned earlier, a single tsunami event can occur and leave its traces in a very wide area, even worldwide. Therefore, tsunami sediments can be excellent time markers (key beds and marker horizons). This is especially true in the case of meteorite-impact tsunamiites that are accompanied by an iridium and other impact-generated-material-bearing clay layer, or in the case of tsunamiites induced by a volcanic eruption, which commonly contain and/or are overlain by volcanic material. The significance of these features of impact tsunamiites are

390 *Chapter 20*

mentioned by Goto (2008, this volume), and volcanic tsunamiites are mentioned by Nishimura (2008, this volume).

2.6 Associated sediments

It is needless to say that a meteorite-impact tsunamiite is characterized by several constituents of cosmic origin such as iridium and by components that prove a high-energy impact shock, such as shocked quartz grains; a volcanism-induced tsunamiite is obviously associated with volcanic material. As to the characteristic features of these tsunamiites, the reader is referred to the previous chapters of this volume.

As mentioned earlier, distinct differences in facies from background sediments characterize common tsunamiites. However, the background sediments are extremely variable, as they depend on their sedimentary conditions and the environments in which they have been deposited. Sandy tsunamiites, for example, can be intercalated between sandy background sediment layers that can closely resemble the sandy tsunamiite layers. In such cases, discrimination between the two kinds of sediments can be difficult. In spite of this problem, it must be said that the peculiarity in a sedimentary succession is generally the most useful key for finding a tsunamiite in the field.

The usage of this key is different for tsunamiites formed because of different trigger mechanisms. Submarine earthquakes happen repeatedly and cyclically in some areas. Under such conditions, seismic tsunamiites commonly occur in groups, and seismites such as liquefied sediments and dykes are, as a rule, associated with seismic tsunamiites. This seems a very critical difference between seismic tsunamiites and impact-induced tsunamiites. It must be remembered, however, that everything in geology has exceptions. Tsunamis can travel across oceans. The seismic tsunami that occurred in Washington, on the western coast of North America in 1700 CE, traveled across the Pacific Ocean and attacked the Japanese coast (Atwater et al., 2005; Satake et al., 1996; Tsuji et al., 1998). In 1960, another seismic tsunami, which occurred in Chile, also attacked the Japanese coast and formed correlatable tsunami deposits. That event and its sediments were studied in detail by Kon'no (1961). On the other hand, it is also logical that proximal-impact tsunamiites are accompanied by remarkable seismic sedimentary signatures, because an earthquake must have been induced by the meteorite impact.

3. Sedimentary structures in tsunamiite beds

Many types of sedimentary structures are known in clastic sedimentology. In this section, occurrence, or nonoccurrence, of some of these structures will be referred to and a few well-known sedimentary structures will be examined from the point of view of possible indicators for tsunami-induced sediments. It must be noted here again that sometimes the passing of a tsunami wave is revealed only by a giant erosion surface without any accompanying sediments.

3.1 Sedimentary structures in a sedimentary set

The Bouma sequence of the sedimentary structures in turbidites (Bouma, 1962) is generally known in sedimentary geology. Is any comparable type of stacked layers with their own primary sedimentary structures unique to tsunamiites? This is an interesting and important question that is still under discussion. For an answer to this question, we must keep in mind the pulses of a tsunami and the repeated arrival of high-density currents separated by a long stagnant stage.

According to the contribution by Fujiwara and his coworkers, a tsunamiite bed consist of several units (or subunits) (Fujiwara, 2004; Fujiwara and Kamataki, 2008, this volume). These units are attributed to run-up in their early stage, larger waves that deposit outsized clasts in the middle stage, and relatively small waves with long stagnant intervals in the later stage, with subsequent foundering of wood and plant debris derived from the land. Each unit consists of a couplet of a coarse-grained unit and a mud unit. The period of tsunami waves is in between that of storm waves and of tide waves and is long enough for the deposition of mud layers (mud drapes) during its relatively calm, stagnant stage.

Stacked units capped by muddy laminae are frequently present in some recent tsunamiites (1964, Alaska; 1703, Kanto; 1960, Chile; see Fujiwara and Kamataki, 2008, this volume). A tsunamiite stack from the Miocene upper-bathyal sediments mentioned earlier consists of several units of different lithofacies, including conglomerates, laminated sandstones with antidunes and inclined convolutions, and so on, indicating various flow directions. The successive units in the K/T-boundary tsunamiites described in detail by Smit et al. (1996) also reveal a sequence of units in shallow-to-deep environments, as mentioned earlier. The couplets with a fine-grained drape of the stacked units can thus be regarded as one of the most noteworthy characteristics of submarine tsunamiites. Fujiwara et al. (2003b) clarified that the vertical grain-size distribution shows a saw-toothed pattern in such stacked layers, as mentioned earlier.

Although the aforementioned model is also applicable to tsunamiites deposited in coastal lakes (Fig. 20.2) and on coastal lowlands, there are many exceptions (Fujiwara, 2008, this volume). For example, only one (single) tsunamiite layer can indicate a deep entrance of a tsunami into an alluvial plain, as in the case of the Sendai Plain in Japan (Minoura et al., 2001). Both the stacked units and the single layer from an inland-inundating tsunamiite have recently been exhibited dramatically by the Indian Ocean tsunami deposits in many coastal areas (see other chapters of this volume).

A relatively simple inner structure, which shows no sign of stacking, in the tsunami-induced sediment gravity flow sediments, a sort of "tsunamiite," was described by Cita and Aoisi (2000) and Cita (2008). This simple structure is discussed sedimentologically in fairly great detail by Shiki and Cita (2008, this volume). In short, the faint grading of grain size (and associated carbonate contents) is the only observable internal feature in the beds.

In any case, a complete inventory of the stacked layers is one of the most important ways to find and distinguish tsunamiites in the field.

3.2 Grading in a layer and fining upward through stacked layers

Normal grading is very common in tsunamiites, as described in many other chapters of this book. Reverse grading also happens occasionally in some submarine and onshore tsunamiite beds.

Tsunamiites in a shallow sea caused by a single event such as an earthquake or a meteorite impact usually form one characteristic stack of several units, as mentioned previously. The units are generally covered by muddy drape layers that are deposited during a relatively calm stage. Even when deposited in deep water, the tsunami-related sediment may form a markedly fining upward sequence if the tsunami is large enough. A good example of this is the K/T-boundary one. In contrast, a turbiditic Mediterranean homogenite consists of a coarse-grained graded unit and a fine-grained, distinctly homogeneous and thick unit that indicates deposition from a suspension cloud.

Internal structures in thin tsunamiite beds are uncommon. The beds usually consist of only a lower sandy division and an upper muddy division, similar to a thin turbidite bed. The sandy division of thin homogenite beds is generally structureless or faintly graded and passes into a muddy division without sharp boundary. Such a tsunamiite, especially if it is a submarine tsunamiite, can hardly be distinguished from a thin turbidite when only a single bed is observed. The sedimentary mechanisms of these two types of sediments are not very different if only the hydrodynamics of the transportation and the deposition of the sediment grains at a certain point are considered. Seismic-induced turbidites, however, generally do not occur alone but happen in a thick succession over a whole area where earthquakes are common, as pointed out earlier. An overall study of the sedimentary succession should make it possible to distinguish between these two different types of deposits that are caused by different triggers.

As to the deep-sea tsunamiites, they are really a kind of "sediment gravity flow" deposit with some possible exceptions in the case of gigantic impact tsunamiites. It is logical that they are graded in their sandy division. However, an actual example, namely the Mediterranean homogenites, does not reveal any sign of other divisions of the Bouma sequence.

3.3 Current structures

Many kinds of current structures can be present in tsunamiites (Figs. 20.2 and 20.3). They are internal and surface structures, such as parallel lamination, cross-lamination, convolutions, current-ripple marks, dunes and antidunes, and some sole structures, such as giant scour marks, current marks, and so on. A few of these current structures and their implication will be referred to here in some detail.

Imbrication of gravels and shells can develop because of a tsunami in a coastal area (see the photos in Scheffers, 2008, this volume). The structure can also develop in submarine environments in some exceptional cases. For example, the imbrication of cobbles and gravel clasts evidences transportation by a tsunami current under the upper-bathyal conditions in the Miocene of central Japan, as has been reported and discussed repeatedly (Yamazaki and Shiki, 1988; Yamazaki et al., 1989; Shiki and Yamazaki, 1996). It is interesting that the cobbles in Scheffers's photo dip seaward as in the case of storm beach sediments. On the other hand, the imbricated cobbles of the backwash tsunami bottom-current under the upper-bathyal conditions dip up-current, very similar to river sediments.

As mentioned earlier, herringbone structures, which are considered as a common sign of tidal sediments, may be a feature of some tsunamiites because of the shuttle movement of a tsunami current. Unlike tidal sediments, however, herringbone structures cannot develop successively in thick layers in a tsunamiite bed. A thin muddy drape layer is generally intercalated between two sandy layers that show ripple cross-lamination into two opposite directions (Takashimizu and Masuda, 2000; see also Fujiwara, 2008, this volume).

As mentioned earlier also, hummocky and swaley cross-stratification (HCS and SCS) are fairly common, and even wave-ripple marks have been reported by Fujiwara and his colleagues from some tsunamiite beds from Holocene deposits in the Boso Peninsula, central Japan (Fujiwara et al., 2003b; Fujiwara and Kamataki, 2008, this volume), and in the K/Pq-boundary impact tsunamiite in Brazil, as reported by Albertao and Martins (2002) and Albertao et al., (2008), this volume. These sedimentary structures often make it difficult to discriminate tsunamiites from the sediments of giant wind-driven waves. HCS has generally been considered as a combined result of wave and current action (e.g., Dott and Bourgeois, 1982; Yokokawa and Masuda, 1991). It should be noted here that the size (length) of a storm-induced wave is very small (short) compared with the length of a tsunami wave. Tsunami water moves like a current but not like a wave even just above the sea floor, as stressed earlier repeatedly. According to this point of view, the genetic interpretation of HCS and wave-ripple marks is still questionable in tsunami sedimentology. There are two hypothetical interpretations.

1. Some high-order (small-scale) waves can occur on the surface of a tsunami wave because of some minute irregular topography of the sea bottom and/or surrounding

394 Chapter 20

coast or because of stormy weather. This interpretation seems speculative. However, it is said that any kind of wave structure in nature can consist of waves of different orders.

2. The so-called hummocky (and swaley-like) structures of tsunamiites are actually hummock-mimics (HCS mimics: Rust and Gibling, 1990) and antidunes in a hydrodynamic sense (Masuda et al., 1993). If this hypothesis is correct, indeed, there is nothing unusual in the occurrence of HCS-like and SCS-like structures in submarine tsunamiites. Hummock-mimics and antidunes are very noticeable sedimentary structures that indicate sedimentation under a very high-flow regimen. This idea, however, cannot be applied to ripple marks at all. Further investigations, including field research, are needed to solve this puzzle concerning the occurrence of storm-wave-induced structures.

Are any well-known sedimentary structures lacking in tsunamiites? Epsilon cross-stratification is a typical example of such a structure. This structure is formed by the lateral accretion of sediments connected with the meandering of a continuous flow (stream) and sediment supply for a relatively long time (order from an hour to some years). Sedimentary conditions like this might occur at the time of backflow of tsunami water from deep inland, but it would be very exceptional.

In some special cases, the passing of a tsunami wave is recorded only by giant erosion surfaces without any sediment deposition as stated earlier. The genetic processes of the giant surface and the meaning of the processes can be recognized and explained only through detailed studies of the sediment succession in the study area.

In short, though there is no golden key for identifying tsunamiites and distinguishing them from other sediments, sedimentary structures do provide useful indications for recognizing possible tsunami-generated sediments.

4. Constituents of tsunamiites

Integrated sedimentological and palaeontological investigation of the constituents of sediments is very useful to identify tsunami deposits. The composition of tsunamiites depends on the sediment sources, even more markedly than many other kinds of sediments that reflect the process of erosion, transportation, and/or deposition. This feature is especially notable in onshore, coastal, and shallow-sea tsunami deposits due to the huge sediment supply from deeper, offshore areas.

Different aspects of the grain-size distribution between run-up sediments and backwash sediments were clarified for the 1993 Hokkaido Nansei-Oki earthquake tsunami deposits (Shigeno and Nanayama, 2002). The mixing of shallow sea and coastal continental sediments through the process of run-up was pointed out on the basis of the bimodal grain-size distribution of the Holocene tsunami deposits of the coast of Kamchatka (Minoura et al., 1996).

The dominant organic remains or fossils such as bivalves, foraminifers, diatoms, fragments of corals, and so on from offshore, including the shallow sea, are characteristics of run-up tsunami wave sediments. On the other hand, wood fragments, and some other material taken from land, can reveal backwash of tsunami waves. For example, strongly mixed molluscan assemblages from various habitats such as a mud bottom and a rocky shore are characteristics of tsunami deposits on the bottom of a bay in the Kanto region, Japan, mentioned earlier (Fujiwara et al., 2003a).

In short, careful layer-by-layer investigation of the clastic grains and the organic constituents in successive layers makes it possible to distinguish between the various sedimentary sources of the layers. Shuttle water movement may thus be proven, just like other features may indicate pulses of the tsunami waves and currents that deposited the sediments under investigation.

The special constituents of impact tsunamiites and volcanic tsunamiites have been dealt with in the previous chapters of this volume.

5. Tsunamiite features in various environments

It must be noted here that the descriptions and explanations of characteristic sedimentary features in specific environments presented in this section concern mainly seismic tsunamiites, except for an additional comment on the deep-sea sediments.

The seismic tsunami deposits that can develop in the various sedimentary environments, such as onshore, in lacustrine and submarine areas, were described in the previous chapters of this present volume. Some additional comments and discussions are given here.

5.1 Coastal plain

As explained by Sugawara et al. (2008, this volume), the intervals between tsunami wave trains become short as the wave trains propagate from the deep sea to a shallow sea. The shape of the tsunami wave changes and shows a gradual rise of the sea surface by a nonbreaking wave, surge of breaker or tidal bore (like a wall of water), depending on the delicate differences in hydrodynamic conditions. Water from each individual tsunami shape, however, can inundate coastal plains very far (Nott and Bryant, 2003; Soeda et al., 2004). This inundation is essentially caused by the long wavelength of the original tsunami and by the enormous volume of water that attends the wave movement, as mentioned earlier. Sedimentary features of coastal plain tsunami deposits have been illustrated by numerous researchers (e.g., Minoura and Nakaya, 1991; Nanayama et al., 2000; Nanayama and Shigeno, 2004; Fujino et al., 2008, this volume; Goto et al., 2008a, this volume). The sedimentary differences between a coastal tsunamiite and a storm deposit were discussed in detail by Nanayama et al. (2000).

396   Chapter 20

A single structureless sand layer may be the only record of an invasion of the most violent tsunami flow deep inland. On the other hand, along a coast (i.e., very close to the sea), a bed caused by a single tsunami event can consist of a stack of a few or more unit layers that document repeated run-up and backwash flows (Fig. 20.2). Various sedimentary structures, like cross-lamination or pebble imbrication, may be present in the units (Nanayama and Shigeno, 2004; Fujiwara, 2008, this volume).

The most apparent sign of tsunami inundation and sedimentation in coastal area is the occurrence of marine material such as offshore biogenic sediments. Such signatures are, however, also common in storm-wave-induced sediments.

The grain sizes of many coastal tsunamiites depend on the supplied sediments, which closely reflect the source material. The most interesting data on the grain sizes of tsunamiites in a coastal lowland were provided by Nanayama and Shigeno (2004). In some cases, giant boulders were removed and transported from a beach and were left onshore, as is documented by Keating et al. (2008). Worldwide examples of tsunami boulder deposits are provided by Scheffers (2008, this volume).

5.2 Coastal lacustrine basin

If there is some topographic barrier that is high enough to prevent an invasion of storm-induced wave deposits, and if the tidal water cannot enter the basin either, the lacustrine basin can provide an excellent and very convenient restricted condition for recognizing and identifying tsunami records. Therefore, a lot of studies have been, and will be, carried out in such modern and prehistorical coastal lakes around the world (e.g., Atwater and Moore, 1992; Clague and Bobrowsky, 1994; Minoura et al., 1994; Smoot et al., 2000; Tsuji et al., 1998). Tsunami deposits in a lacustrine environment are commonly found as a distinct sand layer intercalated within silty to clayey sediments. Kaga and Nanayama (2001) recognized five stacked divisions in a tsunami bed, namely in ascending order, a pebble layer, a sand layer with developed bedforms (sedimentary structures), a clouded rip-up clast layer, a fine alternation of silt and sand lamina, and a silt layer with carbonaceous matter. These units were covered by sediments of the nontsunami background sediments (Fig. 20.2).

The Storegga tsunami deposits of around 7000 BP in western Norway, studied by Bondevik et al. (1997, 2005), are an example of special interest regarding their sedimentary facies successions in several small coastal lakes of different height (altitude) and distance from the sea. The sediments change their facies from basin to basin, and they thin and decrease in grain size as they become positioned more inland. Suspended fine material settled after the withdrawal of the tsunami wave and formed a laminated gyttja layer in the saline and anoxic water in the lakes. These facies contain, or are composed of, various organic fragments or clasts of marine origin. In general, backwash tsunamiites may rest on the run-up tsunamiites.

5.3 Beach

Coastal beaches, including coastal bars, are the sites of daily erosion and deposition. A tsunami can suddenly change the topography and sedimentary bodies around these areas catastrophically, generally by heavy erosion. A large amount of beach and shore sediment can be eroded and transported by a tsunami both inland and offshore. Even giant boulders, distributed along a beach, can be removed and shifted inland. Readers are referred to the review articles of coastline sediments supplied by Dawson and Shi (2000), Scheffers and Kelletat (2003), and other contributions that exhibit the sedimentary features of the 2004 Indian Ocean tsunami (e.g., Fujino et al., 2008, this volume; Goto et al., 2008a, this volume).

5.4 Nearshore

Sedimentary phenomena caused by tsunami action near the shore and offshore regions are new, just developing, topics of research. As is well known, strong storm waves and currents can cause both erosion and deposition in regions such as coastal sand dunes and barrier zones, shore face, and the zone where ripple marks and dunes develop under common stormy and fair-weather conditions. Shore-face erosion and "ravinement" in sedimentary records have attracted interest in submarine sedimentology and stratigraphy. These phenomena must occur much more extensively and heavily because of tsunami activity.

Erosion surfaces by tsunamis can develop in extensive areas, as suggested by Rossetti et al. (2000). The surfaces extend over a much wider area than the area of common shore-face erosion (by storm waves). On the other hand, the gigantic energy of a tsunami can also cause deposition of huge amounts of material. These deposits may occur even in the shore-face environment where sediments are commonly eroded because of the prevailing high-energy conditions.

In short, large amounts of sediments and/or extended erosion surfaces can be remarkable records of tsunami activity around the region that includes the beach, nearshore and offshore environments.

5.5 Bottom of bays

Tsunamiites in bay floor have been described well; they include restricted bays (Massari and D'Alessandro, 2000), incised valleys (Takashimizu and Masuda, 2000), and the bottom of small bays (e.g., Fujiwara et al., 2000).

A depositional model of tsunamiites in a small bay has been described and discussed in detail by Fujiwara et al. (2003b) and by Fujiwara and Kamataki (2008, this volume).

Special attention was given to bedforms such as the vertical succession of sediment sheets, multiple-bed types that reflect the physical properties of the source sediment current, the reversal of current direction, and the long period of tsunami waves. Sedimentary structures such as imbrication of clastic particles and shell fragments, grading, herringbone structures, antidunes, and hummocky and swaley structures indicate the tsunami flow properties. Mud clasts are generally thought to be ripped up not only from the sea bottom but can also be supplied occasionally from side walls of drainage channels in coastal land areas. For further details, the readers are referred to Fujiwara and Kamataki (2008, this volume).

5.6 Shallow sea including continental shelf

Shallow seas, that is, the offshore and continental-shelf areas, are the environment in which the most characteristics and noteworthy features of tsunamiites, including shuttle movement, can develop (Fig. 20.3). Because of this reason, and because of the higher sedimentary preservation potential in shallow seas than onshore and in coastal areas and wave-dominated nearshore regions, shallow-marine environments can house tsunami

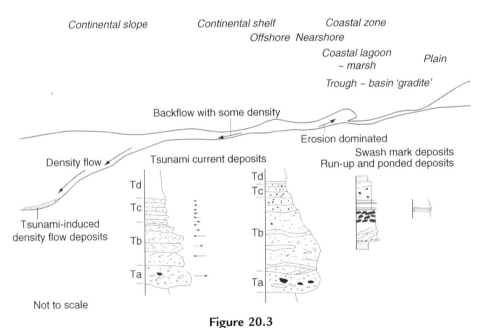

Figure 20.3

Schematic diagram of tsunamiite sedimentology and resulting sediments. Four typical histograms of tsunamiites are schematically illustrated to show the lateral change of the stacking pattern of sedimentary structures in accordance with various sedimentary settings. Ta, Tb, and so on show divisions that may or may not include several units described in the text. As far as the sedimentography of the deep-sea homogenite is concerned, readers are referred to the figures in Shiki and Cita (2008, this volume).

records. Tsunamiites will therefore be identified more often in these areas in future studies, though only a few examples have been studied as yet.

A sedimentary layer formed under an exceptionally high-energy regimen—such as exceptionally coarse sediments, mud clasts, and giant HCS and SCS structures—present among common background sediments can be recognized and may be assigned as tsunamiites there. This is exemplified for the Pliocene sediments in the Joban district, eastern Japan by Sasayama and Saito (2003), and for the Miocene Tanabe Group in central Japan (Matsumoto and Masuda, 2004). The sedimentary structures reveal a gravitation-induced inflow and rapid deposition onto the slightly sloping floor of the continental shelf. The very gentle slope or actually flat basal topography and rapid deposition from a gravity flow may in many cases be more important causes of the tsunamiite deposition on the continental shelf. On the other hand, an erosion surface may develop in some places in this environment also because of the very high-erosion energy of tsunami waves.

5.7 Deep-sea environments

In general, tsunamis neither transport nor deposit sediments in the deep sea, including bathyal areas, as far as submarine earthquakes are concerned. It is deduced, based on the long-wave equation, that the common seismic-tsunami current velocity does not exceed several centimeters per second there (see Sugawara et al., 2008, this volume). However, the energy and the current velocity of a meteorite-impact tsunami can be much larger than that of a common seismic tsunami. Especially in the case of a gigantic oscillation of a tsunami flow, such as that of the K/T-boundary tsunami, the current velocity is enough to erode and transport the deep-sea sediments, as shown by Takayama et al. (2000), Tada et al. (2003) and Goto et al. (2008b, this volume).

The transportation of sediments by tsunami-triggered sediment gravity flows happens commonly in valleys and canyons on the continental slope (Fig. 20.3). The sediments transported by a gravity flow form a turbidite (in the wide sense of its meaning) on the deep-sea fan and the deep-sea floor. Fundamental similarities in the transportation mechanism of the tsunami-induced turbidite and other turbidites make it difficult to distinguish between these sediments. The former can be called "tsunamiite," but the others cannot. However, the shuttle nature of a tsunami may be revealed by faint fluctuations in grain size and composition, as suggested for the Mediterranean homogenites by Shiki and Cita (2008, this volume). Furthermore, the exceptionally structureless character of the homogenites probably reflects the deposition of the fine-grained particles from a diluted suspension cloud generated by a tsunami.

Apparently, many topics and questions remain for future studies of the deep-sea tsunamis and their deposits.

6. Conclusive remarks

Tsunamiites are generated by, or related to, tsunamis that have a gigantic energy that is exceptional for processes on Earth. Owing to this nature, a tsunamiite bed of one gigantic tsunami event can be traced almost all over the world and can provide an excellent key for correlating prehistoric and older strata. On the other hand, tsunamiites only occur sporadically, and they are very variable in occurrence and appearance.

The most noteworthy features that may help to identify tsunami deposits are their large wavelength with a long period and the nature of the water movement, namely the shuttle character of the water current. Because of its large wavelength, tsunami water can propagate deep into coastal plains and leave sandy sheet-flood deposits. Various sedimentary structures and other features can also occur owing to the "current" nature of the tsunami water movement in coastal lakes and, most typically, in shallow-sea sediments. These features are the vertical repetition of coarse sediment layers with mud drapes and various current structures such as imbrication, herringbone structures, and antidunes, which reflect alternating run-up and backwash. Scattered rip-up mud clasts and organic and nonorganic materials form a signature of the long-wave nature of a tsunami and its different sources. Hummocky and swaley structures occur in some submarine tsunamiites, though their tsunami-related mechanism is not understood well.

As to the nature of tsunami water movement in the form of a current, it must be noted that exceptionally gigantic tsunami currents induced by meteorite impacts in the geological past may have affected even ocean-floor sediments.

It should also be remarked that gravity also plays a great and sometimes deadly role on sloping bottoms in seas and lakes and that they erode and transport bottom sediments, thus forming a flow of density-current nature. After a tsunami has ceased its downward movement, it turns over landward because of its shuttle nature. Despite the reversal of tsunami water, however, the density current with its particles continues its downward movement over the bottom. Thus, such deep-sea tsunamiites can be regarded as a sort of turbidite in a broad sense. However, the presence of a thick, homogeneous (structureless) layer may be a special characteristic of the deep-sea tsunamiites in contrast to turbidites, which have various sedimentary structures.

Dawson and Shi (2000) correctly mentioned that the study of tsunamiites is still in its infancy, with many new facts awaiting discovery. Since then, the studies have developed considerably, however, and these are going to develop even more rapidly and successfully. We hope that the information and new ideas in the present volume will help to initiate further discussions and to assist in future research.

Acknowledgments

The authors owe much to Professors Masuda, Minoura, Tsuji, and many other colleagues in sedimentology, both in Japan and abroad, because of their help and support during their tsunamiite studies and the preparation of the present contribution. Discussions and new information obtained during scientific meetings about tsunamis and tsunami deposits, including that held in Seattle in July 2005, were very helpful to improve our knowledge and ideas. The authors thank all the persons who gave them helpful advices and encouragements.

References

Albertão, G.A., Martins Jr., P.P., 2002. Petrographic and geochemical studies in the Cretaceous—Tertiary boundary, Pernambuco—Paraiba basin, Brazil. In: Buffetaut, E., Koeberl, C. (Eds.), Geological and Biological Effects of Impact Events. Springer-Verlag, Berlin, pp. 167—196.

Albertão, G.A., Martins Jr., P.P., Marini, F., 2008. Tsunamiites—conceptual descriptions and a possible case at the cretaceous-tertiary boundary in the Pernambuco Basin, northeastern Brazil. In: Shiki, T., Tsuji, Y., Yamazaki, T., Minoura, K. (Eds.), Tsunamiites—Features and Implications. Elsevier, Amsterdam, pp. 217—250.

Atwater, B.F., Moore, A.L., 1992. A tsunami about 1000 years ago in Puget Sound, Washington. Science 258, 1614—1617.

Atwater, B.F., Musumi-Rokkaku, S., Satake, K., Tsuji, Y., Ueda, K., Yamguchi, D.K., 2005. The Orphan Tsunami of 1700; Japanese Clues to a Parent Earthquake in North America. Published jointly with University of Washington Press, Washington, DC, p. 133. U.S. Geological Survey Professional Paper 1707.

Bondevik, S., Svendsen, J.I., Mangerud, J., 1997. Tsunami sedimentary facies deposited by the Stregga tsunami in shallow marine basins and coastal lakes, Western Norway. Sedimentology 44, 1115—1131.

Bondevik, S., Lovholt, F., Harbitz, C., Mangerud, J., Dawson, A., Svendsen, J.I., 2005. The Storegga slide tsunami; comparing field observations with numerical simulations. Mar. Petrol. Geol. 22, 195—208.

Bouma, A.H., 1962. Sedimentology of Some Flysch Deposits. Elsevier, Amsterdam, p. 168.

Cita, M.B., 2008. Deep-sea homogenites: sedimentary expression of a prehistoric megatsunami in the eastern Mediterranean. In: Shiki, T., Tsuji, Y., Yamazaki, T., Minoura, K. (Eds.), Tsunamiites—Features and Implications. Elsevier, Amsterdam, pp. 185—202.

Cita, M.B., Aoisi, G., 2000. Deep-sea tsunami deposits triggered by the explosion of Santorini (3500yBP), Eastern Mediterranean. Sediment. Geol. 135, 181—203.

Clague, J.J., Bobrowsky, P.T., 1994. Tsunami deposits beneath tidal marsh on Vancouver Island, British Columbia. Geol. Soc. Am. Bull. 106, 1293—1303.

Coleman, P.J., 1968. Tsunamis as geological agents. J. Geol. Soc. Aust. 15, 267—273.

Dawson, A.G., Shi, S., 2000. Tsunami deposits. Pure Appl. Geomorphol. 157, 875—897.

Dott Jr., R.H., Bourgeois, J., 1982. Hummocky stratification: significance of its variable bedding sequence. Bull. Geol. Soc. Am. 93, 663—680.

Fujino, S., Naruse, H., Suphawajruksakul, A., Jarupongsakul, T., Murayama, M., Ichihara, T., 2008. Thickness and grain-size distribution of Indian Ocean tsunami deposits at Khao Lak and Phra Thong Island, South-Western Thailand. In: Shiki, T., Tsuji, Y., Yamazaki, T., Minoura, K. (Eds.), Tsunamiites—Features and Implications. Elsevier, Amsterdam, pp. 123—132.

Fujiwara, O., 2004. Sedimentological and paleontological character of tsunami deposits. Mem. Geol. Soc. Jpn. 58, 35—44 (in Japanese with English abstract).

Fujiwara, O., 2008. Bedforms and sedimentary structures characterizing tsunami deposits. In: Shiki, T., Tsuji, Y., Yamazaki, T., Minoura, K. (Eds.), Tsunamiites—Features and Implications. Elsevier, Amsterdam, pp. 51—67.

402   Chapter 20

Fujiwara, O., Kamataki, T., 2008. Tsunami depositional processes reflecting the waveform in a small bay: interpretation from the grain-size distribution and sedimentary structures. In: Shiki, T., Tsuji, Y., Yamazaki, T., Minoura, K. (Eds.), Tsunamiites—Features and Implications. Elsevier, Amsterdam, pp. 133–152.

Fujiwara, O., Masuda, F., Sakai, T., Irizuki, T., Fuse, K., 2000. Tsunami deposits in Holocene bay mud in Southern Kanto region, Pacific coast of central Japan. Sedimentology 135, 219–230.

Fujiwara, O., Kamataki, T., Fuse, K., 2003a. Genesis of mixed molluscan assemblages in the tsunami deposits distributed in Holocene drowned valleys on the southern Kanto region, East Japan. Quat. Res. 42, 389–412 (in Japanese with English abstract).

Fujiwara, O., Kamataki, T., Tamura, T., 2003b. Grain-size distribution of tsunami deposits reflecting the tsunami waveform—an example from the Holocene drowned valley on the southern Boso Peninsula, East Japan. Quat. Res. 42, 67–81 (in Japanese with English abstract).

Goto, K., 2008. The genesis of oceanic impact craters and impact-generated tsunami deposits. In: Shiki, T., Tsuji, Y., Yamazaki, T., Minoura, K. (Eds.), Tsunamiites—Features and Implications. Elsevier, Amsterdam, pp. 277–297.

Goto, K., Imamura, F., Keerthi, N., Kunthasap, P., Matsui, T., Minoura, K., Ruangrassamee, A., Sugawara, D., Supharatid, S., 2008a. In: Shiki, T., Tsuji, Y., Yamazaki, T., Minoura, K. (Eds.), Distribution and Significance of the 2004 Indian Ocean Tsunami Deposits: Initial Results from Thailand and Sri Lanka, Tsunamiites—Features and Implications. Elsevier, Amsterdam, pp. 105–122.

Goto, K., Tada, R., Tajika, E., Matsui, T., 2008b. Deep-sea tsunami deposits in the proto-Caribbean sea at the Cretaceous-Tertiary boundary. In: Shiki, T., Tsuji, Y., Yamazaki, T., Minoura, K. (Eds.), Tsunamiites—Features and Implications. Elsevier, Amsterdam, pp. 251–275.

Kaga, A., Nanayama, F., 2001. General characteristics of tsunami event deposits from coastal lacustrine sediments. In: Annual Meeting of the Sedimentological Society of Japan, pp. 20–22. Program and Abstracts, (in Japanese).

Keating, B.H., Wanink, M., Helsley, C.E., 2008. Introduction to a tsunami-deposits database. In: Shiki, T., Tsuji, Y., Yamazaki, T., Minoura, K. (Eds.), Tsunamiites—Features and Implications. Elsevier, Amsterdam, pp. 359–381.

Kon'no, E. (Ed.), 1961. Geological Observations of the Sanrilu Coastal Region Damaged by the Tsunami Due to the Chili Earthquake in 1960. 52 Contributions from the Institute of Geology and Paleontology Tohoku University, p. 40 (in Japanese).

Massari, F., D'Alessandro, A., 2000. Tsunami-related scour-and-drape undulations in Middle Pliocene restricted-bay carbonate deposits (Salento, south Italy). Sediment. Geol. 135, 265–281.

Masuda, F., Yokokawa, M., Sakamoto, T., 1993. HCS mimics in Pleistocene, tidal deposits of the Shimosa Group and flood deposits of the Osaka Group, Japan. J. Sed. Soc. Jpn. 39, 27–34 (in Japanese with English abstract).

Matsumoto, D., Masuda, F., 2004. Thick-bedded tsunami deposits of the shallow-marine and deep-sea environments. In: Fujiwara, O., Ikehara, K., Nanayama, F. (Eds.), Earthquake-Induced Event Deposits—From Deep-Sea to on Land, vol. 58. Mem. Geol. Soc. Jpn, pp. 99–110 (in Japanese with English abstract).

Middleton, G.V., Hampton, M.A., 1976. Subaqueous sediment transport and deposition by sediment gravity flows. In: Stanley, D.J., Swift, D.J.P. (Eds.), Marine Sediment Transport and Environmental Management. Wiley, New York, pp. 197–218.

Minoura, K., Nakaya, S., 1991. Traces of tsunami preserved in inter-tidal lacustrine and marsh deposits: some examples from northeast Japan. J. Geol. 99, 265–287.

Minoura, K., Nakaya, S., Uchida, M., 1994. Tsunami deposits in a lacustrine sequence of the Sanriku coast, Northeast Japan. Sediment. Geol. 89, 25–31.

Minoura, K., Gusiakov, V.G., Kurtbatov, A., Takeuti, S., Svendsen, J.I., Bondevik, S., Oda, T., 1996. Tsunami sedimentation associated with the 1923 Kamchatka earthquake. Sediment. Geol. 106, 145–154.

Minoura, K., Imamura, F., Sugawara, D., Kono, Y., Iwashita, T., 2001. The 869 Jogan tsunami deposit and recurrence interval of large-scale tsunami on the Pacific coast of Northeast Japan. J. Nat. Disaster Sci. 23 (2), 83–88.

Miyazaki, M., 1971. Tsunami and flood tide. In: Masuzawa, J. (Ed.), Physical Oceanology. Tokai University Press, Tokyo, pp. 255–325 (in Japanese).

Miyoshi, H., 1968. The tsunami of April 24, 1771. J. Seism. Soc. Jpn. Ser. 2 (21), 314–316 (in Japanese).

Miyoshi, H., Makino, K., 1972. The tsunami of April 24, 1771 (II). J. Seism. Soc. Jpn. Ser. 2 (25), 33–43 (in Japanese with English abstract).

Nanayama, F., Shigeno, K., 2004. An overview of onshore tsunami deposits in coastal lowland and our sedimentological criteria to recognize them. Mem. Geol. Soc. Jpn. 58, 19–33 (in Japanese with English abstract).

Nanayama, F., Shigeno, K., Satake, K., Shimokawa, S., Koitabashi, S., Miyasaka, S., Ishii, M., 2000. Sedimentary differences between the 1993 Hokkaido-naisei-oki tsunami and the 1959 Miyakojima typhoon at Taisei, southwestern Hokkaido, Northern Japan. Sediment. Geol. 135, 255–264.

Nishimura, Y., 2008. Volcanism-induced tsunamis and tsunamiites. In: Shiki, T., Tsuji, Y., Yamazaki, T., Minoura, K. (Eds.), Tsunamiites—Features and Implications. Elsevier, Amsterdam, pp. 163–184.

Nott, J., Bryant, E., 2003. Extreme marine inundations (tsunamis?) of Coastal Western Australia. J. Geol. 111, 691–706.

Rossetti, D.D., Goes, A.M., Truckenbrodt, W., Anaisse, J., 2000. Tsunami-induced large-scale scour-and-fill structures in late Albian to Cenomanian deposite of the Grajau Basin, Northeastern Brazil. Sedimentology 47, 309–323.

Rust, B.R., Gibling, D.A., 1990. Three dimensional antidunes as HCS mimics in a fluvial sandstone: the Pennsylvanian South Bar formation near Sydney, Nova Scotia. J. Sediment. Petrol. 60, 540–548.

Sasayama, T., Saito, T., 2003. Tsunami deposits in shelf successions of the Pliocene Sendai group in Joban district, Japan. In: Annual Meeting Sedimentological Socierty of Japan, pp. 42–43. Abstract, (in Japanese).

Satake, K., Shimizu, K., Tsuji, Y., Ueda, K., 1996. Time and size of a gigantic earthquake in Cascadia inferred from Japanese tsunami record of January 1700. Nature 379, 246–249.

Scheffers, A., 2008. Tsunami boulder deposits. In: Shiki, T., Tsuji, Y., Yamazaki, T., Minoura, K. (Eds.), Tsunamiites—Features and Implications. Elsevier, Amsterdam, pp. 299–317.

Scheffers, A., Kelletat, D., 2003. Sedimentologic and geomorphologic tsunami imprints worldwide—a review. Earth Sci. Rev. 63, 83–92.

Shigeno, K., Nanayama, F., 2002. Tsunami deposit. Earth Sci. 56, 209–210 (in Japanese).

Shiki, T., Cita, M.B., 2008. Tsunami-related sedimentary properties of Mediterranean homogenites as an example of deep-sea tsunamiite. In: Shiki, T., Tsuji, Y., Yamazaki, T., Minoura, K. (Eds.), Tsunamiites—Features and Implications. Elsevier, Amsterdam, pp. 203–215.

Shiki, T., Yamazaki, T., 1996. Tsunami-induced conglomerates in Miocene upper bathyal deposits, Chita Peninsula, central Japan. Sediment. Geol. 104, 175–188.

Smit, J., Roep, T.B., Alvarez, W., Mountanari, A., Claeys, P., Grajales-Nishimura, J.M., Bermudez, J., 1996. Coarse-grained, clastic sandstone complex at the K/T boundary around the Gulf of Mexico: deposition by tsunami waves induced by the Chicxulub impact? Geol. Soc. Am. Spec. Pap. 307, 151–181.

Smoot, J.P., Litwin, R.L., Bischoff, J.L., Lund, S.J., 2000. Sedimentary record of the 1872 earthquake and "tsunami" at Owens Lake, southwest California. Sediment. Geol. 135, 241–254.

Soeda, Y., Nanayama, F., Shigeno, K., Furukawa, F., Kumasaki, N., Ishii, M., 2004. Large prehistorical tsunami traces at the historical site of Kokutaiji temple and the Shionomigawa lowland Eastern Hokkaido. Significance of sedimentological and diatom analysis for identification of past tsunami deposits. Mem. Geol. Soc. Jpn. 58, 63–75 (in Japanese with English Abstract).

Sugawara, D., Minoura, K., Imamura, F., 2008. Tsunamis and tsunami sedimentology. In: Shiki, T., Tsuji, Y., Yamazaki, T., Minoura, K. (Eds.), Tsunamiites—Features and Implications. Elsevier, Amsterdam.

404 Chapter 20

Tada, R., Iturralde-Vinent, M.A., Matsui, T., Tajika, E., Oji, T., Goto, K., Nakano, Y., Takayama, H., Yamamoto, S., Roja-Consuegra, R., Kiyokawa, S., Garcia-Delgado, D., et al., 2003. K/T boundary deposits in the proto-Caribbean basin. Mem. Am. Assoc. Petrol. Geol. 79, 582−604.

Takashimizu, Y., Masuda, F., 2000. Depositional facies and sedimentary successions of earthquake-induced tsunami deposits in upper Pliocene incised valley fills, central Japan. Sediment. Geol. 135, 231−240.

Takayama, H., Tada, R., Matsi, T., Iturralde-Vinent, M.A., Oji, T., Tajika, E., Kiyokawa, S., Garci, D., Okada, H., Hasegawa, T., Toyoda, K., 2000. Origion of the Penalver Formation in Northwestern Cuba and its relationton to K/T boundary impact event. Sediment. Geol. 135, 295−320.

Tsuji, Y., Ueda, K., Satake, K., 1998. Japanese tsunami records from the January 1900 earthquake in the Cascadia subduction zone. Zishin (J. Seismic Soc. Japan) 51, 1−17 (in Japanese with English abstract).

Yamazaki, T., Shiki, T., 1988. Tsunami deposits—an example from the conglomerates of Murozaki group. Mon. Chikyu 10, 511−514 (in Japanese).

Yamazaki, T., Yamaoka, M., Shiki, T., 1989. Miocene offshore tractive current-worked conglomerate—Tsubutegaura, Chita Peninsula, central Japan. In: Taira, A., Masuda, M. (Eds.), Sedimentary Facies and the Active Plate Margin. Terra Scientific Publ., Tokyo, pp. 483−494.

Yokokawa, M., Masuda, F., 1991. Grain fabric of hummocky cross-stratification. J. Geol. Soc. Jpn. 97, 909−916.

CHAPTER 21

Sedimentology of tsunamiites reflecting chaotic events in the geological record — significance and problems

T. Shiki[1], T. Tachibana[2]
[1]Uji, Kyoto, Japan; [2]Soil Engineering Corporation, Higashi-ku, Okayama, Japan

1. Introduction

Environmental change is controlled by agents originating in three main sources: the cosmos, the Earth's interior, and the living organisms. Although the goddess "Gaia" (e.g., Lovelock, 1972, 1990) can live long because of the harmonious coexistence of these agents, she can also be harmed severely by a catastrophic action caused by any of these agents, including impacts of heavenly (cosmic) bodies, eruption of material from the Earth's interior, or a war among human beings.

In the geological past, some tsunamis caused by the impact of giant meteorites in an aqueous environment have been extremely devastating events. Submarine earthquakes and volcanic eruptions, as well as sudden coastal or submarine mass movements, have also been causes of regionally catastrophic tsunami events. These tsunami events have actually been recorded in the geological record in the form of tsunami-induced and tsunami-affected sediments, that is, tsunamiites. Thus, tsunamiites provide potentially valuable information about certain catastrophes and very episodic events that had worldwide effect, throughout the history of the Earth.

In the following sections, we will discuss the significance and usefulness of ancient tsunamiites and try to shed new light on this scientific field. A review of the state of the art of tsunamiite research is, however, not intended, as this is done in a few other chapters of this volume for more precise data processing. Only pre-Quaternary tsunamiites will be dealt with; recent, historical and prehistorical tsunamiites will be mentioned only in special cases.

Tsunamiites. https://doi.org/10.1016/B978-0-12-823939-1.00021-5
Copyright © 2021 Elsevier B.V. All rights reserved.

2. Tsunamiites as records of ancient events

Tsunamis can be triggered by various intense events in and around the sea. These are bolide impacts, submarine earthquakes, sudden coastal mass movements (e.g., submarine slumps), and violent submarine or subaerial volcanism in the sea. These events have been happening since the division of the Earth's surface into sea and land. Since that time, tsunamiites have possibly stored a lot of significant information about events in the Earth's past. The potential information about these events is not restricted to regional and global phenomena, but may also involve the cosmos.

2.1 Studies of impact-induced tsunamiites

The investigation of impact-induced tsunamiites is, for various reasons, particularly significant in studies of tsunamiites with global dimensions: (1) the impact of a giant cosmic body can cause a catastrophic change of the biosphere and the environment all over the Earth, including even the deep sea, and the influence of such an impact does not disappear for a very long time; and (2) the record of a catastrophic event, including a tsunamiite, can be distributed over extensive areas, thus forming a marker horizon for stratigraphic correlation. This characteristic of impact tsunamiites is especially important for the correlation of pre-Phanerozoic strata in which fossil records are absent or of little help. We hope that more investigations of the Precambrian in search for tsunamiites will be carried out in the future.

Bourgeois et al. (1988) reported about possible tsunami deposits at the Cretaceous/Paleogene (K/Pq) boundary in Texas, United States. This was the first explicit sedimentological report of an impact-induced tsunamiite. Since then, a vast number of studies have been carried out regarding possible impact-induced tsunamiites of various ages, though they are mainly focused on those related to the K/Pq boundary (Albertao and Martins, 1996, 2002; Alvarez et al., 1992; Martins et al., 2000; Smit, 1999; Smit et al., 1996; Takayama et al., 2000; and many others). Hassier et al. (2000) described bedforms produced ~2.6 Ga ago by an impact-generated tsunami, in a deep-marine shelf in Hamersley Basin, Western Australia. This is the oldest record of a possible tsunamiite documented up to date. There is no doubt that the study of impact-induced tsunamiites has developed into a highly interesting discipline within the geosciences, as shown by the review by Dypwik and Jansa (2003), and Goto (2008).

There are some problems, however, in the study of pre-Quaternary impact-induced tsunamiites. For example, severe arguments were raised about the K/Pq-boundary tsunamiites, mainly about the ages of the sediments (e.g., Keller et al., 2003; Montgomery et al., 1992). It seems that the debate about the identification of the K/Pq boundary could arise, at least partly, because a classical fundamental biostratigraphic rule was neglected.

This rule states, in short, that a geological age should be based on the disappearance of old fossils but not on the appearance of new taxa. As is well known, the flora had already changed by the late Cretaceous. Most probably, some typically Tertiary faunal elements had also already appeared during the latest Cretaceous. Thus, old (Cretaceous) and new (Paleogene) taxa must have coexisted before the K/Pq mass extinction, and many animal and plant species (of Cenozoic type) survived in spite of the awful changes in environmental conditions after the catastrophe. We suggest that this is why the bolide impact and the related tsunami are misjudged sometimes as earliest Paleogene events. The mass extinction of Mesozoic-type animals actually took place as a result of the meteorite impact, the tsunami, and the catastrophic environmental changes induced by the impact. The K/Pq boundary should therefore be defined based on the mass extinction.

In addition, it must be noted that the timescale of event sedimentation, including that caused by tsunamis, is quite different from that of the classical stratigraphy. A sedimentary layer of a few or more meters thick can be deposited in several minutes, which is less than the time that it took for the affected organisms to die. Actually, the depositional time is negligible in a geochronological sense. Some dinosaurs, for example, may have survived a few days or even several months after the meteorite impact, but they would eventually have died because of the environmental changes, including "cold summer" and lack of food. Their bodies may have been carried to the sea or to some other depositional sites by water currents, and they may have been come to rest there. This explains why their bones can be found in sediment layers above the K/Pq boundary tsunamiite layers.

In short, the classical law of biostratigraphy cannot be applied to event sediments. All "K/Pq boundary tsunami sediment layers," including the Ir-bearing horizon should be regarded as a single "boundary zone."

Despite all this, the mass extinction at the Mesozoic/Paleogene boundary should be examined more precisely in the new light of event sedimentology.

2.2 Earthquake-induced tsunamiites and tectonics in geological time

A number of earthquake-induced Holocene tsunamiites, including prehistorical and historical time, have been investigated and reported by Scheffers and Kelletat (2003). Most studies focus on the periodical occurrence of earthquake-induced tsunamiites. Some general characteristic sedimentary features of such tsunamiites will be given special attention here, based on these studies (e.g., Dawson and Shi, 2000; Fujiwara, 2004; Fujiwara and Kamataki, 2008, this volume; Nanayama et al., 2000; Shigeno and Nanayama, 2002). In some cases, such studies provide significant information about the tectonic history of the region under investigation. Interesting examples concern the uplift

408   Chapter 21

history of the Boso and Miura peninsulas, central Japan (Fujiwara et al., 1997), the topographic change associated with the great earthquake, and tsunamis that occurred in coastal Washington, north-western America (Atwater, 1987, 1997).

The study of ancient, geohistorical, earthquake-induced tsunamiites also provides sedimentary information about regional and local tectonics in the geological past. One example is the finding of the Miocene origin of a fault crossing an island arc. This fault is still active, as shown by the study of the Tsubutegaura tsunamiites, Chita Peninsula, Japan (Shiki and Yamazaki, 1996; Yamazaki et al., 1989). Duringer (1984) reported the possible occurrence of earthquake-induced tsunamiites from the Muschelkalk in mid-Europe. It was new and surprising information for most stratigraphers that submarine earthquakes could occur in the Muschelkalk Sea, which was believed to have been a calm sea surrounded by continents.

Ueda and Miyashiro (1974) were the first to propose a surprising idea "ridge subduction." Kinoshita and Ito (1986) and Kinoshita (1994) discussed in more detail whether two magmatic belts along the south-west Japan and the East Asian margin, ranging in age from Mesozoic to Paleogene, must be ascribed to the subduction of the Farallon-Izanagi and Kura-Pacific ridges. This tectonic and magmatic activity should then have caused violent earthquakes, volcanism, and tsunamis. No research has been carried out, however, to find either tsunami or earthquake evidence in the sedimentary record in connection with this ridge-subduction concept. This interesting subject deserves more attention in the future.

The most significant developments in the Earth Science since the plate-tectonics model have been the proposal of "plume tectonics" (e.g., Foulger et al., 2005; Hansen and Yuen, 1989; Kellogg et al., 1999; Kumazawa and Maruyama, 1994; Maruyama et al., 1994; Maruyama, 1994). As is well known, strong seismic activity happens (in combination with related tsunamis) during rapid plate subduction at the margin of the ocean related to "ocean spreading." This spreading at a gigantic scale is to be regarded essentially as caused by superplume rise from the bottom of the mantle beneath the ocean floor. For example, a lot of tsunami caused by earthquakes must have occurred along the coast of the paleo-Pacific ocean during the Cretaceous period, when a superplume rose up. The search for seismic tsunamiites, including seismic tsunami-induced turbidites of that time, is a very interesting subject for future studies (Shiki et al., 1998).

2.3 Volcanic-eruption-induced tsunamiites

Volcanic explosions along an arc/trench system generally occur in relation to the subduction of an oceanic plate. Tsunamis and related tsunamiites induced by these volcanic eruptions of common scale have been reviewed by Nishimura et al. (1999, Nishimura, this volume). Homogenites from the Bronze Age (3500 BP) in the eastern Mediterranean present a good example of volcanism-related tsunamiites and have been examined by Cita and others (e.g., Cita and Aloisi, 2000; Cita et al., 1996) and in the present volume.

The most famous gigantic volcanic tsunami in history was caused by the Krakatau eruption in the Sunda Strait in 1883. Its sediments were investigated by Carey et al. (2001) and by Van der Bergh et al. (2003). Volcanic tsunamiites that originated in the prehistory and before, however, have not yet been investigated and are left for future studies.

2.4 Submarine slides and tsunamiites

Submarine and subaerial slides and slumps can be triggered by bolide impacts, earthquakes, volcanic eruptions, and so on. A slide or slump into a water body can generate a tsunami. The 1958 Alaska tsunami, which was generated by the failure of a bay's wall due to an earthquake, is the most gigantic example (Miller, 1960). Several historic tsunamis generated by sector collapse of volcanoes have been documented in Japan. For example, a gigantic tsunami was generated by the collapse of the forepart of Mt. Unzen, in western Kyushu, in 1792 (Aida, 1975; Nishimura, 2008, this volume).

The run-up height distribution of some of these tsunamis has been discussed. However, precise sedimentological studies of slump-induced tsunamiites are rather rare. Dawson et al. (1988, 1991) reported characteristics of the sand layer attributed to the Storegga tsunami in north-eastern Scotland. This tsunami was generated by one of the world's largest submarine slides, the second Storegga slide, which took place on the Norwegian continental slope, $\sim$ 7000 BP (Bondevik et al., 1997).

In 2002, a scientific meeting was held in Amsterdam to discuss submarine slides and tsunamis. At the meeting, many presentations concerned both the identification of slide-generated tsunamis recorded in historical catalogs and numerical models of tsunami generation by submarine slumps and slides.

A new development in the study of submarine slide-induced tsunamis is the application of the results of the investigations into slide-generated tsunamis to the studies of both modern and ancient tsunamiites. However, the distinction between slide-generated tsunami sediments and tsunamiites caused directly by the tectonic rise or the subsidence of the sea floor is not easy, especially for areas far away from the slide mass.

3. Patterns of tsunamiite occurrence in time

Patterns in time, such as linear patterns, rhythms, episodic changes, explosive developments, chaotic changes, fluctuations, and fractals, thus representing various levels or orders, are found throughout in nature (e.g., Kruhl, 1994; Seydel, 1985; see Shiki, 1996). Sedimentary records of some of these phenomena have attracted much attention from geologists in the past few decades. Sequence stratigraphy and event sedimentology, including studies of meteorite-impact records, seismites, and tempestites, are clear examples (e.g., Dott, 1983, 1992, 1996; Einsele et al., 1991, 1996; Haq et al., 1987; Shiki,

1996; Vail et al., 1977). The sedimentary records of submarine tsunamis should reveal similar features in their time pattern.

Bolide impacts, submarine earthquakes, and violent volcanism in the sea are characterized by (1) a sudden, catastrophic nature and spectacle and (2) a temporary rise in frequency. The tsunami records of such events should reveal similar patterns in time. This could be helpful in carrying out more efficient research of tsunami records in ancient sediments. In the following, we will examine specifically the occurrence patterns of impact- and earthquake-induced tsunamiites, and we will provide our view with respect to tsunamiite occurrences in sedimentary successions.

3.1 Meteorite-impact frequency and tsunamiites

It was proposed by Furumoto (1991) that the meteorite-impact frequency (and probability) is of the order of 1.9 billion years (Fig. 21.1). Another cyclicity of a few hundreds of

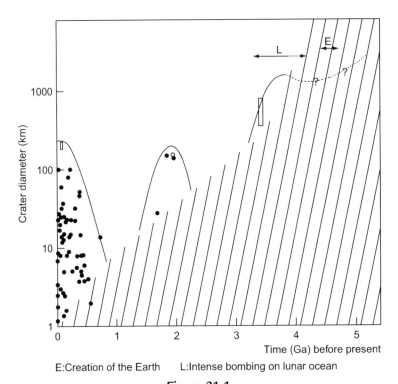

Figure 21.1

Scheme of the meteorite-impact frequency on the Earth—Moon system inferred from the age-size distribution of craters on the Earth, age of ocean genesis on the moon, and age of the solar system. *Solid circles*: craters; *open circle*: possible crater; *open rectangles*: inferred craters; *shaded area*: disappeared record. AE = Ga; E = creation of the Earth; L = intense bombing of lunar oceans. *Based on Furumoto, M., 1991. A rhythm of impact of cosmic bodies on the earth and moon. Month. Chikyu (Earth) 13, 531—535. (in Japanese).*

millions of years has been suggested by Alvarez and Muller (1984). Furthermore, a few examples of "end-of-a-geological-period meteorite-impact tsunamiites" have been recorded. For instance, Friedman (1998) reported on the conglomerates of the end-Cambrian impact and tsunami event in the Taconic sequence of eastern New York State, United States, and Schnyder et al. (2005) described a possible tsunami deposit of possible impact origin at the Jurassic/Cretaceous boundary in the Purbeckian of France. It could be deduced from these data, including the famous K/Pq-boundary tsunamiites record mentioned earlier, that these end-of-a-geological-period meteorite impacts and tsunamis reveal a cyclicity of impact frequency of a few or several million years.

These patterns, especially the 1.9 billion year cyclicity, can be explained by the movement of our solar system through the galaxy and by the movement of numerous small meteoritic bodies in our solar system as has already been suggested by Niitsuma (1973). It is said that the oceans on the moon were generated by the impact of gigantic cosmic bodies at around 3.8 Ga. According to Furumoto (1991), there are several craters on Earth that show another high frequency, namely of ~ 1.9 Ga. The impact at the K/Pq boundary (65 Ma) is within or top of the latest high-frequency phase of the cyclicity.

Sedimentary records of a gigantic tsunami induced by the K/Pq-boundary impact have been found at many locations and are well documented, as mentioned earlier and by Goto (2008, this volume). However, no evidence of the impact-induced tsunami for the 1.9 Ga high-frequency phase has been found as yet. We suggest that special features, such as remarkable sedimentary structures in the iron deposits of the time, should be looked for and carefully studied, including research for possible iridium enrichment or impact-related spherules.

3.2 Earthquake-induced tsunamiite occurrences and sea-level change

Earthquakes happen eventually because of a gradual accumulation of stress in the Earth's crust. Theoretically, it is impossible to predict the precise time of a failure. However, faulting and earthquakes do occur periodically. More precisely, (1) the probability of an earthquake to occur changes cyclically; (2) it is very well possible that the periods with a high earthquake frequency alternate with the intervals of low earthquake frequency.

The frequency of earthquake-induced tsunamiites (and the possibly cyclical variations in this frequency) should reflect—or at least be related to—the episodic occurrence of earthquakes and to the cyclical local or the regional seismicity in the stratigraphic records during the interval concerned. It should also tell about the tectonic history of the area. Remarkable prehistorical examples of such cyclical tsunamis were shown by the studies of tsunamiites in a drowned valley on the Boso Peninsula, east Japan (e.g., Fujiwara and Kamataki, 2008, this volume; Fujiwara et al., 2000), as well as by coastal tsunamiites in western Washington, north-western America (e.g., Atwater, 1997). A lot of recent research of onshore tsunami deposits along the Pacific coast of Kamchatka was also directed to find out the periodicities of tsunamis and strong earthquakes (e.g., Pinegina and Bourgeois, 2001).

As far as ancient (pre-Quaternary) tsunamis are concerned, however, only few examples of cyclical occurrences are known. One distinct example is the occurrence of tsunamiites in the Miocene upper-bathyal sediments of the Chita Peninsula, central Japan.

The question now arises what is the relationship between earthquake-induced tsunamiites and sea-level changes. In fact, the occurrence of an earthquake (with, in many cases, a related tsunami) reflects essentially only tectonism in the area concerned. For example, repeated occurrences of earthquake-induced tsunamis were shown by studies of tsunamiites along the active subduction zone of the western margin of the Pacific Plate, either at a lowstand stage or during highstand. These occurrences happened both before and after the Holocene maximum highstand (e.g., Fujiwara et al., 1999, 2000).

From a more general point of view, however, some relationship appears between eustatic sea-level changes and seismicity, especially in subduction areas, because both phenomena are related to global geotectonics. Global geotectonics thus can be the common cause of these two phenomena.

First-order eustatic changes with a cyclicity of ~400 Ma, called the "Chelegenic Cycle" by Sutton (1963) and global tectonics including extensive magmatism and sea-floor spreading show a close relationship, as was discussed by Worsley et al. (1984) and by many others (see Miall, 1990). The rising sea level coincides with times of magmatic events, including granite emplacement and continental breakup (and is correlatable with worldwide warm time spans), whereas falling sea levels coincide with continental aggradation (e.g., Vail et al., 1977, 1991) (Fig. 21.2). These global tectonics result in changes of seismicity.

In the context of the aforementioned, we want to mention that the continental breakup has been assigned to various causes, such as combined tidal effects caused by the gravitational pull of cosmic bodies in our galaxy and the upheaval of hot plumes caused by heat accumulation beneath the supercontinent that covers the upper mantle (Vail et al., 1991). The mechanism of these possible processes is beyond the scope of this volume. However, it must be noted that the upheaval of the hot superplume induces the mechanical and thermal elevation of young oceanic crust and sea floor (e.g., Maruyama et al., 1994, see aforementioned) with a cyclicity of about 400 Ma. Furthermore, the elevation of sea floor can be the cause of a global sea-level rise that makes the first-order sedimentary sequence (see, among others, Miall, 1990).

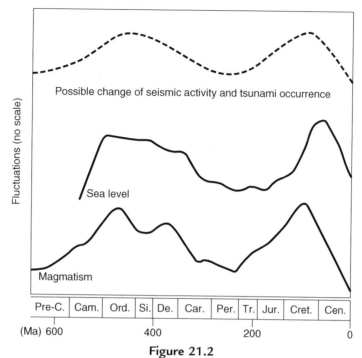

Figure 21.2
Possible changes in the frequency of earthquakes and earthquake-induced tsunamis and tsunamiites (*dashed line*) since the latest Precambrian. The *solid lines* show the first order of cyclicity of magmatism and sea-level change inferred or suggested by many authors. The frequency of earthquake-induced tsunamis roughly reflects these cyclical features because of causal relationship between such tsunamis due to subduction tectonics and uplift by a superplume.

In short, the cyclic upheaval of superplume induces severe subduction tectonics, submarine earthquakes, and tsunamis. On the other hand, the upheaval causes sea-level rise of the first order. Consequently, many seismic tsunamis and tsunamiites can occur during the first-order highstand sea-level time that happens with a cyclicity of about 400 Ma. For example, a lot of earthquakes and resulting tsunamis must have occurred along the coast of the paleo-Pacific Ocean during the Cretaceous time, when a superplume rose up and the sea level became high as mentioned earlier.

A similar relationship may, or may not, come about in some higher orders (smaller in scale) of sedimentary sequences. For example, Vail et al. (1977) showed in their global diagram second-order cycles of 30–60 Ma and third-order cycles of about 10 Ma. It is still being doubted whether these cycles have some connection with the frequency of seismic activity and whether these seismic features can be detected using tsunami records.

Strong tectonic activity and transgression, which imply a high sea level, during the Miocene (especially early to middle Miocene), are common concepts in Japanese geology, based on

414   Chapter 21

classical stratigraphy and structural geology. The high sea level seems contemporaneous with the relatively high global sea level (see Vail et al., 1977). On the other hand, thick accretionary fore-arc sediments were deposited during the Miocene along the Pacific and Philippine Sea coast of the north-eastern and south-western Japanese islands, showing active subduction (and accretion) tectonics. Consequently, a high frequency of tsunamiites is to be expected in the Miocene transgressive sediments around the Japanese islands. Many seismites, including sedimentary dikes associated with these tsunamiites, have been found in such Miocene sedimentary successions. The Tsubutegaura tsunamiites of the Chita Peninsula mentioned earlier and the tsunamiites documented from the Tanabe district (Matsumoto and Masuda, 2004), both in central Japan, are good examples of some tsunamiites associated with seismites, related to the Miocene transgression.

There is also the question concerning much higher-order sedimentary sequences and tsunamiite occurrences. An example of a periodical change in tsunamiite deposition frequency is evidenced by the interbedded succession of tsunamiite-rich units and background sediments in the Tsubutegura succession (Fig. 21.3). The "periodicity" of this tsunamiite "occurrence frequency" may reflect some periodical changes of tectonic activity that occurred along the Pacific (Philippine Sea) coast of the Japanese arc and trench. Sandy and conglomeratic tsunamiite layers alternate with (or are accompanied by) nontsunamiite layers, namely mudstone—sandstone layers and sandy seismites; they indicate periodically repeated tsunami activity (Shiki and Yamazaki, 1996). Furthermore, it has to be noted that the alternating tsunamiite and nontsunamiite layers form relatively thin sedimentary sets that are intercalated in thick muddy background beds. The mudstones in the "non-tsunamiite layers," which alternate with the tsunamiite layers, can also be regarded as "background," as they are similar in their facies to the "thick muddy background beds." No important facies change, indicating an environmental change related with sea-level change, has been observed among these "background" beds of different stratigraphic horizons. That is to say, the tsunamiite-rich unit exhibits a high-frequency seismic activity around the area (Tachibana et al., 2000). In short, cycles or periodicities of two orders are evident in this succession. However, the relationship between the "tsunamiite unit occurrence periodicity" of these orders and the sequence stratigraphy is not clear. It seems that the scale of the periodicity of the lower (bigger) order, that is, the periodicity of "occurrence frequency" in time and space, is too small in comparison to the known various orders of sequence stratigraphy that depend on the global sea-level change.

As to the fourth- and fifth-order cyclicities, there seems to be no relationship between earthquake-induced tsunamiite frequency and sedimentary sequence so that the correlation is weak or nonexistent. These cycles are believed to be glacioeustatic (Haq et al., 1987; Miall, 1990; Vail et al., 1977).

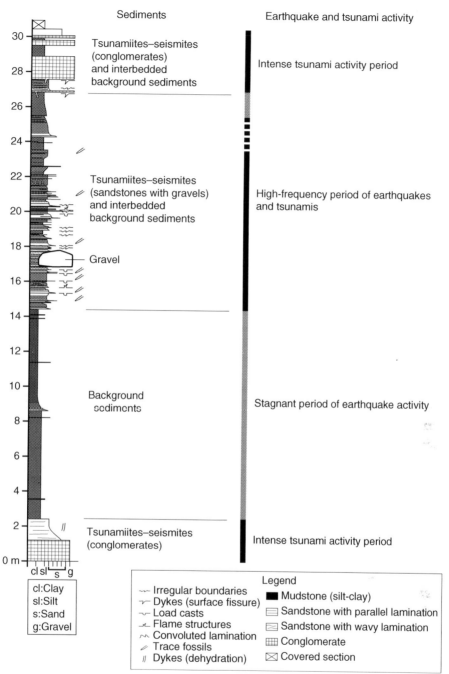

Figure 21.3

Representative columnar section showing two orders of earthquake-induced tsunamiite and nontsunamiite cycles. Tsunamiite—seismite layers contain fine-grained background deposits that jointly form a sedimentary unit exhibiting high-frequency seismic activity. This unit alternates with thick units of fine-grained background sediments. Any remarkable facies difference that reveals environmental change is not observed among these "background" sediments of different stratigraphic horizons. Miocene upper bathyal sediments, Tsubutegaura, Chita Peninsula, central Japan. *Based on Tachibana, T., Shiki, T., Yamazaki, T., 2000. Miocene tsunamiite and seismite succession, Tsubutegaura, Chita peninsula, Japan. Monthly Chikyu (Earth) 24, 724–729. (in Japanese).*

416   Chapter 21

It should be emphasized here that the arrival frequency of tsunamis at a specific coast is higher than the frequency of earthquakes near the coast, because tsunamis can come from many and remote sources. This was clear for the 1960 Chile tsunami, which attacked the Sanriku Coast of north-west Japan (Kato et al., 2002; Kon'no et al., 1961); it was also, and dramatically, exhibited by the damaged coast of Sri Lanka and India by the 2004 Indian Ocean tsunami.

Summarizing, though the frequency of earthquake-induced tsunamiites within a short-time range may not indicate any correlation with sea-level change or sedimentary sequences, it may reveal periods of sea-level highstand of low-order cycles. It has to be pointed out also that, though there is new interest in sequence stratigraphy in recent years (Porebski and Steel, 2003; Print and Nummedal, 2000), very little is discussed regarding the relationship between frequency change of tsunami events and the newly developed sequence stratigraphy. This relationship is a new and interesting subject in tsunamiite sedimentology and should be investigated in the future.

3.3 Slump-induced tsunamiites and sea-level change

It is commonly believed that submarine slides and slumps occur more frequently in a lowstand phase, especially in its beginning, than during highstand, because the fall of the sea level should degrade regional slope stability. Besides that, submarine slides can induce tsunamis, and tsunamis play a role in transporting and depositing materials. Therefore, we may assume that the probability of occurrence of slump-induced tsunamis tends to rise during the times of sea-level fall.

It must be noted, however, that tectonic activity is a more fundamental cause of tsunami generation. The relationship between tsunamiite generation by slumping and sea-level change may be masked by the stochastic nature of seismicity over a short time span and also by the strong effect of tectonic movements of big, low-order character. It is therefore suggested that the aforementioned relationship may be revealed only faintly, and only in some special cases of the fourth- to fifth-order cyclicity, but they may not be revealed in low-order frequency sequences. This question is also left for future study.

4. Preservation potential of tsunamiites

Difficulties and problems in tsunamiite sedimentology are discussed in detail in other chapters of this volume. In review, they are as follows: (1) a lack of geophysical studies on tsunamis, (2) the great variability of tsunamiites, and (3) the ephemeral nature of tsunamiites.

Any ancient tsunamiite deposited in any environment provides valuable information about a catastrophic or very episodic event that has occurred around the globe. Recent studies of many modern run-up (onshore) tsunamiites have clarified numerous interesting facts concerning the special features of such tsunamiites, as demonstrated in the previous chapters. Now, it is possible to apply the many achievements of research on modern tsunamiites to the study of ancient occurrences. However, one significant problem remains, namely that coastal sediments, including run-up tsunamiites, can disappear by erosion due to various natural agents. This means that their record is incomplete in many cases. In other words, ancient sediments rarely reveal only run-up events and then, perhaps, only partially.

In contrast, submarine tsunamiites have a much higher chance of preservation. In this context, we want to stress that submarine tsunamiites can present useful sedimentological information about tsunami events, regardless of whether they were impact-, earthquake-, or slide-induced, in or close to the sea.

It seems that earthquake-induced tsunamiites have a higher preservation potential in a bay or offshore basin with subsiding basement than onshore deposits do. Some good examples are the prehistorical tsunamiites on the Boso Peninsula, the Miocene Tsubutegaura tsunamiites, and the Miocene tsunamiites in the Tanabe district, Japan (Matsumoto and Masuda, 2004), mentioned earlier. Other possible tsunamiites have also been reported from the Paleogene bay sediments in the Joban "coal-seam-bearing" deposits in eastern Japan (Kamata, 1996) and from the Purbeckian in France (Schnyder et al., 2005).

There are a few examples of tsunamiites found in a shoreface environment. The tsunamiites from the Miocene Tanabe Group mentioned earlier and those from the Cretaceous Miyako Group (Fujino et al., 2006) are good examples. One great problem regarding the identification of tsunamiites in a shoreface environment is how to distinguish the effect of giant storm waves from that of tsunamis. In contrast with the shoreface environment, "shallow" seas, which are deep enough not to be affected by storm-generated waves and currents, are therefore interesting and deserve more future research regarding possible tsunami records. Sediments that are exceptionally thick and coarse-grained compared with other layers developed commonly in these environments may be tsunami-induced deposits. Actually, Pliocene tsunami deposits have been found in the shelf sediments of the Joban district, north-east Japan (Sasayama and Saito, 2003), and tsunamiites of some depth were also documented from the Mio-Pliocene sediments, Miyazaki, South Kyushu, Japan (Matsumoto and Masuda, 2004).

All tsunamiites referred to earlier in this section are earthquake-induced. It is to be pointed out that a submarine, offshore environment of sufficient depth offers possibilities for finding impact-induced tsunamiites. In fact, the depositional environment of the oldest-known impact-generated tsunamiite, from Western Australia, was assigned to a submarine environment deeper than storm wave base (Hassier et al., 2000). In addition, some K/Pq-boundary impact-induced tsunamiite beds also occur among the beds of similar submarine

418 Chapter 21

conditions, that is, middle to the outer shelf or somewhat deeper environments (Albertao and Martins, 1996; Bourgeois et al., 1988).

In summary, looking for in-bay and offshore basin sediments can be an efficient approach for finding a tsunamiite, as far as the present knowledge of ancient tsunamiites is concerned. In addition, we suggest that research targets may be settled on beds with some peculiar, exceptional sedimentary features in ancient molasses-type sediments (in their classical geological sense) because of their possible relatively high preservation of characteristic features of tsunamiites.

5. Conclusive remarks and future studies

The study of ancient, geohistorical, tsunamiites is still in its initial development. There are many questions and problems regarding future studies in this field, but future research is potentially truly promising.

The study of the sediments is very important for a better understanding of tsunami-recurrence patterns around the globe. Such research may, for instance, provide additional data on cyclical frequency changes of meteorite impacts and could help to unravel Earth's relative movement through our galaxy. Studies of seismic cyclicities, which have started for historical and prehistorical times, could be extended to include ancient cyclicities and may present important information about subduction tectonics and plume tectonics, including ridge subduction on both the western and the eastern side of the pre-Pacific Oceans during the Mesozoic and Paleozoic.

Further development of tsunamiite sedimentology will result in better identification criteria for ancient tsunamiites and will provide more extensive information for the investigation of the whole history of our Gaia.

Acknowledgments

The authors thank Professor Fujio Masuda, Kyoto University, for his stimulating advice and encouragement. This contribution was improved by the helpful comments from Prof. K. Goto, Professor van Loon, and other geologists who kindly reviewed our manuscript.

References

Aida, I., 1975. Numerical experiment of tsunami which followed 1792 collapse of Mt. Maruyama, Shimabara. Jishin (Earthquake) 2, 449–460 (in Japanese).

Albertao, G.A., Martins Jr., P.P., 1996. A possible tsunami deposit at the Cretaceous—tertiary boundary in Pernambuco, northeastern Brazil. Sediment. Geol. 104, 189–201.

Albertao, G.A., Martins Jr., P.P., 2002. Petrographic and geochemical studies in the cretaceous—tertiary boundary, Pernambuco—Paraiba basin, Brazil. In: Buffetaut, E., Koebert, C. (Eds.), Geological and

Biological Effects of Impact Events—Proceedings of the European Science Foundation "Impact Project" Meeting. Springer-Verlag, Berlin, pp. 167—196.

Alvarez, W., Muller, R.A., 1984. Evidence from crater ages for periodic impacts on the Earth. Nature 308, 718—780.

Alvarez, W., Smit, J., Lowrie, W., Asaro, F., Margolis, S.V., Claeys, P., Kastner, M., Hildebrand, A.R., 1992. Proximal impact deposits at the K-T boundary in the Gulf of Mexico, a restudy of DSDP Leg 77 sites 536 and 540. Geology 20, 697—700.

Atwater, B.F., 1987. Evidence for great earthquakes along the outer coast off Washington State. Science 236, 942—944.

Atwater, B.F., 1997. Coastal evidence for great earthquakes in western Washington. U.S. Geol. Surv. Prof. Pap. 1560, 77—90.

Bondevik, S., Sbendsen, J., Mangerud, J., 1997. Tsunami sedimentary facies deposited by the Storegga tsunami in shallow marine basins and coastal lakes, western Norway. Sedimentology 44, 1115—1131.

Bourgeois, J., Hansen, T.A., Wiberg, P.L., Kauffman, E.G., 1988. A tsunami deposit at the Cretaceous—tertiary boundary in Texas. Science 241, 567—570.

Carey, S., Morelli, D., Sigurdsson, H., Bronto, S., 2001. Tsunami deposits from major explosive eruptions: an example from the 1883 eruption of Krakatau. Geology 29, 347—350.

Cita, M.B., Aloisi, G., 2000. Deep-sea tsunami deposits triggered by the explosion of Santorini (3500yBP), Eastern Mediterranean. Sediment. Geol. 135, 18—203.

Cita, M.B., Camerlenghi, A., Rimoldi, B., 1996. Deep sea tsunami deposits in the Eastern Mediterranean: new evidence and depositional models. Sediment. Geol. 104, 155—173.

Dawson, A.G., Shi, S., 2000. Tsunami deposits. Pure Appl. Geophys. 157, 875—897.

Dawson, A.G., Long, D., Smith, D.E., 1988. The Storegga slides: evidence from Eastern Scotland for a possible tsunami. Mar. Geol. 19, 706—709.

Dawson, A.G., Foster, L.D.L., Shi, S., Smith, D.E., Long, D., 1991. The identification of tsunami deposition coastal sediment sequences. Int. J. Tsunami Soc. 9, 73—82.

Dott Jr., R.H., 1983. Episodic sedimentation—how normal is average? How rare is rare? Does it matter? J. Sediment. Petrol. 59, 5—23.

Dott Jr., R.H., 1992. T.C. Chamberlin's hypothesis of diastrophic control of worldwide change of sea level: a precursor of sequence stratigraphy. In: Dott Jr., R.H. (Ed.), Eustacy: The Historical Ups and Downs of a Major Geological Concept, Geol. Soc. Am. Mem, vol. 180, pp. 31—41.

Dott Jr., R.H., 1996. Episodic event deposits versus stratigraphic sequences—shall the twain never meet? Sediment. Geol. 104, 243—247.

Duringer, P., 1984. Tempêtes et tsunamis: Des dépôts de vague de haute énergie intermittente dass le Muschelkalk supérieur (Trias germanique) de l'est de la France. Bull. Soc. Géol. France 7, 1177—1185.

Dypwik, H., Jansa, L.F., 2003. Sedimentary signatures and processes during marine bolide impacts: a review. Sediment. Geol. 161, 309—337.

Einsele, G., Ricken, W., Seilacher, A., 1991. Cycles and events in stratigraphy—basic concepts and terms. In: Einsele, G., Ricken, W., Seilacher, A. (Eds.), Cycles and Events in Stratigraphy. Springer, Berlin, pp. 1—19.

Einsele, G., Chough, S.K., Shiki, T., 1996. Depositional events and their records—an introduction. Sediment. Geol. 104, 1—9.

Foulger, G.R., Natland, J.H., Presnall, D.C., Anderson, D.L., 2005. Plates, plumes and paradigms. Geol. Soc. Am. Spec. Pap. 388, 881.

Friedman, G.M., 1998. Conglomerates of Terminal Cambrian Ageeitness Bolide Impact and Tsunami Event in Taconic Sequence of Eastern New York State, USA, 15th Sediment. Congr. Alicante, Abstract 348.

Fujino, S., Masuda, F., Tagomori, S., Matsumoto, D., 2006. Structure and depositional process of gravelly tsunami deposit in shallow marine setting: lower Cretaceous, Japan. Sediment. Geol. 187 (3—4), 127—138.

Fujiwara, O., 2004. Sedimentological and paleontological characteristics of tsunami deposits. Mem. Geol. Soc. Jpn. 58, 35—44 (in Japanese with English abstract).

420 Chapter 21

Fujiwara, O., Kamataki, T., 2008. Tsunami depositional processes reflecting the waveform in a small bay—interpretation from the grain-size distribution and sedimentary structures. In: Shiki, T., Tsuji, Y., Yamazaki, T., Minoura, K. (Eds.), Tsunamiites—Features and Implications. Elsevier, Amsterdam, pp. 133—152.

Fujiwara, O., Masuda, F., Sakai, T., Fuse, K., Saito, A., 1997. Tsunami deposits in Holocene bay floor mud and the uplift history of the Boso and Miura peninsulas. Quat. Res. 42, 67—81 (in Japanese with English abstract).

Fujiwara, O., Masuda, F., Sakai, T., Irizuki, T., Fuse, K., 1999. Bay-floor deposits formed by great earthquake during the past 10,000 yrs, near the Sagami trough, Japan. Quat. Res. 38, 489—501 (in Japanese with English abstract).

Fujiwara, O., Masuda, F., Sakai, T., Irizuki, T., Fuse, K., 2000. Tsunami deposits in Holocene bay mud in southern Kanto region, Pacific coast of central Japan. Sediment. Geol. 135, 219—230.

Furumoto, M., 1991. A rhythm of impact of cosmic bodies on the earth and moon. Month. Chikyu (Earth) 13, 531—535 (in Japanese).

Goto, K., 2008. The genesis of oceanic impact craters and impact-generated tsunami deposits. In: Shiki, T., Tsuji, Y., Yamazaki, T., Minoura, K. (Eds.), Tsunamiites—Features and Implications. Elsevier, Amsterdam, pp. 277—297 (this volume).

Hansen, U., Yuen, D.J., 1989. Dynamical influences from thermal-chemical instability at the core-mantle boundary. Geophys. Res. Lett. 16, 629—632.

Haq, B.U., Hardenbol, J., Vail, P.R., 1987. Chronology of fluctuating sea levels since the Triassic (250 million years ago to present). Science 235, 1156—1167.

Hassier, S.W., Robey, H.F., Simonson, B.M., 2000. Bedforms produced by impact-induced tsunami, ~2.6 Ga in the Hammereley basin, Western Australia. Sediment. Geol. 135, 283—294.

Kamata, Y., 1996. Further subjects on researches of the tertiary system developed in the Joban district in Fukushima and Ibaragi prefecture, Japan. No. 2: formation of the Neogene system in the Joban district. Res. Rep. Taira Geol. Club 20, 1—9.

Kato, Y., Jahana, K., Shiroma, M., Watanabe, Y., 2002. The Chilian tsunami of 1960 in the Okinawa islands (summary). Hist. Earthquake 18, 188—189 (in Japanese).

Keller, G., Stinnesbeck, W., Adatte, T., Stuben, D., 2003. Multiple impacts across the cretaceous—tertiary boundary. Earth Sci. Rev. 62, 327—363.

Kellogg, L.H., Hager, B.H., Van der Hilst, R.D., 1999. Compositional stratification in the deep mantle. Science 283, 1881—1884.

Kinoshita, O., 1994. Migration of igneous activities related to ridge subduction in southwest Japan and East Asia continental margins from the Mesozoic to the Paleogene period. Tectonophysics 245, 25—35.

Kinoshita, O., Ito, H., 1986. Migration of Cretaceous igneous activity in Southwest Japan related to ridge subduction. J. Geol. Soc. Jpn. 92, 723—735 (in Japanese with English abstract).

Kon'no, E., Iwai, J., Takayanagi, Y., Nakagawa, H., Onuki, Y., Shibata, T., Mi, H., Kitamura, N., Kotaka, T., Kataoka, J., 1961. Geological observations of the Sanriku coastal region damaged by the tsunami due to the Chili earthquake in 1960. Contrib. Inst. Geol. Paleontol. 52, 1—40.

Kruhl, J.H., 1994. Fractals and Dynamic Systems in Geoscience. Springer-Verlag, Berlin, p. 421.

Kumazawa, M., Maruyama, S., 1994. Whole earth tectonics. J. Geol. Soc. Jpn. 100, 81—102.

Lovelock, J.E., 1972. Gaia as seen through the atmosphere. Atmos. Environ. 6, 579—580.

Lovelock, J.E., 1990. Hands up for the Gaia hypothesis. Nature 344, 100—102.

Martins Jr., P.P., Albertao, G.A., Haddad, R., 2000. The Cretaceous—tertiary boundary in the context of impact geology and sedimentary record—an analytical review of 10 years of researches in Brazil. Rev. Bras. Geosci. 30, 456—461.

Maruyama, S., 1994. Plume tectonics. J. Geol. Soc. Jpn. 100, 24—29.

Maruyama, S., Kumazawa, M., Kawakami, S., 1994. Towards a new paradigm on the Earth's dynamics. J. Geol. Soc. Jpn. 100, 1—3.

Matsumoto, D., Masuda, F., 2004. Thick-bedded tsunami deposits of the shallow-marine and deep-sea environments. Mem. Geol. Soc. Jpn. 58, 99—110 (in Japanese with English abstract).

Miall, A.D., 1990. Regional and global stratigraphic cycles. In: Miall, A.D. (Ed.), Principles of Sedimentary Basin Analysis, second ed. Springer-Verlag, Berlin, p. 668.

Miller, D.J., 1960. Shorter contribution to general geology—giant waves in Lituya Bay, Alaska. U.S. Geol. Surv. Prof. Pap. 354-c, 51—83.

Montgomery, H., Pessagno, E., Soegaard, K., Smith, C., Munoz, I., Pessagno, J., 1992. Misconceptions concerning the cretaceous/tertiary boundary at the Brazos river, falls County, Texas. Earth Planet Sci. Lett. 109, 593—600.

Nanayama, F., Smidgeon, K., Satake, K., Shimokawa, K., Koitabashi, S., Miyasaka, S., Ishi, M., 2000. Sedimentary differences between the 1993 Hokkaido-nansei-oki tsunami and the 1959 Miyakojima typhoon at Taisei, southwestern Hokkaido, northern Japan. Sediment. Geol. 135, 255—264.

Niitsuma, N., 1973. Rotation of the galaxy and history of the Earth. Kagaku (Science) 43, 650—651 (in Japanese).

Nishimura, Y., 2008. Volcanism-induced tsunamis and tsunamiites. In: Shiki, T., Tsuji, Y., Yamazaki, T., Minoura, K., et al. (Eds.), Tsunamiites—Features and Implications. Elsevier, Amsterdam, pp. 163—184 (this volume).

Nishimura, Y., Miyaji, N., Suzuki, M., 1999. Behavior of historic tsunamis of volcanic origin as revealed by onshore tsunami deposits. Phys. Geochem. Earth 24, 985—989.

Pinegina, T.K., Bourgeois, J., 2001. Historical and paleo-tsunami deposits on Kamchatka, Russia: long-term chronologies and long-distance correlations. Nat. Hazards Earth Syst. Sci. 1, 177—185.

Print, A.G., Nummedal, D., 2000. The falling stage systems tract: recognition and importance in sequence stratigraphic analysis. In: Hunt, D., Gawthorpe, R.I. (Eds.), "Sedimentary Response to Forced Regression" 172 Special Publ. Geol. Soc. London, pp. 1—17.

Porebski, S.J., Steel, R.J., 2003. Shelf-margin deltas: their stratigraphic signature and relation to deepwater sands. Earth Sci. Rev. 62, 283—326.

Sasayama, T., Saito, T., 2003. Tsunami deposits in shelf succession of the Pliocene Sendai Group in Joban district, Japan. In: Annual Meeting Sedimentological Soc. Japan Abstract, pp. 42—43.

Scheffers, A., Kelletat, D., 2003. Sedimentologic and geomorphologic tsunami imprints worldwide—a review. Earth Sci. Rev. 63, 83—92.

Schnyder, J., Baudin, F., Deconinck, J.-F., 2005. A possible tsunami deposit around the Jurassic—cretaceous boundary in the Boulonnais area (Northern France). Sediment. Geol. 177, 209—227.

Seydel, R., 1985. From Equilibrium to Chaos, Practical Bifurcation and Stability Analysis. Elsevier, Amsterdam, p. 367.

Shigeno, K., Nanayama, F., 2002. Tsunami deposit. Earth Sci. 56, 209—211 (in Japanese).

Shiki, T., 1996. Reading of the trigger records of sedimentary events—a problem for future studies. Sediment. Geol. 104, 249—255.

Shiki, T., Yamazaki, T., 1996. Tsunami-induced conglomerates in Miocene upper bathyal sediments, Chita Peninsula, central Japan. Sediment. Geol. 104, 175—188.

Shiki, T., Cita, M.B., Jones, B.G., 1998. Sedimentary features of seismites, seismo-turbidites and tsunamiites. In: Abstracts 15th Sediment. Congr. 721. Alicante.

Smit, J., 1999. The global stratigraphy of the Cretaceous—tertiary boundary impact ejecta. Annu. Rev. Earth Planet Sci. 27, 75—113.

Smit, J., Roep, T.B., Alvarez, W., Montanari, A., Clacys, P., Grajales-Nishimura, J.M., Bermudez, J., 1996. Coarse-grained, clastic sandstone complex at the K/T boundary around the Gulf of Mexico: deposition by tsunami waves induced by the Chicxulub impact? Geol. Soc. Am. Spec. Pap. 307, 151—182.

Sutton, J., 1963. Long-term cycles in the evolution of the continents. Nature 198, 731—735.

Tachibana, T., Shiki, T., Yamazaki, T., 2000. Miocene tsunamiite and seismite succession, Tsubutegaura, Chita Peninsula, Japan. Monthly Chikyu (Earth) 24, 724—729 (in Japanese).

Takayama, H., Tada, R., Matui, T., Iturralde-Vinent, M.A., Oji, T., Tjika, E., Kiyokawa, S., Garcia, D., Okada, H., Hasegawa, T., Toyoda, K., 2000. Origin of the Penalver Formation in northwestern Cuba and its relation to K/T boundary impact event. Sediment. Geol. 135, 295—320.

Ueda, S., Miyashiro, A., 1974. Plate tectonics and the Japanese Island: a synthesis. Geol. Soc. Am. Bull. 85, 1159–1170.

Vail, P.R., Mitchum Jr., R.M., Thompson III, S., 1977. Seismic stratigraphy and global changes of sea-level, part four; global cycles of relative changes of sea-level. Am. Assoc. Pet. Geol. Mem. 26, 83–98.

Vail, P.R., Andemard, F., Balman, S.A., Eisner, P.N., Perez-Cruz, C., 1991. The stratigraphic signatures of tectonics, eustacy, and sedimentology—an overview. In: Einsele, G., Ricken, W., Seilacher, A. (Eds.), Cycles and Events in Stratigraphy. Springer-Verlag, Berlin, pp. 617–659.

Van der Bergh, G.D., Boer, W., De Haas, W., Van Weering Tj, C.E., Van Wijhe, R., 2003. Shallow marine tsunami deposited in Teluk Banten (NW Java, Indonesia), generated by 1883 Krakatau eruption. Mar. Geol. 197, 12–34.

Worsley, W., Nance, D., Moody, J.B., 1984. Global tectonics and eustacy for the past 2 billion years. Mar. Geol. 58, 373–400.

Yamazaki, T., Yamaoka, M., Shiki, T., 1989. Miocene offshore tractive current worked conglomerates—Tsubutegaura, Chita peninsula, central Japan. In: Taira, A., Masuda, F. (Eds.), Sedimentary Facies and the Active Plate Margin. Terra Publ, Tokyo, pp. 483–494.

Selected bibliography

Only the papers and books published since the beginning of 2006, when the compilation in the first version of the book was arranged, are now listed in this bibliography. The study of tsunamiites, however, has grown explosively since the publication of the first edition of this book. It is therefore impossible to include every article and writing published since that time. We have only selected those articles which are most interesting to the editors and which are written in English. It is regrettable that the works written in languages other than English cannot be included, except a few printing. Furthermore, many papers which are cited in the individual chapters in the volume may have been omitted from this bibliography for the sake of keeping the numbers of pages to a minimum.

References

Abrantes, F., Alt-Epping, U., Lebreiro, S., Voelker, A., Schneider, R., 2008. Sedimentological record of tsunamis on shallow-shelf areas: the case of the 1969 AD and 1755 AD tsunamis on the Portuguese Shelf off Lisbon. Mar. Geol. 249, 283—293.

Arai, K., Naruse, H., Miura, R., Kawamura, K., Hino, R., Inazu, D., Yokokaw, M., Izumi, N., Nurayama, M., Kasaya, T., 2013. Tsunami-generated turbidity current of the 2011 Tohoku—Oki earthquake. Geology 41, 1195—1198.

Biolchi, S., Furlani, S., Antonioli, F., Baldassini, N., Causon Deguara, J., Devoto, S., Di Stefano, A., Evans, J., Gambin, T., Gauci, R., Mastronuzzi, G., Monaco, C., Scicchitano, G., 2016. Boulder accumulations related to extreme wave events on the eastern coast of Malta. Nat. Hazards Earth Sys. 16, 737—756.

Chagué-Goff, C., Andrew, A., Szczucinski, W., Goff, J., Nishimura, Y., 2012a. Geochemical signatures up to the maximum inundation of the 2011 Tohoku-oki tsunami - implication for the 869 AD Jogan and other paleotsunamis. Sediment. Geol. 282, 65—77.

Chagué-Goff, C., Szczucinski, W., Shinozaki, T., 2017. Applications of geochemistry in tsunami research: a review. Earth Sci. Rev. 165, 203—244.

Cox, R., Zentner, D.B., Kirchner, B.J., Cook, M.S., 2012. Boulder ridges on the Aran Islands (Ireland): recent movements caused by storm waves, not tsunamis. J. Geol. 120, 249—272.

Cox, R., Jahn, K.L., Watkins, O.G., Cox, P., 2018. Extraordinary boulder transport by storm waves (west of Ireland, winter 2013—2014), and criteria for analysing coastal boulder deposits. Earth Sci. Rev. 177, 623—636.

Dawson, A.G., Stewart, I., 2007. Tsunami deposits in the geological record. Sediment. Geol. 200, 166—183.

Dewey, J.F., Ryan, P.D., 2017. Storm, rogue wave, or tsunami origin for megaclast deposits in western Ireland and North Island, New Zealand? Proc. Natl. Acad. Sci. U.S.A. 114, E10639—E10647.

Engel, M., May, S.M., 2012. Bonaire's boulder fields revisited: evidence for Holocene tsunami impact on the Leeward Antilles. Quat. Sci. Rev. 24, 126—141.

424 Selected bibliography

Engel, M., Brückner, H., Wennrich, V., Scheffers, A., Dieter, K., Vott, A., Schabita, F., Danp, G., Wiylershauser, T., 2010. Coastal stratigraphies of eastern Bonaire (Netherlands Antilles): new insights into the palaeo-tsunami history of the southern Caribbean. Sediment. Geol. 231, 14—30.

Etienne, S., Buckley, M., Paris, R., Nandasena, A.K., Clark, K., Strotz, L., Chagué-Goff, C., Goff, J., Richmond, B., 2011. The use of boulders for characterising past tsunamis: lessons from the 2004 Indian Ocean and 2009 South Pacific tsunamis. Earth Sci. Rev. 107, 76—90.

Feldens, P., Schwarzer, K., Sakuna, D., Szczucinski, W., Somgpongchaiykul, P., 2012. Sediment distribution on the inner continental shelf off Khao Lak (Thailand) after the 2004 Indian ocean tsunami. Earth Planets Space 64, 875—887.

Fritz, H.M., Borrero, J.C., Synolakis, C.E., Okal, E.A., Weiss, R., Titov, V.V., Jaffe, B.E., Foteinis, S., Lynett, P.J., Chan, I.-C., Liu, P.L.-F., 2011a. Insights on the 2009 South Pacific tsunami in Samoa and Tonga from field surveys and numerical simulations. Earth Sci. Rev. 107, 66—75.

Fujiwara, O., Kamataki, T., 2007. Identification of tsunami deposits considering the tsunami waveform: an example of subaqueous tsunami deposits in Holocene shallow bay on southern Boso Peninsula, central Japan. Sediment. Geol. 200, 295—313.

Gelfenbaum, G., Jaffe, B., 2003. Erosion and sedimentation from the 17 July, 1998 Papua New Guinea tsunami. Pure Appl. Geophys. 160, 1969—1999.

Goodman-Tchemov, B.N., Austin Jr., J.A., 2015. Deterioration of Israel's Caesarea Maritima's ancient harbor linked to repeated tsunami events identified in geophysical mapping of offshore stratigraphy. J. Archaeol. Sci. Rep. 3, 444—454.

Goodman-Tchemov, B.N., Dey, H.W., Reinhardt, E.G., McCoy, F., Mart, Y., 2009. Tsunami waves generated by the Santorini eruption reached Eastern Mediterranean shores. Geology 37, 943—946.

Goto, K., Chavanich, S.A., Imamura, F., Kunthasap, P., Matsui, T., Minoura, K., Sugawara, D., Yanagisawa, H., 2007. Distribution, origin and transport process of boulders deposited by the 2004 Indian Ocean tsunami at Pakarang Cape, Thailand. Sediment. Geol. 202, 821—837.

Goto, K., Kawana, T., Imamura, F., 2010a. Historical and geological evidences of boulders deposited by tsunamis, southern Ryukyu Islands, Japan. Earth Sci. Rev. 102, 77—99.

Goto, K., Miyagi, K., Kawamata, H., Imamura, F., 2010b. Discrimination of boulders deposited by tsunamis and storm waves at Ishigaki Island, Japan. Mar. Geol. 269, 34—45.

Goto, K., Okada, K., Imamura, F., 2010c. Numerical analysis of boulder transport by the 2004 Indian Ocean tsunami at Pakarang Cape, Thailand. Mar. Geol. 268, 97—105.

Goto, K., Chagué-Goff, C., Fujino, S., Goff, J., Jaffe, B., Nishimura, Y., Richmond, B., Sugawara, D., Szczuciński, W., Tappin, D.R., Witter, R., Yulianto, E., 2011. New insights of tsunami hazard from the 2011 Tohoku-oki event. Mar. Geol. 290, 46—50.

Goto, K., Chagué-Goff, C., Goff, J., Jaffe, B., 2012a. The future of tsunami research following the 2011 Tohoku-oki event. Sediment. Geol. 282, 1—13.

Goto, K., Fujima, K., Sugawara, D., Fujino, S., Imai, K., Tsudaka, R., Abe, T., Haraguchi, T., 2012b. Field measurements and numerical modeling for the run-up heights and inundation distances of the 2011 Tohoku-oki tsunami at Sendai Plain, Japan. Earth Planets Space 64, 1247—1257.

Goto, K., Sugawara, D., Ikema, S., Miyagi, T., 2012c. Sedimentary processes associated with sand and boulder deposits formed by the 2011 Tohoku-oki tsunami at Sabusawa Island, Japan. Sediment. Geol. 282, 188—198.

Goto, K., Fujino, S., Sugawara, D., Nishimura, Y., 2014a. The current situation of tsunami geology under new policies for disaster countermeasures in Japan. Episodes 37, 258—264.

Goto, K., Hashimoto, K., Sugawara, D., Yanagisawa, H., Abe, T., 2014b. Spatial thickness variability of the 2011 Tohoku-oki tsunami deposits along the coastline of Sendai Bay. Mar. Geol. 358, 38—48.

Goto, K., Satake, K., Sugai, T., Ishibe, T., Harada, T., Murotaki, S., 2015. Historical tsunami and storm deposits during the last five centuries on the Sanriku coast, Japan. Mar. Geol. 367, 105—117.

Goto, K., Hongo, C., Watanabe, M., Miyazaki, K., Hisamatsu, A., 2019a. Large tsunamis reset growth of massive corals. Prog. Earth Planet. Sci. 6, 14.

Hawkes, A.D., Bird, M., Cowie, S., Grundy-Warr, C., Horton, B.P., Shau Hwai, A.T., Law, L., Macgregor, C., Nott, J., Ong, J.E., Rigg, J., Robinson, R., Tan-Mullins, M., Sa, T.T., Yasin, Z., Aik, L.W., 2007. Sediments deposited by the 2004 Indian ocean tsunami along the Malaysia-Thailand peninsula. Mar. Geol. 242, 169−190.

Hoffmeister, D., Ntageretzis, K., Aasen, H., Curdt, C., Hadler, H., Willershäuser, T., Bareth, G., Brückner, H., Vött, A., 2014. 3D model-based estimations of volume and mass of high-energy dislocated boulders in coastal areas of Greece by terrestrial laser scanning. Z. Geomorphol. Suppl. Issues 58 (3), 115−135.

Ikehara, K., Irino, T., Usami, K., Jenkins, A., Omura, A., Ashi, J., 2014. Possible submarine tsunami deposits on the outer shelf of Sendai Bay, Japan resulting from the 2-11 earthquake tsunami off the Pacific coast of Tohoku. Mar. Geol. 358, 120−127.

Imamura, F., Goto, K., Ohkubo, S., 2008. A numerical model for the transport of a boulder by tsunami. J. Geophys. Res. Oceans 113, C01008.

Inoue, T., Goto, K., Nishimura, Y., Watanabe, M., Iijima, Y., Sugawara, D., 2017. Paleo-tsunami history along the northern Japan trench: evidence from Noda Village, northern Sanriku coast, Japan. Prog. Earth Planet. Sci. 4 (1), 1−15.

Ioki, K., Tanioka, Y., 2016. Re-estimtion fault model of the 17th century great earthquake off Hokkaido using tsunami deposit data. Earth Planet Sci. Lett. 433, 133−136.

Ishimura, D., 2017. Re-examination of the age of historical and paleo-tsunami deposits at Koyadori on the Sanriku Coast, Northeast Japan. Geosci. Lett. 4, 11.

Ishimura, D., Miyauchi, T., 2015. Historical and paleo-tsunami deposits during the last 4000 years and their correlations with historical tsunami events in Koyadori on the Sanriku Coast, northeastern Japan. Prog. Earth Planet. Sci. 2 (1), 16.

Ishizawa, T., Goto, K., Yokoyama, Y., Miyairi, Y., Sawada, C., Nishimura, Y., Sugawara, D., 2017. Sequential radiocarbon measurement of bulk peat for high-precision dating of tsunami deposits. Quat. Geochronol. 41, 202−210.

Ishizawa, T., Goto, K., Yokoyama, Y., Miyairi, Y., Sawada, C., Takada, K., 2018. Reducing the age range of tsunami deposits by 14C dating of rip-up clasts. Sediment. Geol. 364, 334−341.

Jaffe, B.E., Goto, K., Sugawara, D., Richmond, B.M., Fujino, S., Nishimura, Y., 2012. Flow speed estimated by inverse modeling of sandy tsunami deposits: results from the 11 March 2011 tsunami on the coastal plain near the Sendai Airport, Honshu, Japan. Sediment. Geol. 282, 90−109.

Jaffe, B., Goto, K., Sugawara, D., Gelfenbaum, G., La Selle, S., 2016. Uncertainty in tsunami sediment transport modeling. J. Disaster Res. 11 (4), 647−661.

Jagodziński, R., Sternal, B., Szczuciński, W., Chagué-Goff, C., Sugawara, D., 2012. Heavy minerals in the 2011 Tohoku-oki tsunami deposits—insights into sediment sources and hydrodynamics. Sediment. Geol. 282, 57−64.

Kelletat, D., Scheffers, S.R., Scheffers, A., 2007. Field signatures of the S-E Asian mega-tsunami along the west coast of Thailand compared to Holocene paleo-tsunami from the Atlantic region. Pure Appl. Geophys. 164, 413−431.

Lau, A.Y.A., Etienne, S., Terry, J.P., Switzer, A.D., Lee, Y.S., 2014. A preliminary study of the distribution, sizes and orientation of large reef-top coral boulders deposited by extreme waves at Makemo Atoll, French Polynesia. J. Coastal Res. Spec. Iss. 70, 272−277.

Lorang, M.S., 2011. A wave-competence approach to distinguish between boulder and megaclast deposits due to storm waves versus tsunamis. Mar. Geol. 283, 90−97.

Maouche, S., Morhange, C., Meghraoui, M., 2009. Large boulder accumulation on the Algerian coast evidence tsunami events in the western Mediterranean. Marine Geol. 262, 96−104.

Matsumoto, D., Naruse, H., Fujino, S., Surphawajruksakul, A., Jarupongsakul, T., Sakakura, N., Murayama, M., 2008. Truncated flame structures within a deposit of the Indian Ocean Tsunami: evidence of synsedimentary deformation. Sedimentology 55, 1559−1570.

Matsumoto, D., Shimamoto, T., Hirose, T., Gunatilake, J., Wickramasooriya, A., Delile, J., Young, S., Rathnayake, C., Ranasooriya, J., Murayama, M., 2010. Thickness and grain-size distribution of the 2004 Indian Ocean tsunami deposits in Periya Kalapuwa Lagoon, eastern Sri Lanka. Sediment. Geol. 230, 95–104.

Medina, F., Mhammdi, N., Chiguer, A., Akil, M., Jaaidi, E., 2011. The Rabat and Larache boulder fields; new examples of high-energy deposits related to storms and tsunami waves in north-western Morocco. Nat. Hazards 59, 725–747.

Mhammdi, N., Medina, F., Kelletat, D., Ahmamou, M., Aloussi, L., 2008. Large boulders along the Rabat coast (Morocco); possible emplacement by the November 1st, 1755 A.D. tsunami. Sci. Tsunami Hazards 27, 17–30.

Milker, Y., Wilken, M., Schumann, J., Sakuna, D., Feldens, P., Schwarzer, K., Schmiedl, G., 2013. Sediment transport on the inner shelf off Khao Lak (Andaman Sea, Thailand) during the 2004 Indian Ocean tsunami and former storm events: evidence from foraminiferal transfer functions. Nat. Hazards Earth Syst. Sci. 13, 3113–3128.

Mori, N., Takahashi, T., The 2011 Tohoku Earthquaske Joint Survey Group, 2012. Nationwide post event survey and analysis of the 2011Tohoku earthquake tsunami. Coast Eng. J. 54 (1), 1250001-1-1250001-27.

Morton, R.A., Richmond, B.M., Jaffe, B.E., Gelfenbaum, G., 2008. Coarse-clast ridge complexes of the Caribbean: a preliminary basis for distinguishing tsunami and storm wave origins. J. Sediment. Res. 78, 624–637.

Namegaya, Y., Satake, K., 2014. Reexamination of the A.D. 869 Jogan earthquake size from tsunami deposit distribution, simulated flow depth, and velocity. Geophys. Res. Lett. 41, 2297–2303.

Nandasena, N.A.K., Paris, R., Tanaka, N., 2011a. Reassessment of hydrodynamic equations: minimum flow velocity to initiate boulder transport by high energy events (storms, tsunamis). Mar. Geol. 281, 70–84.

Nandasena, N.A.K., Paris, R., Tanaka, N., 2011b. Numerical assessment of boulder transport by the 2004 Indian ocean tsunami in Lhok nga, west Banda Aceh (Sumatra, Indonesia). Comput. Geosci. 37, 1391–1399.

Papadopoulos, G.A., Gràcia, E., Urgeles, R., Sallares, V., De Martini, P.M., Pantosti, D., González, M., Yalciner, A.C., Mascle, J., Sakellariou, D., Salamon, A., Tinti, S., Karastathis, V., Fokaefs, A., Camerlenghi, A., Novikova, T., Papageorgiou, A., 2014. Historical and pre-historical tsunamis in the Mediterranean and its connected seas: geological signatures, generation mechanisms and coastal impacts. Mar. Geol. 354, 81–109.

Paris, R., Lavigne, F., Wassmer, P., Sartohadi, J., 2007. Coastal sedimentation associated with the December 26, 2004 tsunami in Lhok nga, west Banda Aceh (Sumatra, Indonesia). Mar. Geol. 238, 93–106.

Paris, R., Wassmer, P., Sartohadi, J., Lavigne, F., Barthomeuf, B., Desgages, E., Grancher, D., Baumert, P., Vautier, F., Brunstein, D., Gomez, C., 2009. Tsunamis as geomorphic crises: lessons from the December 26, 2004 tsunami in Lhok nga, west Banda Aceh (Sumatra, Indonesia). Geomorphology 104, 59–72.

Paris, R., Fournier, J., Poizot, E., Etienne, S., Morin, J., Lavigne, F., Wassmer, P., 2010. Boulder and fine sediment transport and deposition by the 2004 tsunami in Lhok Nga (western Banda Aceh, Sumatra, Indonesia): a coupled offshore-onshore model. Mar. Geol. 268, 43–54.

Pilarczyk, J.E., Hoton, B.P., Witter, R.C., Vane, C.H., Chgue-Goff, C., Goff, J.R., 2012. Sedimentary and foramuniferal evidence of the 2011 Tohoku-oki tsunami on the Sendai coastal plain, Japan. Sediment. Geol. 282, 78–89.

Polonia, A., Bonatti, E., Camerlenghi, A., Lucchi, R.G., Panieri, G., Gasperini, I., 2013. Mediterranean megaturbidite triggered by the AD 365 Cretan earthquake and tsunami. Sci. Rep. 3, 128.

Polonia, A., Vaiani, S.C., Nelson, C.H., Romano, S., Gasparotto, G., Gasprotto, L., 2015. Generation of Tsunamite seismic-turbidites in the Ionian Sea (Mediterranean Basin). In: AGU Fall Meeting Abstracts.id. NH21E-04.

Quintela, M., Costa, P.J.M., Fatela, F., Drago, T., Hoska, N., Andrade, C., Freitas, M.C., 2016. The AD 1755 tsunami deposits onshore and offshore of Algarve (south Portugal): sediment transport interpretations based on the study of Foraminifera assemblages. Quatern. Int. 408, 123–138.

Richmond, B., Szczuciński, W., Chagué-Goff, C., Goto, K., Sugawara, D., Witter, R., Tappin, D.R., Jaffe, B., Fujino, S., Nishimura, Y., Goff, J., 2012. Erosion, deposition and landscape change on the Sendai coastal plain, Japan, resulting from the March 11, 2011 Tohoku-oki tsunami. Sediment. Geol. 282, 27–39.

Sakuna, D., Szczucinski, W., Feldens, P., Schwarzer, K., Khokiattiwong, S., 2012. Sedimentary deposits left by the 2004 Indian Ocean tsunami on the inner continental shelf offshore of Khao Lak, Andaman Sea (Thailand). Earth Planets Space 64, 931–943.

Sakuna, D., Feldens, P., Schwarzer, K., Khokiattiwong, S., Stattegger, K., 2015. Internal structure of event layers preserved on the Andaman Sea continental shelf, Thailand: tsunami vs. storm and flashflood deposits. Nat. Hazards Earth Syst. Sci. 15, 1181–1199.

Sawai, Y., Fujii, Y., Fujiwara, O., Kamataki, T., Komatsubara, J., Okamura, Y., Shishikura, M., 2008. Marine incursion of the past 1500 years and evidence of tsunamis at Suijinmuura, a coastal lake facing the Japan Trench. Holocene 18, 517–528.

Sawai, Y., Jankaew, K., Martin, M.E., Prendergast, A., Choowong, M., Charoentitirat, T., 2009a. Diatom assemblages in tsunami deposits associated with the 2004 Indian Ocean tsunami at Phra Thong Island, Thailand. Mar. Micropaleontol. 73, 70–79.

Sawai, Y., Namegaya, Y., Okamura, Y., Satake, K., Shishikura, M., 2012. Challenges of anticipating the 2011 Tohoku earthquake and tsunami using coastal geology. Geophys. Res. Lett. 39, L21309.

Scheffers, A., 2006. Sedimentary impacts of Holocene tsunami events from the intraAmerican seas and southern Europe. Z. Geomorphol. 146, 7–37.

Scheffers, A., Kinis, S., 2014. Stable imbrication and delicate/unstable settings in coastal boulder deposits: indicators for tsunami dislocation? Quatern. Intern. 332, 73–84.

Scheffers, A., Kelletat, D., Vött, A., May, S.M., Scheffers, S., 2008b. Late Holocene tsunami traces on the western and southern coastlines of the Peloponnesus (Greece). Earth Planet Sci. Lett. 269 (1–2), 271–279.

Scheffers, S.R., Haviser, J., Browne, T., Scheffers, A., 2009. Tsunamis, hurricanes, the demise of coral reefs and shifts in prehistoric human populations in the Caribbean. Quatern. Intern. 195, 69–87.

Scheffers, A., Kelletat, D., Haslett, S., Scheffers, S., Browne, T., 2010. Coastal boulder deposits in Galway Bay and the Aran Islands, western Ireland. Z. Geomorphol. Suppl. Issues 54 (Issue 3), 247–279.

Scheffers, A.M., Engel, M., May, S.M., Scheffers, S.R., Joannes-Boyau, R., Hänssler, E., Kennedy, K., Kelletat, D., Brückner, H., Vött, A., Schellmann, G., Schäbitz, F., Radtke, U., Sommer, B., Willershäuser, T., Felis, T., 2014. Potential and limits of combining studies of coarse-and fine-grained sediments for the coastal event history of a Caribbean carbonate environment. In: Martini, I.P. (Ed.), Sedimentary Coastal Zones from High to Low Latitudes: Similarities and Differences, vol. 388. Geol. Soc. London Spec. Pub, pp. 503–531.

Scicchitano, G., Monaco, C., Tortorici, L., 2007. Large boulder deposits by tsunami waves along the Ionian coast of south-eastern Sicily (Italy). Mar. Geol. 238, 75–91.

Shinozaki, T., Fujino, S., Ikehara, M., Sawai, Y., Tamura, T., Goto, K., Sugawara, D., Abe, T., 2015a. Marine biomarkers deposited on coastal land by the 2011 Tohoku-oki tsunami. Nat. Hazards 77, 445–460.

Shinozaki, T., Goto, K., Fujino, S., Sugawara, D., Chiba, T., 2015b. Erosion of a paleo-tsunami record by the 2011 Tohoku-oki tsunami along the southern Sendai Plain. Mar. Geol. 369, 127–136.

Smedile, A., De Martini, P.M., Pantosti, D., Belluoci, L., Del Carto, P., Gasperni, L., Pirrotta, C., Polonia, A., Boschi, B., 2011. Possible tsunami signatures from an integrated study in the Augusta Bay offshore (Eastern Sicily – Italy). Mar. Geol. 281, 1–13.

Smedile, A., De Martini, P.M., Pantosti, D., 2012. Combining inland and offshore palaeotsunami evidence; The Augusta Basy (eastern Sicily, Italy) case study. Nat. Hazards Earth Syst. Sci. 12, 2557–2567.

Sugawara, D., 2017. The geological record of tsunamis in the Anthropocene. In: Yasuda, Y., Hudson, M.J. (Eds.), Multidisciplinary Studies of the Environment and Civilization: Japanese Perspectives. Routledge, New York, pp. 43–56.

Sugawara, D., 2018. Evolution of numerical modeling as a tool for predicting tsunami-induced morphological changes in coastal areas: a review since the 2011 Tohoku Earthquake. In: Santiago-Fandino, V., Sato, S.,

428 Selected bibliography

Maki, N., Iuchi, K. (Eds.), The 2011 Japan Earthquake and Tsunami: Reconstruction and Restoration. Insights and Assessment after 5 Years, Advances in Natural and Technological Hazards Research, vol. 47. Springer, Japan, pp. 451—467.

Sugawara, D., Goto, K., 2012. Numerical modeling of the 2011 Tohoku-oki tsunami in the offshore and onshore of Sendai Plain, Japan. Sediment. Geol. 282, 110—123.

Sugawara, D., Minoura, K., Nemoto, N., Tsukawaki, S., Goto, K., Imamura, F., 2009. Foraminiferal evidence of submarine sediment transport and deposition by backwash during the 2004 Indian Ocean tsunami. Isl. Arc 18, 513—525.

Sugawara, D., Goto, K., Imamura, F., Matsumoto, H., Minoura, K., 2012. Assessing the magnitude of the 869 Jogan tsunami using sedimentary deposits: prediction and consequence of the 2011 Tohoku-oki tsunami. Sediment. Geol. 282, 14—26.

Sugawara, D., Goto, K., Jaffe, B., 2014a. Numerical models of tsunami sediment transport — Current understanding and future directions. Mar. Geol. 352, 295—320.

Sugawara, D., Takahashi, T., Imamura, F., 2014b. Sediment transport due to the 2011 Tohoku-oki tsunami at Sendai: results from numerical modeling. Mar. Geol. 358, 18—37.

Szczuciński, W., 2012. The post-depositional changes of the onshore 2004 tsunami deposits on the Andaman Sea coast of Thailand. Nat. Hazards 60, 115—133.

Szczuciński, W., Kokociński, M., Rzeszewski, M., Chagué-Goff, C., Cachão, M., Goto, K., Sugawara, D., 2012b. Sediment sources and sedimentation processes of 2011 Tohoku-oki tsunami deposits on the Sendai Plain, Japan — Insights from diatoms, nannoliths and grain size distribution. Sediment. Geol. 282, 40—56.

Takashimizu, Y., Urabe, A., Suzuki, K., Sato, Y., 2012. Deposition by 2011 Tohoku-oki tsunami on coastal lowland controlled by beach ridges near Sendai, Japan. Sediment. Geol. 282, 124—141.

Tamura, T., Sawai, Y., Ikehara, K., Nakashima, R., Hara, J., Kanai, Y., 2015. Shallow-marine deposits associated with the 2011 Tohoku-oki tsunami in Sendai Bay, Japan. Quatern. Sci. 30, 293—297.

Tanaka, G., Naruse, H., Yamashita, S., Arai, K., 2012. Ostracodes reveal the sea-bed origin of tsunami deposits. Geophys. Res. Lett. 39, L05406.

Tappin, D.R., 2017. Tsunamis from submarine landslides. Geol. Today 33, 190—200.

Tappin, D.R., Evans, H.M., Jordan, C.J., Richmond, B., Sugawara, D., Goto, K., 2012. Coastal changes in the Sendai area from the impact of the 2011 Tohoku-oki tsunami: interpretations of time series satellite images, helicopter-borne video footage and field observations. Sediment. Geol. 282, 151—174.

Tappin, D.R., Grilli, S.T., Harris, J.C., Geller, R.J., Masterlark, T., Kirby, J.T., Shi, F., Ma, G., Thingbaijam, K.K.S., Mai, P.M., 2014. Did a submarine landslide contribute to the 2011 Tohoku tsunami? Mar. Geol. 357, 344—361.

Terry, J.P., Lau, A.Y.A., Etienne, S., 2013. Reef-platform Coral Boulders - Evidence for High-Energy Marine Inundation Events on Tropical Coastlines. Springer, Briefs in Earth Sciences, p. 105.

Tipmanee, D., Deelaman, W., Pongpiachan, S., Schwarzer, K., Sompongchaiyakul, P., 2012. Using Polycyclic Aromatic Hydrocarbons (PAHs) as a chemical proxy to indicate Tsunami 2004 backwash in Khao Lak coastal area, Thailand. Nat. Hazards Earth Syst. Sci. 12, 1441—1451.

Toyofuku, T., Duros, P., Fontanier, C., Mamo, B., Bichon, S., Buscail, R., Chabaud, G., Deflandre, B., Goubet, S., Gremare, A., Menniti, C., Fujii, M., Kawamura, K., Koho, K.A., Noda, A., Namegaya, Y., Oguri, K., Radakovitch, O., Murayama, M., de Nooijer, L.J., Kurasawa, A., Ohkawara, N., Okutani, T., Sakaguchi, A., Jorissen, F., Reichart, G.J., Kitazato, H., 2014. Unexpected biotic resilience on the Japanese seafloor caused by the 2011 Tohoku-Oki tsunami. Sci. Rep. 4, 7517.

Tyuleneva, N., Braun, Y., Katz, T., Suchkov, B., Goodman-Tchernov, B., 2018. A new chalcolithic-era tsunami event identified in the offshore sedimentary record of Jisr al-Zarka (Israel). Mar. Geol. 396, 67—78.

Udo, K., Sugawara, D., Tanaka, H., Imai, K., Mano, A., 2012. Impact of the 2011 Tohoku earthquake and tsunami on beach morphology along the northern Sendai coast. Coast Eng. J. 54 (Issue 1), 1250009.

Udo, K., Tanaka, H., Mono, A., Takeda, Y., 2013. Beach morphology change of southern Sendai Coast caused by the 2011 Tohoku, Earthquake Tsunami. J. Jpn. Soc. Civil Eng. Ser. B2 (Coastal Eng.) 69 (I), 1391-I—1395 (in Japanese with English abstract).

Usami, K., Ikehara, K., Kanamatsu, T., McHugh, C.M., 2018. Supercycle in great earthquake recurrence along the Japan Trench over the last 4000 years. Geosci. Lett. 5 (1), 1–12.

Weiss, R., Bahlburg, H., 2006. A note on the preservation of offshore tsunami deposits. J. Sediment. Res. 76, 1267–1273.

Williams, D.M., 2010. Mechanisms of wave transport of megaclasts on elevated cliff-top platforms: examples from western Ireland relevant to the storm-wave versus tsunami controversy. Irish J. Earth Sci. 28, 13–23.

Yamada, M., Fujino, S., Goto, K., 2014a. Deposition of sediments of diverse sizes by the 2011 Tohoku-oki tsunami at Miyako City, Japan. Mar. Geol. 358, 67–78.

Yamashita, K., Sugawara, D., Takahashi, T., Imamura, F., Saito, Y., Imato, Y., Kai, T., Uehara, H., Kato, T., Nakata, K., Saka, R., Nishikawa, A., 2016. Numerical simulations of large-scale sediment transport caused by the 2011 Tohoku earthquake tsunami in Hirota Bay, southern Sanriku coast. Coast Eng. J. 58 (4), 1640015.

Yoshikawa, S., Kanamatsu, T., Goto, K., Sakamoto, I., Yagi, M., Fujimaki, M., Imura, R., Nemoto, K., Sakaguchi, H., 2015. Evidence for erosion and deposition by the 2011 Tohoku-oki tsunami on the nearshore shelf of Sendai Bay, Japan. Geo Mar. Lett. 35 (4), 315–328.

Further reading

Aarup, T., Baptista, M.A., Costa, P.J.M., Matias, L.M., Pires, P., Santini, M., Santoro, F., Soddu, P.L., 2013. Tsunamis Preparedness Civil Protection - Good Practice Guide. IOC/UNESCO, p. 73.

Abad, M., Izquierdo, T., Caceres, M., Bernardez, E., Rodriguez-Vidal, J., 2020. Coastal boulder deposit as evidence of an ocean-wide prehistoric tsunami originated on the Atacama Desert coast (northern Chile). Sedimentology 67, 1505–1528.

Abadie, S.M., Harris, J.C., Grilli, S.T., Fabre, R., 2012. Numerical modelling of tsunami waves generated by the flank collapse of the Cumbre Vieja volcano (La Palma, Canary Isles): tsunami source and near field effects. J. Geophys. Res. 117, C05030.

Abbott, D., Bryant, E., Gusiakov, V., Masse, W., 2007. Megatsunami of the world ocean - did they occur in the recent past? EOS Trans. 88. PP42A-04.

Abbott, D., Gusiakov, V., Rambolamanana, G., Breger, D., Mazumder, R., Galinskaya, K., 2017. What are the origins of v-shaped (chevron) dunes in Madagascar? The case of their deposition by a Holocene megatsunami. In: Mazumder, R. (Ed.), Sediment Provenance - Influences on Compositional Change from Source to Sink. Elsevier, the Netherlands, pp. 155–182.

Abe, T., Goto, K., Sugawara, D., 2012. Relationship between the maximum extent of tsunami sand and the inundation limit of the 2011 Tohoku-oki tsunami on the Sendai Plain. Sediment. Geol. 282, 142–150.

Abe, T., Goto, K., Sugawara, D., 2020. Spatial distribution and sources of tsunami deposits in a narrow valley setting - insight from 2011 Tohoku-oki tsunami deposits in northeastern Japan. Prog. Earth Planet. Sci. 7 (1), 1–21.

Adityawan, M.B., Roh, M., Tanaka, H., Mano, A., Udo, K., 2012. Investigation of tsunami propagation characteristics in river and on land induced by the Great East Japan Earthquake 2011. J. Earthq. Tsunami 65 (3), 1250039.

Alam, E., Dominey-Howes, D., 2014. An analysis of the AD 1762 earthquake and tsunami in SE Bangladesh. Nat. Hazards 70, 903–933.

Alberico, I., Budillon, F., Casalbore, D., Di Fiore, V., Iavarone, R., 2018. A critical review of potential tsunamigenic sources as first step towards the tsunami hazard assessment for the Napoli Gulf (Southern Italy) highly populated area. Nat. Hazards 92, 43–76.

Alpar, B., Ünlü, S., Altınok, Y., Özer, N., Aksu, A., 2012. New approaches in assessment of tsunami deposits in Dalaman (SW Turkey). Nat. Hazards 60, 27–41.

Ambraseys, N., 2009. Earthquakes in the Mediterranean and Middle East. A Multidisciplinary Study of Seismicity up to 1900. Cambridge University Press, Cambridge, p. 947.

430 Selected bibliography

Ambraseys, N., Synolakis, C., 2010. Tsunami catalogs for the eastern Mediterranean, revisited. J. Earthq. Eng. 14, 309–330.

Ando, M., Nakamura, M., 2013. Seismological evidence for a tsunami earthquake recorded four centuries ago on historical documents. Geophys. J. Int. 195, 1088–1101.

Ando, M., Kitamura, A., Tu, Y., Ohashi, Y., Imai, T., Nakamura, M., Ikuta, R., Miyairi, Y., Yokoyama, Y., Shishikura, M., 2018. Source of high tsunamis along the southernmost Ryukyu trench inferred from tsunami stratigraphy. Tectonophysics 722, 265–276.

Andrade, C., Borges, P., Freitas, M.C., 2006. Historical tsunami in the Azores archipelago (Portugal). J. Volcanol. Geoth. Res. 156, 172–185.

Andrade, C., Freitas, M.C., Oliveira, M.A., Costa, P.J.M., 2016. On the sedimentological and historical evidences of seismic-triggered tsunamis on the Algarve coast of Portugal. In: Duarte, J. (Ed.), Natural Hazards and Plate Tectonics. AGU/Wiley, p. 335. ISBN 978-1-119-05397-2.

Apotsos, A., Gelfenbaum, G., Jaffe, B., Watt, S., Pack, B., Buckley, M., Stevens, A., 2011a. Tsunami inundation and sediment transport in a sediment – limited embayment in American Samoa. Earth Sci. Rev. 107, 1–11.

Apotsos, A., Jaffe, B., Gelfenbaum, G., 2011b. Wave characteristic and morphologic effects on the onshore hydrodynamic response of tsunamis. Coast. Eng. 58, 1034–1048.

Aranguiz, R., Gonzalez, G., Gonzalez, J., Catalan, P.A., Cienfuegos, R., Yagi, Y., Okuwaki, R., Urra, L., Contreras, K., Del Rio, I., Rojas, C., 2016. The 16 September 2015 Chile tsunami from the post-tsunami survey and numerical modeling perspectives. Pure Appl. Geophys. 173, 333–348.

Araoka, D., Yokoyama, Y., Suzuki, A., Goto, K., Miyagi, K., Miyazawa, K., Matsuzaki, H., Kawahata, H., 2013. Tsunami recurrence revealed by Porites coral boulders in the southern Ryukyu Islands, Japan. Geology 41, 919–922.

Arcos, M.E.M., 2012. The AD 900-930 Seattle fault earthquake with a wider coseismic rupture patch and postseismic submergence: inference from new sedimentary evidence. Bul. Seismol. Soc. Am. 102 (3), 1079–1098.

Armesto, J., Ordonez, C., Alejano, L., Arias, P., 2009. Terrestrial laser scanning used to determine the geometry of a granite boulder for stability analysis purposes. Geomorphology 106, 271–277.

Atwater, B.F., Cisternas, M., Yulianto, E., Prendergast, A.L., Jankaew, K., Eipert, A.A., Fernando, W.I.S., Tejakusuma, I., Schiappacasse, I., Sawai, Y., 2013. The 1960 tsunami on beach-ridge plains near Maullín, Chile: landward descent, renewed breaches, aggraded fans, multiple predecessors. Andean Geol. 40 (3), 393–418.

Atwater, B.F., ten Brink, U.S., Cescon, A.L., Feuillet, N., Fuentes, Z., Halley, R.B., Nunez, C., Reinhardt, E.G., Roger, J.H., Sawai, Y., Spiske, M., Tuttle, M.P., Wei, Y., Weil-Accardo, J., 2017. Extreme waves in the British Virgin Islands during the last centuries before 1500 CE. Geosphere 13, 301–368.

Aung, T.T., Satake, K., Okamura, Y., Shishikura, M., Swe, W., Saw, H., Swe, T.L., Tun, S.T., Aung, T., 2008. Geologic evidence for three great earthquakes in the past 3400 years off Myanmar. J. Earthq. Tsunami 2, 259–265.

Australian Tsunami Research Centre, Donato, S.V., Reinhardt, E.G., Boyce, J.I., Pilarczyk, J.E., Jupp, B.P., 2009. Particle-size distribution of inferred tsunami deposits in Sur Lagoon, Sultanate of Oman. Mar. Geol. 257, 54–64.

Autret, R., Dodet, G., Suanez, S., Roudaut, G., Fichaut, B., 2018. Long-term variability of supratidal coastal boulder activation in Brittany (France). Geomorphology 304, 184–200.

Aydan, O., Tokashiki, N., 2018. Tsunami boulders and their implications on the potential for a mega-earthquake along the Ryukyu Archipelago, Japan. Bull. Eng. Geol. Environ. 78, 3917–3925.

Bahlburg, H., Spiske, M., 2012. Sedimentology of tsunami inflow and backflow deposits: key differences revealed in a modern example. Sedimentology 59, 1063–1086.

Bahlburg, H., Nentwig, V., Kreutzer, M., 2018. The September 16, 2015 Illapel tsunami, Chile - sedimentology of tsunami deposits at the beaches of La Serena and Coquimbo. Mar. Geol. 396, 43–53.

Bai, Y., Yamazaki, Y., Cheung, K.F., 2018. Amplification of drawdown and runup over Hawaii's insular shelves by tsunami N -waves from mega Aleutian earthquakes. Ocean Model. 124, 61–74.

Baranes, H.E., Woodruff, J.D., Wallace, D.J., Kanamaru, K., Cook, T.L., 2016. Sedimentological records of the C.E. 1707 Hōei Nankai trough tsunami in the Bungo channel, southwestern Japan. Nat. Hazards 84, 1185−1205.

Barbano, M.S., Pantosti, D., Gerardi, F., Vaiani, S.C., Gasperini, G., Gasprini, L., Nelson, C.H., 2014. Large boulders along the south-eastern coast of Sicily: storm or tsunami deposits. Mar. Geol. 275, 140−154.

Baumann, J., Chaumillon, E., Bertin, X., Schneider, J.L., Guillot, B., Schmutz, M., 2017. Importance of infragravity waves for the generation of washover deposits. Mar. Geol. 391, 20−35.

Beavan, J., Wang, X., Holden, C., Wilson, K., Power, W., Prasetya, G., Bevis, M., Kautoke, R., 2010. Nearsimultaneous great earthquakes at Tongan megathrust and outer rise in September 2009. Nature 466, 959−963.

Bellanova, P., Bahlburg, H., Nentwig, V., Spiske, M., 2016. Microtextural analysis of quartz grains of tsunami and non-tsunami deposits — a case study from Tirúa (Chile). Sediment. Geol. 343, 72−84.

Bellanova, P., Frenken, M., Reicherter, K., Jaffe, B., Szczucinski, W., Schwarzayer, J., 2020a. Authropogenic pollutants and biomarker for the identification of 2011 Tohoku-oki tsunami deposits (Japan). Mar. Geol. 422, 106117.

Bellanova, P., Frenken, M., Richmond, B., Schwarzbauer, J., La Selle, S., Griswold, F., Jaffe, B., Nelson, A., Reicherter, K., 2020b. Organic geochemical investigation of far-field tsunami deposits of the Kahana Valley, O'ahu, Hawai'i. Sedimentology 67, 1230−1248.

Benner, R., Browne, T., Brückner, H., Kelletat, D., Scheffers, A.M., 2010. Boulder transport by waves: progress in physical modeling. Z. Geomorphol. Suppl. Issues 54 (Issue 3), 127−146.

Bertrand, S., Doner, L., Akçer Ön, S., Sancar, U., Schudack, U., Mischke, S., Çagatay, M.N., Leroy, S.A.G., 2011. Sedimentary record of coseismic subsidence in Hersek coastal lagoon (Izmit bay, Turkey) and the late Holocene activity of the north Anatolian fault. G-cubed 12, 1−17.

Boesl, F., Engel, M., Eco, R.C., Galang, J.B., Gonzalo, L.A., Llanes, F., Quix, E., Helmut Brückner, H., 2020. Digital mapping of coastal boulders - high-resolution data acquisition to infer past and recent transport dynamics. Sedimentology 67, 1393−1410.

Bondevik, S., Stormo, S.K., Skjerdal, G., 2012. Green mosses date the Storegga tsunami to the chilliest decades of the 8.2 ka cold event. Quatanary Sci. Rev. 45, 1−6.

Bony, D., Marriner, N., Morhange, C., Kaniewski, D., Perincek, D., 2011. A high-energy deposit in the Byzantine harbour of Yenikapi, Istanbul (Turkey). Quat. Int. 266 (17), 117−130.

Boulton, S.J., Whitworth, M.R., 2018. Block and Boulder Accumulations on the Southern Coast of Crete (Greece): Evidence for the 365 CE Tsunami in the Eastern Mediterranean, vol. 456. Geological Society, London, pp. 105−125. Special Publications.

Bourgeois, J., 2009. Geological effect of records of tsunamis. Chapter 3. In: Bernard, E.N., Robinson, A.R. (Eds.), The Sea, Tsunami, vol. 15. Harvard University Press, pp. 55−91.

Bourgeois, J., MacInnes, B.T., 2010. Tsunami boulder transport and other dramatic effects of the 15 November 2006 central Kuril Islands tsunami on the island of Matua. Z. Geomorphol. Suppl. Issues 54 (Issue 3), 175−190.

Bourgeois, J., Weiss, R., 2009. "Chevrons" are not megatsunami deposits - a sedimentologic assessment. Geology 37, 403−406.

Bourgeois, J., Pinegina, T.K., Ponomareva, V., Zaretskaia, N., 2006. Holocene tsunamis in the southwestern Bering Sea, Russian Far East, and their tectonic implications. Geol. Soc. Am. Bull. 118, 449−463.

Bradák-Hayashi, B., Tanigawa, K., Hyodo, M., Seto, Y., 2017. Magnetic fabric evidence for rapid, characteristic changes in the dynamics of the 2011 Tohoku-oki tsunami. Mar. Geol. 387, 85−96.

Briggs, R.W., Engelhart, S.E., Nelson, A.R., Dura, T., Kemp, A.C., Haeussler, P.J., Corbett, D.R., Angster, S.J., Bradley, L.A., 2014. Uplift and subsidence reveal a nonpersistent megathrust rupture boundary (Sitkinak Island, Alaska). Geophys. Res. Lett. 41, 2289−2296.

Brill, D., Brückner, H., Jankaew, K., Kelletat, D., Scheffers, A., Scheffers, S., 2011. Potential predecessors of the 2004 Indian ocean tsunami—sedimentary evidence of extreme wave events at ban Bang Sak, SW Thailand. Sediment. Geol. 239, 146−161.

432 **Selected bibliography**

Brill, D., Klasen, N., Brückner, H., Jankaew, K., Kelletat, D., Scheffers, A., Scheffers, S., 2012. Local inundation distances and regional tsunami recurrence in the Indian Ocean inferred from luminescence dating of sandy deposits in Thailand. Nat. Hazards Earth Syst. Sci. 12, 2177–2192.

Brill, D., Reimann, T., Wallinga, J., May, S.M., Engel, M., Riedesel, S., Brückner, H., 2018. Testing the accuracy of feldspar single grains to date late Holocene cyclone and tsunami deposits. Quat. Geochronol. 48, 91–103.

Brill, D., Seeger, K., Pint, A., Reize, F., Hlaing, K.T., Seeliger, M., Opitz, S., Win, K.M.M., Nyunt, W.T., Aye, N., Aung, A., Kyaw, K., Kraas, F., Brückner, H., 2020. Modern and historical tropical cyclone and tsunami deposits at the coast of Myanmar: implications for their identification and preservation in the geological record. Sedimentology 67, 1431–1459.

Bruins, H.J., MacGillivray, J.A., Synolakis, C.E., Benjamini, C., Keller, J., Kisch, H.J., Klügel, A., van der Plicht, J., 2008. Geoarchaeological tsunami deposits at Palaikastro (Crete) and the late Minoan IA eruption of Santorini. J. Archaeol. Sci. 35, 191–212.

Bryant, E., 2014. Tsunami - The Underrated Hazard, second ed. Springer International Publishing, Berlin, Germany, p. 222.

Bryant, E.A., Haslett, S.K., 2007. Catastrophic wave erosion, Bristol channel, United Kingdom: impact of tsunami? J. Geol. 115, 253–269.

Butler, R., 2012. Re-examination of the potential for great earthquakes along the Aleutian Island arc with implications for tsunamis in Hawaii. Seismol. Res. Lett. 83, 29–38.

Butler, R., Burney, D., Walsh, D., 2014. Paleotsunami evidence on Kaua'i and numerical modeling of a great Aleutian tsunami. Geophys. Res. Lett. 41, 6795–6802.

Butler, R., Burney, D.A., Rubin, K.H., Walsh, D., 2017. The orphan Sanriku tsunami of 1586: new evidence from coral dating on Kaua'i. Nat. Hazards 88, 797–819.

Cabioch, G., Montaggioni, L., Frank, N., Seard, C., Salle, E., Payri, C., Pelletier, B., Paterne, M., 2008. Successive reef depositional events along the Marquesas foreslopes (French Polynesia) since 26 ka. Mar. Geol. 254, 18–34.

Carvajal, M., Cisternas, M., Catalan, P.A., 2017a. Source of the 1730 Chilean earthquake from historical records: implications for the future tsunami hazard on the coast of Metropolitan Chile. J. Geophys. Res. Solid Earth 122, 3648–3660.

Carvajal, M., Contreras-Lopez, M., Winckler, P., Sepulveda, I., 2017b. Meteotsunamis occurring along the southwest coast of south America during an intense storm. Pure Appl. Geophys. 147, 3313–3323.

Cascalho, J., Costa, P., Dawson, S., Milne, F., Rocha, A., 2016. Heavy mineral assemblages of the Storegga tsunami deposit. Sediment. Geol. 334, 21–33.

Chagué, C., Sugawara, D., Goto, K., Goff, J., Dudley, W., Gadd, P., 2018. Geological evidence and sediment transport modelling for the 1946 and 1960 tsunamis in Shinmachi, Hilo, Hawaii. Sediment. Geol. 364, 319–333.

Chagué-Goff, C., 2010. Chemical signatures of palaeotsunamis: a forgotten proxy? Mar. Geol. 271, 67–71.

Chagué-Goff, C., Schneider, J.L., Goff, J.R., Dominey-Howes, D., 2011. Expanding the proxy toolkit to help identify past events — lessons from the 2004 Indian Tsunami and the 2009 South Pacific Tsunami. Earth Sci. Rev. 107, 107–122.

Chagué-Goff, C., Goff, J., Nichol, S.L., Dudley, W., Zawadzki, A., Bennett, J.W., Mooney, S.D., Fierro, D., Heijnis, H., Dominey-Howes, D., Courtney, C., 2012b. Multi-proxy evidence for trans-Pacific tsunamis in the Hawai'ian Islands. Mar. Geol. 299–302, 77–89.

Chagué-Goff, C., Niedzielski, P., Wong, H.K.Y., Szczuciński, W., Sugawara, D., Goff, J., 2012c. Environmental impact assessment of the 2011 Tohoku-oki tsunami on the Sendai plain. Sediment. Geol. 282, 175–187.

Chagué-Goff, C., Goto, K., Goff, J., Jaffe, B., 2012d. The 2011 Tohoku-oki tsunami. Sediment. Geol. 282, 1–374.

Chagué-Goff, C., Wong, H.K.Y., Sugawara, D., Goff, J., Nishimura, Y., Beer, J., Szczuciński, W., Goto, K., 2014. Impact of tsunami inundation on soil salinisation: up to one year after the 2011 Tohoku-oki tsunami. In: Kontar, Y.A., Santiago-Fandiño, V., Takahashi, T. (Eds.), Tsunami Events and Lessons Learned - Environmental and Societal Significance, Advances in Natural and Technological Hazards Research, vol. 35. Springer, Japan, pp. 193–214.

Chagué-Goff, C., Goff, J., Wong, H.K., Cisternas, M., 2015. Insights from geochemistry and diatoms to characterise a tsunami's deposit and maximum inundation limit. Mar. Geol. 359, 22–34.

Chagué-Goff, C., Chan, J.C.H., Goff, J., Gadd, P., 2016. Late Holocene record of environmental changes, cyclones and tsunamis in a coastal lake, Mangaia, Cook Islands. Isl. Arc 25, 333–349.

Chagué-Goff, C., Goto, K., Sugawara, D., Nishimura, Y., Komai, T., 2018. Restoration measures after the 2011 Tohoku-oki tsunami and their impact on tsunami research. In: Santiago-Fandino, V., Sato, S., Maki, N., Iuchi, K. (Eds.), The 2011 Japan Earthquake and Tsunami: Reconstruction and Restoration. Insights and Assessment after 5 Years, Advances in Natural and Technological Hazards Research, vol. 47. Springer, Japan, pp. 229–247.

Choowong, M., Murakoshi, N., Hisada, K., Charusiri, P., Charoentitirat, T., Chutakositkanon, V., Phantuwongraj, S., 2008a. Indean ocean tsunami inflow and outflow at Phuket, Thaland. Mar. Geol. 248, 179–192.

Choowong, M., Murakoshi, N., Hisada, K.I., Charoentitirat, T., Charusiri, P., Phantuwongraj, S., Wongkok, P., Choowong, A., Subsayjun, R., Chutakositkanon, V., Jankaew, K., Kanjanapayont, P., 2008b. Flow conditions of the 2004 Indian ocean tsunami in Thailand, inferred from capping bedforms and sedimentary structures. Terra. Nova 20, 141–149.

Cisternas, M., Atwater, B.F., Torrejon, F., Sawai, Y., Machuca, G., Lagos, M., Eiopert, A., Youlton, C., Salgado, I., Kamataki, T., Shishikura, M., Rajendran, C.P., Malik, J.K., Rizal, Y., Husni, M., 2005. Predecessors of the giant 1960 Chile earthquake. Nature 47, 404–407.

Cisternas, M., Garrett, E., Wesson, R., Dura, T., Ely, L.L., 2017. Unusual geologic evidence of coeval seismic shaking and tsunami shows variability in earthquake size and recurrence in the area of the giant 1960 Chile earthquake. Mar. Geol. 385, 101–113.

Coello Bravo, J.J., Martín González, M.E., Hernández Gutiérrez, L.E., 2014. Tsunami deposits originated by a giant landslide in Tenerife (Canary Islands). Vieraea 42, 79–102.

Costa, M., Andrade, C., 2020. Tsunami deposits: present knowledge and future challenges. Sedimentology 67, 1189–1206.

Costa, P.J.M., Dawson, S., 2014. Tsunami sedimentology. In: Meyers, A.R. (Ed.), Encyclopedia of Complexity and System Science. Springer, pp. 1–17.

Costa, P.J.M., Andrade, C., Freitas, M.C., Oliveira, M.A., da Silva, C.M., Omira, R., Baptista, M.A., 2011. Boulder deposition during major tsunami events. Earth Surf. Proc. Land. 36, 2054–2068.

Costa, P.J.M., Andrade, C., Dawson, A.G., Mahaney, W.C., Freitas, M.C., Paris, R., Taborda, R., 2012a. Microtextural characteristics of quartz grains transported and deposited by tsunamis and storms. Sediment. Geol. 275–276, 55–69.

Costa, P.J.M., Andrade, C., Freitas, M.C., Oliveira, M.A., Lopes, V., Dawson, A.G., Moreno, J., Fatela, F., Jouanneau, J.M., 2012b. A tsunami record in the sedimentary archive of the central Algarve coast, Portugal: characterizing sediment, reconstructing sources and inundation paths. Holocene 22, 899–914.

Costa, P.J.M., Andrade, C., Dawson, S., 2014. Geological recognition of on shore tsunami deposits. In: Finki, C.W., Makowski, C. (Eds.), Series: Coastal Research Library, Advances in Coast and Marine Resources Environmental Management and Governances, vol. 8. Springer, p. 28.

Costa, P.J.M., Andrade, C., Cascalho, J., Dawson, A.G., Freitas, M.C., Paris, R., Dawson, S., 2015. Onshore tsunami sediment transport mechanisms inferred from heavy mineral assemblages. Holocene 25, 795–809.

Costa, P.J.M., Costas, S., Gonzalez-Villanueva, R., Oliveira, M.A., Roelvink, D., Andrade, C., Freitas, M.C., Cunha, P.P., Martins, A., Buylaert, J.-P., Murray, A., 2016a. How did the AD 1755 tsunami impact on sand barriers across the southern coast of Portugal? Geomorphology 268, 296–311.

Costa, P.J.M., Oliveira, M.A., Gonzalez-Villanueva, R., Andrade, C., Freitas, M.C., 2016b. Imprints of the AD 1755 tsunami in Algarve, south Portugae. In: Tsunamis and Earthquakes in Coastal Environments: Signature and Restoration. Springer International Publishing, pp. 17–30.

Costa, P.J.M., La Selle, S., Dawson, S., Gelfenbaum, G., Milne, F., Andrade, C., Lira, C.P., Cascalho, J., Freitas, M.C., Jaffe, B., 2017a. The application of microtextural and heavy mineral analysis to discriminate between storm and tsunami deposits. In: Scourse, E.M., Chapman, N.A., Tappin, D.R., Wallis, S.R. (Eds.), Tsunamis: Geology, Hazards and Risks, vol. 456. Geological Society of London Special Publications, pp. 167–190.

Costa, P.J.M., Andrade, C., Freitas, M.C. (Eds.), 2017b. 5th International Tsunami Field Symposium — Volume of Abstracts, p. 131.

Costa, P.J.M., Kim, Y.D., Park, Y.S., Quintele, M., Mahaney, W.C., Dourado, F., Dawson, S., 2017c. Imprints in silica grains induced during a wave experiment: approach to determine microtextural signatures during aqueous transport. J. Sediment. Res. 67 (7), 677–687.

Costa, P.J.M., Gelfenbaum, B., Dawson, S., La Selle, S., Milne, F., Cascalho, J., Ponte Lira, C., Andrade, C., Freitas, M.C., Jaffe, B., 2018. The application of microtextural and heavy mineral analysis to discriminate between storm and tsunami deposits. In: Geological Society, vol. 456. Special Publication, London, pp. 167–190.

Cummins, P.R., 2007. The potential for giant tsunamigenic earthquakes in the northern Bay of Bengal. Nature 449, 75–78.

Cunha, P.P., Buylaert, J.-P., Murray, A.S., Andrade, C., Freitas, M.C., Fatela, F., Munha, J.M., Martins, A.A., Sugisaki, S., 2010. Optical dating of clastic deposits generated by an extreme marine coastal flood: the 1755 tsunami deposits in the Algarve (Portugal). Quat. Geochronol. 5, 329–335.

Cuven, S., Paris, R., Falvard, S., Miot-Noirault, E., Benbakkar, M., Schneider, J.L., Billy, I., 2013. High-resolution analysis of a tsunami deposit: case-study from the 1755 Lisbon tsunami in southwestern Spain. Mar. Geol. 337, 98–111.

Dawson, A.G., Dawson, S., Bondevik, S., Costa, P.J.M., Hill, J., Stewart, I., 2020. Reconciling Storegga tsunami sedimentation patterns with modelled wave heights: a discussion from the Shetland Isles field laboratory. Sedimentology 67, 1344–1353.

De Martini, P.M., Barbano, M.S., Smedile, A., Gerardi, F., Pantosti, D., Del Carlo, P., Pirrotta, C., 2010. A unique 4000 year long geological record of multiple tsunami inundations in the Augusta bay (eastern Sicily, Italy). Mar. Geol. 276, 42–57.

De Martini, P.M., Barbano, M.S., Pantosti, D., Smedile, A., Pirrotta, C., Del Carlo, P., Pinzi, S., 2012. Geological evidence for paleotsunami along eastern Sicily (Italy): an overview. Nat. Hazards Earth Syst. Sci. 12, 2569–2580.

Dey, H., Goodman-Tchernov, B., Sharvit, J., 2014. Archaeolgical evidence for the tsunami o January 18, A.D. 749: a chapter in the history of early Islamic Qaysariyah (Caesarea Mariima). J. Roman Archeol. 27, 357–373.

Dominey-Howes, D., 2007. Geological and historical records of Australian tsunami. Mar. Geol. 239, 99–123.

Donato, S.V., Reinhardt, E.G., Boyce, J.I., Rothaus, R., Vosmer, T., 2008. Identifying tsunami deposits using bivalve shell taphonomy. Geology 36, 199–202.

Donato, S.V., Reihardt, E.G., Boyce, J.I., Pilarczyk, J.E., Jupp, B.P., 2009. Particle-size distribution inferred tsunami deposits in Sur Lagoon, Sultanate of Oman. Mar. Geol. 257, 54–64.

Dunbar, P., McCullough, H., Mungov, G., Varner, J., Stroker, K., 2011. Tohoku earthquake and tsunami data available from the National Oceanic and Atmospheric Administration/National Geophysical Data Center. Geomat. Nat. Hazards Risk 2, 305–323.

Dura, T., Cisternas, M., Horton, B.P., Ely, L.L., Nelson, A.R., Wesson, R.L., Pilarczyk, J.E., 2015. Coastal evidence for Holocene subduction-zone earthquakes and tsunamis in central Chile. Quat. Sci. Rev. 113, 93–111.

Dura, T., Hemphill-Haley, E., Sawai, Y., Horton, B.P., 2016. The application of diatoms to reconstruct the history of subduction zone earthquakes and tsunamis. Earth Sci. Rev. 152, 181–197.

Ely, L.L., Cisternas, M., Wesson, R.L., Dura, T., 2014. Five centuries of tsunamis and land-level changes in the overlapping rupture area of the 1860 and 2010 Chirian earthquakes. Geology 42, 995–998.

Engel, M., Brückner, H., 2011. The identification of palaeo-tsunami deposits — a major challenges in coastal sedimentary research. In: Karius, Hadiler, Decke, von Eynatten, Bruckner, Vott (Eds.), Dynamische Kusten — Prozese, Zusammenhange und Auswirkungen, Coastline Reports, vol. 17. ISSN 0928-2734, ISBN 9.

Engel, M., May, S.M., Scheffers, A., Squire, P., Pint, A., Kelletat, D., Brückner, H., 2015. Prograded foredunes of Western Australia's macrotidal coast - implications for Holocene sea-level change and high-energy wave impacts. Earth Surf. Proc. Land. 40, 726–740.

Engel, M., Oetjen, J., May, S.M., Brückner, H., 2016. Tsunami deposits of the Caribbean — towards an improved coastal hazard assessment. Earth Sci. Rev. 163, 260–296.

Engel, M., Pilarczyk, J., May, S.-M., Brill, D., Garrett (Eds.), 2020. Geological Records of Tsunami and Other Extreme Waves. Elsevier, p. 844.

Falvard, S., Paris, R., 2017. X-ray tomography of tsunami deposits: towards a new depositional model of tsunami deposits. Sedimentology 64, 453–477.

Falvard, S., Norpoth, M., Pint, A., Brill, D., Engel, M., 2016. A mid-Holocene candidate tsunami deposit from the NW Cape (Western Australia). Sediment. Geol. 332, 40–50.

Falvard, S., Paris, R., Belousova, M., Belousov, A., Giachetti, T., Cuven, S., 2018. Scenario of the 1996 volcanic tsunamis in Karymskoye Lake, Kamchatka, inferred from X-ray tomography of heavy minerals in tsunami deposits. Mar. Geol. 396, 160–170.

Ferrer, M., Gonzalez de Vallejo, L., Seisdedos, J., Coello, J.J., García, J.C., Hernández, L.E., Casillas, R., Martn, C., Rodríguez, J.A., Madeira, J., Andrade, C., Freitas, M.C., Lomoschitz, A., Yepes, J., Meco, J., Betancort, J.F., 2013. Güímar and La Orotava mega-landslides (Tenerife) and tsunamis deposits in canary islands. In: Margottini, C., Canuti, P., Sassa, K. (Eds.), Landslide Science and Practice: Volume 5: Complex Environment, pp. 27–33.

Finger, K., 2018. Tsunami-generated rafting of foraminifera across the north Pacific ocean. Aquat. Invasion 13 (Issue 13), 17–30.

Finkler, C., Fischer, P., Kaika, K., Rigakou, D., Metallinou, G., Hadler, H., Vött, A., 2018. Tracing the Alkinoos Harbor of ancient Kerkyra, Greece, and reconstructing its paleotsunami history. Geoarchaeology 33, 24–42.

Fischer, P., Finkler, C., Röbke, B.R., Baika, K., Hadler, H., Willershäuser, H., Rigakou, D., Metallinou, G., Vött, A., 2016. Impact of Holocene tsunamis detected in lagoonal environments on Corfu (Ionian Islands, Greece): geomorphological, sedimentary and microfaunal evidence. Quatern. Int. 401, 4–16.

Font, E., Nascimento, C., Omira, R., Baptista, M.A., Silva, P.F., 2010. Identification of tsunami-induced deposits using numerical modeling and rock magnetism techniques: a study case of the 1755 Lisbon tsunami in Algarve, Portugal. Phys. Earth Planet. Inter. 182, 187–198.

Fritz, H.M., Petroff, C.M., Catalán, P.A., Cienfuegos, R., Winckler, P., Kalligeris, N., Weiss, R., Barrientos, S.E., Meneses, G., Valderas-Bermejo, C., Ebeling, C., Papadopoulos, A., Contreras, M., Almar, R., Dominguez, J.C., Synolakis, C.E., 2011b. Field survey of the 27 February 2010 Chile tsunami. Pure Appl. Geophys. 168 (11), 1989–2010.

Fruergaard, M.S., Piasecki, P., Johannessen, N., Noe- Nygaard, T., Andersen, M., Pejrup, M., Nielsen, L., 2015. Tsunami propagation over a wide, shallow continental shelf caused by the Storegga Slide, southeastern North Sea, Denmark. Geology 43, 1047–1050.

Fujie, G., Kodira, S., Nakamura, Y., Morgan, J.P., Dannowski, A., Thorwant, M., Grevemeyer, I., Miura, S., 2020. Spatial variations of incoming sediments at the northeastern Japan arc and their implications. Geology 48, 614–619.

Fujino, S., Naruse, H., Matsumoto, D., Jarupongsakul, T., Sphawajruksakul, A., Sakakura, N., 2009. Stratigraphic evidence for pre-2004 tsunamis in southwestern Thailand. Mar. Geol. 262, 25–28.

Fujino, S., Naruse, H., Matsumoto, D., Sakakura, N., Suphawajruksakul, A., Jarupongsakul, T., 2010. Detailed measurements of thickness and grain size of a widespread onshore tsunami deposit in Phang-nga Province, southwestern Thailand. Isl. Arc 19 (3), 389–398.

Fujino, S., Goto, K., Tappin, D., Fujiwara, O., 2016. Geological records of storms, tsunamis and other extreme events. Isl. Arc 25 (5), 303–305.

Fujino, S., Kimura, H., Komatsubara, J., Matsumoto, D., Namegaya, Y., Sawai, Y., Shishikura, M., 2018. Stratigraphic evidence of historical and prehistoric tsunamis on the Pacific coast of central Japan: implications for the variable recurrence of tsunamis in the Nankai Trough. Quat. Sci. Rev. 201, 147–161.

Fujiwara, O., 2015. The Science of Tsunami Deposits. University of Tokyo Press, p. 283 (in Japanese).

Fujiwara, O., Tanigawa, K., 2014. Bedform record the flow condition of the 2011 Tohoku-Oki tsunami on the Sendai Plain, northeast Japan. Mar. Geol. 358, 79–88.

436 Selected bibliography

Fujiwara, O., Aoshima, A., Irizuki, T., Ono, E., Obrochta, S.P., Sampei, Y., Sato, Y., Takahashi, A., 2020. Tsunami deposits refine great earthquake rupture extent and recurrence over the past 1300 years along the Nankai and Tokai fault segments of the Nankai Trough, Japan. Quat. Sci. Rev. 277, 105999.

Furumura, T., Imai, K., Maeda, T., 2011. A revised tsunami source model for the 1707 Hoei earthquake and simulation of tsunami inundation of Ryujin Lake, Kyushu, Japan. J. Geophys. Res. 116, B02308.

Furusato, E., Tanaka, N., 2014. Maximum sand sedimentation distance after backwash current of tsunami - simple inverse model and laboratory experiments. Mar. Geol. 353, 128−139.

Garret, E., Shennan, I., Watcham, E.P., Woodroffe, S.A., 2013. Reconstructing paleoseismic deformation, 1: modern analogues from the 1960 and 2010 Chilean great earthquakes. Quatern. Sci. Rev. 75, 11−21.

Garrett, E., Shennan, I., Woodroffe, S.A., Cisternas, M., Hocking, E.P., Gulliver, P., 2015. Reconstructing paleoseismic deformation, 2: 1000 years of great earthquakes at Chucalen, south central Chile. Quatern. Sci. Rev. 113, 112−122.

Garrett, E., Fujiwara, O., Garrett, P., Heyvaert, V.M.A., Shishikura, M., Yokoyama, Y., Hubert-Ferrari, A., Brückner, H., Nakamura, A., De Batist, M., The QuakeRecNankai Team, 2016. A systematic review of geological evidence for Holocene earthquakes and tsunamis along the Nankai-Suruga Trough, Japan. Earth Sci. Rev. 159, 337−357.

Garrett, E., Pilarczyk, J., Brill, D., 2017. Preface to marine geology special issue: geological records of extreme wave events. Mar. Geol. 396, 1−5.

Garrett, E., Fujiwara, O., Riedesel, S., Deforce, W.J., Yokoyama, Y., Schmidt, S., Brückner, H., De Batist, M., Heyvaert, V.M.A., The QuakeRecNankai Team, 2018a. Historical Nankai-Suruga megathrust earthquakes recorded by tsunami and terrestrial mass movement deposits on the Shirasuka coastal Lowlands, Shizuoka Prefecture, Japan. Holocene 28, 968−983.

Garrett, E., Pilarczyk, J.E., Brill, D., 2018b. Geological records of extreme wave events. Mar. Geol. 396, 1−230.

Gelfenbaum, G., Apotsos, A., Stevens, A.W., Jaffe, B., 2011. Effects of fringing reefs on tsunami inundation: American Samoa. Earth Sci. Rev. 107, 12−22.

Giachetti, T., Paris, R., Kelfoun, K., Perez-Torrado, F.J., 2011. Numerical modelling of the tsunami triggered by the Güimar debris avalanche, Tenerife (Canary Islands): comparison with field-based data. Mar. Geol. 284, 189−202.

Gienko, G.A., Terry, J.P., 2014. Three-dimensional modeling of coastal boulders using multi-view image measurements. Earth Surf. Proc. Land. 38, 853−864.

Gisler, G.R., Weaver, R.P., Gittings, M.L., 2006. Sage calculations of the tsunami threat from La Palma. Sci. Tsunami Hazards 24, 288−301.

Goff, J., 2011. Evidence of a previously unrecorded local tsunami, 13 April 2010, Cook Islands: implications for Pacific Island countries. Nat. Hazards Earth Syst. Sci. 11, 1371−1379.

Goff, J., Chagué -Goff, C., 2014. The Australian tsunami database - a review. Prog. Phys. Geogr. 38, 218−240.

Goff, J., Sugawara, D., 2014. Seismic-driving of sand beach ridge formation in northern Honshu, Japan? Mar. Geol. 358, 138−149.

Goff, J., Dudley, W.C., deMaintenon, M.J., Cain, G., Coney, J.P., 2006. The largest local tsunami in 20th century Hawaii. Mar. Geol. 226, 65−79.

Goff, J., Chagué-Goff, C., Dominey-Howes, D., McAdoo, B., Cronin, S., Bonte-Grapetin, M., Nichol, S., Horrocks, M., Cisternas, M., Lamarche, G., Pelletier, B., Jaffe, B., Dudley, W., 2011a. Palaeotsunamis in the Pacific islands. Earth Sci. Rev. 107, 141−146.

Goff, J., Lamarche, G., Pelletier, B., Chagué-Goff, C., Strotz, L., 2011b. Predecessors to the 2009 south Pacific tsunami in the Wallis and Futuna archipelago. Earth Sci. Rev. 107, 91−106.

Goff, J., Chagué-Goff, C., Nicolo, S., Jaffe, B., Dominey-Howesm, D., 2012. Progress in palaeotsunami research. Sediment. Geol. 243−244, 70−88.

Goff, J., Terry, J.P., Chague-Goff, C., Goto, K., 2014. What is a mega-tsunami? Mar. Geol. 358, 12−17.

Goff, J.R., Goto, K., Chagué-Goff, M., Watanabe, M., Gadd, P., 2016a. New Zealand's Most Easterly Palaeotsunami Deposits Supporting Evidence for Major Regionwide Event. AGUFM 2016, NH41-1772.

Goff, J., Knight, J., Sugawara, D., Terry, J.P., 2016b. Anthropogenic disruption to the seismic driving of beach ridge formation: the Sendai coast, Japan. Sci. Total Environ. 544, 18–23.

Goff, J., Goto, K., Chagué, C., Watanabe, M., Gadd, P.S., King, D.N., 2018. New Zealand's most easterly paleotsunami deposit confirms evidence for major trans- Pacific events. Mar. Geol. 404, 158–173.

Goodman-Tchermov, B.N., Reinhadt, E., Mart, Y., 2007. Results of coring survey indicate Eastern Mediterranean paleotsunami events. In: American Geophy. Union, Fall Meeting 2007 Abstract. OS13B-5.

Goodman-Tchermov, B., Katz, T., Shaked, Y., Qupty, N., Kanari, M., Niemi, T., Agnon, A., 2016. Offshore evidence for an undocumented tsunami event in the 'low risk' Gulf Aqaba-Eilat, Northrn Red Sea. PLoS One 11 (1), e045802.

Goto, K., Okada, K., Imamura, F., 2009. Characteristics and hydrodynamics of boulders transported by storm waves at Kudaka Island, Japan. Mar. Geol. 262, 14–24.

Goto, K., Shinozaki, T., Minoura, K., Okada, K., Sugawara, D., Imamura, F., 2010d. Distribution of boulders at Miyara Bay of Ishigaki Island, Japan: a flow characteristic indicator of tsunami and storm waves. Isl. Arc 19 (3), 412–426.

Goto, K., Sugawara, D., Abe, T., Haraguchi, T., Fujino, S., 2012d. Liquefaction as an important local source of the 2011 Tohoku-oki tsunami deposits at Sendai Plain, Japan. Geology 40, 887–890.

Goto, K., Miyagi, K., Imamura, F., 2013. Localized tsunamigenic earthquakes inferred from preferential distribution of coastal boulders on Ryukyu Islands, Japan. Geology 41, 1139–1142.

Goto, K., Ikehara, K., Goff, J., Chagué-Goff, C., Jaffe, B., 2014c. The 2011 Tohoku-oki tsunami- 3 years on. Mar. Geol. 358, 2–11.

Goto, K., Chagué-Goff, C., Goff, J., Ikehara, K., Jaffe, B., 2014d. In the wake of the 2011 Tohoku-oki tsunami - three years on. Mar. Geol. 358, 1–150.

Goto, T., Satake, K., Sugai, T., Ishibe, T., Harada, T., Gusman, A.R., 2017. Effects of topography on particle composition of 2011 tsunami deposits on the ria-type Sanriku coast, Japan. Quat. Int. 456, 17–27.

Goto, K., Satake, K., Sugai, T., Ishibe, T., Harada, T., Gusman, A.R., 2019b. Tsunami history over the past 2000 years on the Sanriku coast, Japan, determined using gravel deposits to estimate tsunami inundation behavior. Sediment. Geol. 382, 85–102.

Gouramanis, C., Switzer, A.D., Polivka, P.M., Bristow, C.S., Jankaew, K., Dat, P.T., Pile, J., Rubin, C.M., Yingsin, L., Ildefonso, S.R., Jol, H.M., 2015. Ground penetrating radar examination of thin tsunami beds - a case study from Phra Thong Island, Thailand. Sediment. Geol. 329, 149–165.

Gracia, F.J., Alonso, C., Benavente, J., Anfuso, G., Del- Rio, L., 2006. The different records of the 1755 tsunami waves along the south Atlantic Spanish Coast. Z. Geomorphol. 146, 195–220.

Grand Pre, C.A., Horton, B.P., Kelsey, H.M., Rubin, C.M., Hawkes, A.D., Daryono, M.R., Rosenberg, G., Culver, S.J., 2012. Stratigraphic evidence for an early Holocene earthquake in Aceh, Indonesia. Quatern. Sci. Rev. 54, 142–151.

Grauert, M., Bjorck, S., Bondevik, S., 2001. Storegga tsunami deposits in a coastal lake on Suduroy, the Faroe Islands. Boreas 30, 263–271.

Grzelak, K., Szczuciński, W., Kotwicki, L., Sugawara, D., 2014. Ecological status of sandy beaches after tsunami events: insights from Meiofauna investigations after the 2011 Tohoku-oki tsunami, Sendai bay, Japan. In: Kontar, Y.A., Santiago-Fandiño, V., Takahashi, T. (Eds.), Tsunami Events and Lessons Learned - Environmental and Societal Significance, Advances in Natural and Technological Hazards Research, vol. 35. Springer, Japan, pp. 177–191.

Gusiakov, V., Abbott, D.H., Bryant, E.A., Masse, W.B., Breger, D., Beer, T., 2010. Mega Tsunami of the world oceans: chevron Dune formation, micro-ejecta, and rapid climate change as the evidence of recent oceanic bolide impacts geophysical hazards. In: Mulder, E.F.J., Derbyshire, E. (Eds.), Geophysical Hazards - Minimizing Risk, Maximizing Awareness. Springer, the Netherlands, pp. 197–227.

Hadler, H., Willershäuser, T., Ntageretzis, K., Henning, P., Vött, A., 2012. Catalogue entries and non-entries of earthquake and tsunami events in the Ionian Sea and the Gulf of Corinth (eastern Mediterranean, Greece) and their interpretation with regard to palaeotsunami research. Beiträge der 29, 1–15.

438 Selected bibliography

Hadler, H., Baika, K., Pakkanen, J., Evangelistis, D., Fischer, P., Ntageretzis, K., Röbke, B., Willershäuser, T., Vött, A., 2015a. Palaeotsunami impact on the ancient harbour site Kyllini (western Peloponnese, Greece) based on a geomorphological multi-proxy approach. Z. Geomorphol. Suppl. Issues 59 (4), 7–41.

Hadler, H., Vött, A., Fischer, P., Ludwig, S., Heinzelmann, M., Rohn, C., 2015b. Temple-complex post-dates tsunami deposits found in the ancient harbour basin of Ostia (Rome, Italy). J. Archaeol. Sci. 61, 78–89.

Hadler, H., Fischer, P., Obrocki, L., Heinzelmann, M., Vött, A., 2020. River channel evolution and tsunami impacts recorded in local sedimentary archives - the 'Fiume Morto' at Ostia Antica (Tiber River, Italy). Sedimentology 67, 1309–1343.

Hill, J., Collins, G.S., Avdis, A., Kramer, S.C., Piggot, M.D., 2014. How does multiscale modelling and inclusion of realistic palaeobathymetry affect numerical simulation of the Storegga Slide tsunami? Ocean Model. 83, 11–25.

Hoffmann, G., Reicherter, K., Wiatr, T., Grützner, C., Rausch, T., 2013. Block and boulder accumulations along the coastline between Fins and Sur (Sultanate of Oman): tsunamigenic remains? Nat. Hazards 65, 851–873.

Hoffmann, N., Master, D., Goodman-Tchernov, B., 2018. Possible tsunami inundation identified amongst 4-5th century BCE archaeological deposits at Tel Ashkelon, Israel. Mar. Geol. 396, 150–159.

Hoffmeister, D., Tilly, N., Curdt, C., Ntageretzis, K., Bareth, G., Vött, A., 2013. Monitoring annual changes of the coastal sedimentary budget in western Greece by terrestrial laser scanning. Z. Geomorphol. Suppl. Issues 57 (4), 47–67.

Hoffmeister, D., Curdt, C., Bareth, G., 2020. Monitoring the sedimentary budget and dislocated boulders in western Greece - results since 2008. Sedimentology 67, 1411–1430.

Hong, I., Dura, T., Ely, L.L., Horton, B.P., Nelson, A.R., Cisternas, M., Nikitina, D., Wesson, R.L., 2017. A 600-year-long stratigraphic record of tsunamis in south-central Chile. Holocene 27 (1), 39–51.

Hori, K., Kuzumoto, R., Hirouchi, D., Umitsu, M., Janjirawuttikul, N., Patanakanog, B., 2007. Horizontal and vertical variation of 2004 Indian tsunami deposits: an example of two transects along the western coast of Thailand. Mar. Geol. 239, 163–172.

Horton, B., Sawai, Y., Hawkes, A., Witter, R., 2011. Sedimentology and paleontology of a tsunami deposit accompanying the great Chilean earthquake of February 2010. Mar. Micropaleontol. 79, 132–138.

Hunt, J.E., Wynn, R.B., Talling, P.J., Masson, D.G., 2013. Turbidite record of frequency and source of large volume (>100 km^3) Canary Island landslides in the last 1.5 Ma: implications for landslide triggers and geohazards. Geochem. Geophys. Geosyst. 14, 2100–2123.

Ikehara, k., Usami, K., 2017. Submarine earthquake- and tsunami- induced event deposits. Synthesiology, English Edition 11 (Issue 1), 12–22.

Ikehara, K., Kanamatsu, T., Nagahoahi, Y., Strasser, M., Fink, H., Usami, K., Irino, T., Wefer, G., 2016. Documenting large earthquakes similar to the 2011 Tohoku-oki earthquake from sediments deposited to the Japan Trench over the past 1500 years. Earth Planet Sci. Lett. 445, 48–56.

Ikehara, K., Usami, K., Katamatsu, T., Danbara, T., Yamashita, T., 2017. Three important Holocene tephras off the Pacific region, Northeast Japan: implication for correlating onshore and offshore event deposits. Quat. Int. 456, 138–153.

Irizuki, T., Fujiwara, O., Yoshioka, K., Suzuki, A., Tanaka, Y., Nagao, M., Kawagata, S., Kawano, S., Nishimura, O., 2019. Geochemical and micropaleontological impacts caused by the 2011 Tohoku-oki tsunami in Matsushima Bay, northeastern Japan. Mar. Geol. 407, 261–274.

Ishimura, D., Yamada, K., 2019. Palaeo-tsunami inundation distances deduced from roundness of gravel particles in tsunami deposits. Sci. Rep. 16, 10251.

Ishizawa, T., Goto, K., Yokoyama, Y., Goff, J., 2020. Dating tsunami deposits: present knowledge and challenges. Earth Sci. Rev. 200, 102971.

Jaffe, B.E., Gelfenbuam, G., 2007. A simple model for calculating tsunami flow speed from tsunami deposits. Sediment. Geol. 200, 347–361.

Jaffe, B.E., Gelfenbaum, G., Buckley, M.L., Steve Watt, S., Apotsos, A., Stevens, A.W., Richmond, B.M., 2010. The Limit of Inundation of the September 29, 2009, Tsunami on Tutuila, American Samoa. US Geological Survey Open-File Report 2010−1018, p. 27.

Jaffe, B., Buckley, M., Richmond, B., Strotz, L., Etienne, S., Clark, K., Watt, S., Gelfenbaum, G., Goff, J., 2011. Flow speed estimated by inverse modeling of sandy sediment deposited by the 29 September 2009 tsunami near Satitoa, East Upolu, Samoa. Earth Sci. Rev. 107, 23−37.

Jankaew, K., Atwater, B.F., Sawai, Y., Choowong, M., Charoentitirat, T., Martin, M.E., Prendergast, A., 2008. Medieval forewarning of the 2004 Indian Ocean tsunami in Thailand. Nature 455, 1228−1231.

Jonathan, M.P., Srinivasalu, S., Thangadurai, N., Rajeshwara-Rao, N., Ram-Mohan, V., Narmatha, T., 2012. Offshore depositional sequence of 2004 tsunami from Chennai, SE coast of India. Nat. Hazards 62, 1155−1168.

Judd, K., Chagué-Goff, C., Goff, J., Gadd, P., Zawadzki, A., Fierro, D., 2017. Multi-proxy evidence for small historical tsunamis leaving little or no sedimentary record. Mar. Geol. 385, 204−215.

Kain, C.L., Gomez, C., Hart, D.E., Chagué-Goff, C., Goff, J., 2015. Analysis of environmental controls on tsunami deposit texture. Mar. Geol. 368, 1−14.

Kain, C., Wassmer, P., Goff, J., Chagué-Goff, C., Gomez, C., Hart, D.E., Fierro, D., Jacobsen, G., Zawadzki, A., 2017. Determining flow patterns and emplacement dynamics from tsunami deposits with no visible sedimentary structure. Earth Surf. Proc. Land. 42, 763−780.

Kawakami, G., Nishina, K., Kase, Y., Tajika, J., Hayashi, K., Hirose, W., Sagayama, T., Watanabe, T., Ishimaru, S., Koshimizu, K., Takahashi, R., Hirakawa, K., 2017. Stratigraphic records of tsunami along the Japan sea, southwest Hokkaido, northern Japan. Isl. Arc 26, 1−18.

Keating, B.H., 2008. The 1946 tsunami at Kahuku, NE Oahu, Hawaii. In: Wallendorf, L., Ewing, L., Jones, C., Jaffe, B. (Eds.), Solutions to Coastal Disasters 2008: Tsunamis. American Society of Civil Engineers, Reston, VA, pp. 157−168.

Kelletat, D., 2008. Comments to Dawson, A.G. and Stewart, I. (2007) Tsunami deposits in the geological record (Sediment. Geol., 200, 166−183). Sediment. Geol. 211, 87−91.

Kelsey, H., Engelhart, S.E., Pilarczyk, J.E., Horton, B.P., Rubin, C.M., Daryono, M., Ismail, N., Hawkes, A.D., Bernhardt, C., Cahill, N., 2015. Accommodation space, relative sea level, and the archiving of paleo-earthquakes along subduction zones. Geology 43, 675−678.

Kempf, P., Moermaut, J., Van Daele, M., Vermassen, F., Vandoorne, W., Pino, M., Urrutia, R., Schmidt, S., Garret, E., De Batist, M., 2015. The sedimentary record of the 1960 tsunami in two coastal Lakes on Isle de Chiloe, south central Chile. Sediment. Geol. 328, 73−88.

Kempf, P., Moernaut, J., Van Daele, M., Vandoorne, W., Pino, M., Urrutia, R., De Batist, M., 2017. Coastal lake sediments reveal 5500 years of tsunami history in south central Chile. Quat. Sci. Rev. 161, 99−116.

Khan, A.S., Ravikuman, B., Lyla, S., Manokaran, S., 2018. Impact of the 2004 tsunami on the macrofauna of the continental slope of the southeast coast of India. Mar. Ecol. 39 (5), e12527.

Kioka, A., Schwestermann, T., Moernaut, J., Ikehara, K., Kanamatsu, T., Eglinton, T., Strasser, M., 2019. Event stratigraphy in a hadal oceanic trench: the Japan Trench as sedimentary archive recording recurrent giant subduction zone earthquakes and their role in organic carbon export to the deep sea. Front. Earth Sci. 7.

Kitamura, A., Fujiwara, O., Shinohara, K., Akaike, S., Masuda, T., Ogura, K., Urano, Y., Kobayashi, K., Tamaki, C., Mori, H., 2013. Identifying possible tsunami deposits on the Shizuoka Plain, Japan and their correlation with earthquake activity over the past 4000 years. Holocene 23, 1684−1698.

Kitamura, A., Ito, M., Ikuta, R., Ikeda, M., 2018a. Using molluscan assemblages from paleotsunami deposits to evaluate the influence of topography on the magnitude of late Holocene mega-tsunamis on Ishigaki Island, Japan. Prog. Earth Planet. Sci. 5, 41.

Kitamura, A., Ito, M., Sakai, S., Yokoyama, Y., Miyairi, Y., 2018b. Identification of tsunami deposits using a combination of radiometric dating and oxygen-isotope profiles of articulated bivalves. Mar. Geol. 403, 57−61.

440 Selected bibliography

Kitamura, A., Ina, T., Suzuki, D., Tsutahara, K., Sugawara, D., Yamada, K., Aoshima, A., 2019. Geologic evidence for coseismic uplift at ~AD 400 in coastal lowland deposits on the Shimizu Plain, central Japan. Prog. Earth Planet. Sci. 6, 57.

Koiwa, N., Kasai, M., Kataoka, S., Isono, T., 2014. Examination of relation with tsunami behavior reconstructed from on-shore sequence photographs, topography, and sedimentary deposits from the 2011 Tohoku-oki tsunami on the Kamikita Plain, Japan. Mar. Geol. 358, 107−119.

Komatsubara, J., Fujiwara, O., 2007. Overview of Holocene tsunami deposits along the Nankai, Suruga, and Sagami Troughs, Southwest Japan. Pure Appl. Geophys. 164, 493−507.

Komatsubara, J., Fujiwara, O., Takada, K., Sawai, Y., Than, T.A., Kamataki, T., 2008. Historical tsunamis and storms recorded in a coastal lowland, Shizuoka prefecture, along the Pacific Coast of Japan. Sedimentology 55, 1703−1716.

Kon, S., Nakamura, N., Nishimura, Y., Goto, K., Sugawara, D., 2017. Inverse magnetic fabric in unconsolidated sandy event deposits in Kiritappu Marsh, Hokkaido, Japan. Sediment. Geol. 349, 112−119.

Kortekaas, S., Dawson, A.G., 2007. Distinguishing tsunami and storm deposits: an example from Martinhal, SW Portugal. Sediment. Geol. 200, 208−221.

Koster, B., Reicherter, K., 2014. Sedimentological and geophysical properties of a ca. 4000 years old tsunami deposit in southern Spain. Sediment. Geol. 314, 1−16.

Koster, B., Hadler, H., Vött, A., Reicherter, K., 2013. Application of GPR for visualising spatial distribution and internal structures of tsunami deposits−case studies from Spain and Greece. Z. Geomorphol. Suppl. Issues 57 (4), 29−45.

Koster, B., Hoffmann, G., Grutzner, C., Reicherter, K., 2014. Ground penetrating radar facies of inferred tsunami deposits on the shores of the Arabian Sea (Northern Indian Ocean). Mar. Geol. 351, 13−24.

Koster, B., Vött, A., Mathes-Schmidt, M., Reicherter, K., 2015. Geoscientific investigations in search of tsunami deposits in the environs of the Agoulinitsa peatland, Kaiafas Lagoon and Kakovatos (Gulf of Kyparissia, western Peloponnese (Greece). Z. Geomorphol. Suppl. Issues 59 (Issue 4), 125−156.

Kumamoto, S., Goto, T., Sugai, T., Omori, T., Satake, K., 2018. Geological evidence of Tsunamis in the past 3800 years at a coastal lowland in the central Fukushima Prefecture, Japan. Mar. Geol. 404, 37−146.

Kunz, A., Frechen, M., Ramesh, R., Urban, B., 2010. Revealing the coastal event-history of the Andaman islands (Bay of Bengal) during the Holocene using radiocarbon and OSL dating. Int. J. Earth Sci. 99, 1741−1761.

Kuriyama, Y., Chida, Y., Uno, Y., Honda, K., 2020. Numerical simulation of sedimentation and erosion caused by 2011 Tohoku tsunami in Oarai Port, Japan. Mar. Geol. 427, 106225.

Kuwatani, T., Nagata, K., Okada, M., Watanabe, T., Ogawa, Y., Komai, T., Tsuchiya, N., 2014. Machine-learning techniques for geochemical discrimination of 2011 Tohoku tsunami deposits. Sci. Rep. 4, 7077.

La Selle, S.M., Richmond, B.M., Griswold, F.R., Lunghino, B.D., Jaffe, B.E., Kane, H.H., Bellanova, P., Arcos, M.E., Nelson, A.R., Chague, C., Bishop, J.M., Gelfenbaum, G., 2019. Core Logs, Scans, Photographs, Grain Size, and Radiocarbon Data From Coastal Wetlands on the Hawaiian Islands of Kauai, Oahu, and Hawaii. U.S. Geological Survey data release.

La Selle, S., Richmond, B.M., Jaffe, B.E., Nelson, A.R., Griswold, F.R., Arcos, M.E.M., Chague, C., Bishop, J.M., Bellanova, P., Kane, H.H., Lunghino, B.D., Gelfenbaum, G., 2020. Sedimentary evidence of prehistoric distant-source tsunamis in the Hawaiian Islands. Sedimentology 67, 1249−1273.

Lakhsmi, C.S.V., Srinivasan, P., Murthy, S.G.N., Trivedi, D., Nair, R.R., 2010. Granularity and textural analysis as a proxy for extreme wave events in southeast coast of India. J. Earth Syst. Sci. 119, 297−305.

Lamarche, G., Pelletier, B., Goff, J., 2010. Impact of the 29 September 2009 south Pacific tsunami on Wallis and Futuna. Mar. Geol. 271, 297−302.

Lario, J., Bardají, T., Silva, P.G., Zazo, C., Goy, J.L., 2016. Improving the coastal record of tsunamis in the ESI-07 scale: tsunami environmental effects scale (TEE-16 scale). Geol. Acta 14 (2), 179−193.

Lario, J., Spencer, C., Bardají, T., Marchante, A., Gardunomonroy, V.H., Macias, J., Ortega, S., 2020. An extreme wave event in eastern Yucatan, Mexico: evidence of a palaeotsunami event during the Mayan times. Sedimentology 67, 1481−1504.

Lau, A.Y.A., Terry, J.P., Ziegler, A.D., Switzer, A.D., Lee, Y., Etienne, S., 2016. Understanding the history of extreme wave events in the Tuamotu Archipelago of French Polynesia from large carbonate boulders on Makemo Atoll, with implications for future threats in the central South Pacific. Mar. Geol. 380, 174−190.

Lau, A.Y.A., Terry, J.P., Ziegler, A., Pratap, A., Harris, D., 2018. Boulder emplacement and remobilisation by cyclone and submarine landslide tsunami waves near Suva City, Fiji. Sediment. Geol. 364, 242−257.

Lay, T., Ammon, C.J., Kanamori, H., Rivera, L., Koper, K.D., Hutko, A.R., 2010. The 2009 Samoa-Tonga great earthquake triggered doublet. Nature 466, 964−968.

Le Bas, T.P., Masson, D.G., Holtom, R.T., Grevemeyer, I., 2007. Slope failures of the flanks of the southern Cape Verde Islands. In: Lykousis, V., Sakellariou, D., Locat, J. (Eds.), Submarine Mass Movements and Their Consequences, Advances in Natural and Technological Hazards Research, vol. 27. Springer, Dordrecht, the Netherlands, pp. 337−345.

Li, L., Qiu, Q., Huang, Z., 2012. Numerical modeling of the morphological change in Lhok Nga, west Banda Aceh, during the 2004 Indian Ocean tsunami: understanding tsunami deposits using a forward modeling method. Nat. Hazards 64, 1549−1574.

Lopez, G.I., Goodman-Tchernov, B.N., Porat, N., 2018. OSL over-dispersion: a pilot study for the characterization of extreme events in the shallow marine realm. Sediment. Geol. 378, 35−51.

Lorito, S., Tiberti, M.M., Basili, R., Piatanesi, A., Valensise, G., 2008. Earthquake generated tsunamis in the Mediterranean Sea: scenarios of potential threats to southern Italy. J. Geophys. Res. Solid Earth 113, 1−14.

Løvholt, F., Bungum, H., Harbitz, C.B., Glimsdal, S., Lindholm, C.D., Pedersen, G., 2006. Earthquake related tsunami hazard along the western coast of Thailand. Nat. Hazard. Earth Syst. Sci. 6, 979−997.

Løvholt, F., Bondevik, S., Laberg, J.S., Kim, J., Boylan, N., 2017. Some giant submarine landslides do not produce large tsunamis. Geophys. Res. Lett. 44 (16), 8463−8472.

Luczynsky, P., 2012. Record of possible palaeotsunami 9n the Upper Silurian of Podolia (Ukraina). In: Abstract Book of the 29th Meeting of the Intern. Asso. of Sedimentologists, vol. 441. Shiadming.

Machnnes, B.T., Bourgeois, J., Kravchunovskaya, E.A., Pinegina, T.K., 2009. Tsunami geomorphology, erosion and deposition from the 15 November 2006 Kuril Island tsunami. Geology 37, 995−998.

Macías, J., Mercado, A., González-Vida, J.M., Ortega, S., Castro, M.J., 2016. Comparison and computational performance of Tsunami-HySEA and MOST models for LANTEX 2013 scenario: impact assessment on Puerto Rico coasts. Pure Appl. Geophys. 173 (12), 3973−3997.

Macías, J., Castro, M.J., Ortega, S., Escalante, C., González-Vida, J.M., 2017. Performance benchmarking of tsunami-HySEA model for NTHMP's inundation mapping activities. Pure Appl. Geophys. 174 (8), 3147−3183.

MacInnes, B.T., Weiss, R., Bourgeois, J., Pinegina, T.K., 2010. Slip distribution of the 1952 Kamchatka Great Earthquake based on near-field tsunami deposits and historical records. Bull. Seismol. Soc. Am. 100, 1695−1709.

Madeira, J., Ramalho, R.S., Hoffmann, D.L., Mata, J., Moreira, M., 2020. A geological record of multiple Pleistocene tsunami inundations in an oceanic island: the case of Maio, Cape Verde. Sedimentology 67, 1529−1552.

Mamo, B., Strotz, L., Dominey-Howes, D., 2009. Tsunami sediments and their foraminiferal assemblages. Earth Sci. Rev. 96, 263−278.

Maramai, A., Brizuela, B., Graziani, L., 2014. The Euro-mediterranean tsunami catalogue. Ann. Geophys. 57, S0435.

Marriner, N., Kaniewski, D., Morhange, C., Flaux, C., Giaime, M., Vacci, M., 2017. Tsunamis in the geological records: making waves with a cautionary tale from the Mediterranean. Sci. Adv. 3 (10), e1700485.

Martin, M.F., Weiss, R., Bourgeois, J., Pinegina, T.K., Houston, H., Titov, V.V., 2008. Combining constraints from tsunami modeling and sedimentology to untangle the 1969 Ozernoi and 1971 Kamchatskii tsunamis. Geophys. Res. Lett. 5 (1), 101610.

Massari, F., D'Alessandro, A., Davaud, E., 2009. A coquinoid tsunamite from the Pliocene of Salento (SE Italy). Sediment. Geol. 221, 7–18.

Masson, D.G., Le Bas, T.P., Grevemeyer, I., Weinrebe, W., 2008. Flank collapse and large-scale landsliding in the Cape Verde islands, off west Africa. Geochem. Geophys. Geosyst 9, Q07015.

Mastronuzzi, G., Pignatelli, C., Sanso, P., 2006. Boulder fields: a valuable morphological indicator of paleotsunami in the Mediterranean Sea. Z. Geomorphol. NF Suppl. 146, 173–194.

Mastronuzzi, G., Pignatelli, C., Sanso, P., Selleri, G., 2007. Boulder accumulations produced by the 20th of February, 1743 tsunami along the coast of southeastern Salento (Apulia region, Italy). Mar. Geol. 242, 191–205.

Mastronuzzi, G., Brückner, H., Sanso, P., Vött, A. (Eds.), 2010. Tsunami Fingerprints in Different Archives - Sediments, Dynamics and Modelling Approaches Proceedings of the 2nd International Tsunami Field Symposium, Septembaer 22–28, 2008, in Ostuni (Italy) and Lefkada (Greece).

Mathes-Schmidt, M., Schwarzbauer, J., Papanikolaou, L.D., Syberberg, F., Thiele, A., Wittkopp, F., Reicherter, K.R., 2013. Geochemical and micropalaeontological investigations of tsunasmigenetic layers along the Thracian coast (north Aegean sea, Greece). Z. Geomorphol. Suppl. Issues 57 (Issue 5), 7.

Matsumoto, D., Sawai, Y., Tanigawa, K., Fujiwara, O., Namegawa, Y., Shisgikura, M., Kimura, H., 2016. Tsunami deposit associated with the 2011 Tohoku-oki tsunami in the Hasunuma site of the Kujukuri coast plain, Japan. Isl. Arc 25, 369–385.

Matthias, May, S., Brill, D., Engel, M., Schefers, A., Pint, A., Opitz, S., Wennrich, V., Squire, R., Kelletat, D., Bruckner, H., 2015. Traces of historical tropical cyclones and tsunamis in the Ashburton Delta (north-west Australia). Sedimentology 62, 1346–1542.

May, S.M., Willershäuser, T., Vött, A., 2010. Boulder transport by high-energy wave events at Cap Bon (NE Tunisia). Coastline Rep. 16, 1–10.

May, S.M., Vött, A., Brückner, H., Smedile, A., 2012. The Gyra washover fan in the Lefkada Lagoon, NW Greece—possible evidence of the 365 AD Crete earthquake and tsunami. Earth Planets Space 64 (10), 859–874.

May, S.M., Brill, D., Engel, M., Scheffers, A., Pint, A., Opitz, S., Wennrich, V., Sqdire, P., Klletat, D., Bruckner, H., 2015. Traces of historical tropical cyclones and tsunamis in the Ashburton Delta (north-west Australia). Sedimentology 62 (6), 1546–1572.

McAdoo, B., Fritz, H., Jackson, K., Kalligeris, N., Kruger, J., Bonte-Grapentin, M., Moore, A., Rafiau, W., Billy, D., Tiano, B., 2008. Solomon Islands tsunami, one year later. EOS 89, 169–176.

McCloskey, T.A., Bianchette, T.A., Liu, K., 2015. Geological and sedimentological evidence of a large tsunami occurring - 1100 year BP from a small lake along the bay of La Paz in Baja California Sur, Mexico. J. Mar. Sci. 3, 1544–1567.

McHugh, C.M.G., Seeber, L., Cormie, M.-H., Dutton, J., Cagatay, N., Polonia, A., Ryan, W.B.F., Corur, N., 2006. Submarine earthquake geology along the North Anatolia fault in the Marmara Sea, Turkey: a model for transform basin sedimentation. Earth Planet Sci. Lett. 248, 661–684.

McKenna, J., Jackson, D.W., Cooper, J., Andrew, G., 2011. In situ exhumation from bedrock of large rounded boulders at the Giant's Causeway, Northern Ireland: an alternative genesis for large shore boulders (mega-clasts). Mar. Geol. 283, 25–35.

Mikami, T., Suzuki, H., Kawabe, T., Shiki, T., Tachibana, T., 2012. Diversity of the tsunami deposits and the facies formed by the 2011 Tohoku Earthquake Tsunami. In: Abstract Book of the 29th Meeting of the Intern. Asso. Sedimentoloists, Achiadming.

Miller, S., Rowe, D.A., Brown, L., Mandal, A., 2014. Wave-emplaced boulders: implications for development of "prime real estate" seafront, North Coast Jamaica. Bull. Eng. Geol. Environ. 73, 109–122.

Minoura, K., Sugawara, D., Yamanoi, T., Yamada, T., 2015. Aftereffects of subduction-zone earthquakes: potential tsunami hazards along the Japan sea coast. Tohoku J. Exp. Med. 237, 91–102.

Monecke, K., Finger, W., Klarer, D., Kongko, W., McAdoo, B.G., Moore, A.L., Sudrajat, S.U., 2008. A 1,000-year sediment record of tsunami recurrence in northern Sumatra. Nature 455, 1232–1234.

Moore, A.L., McAdoo, B.G., Ruffman, A., 2007. Landward fining from multiple sources in a sand sheet deposited by the 1929 Grano Banks tsunami, Newfoundland. Sediment. Geol. 200, 336–346.

Moore, A., Goto, J., McAdoo, B.G., Fritz, H.M., Gusman, A., Kalligeris, N., Synolakis, C.E., 2011. Sedimentary deposits from the 17 July 2006 Western Java Tsunami, Indonesia: use of grainsize analyses to assess tsunami flow depth, speed, and traction carpet characteristics. Pure Appl. Geophys. 168, 1951–1961.

Moreira, S., Costa, P.J.M., Andrade, C., Lira, C.P., Freitas, M.C., Oliveira, M.A., Reichart, G.-J., 2017. High resolution geochemical and grain-size analysis of the AD 1755 tsunami deposit: insights into the inland extent and inundation phases. Mar. Geol. 390, 94–105.

Mori, N., Takahashi, T., Yasuda, T., Yanagisawa, H., 2011. Survey of 2011 Tohoku earthquake tsunami inundation and run-up. Geophys. Res. Lett. 38, L00G14.

Morton, R.A., Richmond, B.M., Jaffe, B.E., Gelfenbaum, G., 2006. Reconnaissance Investigation of Caribbean Extreme Wave Deposits - Preliminary Observations, Interpretations, and Research Directions. US Geological Survey, p. 41. USGS Open-File Report 2006-1293.

Morton, R.A., Gelfenbaum, G., Jaffe, B.E., 2007. Physical criteria for distinguishing sandy deposits and storm deposits using modern examples. Sediment. Geol. 200, 184–207.

Morton, E.A., Gelfenbaum, G., Buckley, M.L., Richmond, B.M., 2011. Geological effects and implications of the 2010 tsunami along the central coast of Chile. Sediment. Geol. 242 (11–41), 34–51.

Mulder, T., Zaragosi, S., Razin, P., Frrelaud, C., Lanfumey, V., Bavoi, F., 2009. A new conceptual model for the deposition process of homogenite: application to a cretaceous megaturbidite of the western Pyrenees (Basque region, SW France). Sediment. Geol. 222, 263–273.

Nakamura, M., 2009. Fault model of the 1771 Yaeyama earthquake along the Ryukyu Trench estimated from the devastating tsunami. Geophys. Res. Lett. 36, L19307.

Nakamura, Y., Nishimura, Y., Putra, P.S., 2012. Local variation of inundation, sedimentary characteristics, and mineral assemblages of the 2011 Tohoku-oki tsunami on the Misawa coast, Aomori, Japan. Sediment. Geol. 282, 216–227.

Namegaya, Y., Yata, T., 2014. Tsunamis which affected the Pacific coast of eastern Japan in medieval times inferred from historical documents. J. Seismol. Soc. Jpn. 66, 73–81.

Namegaya, Y., Satake, K., Shishikura, M., 2011. Fault models of the 1703 Genroku and 1923 Taisho Kanto earthquakes inferred from coastal movements in the southern Kanto area. Ann. Rep. Active Fault Paleoearthq. Res. 11, 107–120.

Nanayama, F., Shigeno, K., 2006. Inflow and outflow facies from the 1993 tsunami in southwest Hokkaido. Sediment. Geol. 187, 139–158.

Nanayama, F., Furukawa, R., Shigeno, K., Makino, A., Soeda, Y., Igarashi, Y., 2007. Nine unusually large tsunami deposits from the past 4000 years at Kiritappu marsh along the southern Kuril Trench. Sediment. Geol. 200, 275–294.

Nandasena, N.A.K., Sasaki, Y., Tanaka, N., 2012. Modeling field observations of the 2011 Great East Japan tsunami: efficacy of artificial and natural structures on tsunami mitigation. Coast. Eng. 67, 1–13.

Nandasena, N.A.K., Tanaka, N., Sasaki, Y., Osada, M., 2014. Reprint of "Boulder transport by the 2011 Great East Japan tsunami: comprehensive field observations and whither model predictions?". Mar. Geol. 358, 49–66.

Naruse, H., Fujino, S., Suphawajrukskul, A., Jarupongsakul, T., 2010. Features and formation process of multiple deposition layers rom the 2004 Indian Ocean Tsunami at Ban Nam Km, southern Thailand. Isl. Arc 19, 399–411.

Naruse, H., Arai, K., Matsumoto, D., Takahashi, H., Yamashita, S., Tanaka, G., Murayama, M., 2012. Sedimentary feature observed in the tsunami deposits at Rikuzentakada city. Sediment. Geol. 282, 199–215.

444 Selected bibliography

Nelson, A.R., Briggs, R.W., Dura, T., Engelhart, S.E., Gelfenbaum, G., Bradley, L.A., Forman, S.L., Vane, C.H., Kelley, K.A., 2015. Tsunami recurrence in the eastern Alaska-Aleutian arc: a Holocene stratigraphic record from Chirikof Island, Alaska. Geosphere 11 (4), 1172−1203.

Nentwig, V., Tsukamoto, S., Frechen, M., Bahlburg, H., 2015. Reconstructing the tsunami record in Tirua, Central Chile beyond the historical record with quartz-based SAROSL. Quatern. Geochronol. 30, 299−305.

Nichol, S.L., Kench, P.S., 2008. Sedimentology and preservation potential of carbonate sand sheets deposited by the December 2004 Indian Ocean tsunami: south Baa Atoll, Maldives. Sedimentology 55, 1173−1187.

Niwa, Y., Sugai, T., Matsushima, Y., Tod, S., 2017. Subsidence along the central to southern Sanriku coast, Northeast Japan, near the source region of the 2011 Tohoku-0ki earthquake, estimated from the Holocene sedimentary succession along a ria coast. Quatan. Int. 456, 1−16.

Ntageretzis, K., Vött, A., Emde, K., Fischer, P., Hadler, H., Röbke, B.R., Willershäuser, T., 2015a. Palaeotsunami record in near-coast sedimentary archives in southeastern Lakonia (Peloponnese, Greece). Z. Geomorphol. Suppl. Issues 59 (4), 275−299.

Ntageretzis, K., Vött, A., Fischer, P., Hadler, H., Emde, K., Röbke, B.R., Willershäuser, T., 2015b. Palaeotsunami history of the Elos plain (Evrotas river delta, Peloponnese, Greece). Z. Geomorphol. Suppl. Issues 59 (4), 253−273.

Ntageretzis, K., Vött, A., Fischer, P., Hadler, H., Emde, K., Röbke, B.R., Willershäuser, T., 2015c. Traces of repeated tsunami landfall in the vicinity of Limnothalassa Moustou (Gulf of Argolis−Peloponnese, Greece). Z. Geomorphol. Suppl. Issues 59 (4), 301−317.

Obrocki, L., Vött, A., Wilken, D., Fischer, P., Willershäuser, T., Koster, B., Lang, F., Papanikolaou, I., Rabbel, W., Reicherter, K., 2020. Tracing tsunami signatures of the AD 551 and AD 1303 tsunamis at the Gulf of Kyparissia (Peloponnese, Greece) using direct push in situ sensing techniques combined with geophysical studies. Sedimentology 67, 1274−1308.

Oetjen, J., Engel, M., Brückner, H., Pudasaini, S.P., Schüttrumpf, H., 2017. Enhanced field observation-based physical and numerical modelling of tsunami-induced boulder transport. Phase I: physical experiments. Coast. Eng. Proc. 35. Management 4.

Ohta, K., Naruse, H., Yokokawa, M., Vipaelli, E., 2017. New bedform phase diagrams and discriminant function for formative conditions of bedforms in open-channel flows. J. Geophys. Res. Earth Surf. 122, 2139−2158.

Oliveira, M.A., Andrade, C., Freitas, M.C., Costa, P., 2008. In: Using the Historical Record and Geomorphological Setting to Identify Tsunami Deposits in the Southern Coast of Algarve (Portugal) 2nd International tsunami Field Symposium. IGCP Project, vol.495.

Omira, R., Quartau, R., Ramalho, I., Baptista, M.A., Mitchell, N.C., 2016. The tsunami effects of a collapse of a volcanic island on a semienclosed basin: the Pico-Sao Jorge Channel in the Azores Archipelago. In: Duarte, J.C., Schellart, W.P. (Eds.), Plate Boundaries and Natural Hazards, Geophysical Monograph, vol. 219. American Geophysical Union, John Wiley & Sons, Inc., Washington, DC, pp. 271−287.

Ortlieb, L., Vargas, G., Saliege, J.F., 2011. Marine radiocarbon reservoir effect along the northern Chile-southern Peru coast (14-24°S) throughout the Holocene. Quatern. Res. 75, 91−103.

Papadopoulos, G.A., 2015. Tsunamis in the European- Mediterranean Region. From Historical Record to Risk Mitigation. Elsevier, Amsterdam, p. 271.

Paris, R., Giachetti, T., Chevalier, J., Guillou, H., Frank, N., 2011. Tsunami deposits in Santiago Island (Cape Verde archipelago) as possible evidence of a massive flank failure of Fogos volcano. Sediment. Geol. 239, 129−145.

Paris, R., Wassmer, P., Lavigne, F., Belousov, A., Belousova, M., Iskandarsyah, Y., Benbakkar, M., Ontowirjo, B., Mazzoni, N., 2014. Coupling eruption and tsunami records: the Krakatau 1883 case-study, Indonesia. Bull. Volcanol. 76, 814.

Paris, R., Coello Bravo, J.J., Martín González, M.E., Kelfoun, K., Nauret, F., 2017. Explosive eruption, flank collapse and mega-tsunami at Tenerife ca. 170 ky ago. Nat. Commun. 8, 15246.

Paris, R., Ramalho, R.S., Madeira, J., Ávila, S., May, S.M., Rixhon, G., Engel, M., Brückner, H., Herzog, M., Schukraft, G., Perez-Torrado, F.J., Rodriguez-Gonzales, A., Carracedo, J.C., Giachetti, T., 2018. Mega-tsunami conglomerates and flank collapses of ocean volcanoes. Mar. Geol. 395, 168−187.

Paris, R., Falvard, S., Chague, C., Goff, J., Etienne, S., Doumalin, P., 2020. Sedimentary fabric characterized by X-ray tomography: a case-study from tsunami deposits on the Marquesas Islands, French Polynesia. Sedimentology 67 (3), 1207−1229.

Perez-Torrado, F.J., Paris, R., Cabrera, M.C., Schneider, J.L., Wassmer, P., Carracedo, J.C., Rodriguez Santana, A., Santana, F., 2006. The Agaete tsunami deposits (Gran Canaria): evidence of tsunamis related to flank collapses in the Canary Islands. Mar. Geol. 227, 137−149.

Periancz, R., Abril, J.M., 2014. Modelling tsunamis in the eastern Mediterranean sea. Application in the Minoan Santorini tsunami sequence as a potential scenario for the hidtolical exous. J. Mar. Syst. 139, 91−102.

Peters, R., Jaffe, B., 2010. Identification of Tsunami Deposits in the Geologic Record; Developing Criteria Using Recent Tsunami Deposits, p. 39. U.S. Geological Survey Open-File Report 1239.

Peters, R., Jaffe, B., Geifenbaum, G., 2007. Distribution and sedimentary characteristics of tsunami deposits along the Cascadia margin North America. Sediment. Geol. 200, 372−386.

Peterson, C.D., Cruikshank, K.M., Schlichuring, R.B., Braunsten, S., 2010. Distal run-up records of Latest Holocene palaeotsunami inundation in alluvial flood plain: Neskowin and Beaver, Oregon, Central Cascadia Margin, West Coast U.S.A. J. Coastal Res. 26. (264), 622−634.

Peterson, C.D., Carver, G.A., Cruikshank, K.M., Abramson, H.F., Garrison-Laney, C.E., Dengler, L.A., 2011. Evaluation of the use of paleotsunami deposits to reconstruct inundation distance and runup heights associated with prehistoric inundation events, Crescent Cuty, southern Cascadia margin. Earth Surface Proc. Land. 36, 967−980.

Pham, D.T., Gouramanis, C., Switzer, A.D., Rubin, C.M., Jones, B.G., Jankaew, K., Carr, P.F., 2017. Elemental and mineralogical analysis of marine and coastal sediments from Phra Thong Island, Thailand: insights into the provenance of coastal hazard deposits. Mar. Geol. 385, 274−292.

Phantuwongrai, M., Choowong, M., 2012. Tsunami versus storm deposits from Thailand. Nat. Hazards 63, 31−58.

Pignatelli, C., Sanso, P., Mastronuzzi, G., 2009. Evaluation of tsunami flooding using geomorphologic evidence. Mar. Geol. 260, 6−18.

Pilarczyk, J.E., Reinhardt, E.G., 2012a. Homotrema rubrum (Lamarck) taphonomy as in overwash indicator in Marine Ponds on Anegada, British Virgin Islands. Nat. Hazards 63, 85−100.

Pilarczyk, J.E., Reinhardt, E.G., 2012b. Testing foraminiferal taphonomy as a tsunami indicator in a shallow and system lagoon; Sur, Sultanate of Oman. Mar. Geol. 295−298, 128−136.

Pilarczyk, I.E., Dura, T., Horton, B.P., Engelhart, S.E., Kemp, A.C., Sawai, Y., 2014. Microfossils from coastal environments as indicators of paleo-earthquakes tsunamis and storms. Palaeogeogr. Palaeclimatol. Palaeoecol. 413, 144−157.

Pilarczyk, J.E., Sawai, Y., Matsumoto, D., Namegaya, Y., Nishida, N., Ikehara, K., Fujiwara, O., Gouramanis, C., Dura, T., Horton, B.P., 2020. Constraining sediment provenance for tsunami deposits using distributions of grain size and foraminifera from the Kujukuri coastline and shelf, Japan. Sedimentology 67 (3), 1373−1392.

Pinegina, T.K., Bourgeois, J., 2020. Tephrostratigraphy and tephrochronology. In: Engel, M., et al. (Eds.), Geological Record of Tsunamis and Other Extensive Waves. Elsevier, pp. 745−759.

Pinegina, T.K., Bourgeois, J., Bazanova, L.L., Zelenin, E.A., Krasheninnikov, S.P., Portnyagin, M.V., 2020. Coseismic coastal subsidence associated with unusually wide rupture of prehistoric earthquakes on the Kamuchatka subduction zone: a record in buried erosional scarps and tsunami deposits. Quat. Sci. Rev. 233, 106171.

Piotrowski, A., Szczucinski, W., Sydor, P., Kotrys, B., Rzodkiewicz, M., Krzyminska, J., 2017. Sedimentary evidence of extreme storm surge or tsunami events in the southern Baltic Sea (Rogowo area, NW Poland). Geol. Q. 61, 973−986.

446 Selected bibliography

Pistolesi, M., Bertagnini, A., DiRoberto, A., Ripepe, M., Rosi, M., 2020. Tsunami and tephra deposits record interaction between past eruptive activity and landslides at Stromboli volcano, Italy. Geology 48, 436–440.

Polonia, A., Nelson, H., Romano, S., Vaiani, S.C., Colizza, E., Gasparotto, G., Gaspeparoto, L., 2016a. Seismic shaking tsunami wave erosion and generation of seismo-turbidites in the Ionian Sea. In: EGU General Assembly Conference Abstracts, vol. 18. Id. EPS2016-16602.

Polonia, A., Vaiani, S.C., de Lange, G.J., 2016b. Did the A.D. 365 Crete earthquake/tsunami trigger synchronous giant turbidity currents in the Mediterranan Sea? Geology 44 (3), 191–194.

Prendergast, A., September 2006. Echoes of Ancient Tsunamis. Australian Government Geoscience Australia. AUSGEO News, p. 5. Issue No. 83.

Priest, G.R., Witter, R.C., Zhang, Y.J., Goldfinger, C., Wang, K., Allan, J.C., 2017. New constraints on coseismic slip during southern Cascadia subduction zone earthquakes over the past 4600 years implied by tsunami deposits and marine turbidites. Nat. Hazards 88, 285–313.

Pufahl, P., Kane, I., Porta, G.D., Costa, P., Andrade, C., 2020. Tsunami geoscience: present geomorphologie, supplementary issues 54, issue 3 knowledge and future challenges. Sedimentology 67 (3), 1189–1600.

Putra, P.S., 2018. Tsunami sediments and their grain size characteristics. IOP Conf. Ser. Earth Environ. Sci. 118, 012035.

Putra, P.S., Nishimura, Y., Nakamura, Y., Yulianto, E., 2013. Source and transportation mode of the 2011 Tohoku-0ki tsunami deposits on the central east, Japan coast. Sediment. Geol. 294, 282–293.

Ramalho, R.S., Winckler, G., Madeira, J., Helffrich, G.R., Hipolito, A.R., Quartau, R., Adena, K., Schaefer, J.M., 2015. Hazard potential of volcanic flank collapses raised by new megatsunami evidence. Sci. Adv. 1, e1500456.

Ramirez-Herrra, M.T., Lagos, M., Huchnson, I., Kostogolodov, V., Machain, M.L., Caballero, M., Goguitchaichvili, A., Aguilar, B., Chagué-Goff, C., Goff, J., Ruiz-Fernández, A.-C., Ortiz, M., Nava, H., Bautista, F., Lopez, G.I., Quintana, P., 2012. Extreme wave deposits on the Pacific coast of Mexico: tsunamis or storms? A multi-proxy approach. Geomorphology 139–140, 360–371.

Rasmussen, H., Bondevik, S., Corner, G.D., 2018. Holocene relative sea level history and Storegga tsunami run-up in Lyngen, northern Norway. J. Quat. Sci. 33, 393–408.

Razjigaeva, N.G., Ganzey, L.A., Grebennikova, T.A., Arslanov, K.A., Ivanova, E.D., Ganzey, K.S., Kharlamov, A.A., 2018. Historical tsunami records on Russian Island, the Sea of Japan. Pure Appl. Geophys. 175, 1507–1523.

Reicherter, K., Vonberg, D., Koster, B., Fernander-Steeger, T., Grrutzne, C., Mathes-Scmidt, M., 2010. The sedimentary inventory of tsunamis along the southern Gulf of Cadiz (southwestern Spain). Z. Geomorphol. Suppl. Issue 3, 147–173.

Reymond, D., Hyvernaud, O., Okal, E.A., 2013. The 2010 and 2011 tsunamis in French Polynesia: operational aspects and field surveys. Pure Appl. Geophys. 170, 1169–1187.

Rhodes, B., Tuttle, M., Horton, B., Doner, L., Kelsey, H., Nelson, A., Cisternas, M., 2006. Paleotsunami research. EOS Trans. Am. Geophys. Union 87, 205–209.

Richmond, B.M., Buckley, M., Etienne, S., Chagué-Goff, C., Clark, K., Goff, J., Dominey-Howes, D., Strotz, L., 2011a. Deposits, flow characteristics, and landscape change resulting from the September 2009 South Pacific tsunami in the Samoan islands. Earth Sci. Rev. 107, 38–51.

Richmond, B.M., Watt, S., Buckley, M., Jaffe, B.E., Gelfenbaum, G., Morton, R.A., 2011b. Recent storm and tsunami coarse-clast deposit characteristics, southeast Hawai'i. Mar. Geol. 283, 79–89.

Riedesel, S., Brill, D., Roberts, H.M., Duller, G.A.T., Garrett, E., Zander, A.M., King, G.E., Tamura, T., Burow, C., Cunningham, A., Seeliger, M., De Batist, M., Heyvaert, V.M.A., Fujiwara, O., Brückner, H., The QuakeRecNankai Team, 2018. Single-grain feldspar luminescence chronology of historical extreme wave event deposits recorded in a coastal lowland, Pacific coast of central Japan. Quat. Geochronol. 45, 37–49.

Riha, k., Khupka, A., Costa, P.J.M., 2019. Image analysis applied to quartz grain microtextural provenance studies. Comput. Geosci. 25, 98–108.

Riou, B., Chaumillon, E., Schneider, J.L., Corrège, T., Chagué, C., 2020. The sediment-fill of Pago Pago Bay (Tutuila Island, American Samoa): new insights on the sediment record of past tsunamis. Sedimentology 67 (3), 1577−1600.

Rixhon, G., May, S.M., Engel, M., Mechernich, S., Schroeder- Ritzrau, A., Frank, N., Fohlmeister, J., Boulvain, F., Dunai, T., Brückner, H., 2018. Multiple dating approach (14C, 230Th/U and 36Cl) of tsunami-transported reef-top boulders on Bonaire (Leeward Antilles) - current achievements and challenges. Mar. Geol. 396, 100−113.

Röbke, B.R., Vött, A., 2017. The tsunami phenomenon. Prog. Oceanogr. 159, 296−322.

Röbke, B.R., Schüttrumpf, H., Fler, T.W., Fischer, P., Hadler, H., Ntageretzis, K., Willershäuser, T., Vött, A., 2013. Tsunami inundation scenarios for the Gulf of Kyparissia (western Peloponnese, Greece) derived from numerical simulation and geo-scientific field evidence. Z. Geomorphol. Suppl. Issues 57 (Issue 4), 69−104.

Röbke, B.R., Vött, A., Willershäuser, T., Fischer, P., Hadler, H., 2015. Considering coastal palaeogeographical changes in a numerical tsunami model−a progressive base to compare simulation results with field traces from three coastal settings in western Greece. Z. Geomorphol. Suppl. Issues 59 (Issue 4), 157−188.

Röbke, B.R., Schüttrumpf, H., Vött, A., 2016. Effects of different boundary conditions and palaeotopographies on the onshore response of tsunamis in a numerical model−A case study from western Greece. Continent. Shelf Res. 124, 182−199.

Roig-Munar, F.X., Vilaplana, J.M., Rodriguez-Perea, A., Martin-Prieto, J.A., Gelabert, B., 2018. Tsunamis boulders on the rocky shores of Minorca (Balearic Islands). Nat. Hazards Earth Syst. Sci. 18, 1985−1998.

Romundset, A., Bondevik, S., 2011. Propagation of the Storegga tsunami into ice-free lakes along the southern shores of the Barents Sea. J. Quat. Sci. 26, 457−462.

Rubin, C., Horton, B., Sieh, K., Pilarczyk, J., Daly, P., Ismail, N., Parnell, A., 2017. Highly variable recurrence of tsunamis in the 7,400 years before the 2004 Indian Ocean tsunami. Nat. Commun. 8, 1−12.

Ruiz, F., Abad, M., Cáceres, L.M., Rodríguez Vidal, J., Carretero, M.I., Pozo, M., González-Regalado, M.L., 2010. Ostracods as tsunami tracers in Holocene sequences. Quatern. Res. 73, 130−135.

Ryan, W.B.F., Carbotte, S.M., Coplan, J., O'Hara, S., Melkonian, A., Arko, R., Weissel, R.A., Ferrini, V., Goodwillie, A., Nitsche, F., Bonczkowski, J., Zemsky, R., 2009. Global multi-resolution topography (GMRT) synthesis data set. Geochem. Geophys. Geosyst. 10, Q03014.

Salamon, A., Rockwell, T., Ward, S.N., Guidoboni, E., Comastri, A., 2007. Tsunami hazard evaluation of the eastern Mediterranean historical analysis and selected modeling. Bull. Seismol. Soc. Am. 97, 705−724.

Salem, E.S.M., 2009. Paleo-tsunami deposits on the Red beach, Egipt. Arab. J. Geosci. 2, 185−197.

Satake, K., Atwater, B., 2007. Long-term perspectives on giant earthquakes and tsunamis at subduction zones. Annu. Rev. Earth Planet Sci. 35, 349−374.

Satake, K., Aung, T.T., Sawai, Y., Okamura, Y., Win, K.S., Swe, W., Swe, C., Swe, T.L., Tun, S.T., Soe, M.M., Oo, T.Z., Zaw, S.H., 2006. Tsunami heights and damage along the Myanmar coast from the December 2004 Sumatra-Andaman earthquake. Earth Planet Space 58, 143−152.

Sawai, Y., Kamataki, T., Shishikura, M., Nasu, H., Okamura, Y., Satake, K., Thomson, K.H., Matsumoto, D., Fujii, Y., Komatsubara, J., Aung, T.T., 2009b. Aperiodic recurrence of geologically recorded tsunamis during the past 5500 years in eastern Hokkaido, Japan. J. Geophy. Res. 114, B01319.

Sawai, Y., Namegaya, Y., Tamura, T., Nakashima, R., Tanigawa, K., 2015. Shorter intervals between great earthquakes near Sendai: Scour ponds and a sand layer attributable to A.D. 1454 overwash. Geophys. Res. Lett. 42, 4795−4800.

Sawai, Y., Tanigawa, K., Tamura, T., Namegaya, Y., 2016. Medieval coastal inundation revealed by a sand layer on the Ita lowland adjacent to the Suruga Trough, central Japan. Nat. Hazards 80, 505−519.

Scheffers, A., Kelletat, D., Scheffers, S.R., Abbott, D.H., Bryant, E.A., 2008a. Chevrons-enigmatic sedimentary coastal features. Z. Geomorphol. 52, 375−402.

Scheffers, S.R., Scheffers, A., Kelletat, D., Bryant, E.A., 2008c. The Holocene paleo-tsunami history of West Australia. Earth Planet Sci. Lett. 270, 137−146.

448 Selected bibliography

Schichting, R.B., Perterson, C.D., 2006. Mapped overland distance of paleotsunami high velocity inundation on back-barrer wetland of the central Cascadia Margin, USA. J. Geol. 114, 577−592.

Schneider, J.-L., Chagué-Goff, C., Bouchez, J.-L., Goff, J., Sugawara, D., Goto, K., Jaffe, B., Richmond, B., 2014. Using magnetic fabric to reconstruct the dynamics of tsunami deposition on the Sendai Plain, Japan — The 2011 Tohoku-oki tsunami. Mar. Geol. 358, 89−106.

Schneider, B., Hoffman, G., Falkenroth, M., Grade, J., 2019. Tsunami and storm sediments in Oman: characterizing extreme wave deposits using terrestrial laser scanning. J. Coast Conserv. 23, 801−815.

Sciccitano, G., Costa, B., Stefano, A., Longhitano, S.G., Monaco, C., 2010. Tsunami and storm deposits preserved within a ria-type rocky coastal setting (Siracusa, SE Sicily). Z. Geomorphol. Suppl. Issues 54 (3), 51−77.

Scicchitano, G., Pignatelli, C., Spampinato, C.R., Piscitelli, A., Milella, M., Monaco, C., Mastronuzzi, G., 2012. Terrestrial Laser Scanner techniques in the assessment of tsunami impact on the Maddalena peninsula (south-eastern Sicily, Italy). Earth Planets Space 64, 889−903.

Scourse, E.M., Chapman, N.A., Tappin, D.R., Wallis, S.R. (Eds.), 2018. Tsunamis: Geology, Hazards and Risks, vol. 456. Geological Society, London, Special Publications.

Seardino, G., Piscittell, A., Milella, M., Sanso, R., Mastronuzzi, G., 2020. Tsunami fingerprints along the Mediteranean coasts. Rendicanti Lincei. Sci. Fisch Nat. 1−17.

Seike, K., Shirai, K., Kogure, Y., 2013. Disturbance of shallow marine soft-bottom environments and megabenthos assemblages by a huge tsunami induced by the 2011 M9.0 Tohoku-Oki earthquake. PLoS One 8, e65417.

Seike, K., Kitahashi, T., Noguchi, T., 2016. Sedimentary features of Onagawa Bay, northeastern Japan after the 2011 off the Pacific coast of Tohoku earthquake: sediment mixing by recolonized benthic animals decreases the preservation potential of tsunami deposits. J. Ocean. 72, 141−149.

Seike, K., Kobayashi, G., Kogure, K., 2017. Postdepositional alteration of shallow-marine tsunami-induced sand layers: a comparison of recent and ancient tsunami deposits, Onagawa Bay, northeastern Japan. Isl. Arc 26, e12174.

Sepic, J., Vilibic, I., Rabinovich, A.B., Monserrat, S., 2015. Widespread tsunami-like waves of 23−27 June in the Mediterranean and Black Seas generated by highaltitude atmospheric forcing. Sci. Rep. 5, 11682.

Sepic, J., Vilibic, I., Rabinovich, A., Tinti, S., 2018. Meteotsunami ("Marrobbio") of 25−26 June 2014 on the Southwestern Coast of Sicily, Italy. Pure Appl. Geophys. 175, 1573−1593.

Shah-Hosseini, M., Morhange, C., Naderi, B.A., Narriner, N., Lahijani, H., Hamzeh, M., Sabatier, F., 2011. Coastal boulders as evidence for high-energy waves on the Iranian coast of Makran. Mar. Geol. 290, 17−28.

Shah-Hosseini, M., Saleem, A., Mahmoud, A.M.A., Morhange, C., 2016. Coastal boulder deposits attesting to large wave impacts on the Mediterranean coast of Egypt. Nat. Hazards 83, 849−865.

Shanmugam, G., 2012. Process-sedimentological challenges in distinguishing paleo-tsunami deposits. Nat. Hazards 63, 5−30.

Shaw, C.E., Benson, L., 2015. Possible tsunami deposits on the Caribbean coast of the Yucatan Peninsula. J. Coastal Res. 31, 1306−1316.

Shaw, B., Ambraseys, N.N., England, P.C., Floyd, M.A., Gorman, G.J., Higham, T.F.G., Jackson, J.A., Nocquet, J.-M., Pain, C.C., Piggott, M.D., 2008. Eastern Mediterranean tectonics and tsunami hazard inferred from the AD 365 earthquake. Nat. Geosci. 1, 268−276.

Shimada, Y., Fujino, S., Sawai, Y., Tanigawa, K., Matsumoto, D., Momohara, A., Saito-Kato, M., Yamada, M., Hirayama, E., Suzuki, T., Chagué, C., 2019. Geological record of prehistoric tsunamis in Mugi Town, facing the Nankai Trough, western Japan. Prog. Earth Planet. Sci. 6 (33).

Shimazaki, K., Kim, H.Y., Chiba, T., Satake, K., 2011. Geological evidence of recent great Kanto earthquakes at the Miura Peninsula, Japan. J. Geophys. Res. 116, B12408.

Shinozaki, T., Sawai, Y., Ito, K., Hara, J., Matsumoto, D., Tanigawa, K., Pilarczysk, J., 2020. Recent and historical tsunami deposits from Lake Tokotan, eastern Hokkaido, Japan, inferred from nondestructive, grain-size, and radioactive cesium analyses. Nat. Hazards 103, 713−730.

Shishikura, M., 2014. History of the paleo-earthquakes along the Sagami Trough, central Japan: review of coastal paleoseismological studies in the Kanto region. Episodes 37, 246–257.

Sieh, K., Daly, P., McKinnon, E.E., Pilarczyk, J.E., Chiang, H.W., Horton, B., Rubin, C.M., Shen, C.C., Ismail, N., Vane, C.H., Feener, R.M., 2015. Penultimate predecessors of the 2004 Indian Ocean tsunami in Aceh, Sumatra: stratigraphic, archaeological, and historical evidence. J. Geophys. Res. Solid Earth 120, 308–325.

Smedile, A., De Martini, P.M., Bellucci, L., Gasperini, L., Segnotti, L., Del Carlo, P., Plonia, A., Pantosti, D., Barbano, M.S., Gerardi, F., 2008. Paleotsunami deposits in the Augusta Bay area (Eastern Sicily, Italy): preliminary results from offshore data. Comput. Geosci. 34 (2), 103–114.

Smedile, A., Molisso, F., Chague, C., Iorio, M., De Martini, P.M., Pinzi, S., Collins, P.E.F., Sagnotti, L., Pantosti, D., 2020. New coring study in Augusta Bay expands understanding of offshore tsunami deposits (Eastern Sicily, Italy). Sedimentology 67, 1553–1576.

Soria, J.L.A., Switzer, A., Pilarczyk, J., Siringan, F., Khan, N., Fritz, H., 2017. Typhoon Haiyan overwash sediments from Leyte Gulf coastlines show local spatial variations with hybrid storm and tsunami signatures. Sediment. Geol. 358, 121–138.

Spiske, M., Bahlburg, H., 2011. A quasi-experimental setting of coarse clast transport by the 2010 Chile tsunami (Bucalemu, Central Chile). Mar. Geol. 289, 72–85.

Spiske, M., Böröcz, Z., Bahlburg, H., 2008. The role of porosity in discriminating between tsunami and hurricane emplacement of boulders—a case study from the Lesser Antilles, southern Caribbean. Earth Planet Sci. Lett. 268 (3–4), 384–396.

Spiske, M., Piepenbreier, J., Benavente, C., Bahlburg, H., 2013a. Preservation potential of tsunami deposits on arid siliciclastic coasts. Earth Sci. Rev. 126, 58–73.

Spiske, M., Piepenbreier, J., Benavente, C., Kunz, A., Bahlburg, H., Steffahn, J., 2013b. Historical tsunami deposits in Peru - sedimentology, inverse modeling and optically stimulated luminescence dating. Quatern. Int. 305, 31–44.

Spiske, M., Garcia Garcia, A.-M., Tsukamoto, S., Schmidt, V., 2020a. High-energy inundation events versus long-term coastal processes - room for misinterpretation. Sedimentology 67, 1460–1480.

Spiske, M., Tang, H., Bahlburg, H., 2020b. Post-depositional alteration of onshore tsunami deposits – implications for the reconstruction of past events. Earth Sci. Rev. 202, 103068.

Srinivasalu, S., Thangadurai, N., Switzer, A., Mohan, V.R., Ayyamperumal, T., 2007. Erosion and sedimentation in Kalpakkam (N Tamil Nadu, India) from the 26th December 2004 tsunami. Mar. Geol. 240, 65–75.

Stephenson, W.J., Naylor, L.A., 2011. Geological controls on boulder production in a rock coast setting: insights from South Wales, UK. Mar. Geol. 283, 12–24.

Sugawara, D., Takahashi, T., 2014. Numerical simulation of coastal sediment transport by the 2011 Tohoku-oki earthquake tsunami. In: Kontar, Y.A., Santiago-Fandino, V., Takahashi, T. (Eds.), Tsunami Events and Lessons Learned, Advances in Natural and Technological Hazards Research. Springer, pp. 99–112.

Sugawara, D., Imamura, F., Goto, K., Matsumoto, H., Minoura, K., 2013. The 2011 Tohoku-oki earthquake tsunami: similarities and differences between the 869 Jogan Tsunami on the Sendai Plain. Pure Appl. Geophys. 170, 831–843.

Sugawara, D., Jaffe, B., Goto, K., Gelfenbaum, G., La Selle, S., 2015. Exploring hybrid modeling of tsunami flow and deposit characteristics. In: Wang, P., Rosati, D., Cheng, J. (Eds.), The Proceedings of the Coastal Sediments 2015.

Sugawara, D., Yu, N.T., Yen, J.Y., 2019. Estimating a tsunami source by sediment transport modeling: a primary attempt on a historical/1867 normal-faulting tsunami in Northern Taiwan. J. Geophys. Res. Earth Surface 124, 1675–1700.

Switzer, A.D., Burston, J.M., 2010. Competing mechanisms for boulder deposition on the southeast Australian coast. Geomorphology 114, 42–54.

450 Selected bibliography

Switzer, A.D., Jones, B.G., 2008. Large-scale washover sedimentation in a freshwater lagoon from the southeast Australian coast: sea-level change, tsunami or exceptionally large storm? Holocene 18, 787–803.

Szczucinski, W., Chaimanee, N., Niedzielski, P., Rachlewicz, G., Saisuttichai, D., Tepsuwan, T., Lorenc, S., Siepak, J., 2006. Environmental and geological impacts of the 26 December 2004 tsunami in coastal zone of Thailand - overview of short and long-term effects. Pol. J. Environ. Stud. 15, 793–810.

Szczucinski, W., Niedzielski, P., Kozak, L., Frankowski, M., Ziola, A., Lorenc, S., 2007. Effects of rainy season on mobilization of contaminants from tsunami deposits left in a coastal zone of Thailand by the 26 December 2004 tsunami. Environ. Geol. 53, 253–264.

Szczuciński, W., Rachlewicz, G., Chaimanee, N., Saisuttichai, D., Tepsuwan, T., Lorenc, S., 2012a. 26 December 2004 tsunami deposits left in areas of various tsunami run up in coastal zone of Thailand. Earth Planets Space 64, 5.

Szczuciński, W., Pawłowska, J., Lejzerowicz, F., Nishimura, Y., Kokociński, M., Majewski, W., Nakamura, Y., Pawlowski, J., 2016. Ancient sedimentary DNA reveals past tsunami deposits. Mar. Geol. 381, 29–33.

Tachibana, T., 2013. Lonestones as indicators of tsunami deposits in deep-sea sedimentary rocks of the Miocene Morozaki Group, central Japan. Sediment. Geol. 289, 62–73.

Tachibana, T., Tsuji, Y., 2011. Geological and hydrodynamical examination of the bathyal tsunamigenic origin of Miocene conglomerates in Chita peninsula, Central Japan. Pure Appl. Geophys. 168 (6–7), 997–1014.

Takashimizu, Y., Kawamura, Y., Rodriguez-Tovar, F.J., Dorador, J., Ducassou, E., Hemandez-Molina, F.J., Stow, D.A., Alvarez-Zarikan, 2016. Reworked tsunami deposits by bottom current: circumstantial evidences from Late Pleistcene to early Holocene in the Gulf of Cadiz. Mar. Geol. 377, 95–602.

Takeda, H., Goto, K., Goff, J., Matsumoto, H., Sugawara, D., 2018. Could tsunami risk be under-estimated using core-based reconstructions? Lessons from ground penetrating radar: GPR - a key first step for palaeotsunami research. Earth Surf. Process. Landforms 43 (4), 808–816.

Tang, H., Wang, J., Weiss, R., Xiao, H., 2018. TSUFLIND-EnKF: inversion of tsunami flow depth and flow speed from deposits with quantified uncertainties. Mar. Geol. 396, 16–25.

Tanigawa, K., Sawai, Y., Shishikura, M., Namegaya, Y., Matsumoto, D., 2014. Geological evidence for an unusually large tsunami on the Pacific coast of Aomori, northern Japan. J. Quatern. Sci. 29, 200–208.

Tanigawa, K., Shishikura, M., Fujiwara, O., Namegaya, Y., Matsumoto, D., 2018. Mid-to late-Holocene marine inundations inferred from coastal deposits facing the Nankai Trough in Nankoku, Kochi Prefecture, southern Japan. Holocene 28, 867–878.

Sedimentary features of tsunami deposits - their origin, recognition and discrimination: an introduction. In: Tappin, D.R. (Ed.), Sediment. Geol. 200, 151–388.

Tsuji, I., 2013. Catalog of distant tsunamis reaching Japan from Chile and Peru. Rep. Tsunami Eng. 30, 61–68.

Uchida, J., Fujiwara, O., Hasegawa, S., Kamataki, T., 2010. Sources and depositional processes of tsunami deposits: analysis using foraminiferal tests and hydrodynamic verification. Isl. Arc 19, 427–442.

Uchiyama, S., Machida, J., Hoyanagi, K., 2019. Tsunami deposits in Holocene estuary sediments, based on lithologic study of core from two drilling sites in Odaka district, Minamisoma City, Fukushima Prefecture, Northeast Japan. J. Geol. Soc. Jpn. 78 (1), 3–14.

Usumi, K., Ikehara, K., Jenkins, R.G., Ashi, J., 2017. Benthic foraminiferal evidence of deep-sea sediment transport by the 2011 Tohoku-oki earthquake and tsunami. Mar. Geol. 384, 214–224.

Vaela, A.N., Richiano, S., Poire, D.G., 2011. Tsunami vs. storm origin for shell bed deposits in a lagoon environment: an example from the Upper Cretaceous of Southern Patagonia, Argentina. Lat. Am. J. Sedimentol. Basin Anal. 18, 63–85.

Vött, A., Kelletat, D., 2015a. Holocene palaeotsunami landfalls and neotectonic dynamics in the western and southern Peloponnese. Z. Geomorphol. Suppl. Issues 59 (issue 4), 1–5.

Vött, A., Kelletat, D., 2015b. Holocene Palaeotsunami landfalls and neotectonic dynamics in the Western and Southern Peloponnese (Greece): the past as a key to the present and predictor of the future. Z. Geomorphol. Suppl. Issues 59 (Issue 4).

Vött, A., Brückner, H., Brockmüller, S., Handl, M., May, S.M., Gaki-Papanastassiou, K., Herd, R., Lang, F., Maroukian, H., Nelle, O., Papanastassiou, D., 2009a. Traces of Holocene tsunamis across the sound of Lefkada, NW Greece. Glob. Planet Change 66, 112—128.

Vött, A., Brückner, H., May, S.M., Sakellariou, D., Nelle, O., Lang, F., Kapsimalis, V., Jahns, S., Herd, R., Handl, M., Fountoulis, I., 2009b. The Lake Voulkaria (Akarnania, NW Greece) palaeoenvironmental archive—a sediment trap for multiple tsunami impact since the mid-Holocene. Z. Geomorphol. Suppl. Issues 53 (1), 1—37.

Vött, A., Bareth, G., Brückner, H., Curdt, C., Fountoulis, I., Grapmayer, R., Hadler, H., Hoffmeister, D., Klasen, N., Lang, F., Masberg, P., May, S.M., Ntageretzis, K., Sakellariou, D., Willershäuser, T., 2010. Beachrock-type calcarenitic tsunamites along the shores of the eastern Ionian Sea (western Greece)—case studies from Akarnania, the Ionian Islands and the western Peloponnese. Z. Geomorphol. Suppl. Issues 54 (3), 1—50.

Vött, A., Lang, F., Brückner, H., Gaki-Papanastassiou, K., Maroukian, H., Papanastassiou, D., Giannikos, A., Hadler, H., Handl, M., Ntageretzis, K., Willershäuser, T., Zander, A., 2011. Sedimentological and geoarchaeological evidence of multiple tsunamigenic imprint on the Bay of Palairos-Pogonia (Akarnania, NW Greece). Quat. Int. 242 (1), 213—239.

Vött, A., Reicherter, K., Papanikolaou, I., 2013. Reconstruction and modeling of Palaeotsunami events -multi-proxy approaches, geophysical studies, numerical simulations. Z. Geomorphol. Suppl. Issues 57 (Issue 4).

Vött, A., Fischer, P., Röbke, B.R., Werner, V., Emde, K., Finkler, C., Hadler, H., Handl, M., Ntageretzis, K., Willershäuser, T., 2015. Holocene fan alluviation and terrace formation by repeated tsunami passage at Epitalio near Olympia (Alpheios River valley, Greece). Z. Geomorphol. Suppl. Issues 59 (4), 81—123.

Vött, A., Bruins, H.J., Gawehn, M., Goodman-Tchernov, B.N., De Martini, P.M., Kelletat, D., Mastronuzzi, G., Reicherter, K., Röbke, B.R., Scheffers, A., Willershäuser, T., Avramidis, P., Bellanova, P., Costa, P.J.M., Finkler, C., Hadler, H., Koster, B., Lario, J., Reinhardt, E., Mathes-Schmidt, M., Ntageretzis, K., Pantosti, D., Papanikolaou, I., Sansò, P., Scicchitano, G., Smedile, A., Szczuciski, W., 2019. Publicity waves based on manipulated geoscientific data suggesting climatic trigger for majority of tsunami findings in the Mediterranean — response to 'Tsunamis in the geological records: making waves with a cautionary tale from the Mediterranean' by Marrier et al. (2017). Z. Geomorphol. Suppl. Issues. 62 (Issue 2), 7—45.

Wagner, B., Bennike, O., Klug, M., Cremer, H., 2007. First indication of Storegga tsunami deposits from East Greenland. J. Quat. Sci. 22, 321—325.

Wallis, S.R., Fujiwara, O., Goto, K., 2018. Geological Studies in Tsunami Research since the 2011 Tohoku Earthquake, vol. 456. Geological Society London Spec. Publication, pp. 39—53.

Wassmer, P., Schneider, J.-L., Fonfrege, A.V., Lavigne, F., Paris, R., Gomez, C., 2010. Use of anisotropy of magnetic susceptibility (AMS) in the study of tsunami deposits: application to the 2004 deposits on the eastern coast of Banda Aceh, North Sumatra, Indonesia. Mar. Geol. 275, 255—272.

Watanabe, M., Goto, K., Imamura, F., Hongo, C., 2016. Numerical identification of tsunami boulders and estimation of local tsunami size at Ibaruma reef of Ishigaki Island, Japan. Isl. Arc 25, 316—332.

Watanabe, M., Goto, K., Bricker, J.D., Imamura, F., 2018. Are inundation limit and maximum extent of sand useful for differentiating tsunamis and storms? An example from sediment transport simulations on the Sendai Plain, Japan. Sediment. Geol. 364, 204—216.

Watanabe, M., Goto, K., Imamura, F., Kennedy, A., Sugawara, D., Nakamura, N., Tonosaki, T., 2019. Modeling boulder transport by coastal waves on cliff topography: case study at Hachijo Island, Japan. Earth Surf. Proc. Land. 44 (15), 2939—2956.

Weiss, R., 2008. Sediment grains moved by passing waves: tsunami deposits in deep water. Mar. Geol. 250, 251—257.

Weiss, R., 2012. The mystery of boulders moved by tsunamis and storms. Mar. Geol. **295**—298, 28—33.

Weiss, R., Bourgeois, J., 2012. Understanding sediment-reducing tsunami risk. Science 336, 1117—1118.

Werner, V., Baika, K., Fischer, P., Hadler, H., Obrocki, L., Willershäuser, T., Tzigounaki, A., Tsigkou, A., Reicherter, K., Papanikolaou, I., Emde, K., Vött, A., 2018. The sedimentary and geomorphological imprint of the AD 365 tsunami on the coasts of southwestern Crete (Greece) - examples from Sougia and Palaiochora. Quatern. Int. 473, 66–90.

Willershäuser, T., Vött, A., Hadler, H., Henning, P., Ntageretzis, K., 2012. Evidence of high-energy impact near Kato Samiko, Gulf of Kyparissia (western Peloponnese, Greece), during history. Beiträge der 29, 26–36.

Willershäuser, T., Vött, A., Hadler, H., Fischer, P., Röbke, B., Ntageretzis, K., Kurt, E., Brückner, H., 2015. Geo-scientific evidence of tsunami impact in the Gulf of Kyparissia (western Peloponnese, Greece). Z. Geomorphol. Suppl. Issues 59 (4), 43–80.

Witter, R.C., Briggs, R.W., Engelhart, S.E., Gelfenbaum, G., Koehler, R.D., Barnhart, W.D., 2014. Little late Holocene strain accumulation and release on the Aleutian megathrust below the Shumagin Islands, Alaska. Geophys. Res. Lett. 41, 2359–2367.

Witter, R.C., Carver, G.A., Briggs, R.W., Gelfenbaum, G., Koehler, R.D., La Selle, S.P., Bender, A.M., Engelhart, S.E., Hemphill-Haley, E., Hill, T.D., 2016. Unusually large tsunamis frequent a currently creeping part of the Aleutian megathrust. Geophys. Res. Lett. 43, 76–84.

Witter, R.C., Briggs, R.W., Engelhart, S.E., Gelfenbaum, G., Koehler, R.D., Nelson, A.R., Wallace, K.L., 2018. Evidence for frequent, large tsunamis spanning locked and creeping parts of the Aleutian megathrust. G S A Bull. 131, 707–729.

Yamada, K., Terakura, M., Tsukawaki, S., 2014b. The impact on bottom sediments and ostracods in the Khlong Thom River mouth following the 2004 Indian Ocean Tsunami. Paleontol. Res. 18, 104–117.

Yamada, M., Fujino, S., Chiba, T., Goto, K., Goff, J., 2020. Redeposition of volcaniclastic sediments by a tsunami 4600 years ago at Kushima City, south-eastern Kyushu, Japan. Sedimentology 67, 1354–1372.

Yanagisawa, H., Goto, K., Sugawara, D., Kanamaru, K., Iwamoto, N., Takamori, Y., 2016. Tsunami earthquake can occur elsewhere along the Japan Trench—historical and geological evidence for the 1677 earthquake and tsunami. J. Geophys. Res. Solid Earth 121, 3504–3516.

Yawsangratt, S., Szczucinski, W., Chaimanee, N., Chatprasert, S., Majewski, W., Lorenc, S., 2012. Evidence of probable paleotsunami deposits on KHO Island, Phang, Nga Province, Thailand. Nat. Hazards 63, 151–163.

Yoshii, T., Tanaka, S., Matsuyama, M., 2017. Tsunami deposits in a super-large wave flume. Mar. Geol. 391, 98–107.

Yoshii, T., Tanaka, S., Matsuyama, M., 2018. Tsunami inundation, sediment transport, and deposition process of tsunami deposits on coastal lowland inferred from the Tsunami Sand Transport Laboratory Experiment (TSTLE). Mar. Geol. 400, 107–118.

Index

Note: 'Page numbers followed by "*f*" indicate figures and "*t*" indicate tables.'

A

Age of deposition, 358
Alaska earthquake (1964 CE),
 59–60
Alaska tsunami (1958), 409
Amphistegina, 213–214
Ancient tsunamis, 109
 quantitative evaluation of,
 43–46
Andaman–Sumatra tsunami,
 360–362
Aniakchak tsunami deposits
 (3500 BP), 249–250
Ansei-Tokai earthquake tsunami
 (1854), 32, 33f
Anthropogenic materials, 89
Antidunes, 28–29, 57
Ariadnaesporites sp., 277
Asthenosphere, 11
Atlantic Ocean, 346–347
Augias turbidite, 214, 226

B

B-type (mega)-turbidite, 5
Back-arc basin, 6
Backwash, tsunami, 28–29,
 162–163, 184, 186
 conceptual model, 184f
 current of tsunamis, 233–234
 sedimentological features
 associated with, 38–40
Bahamas carbonate platform, 307
Bang Sak beach, Thailand,
 135–139
 map, 131f
 tsunami deposits, 137f
 and sedimentary structures,
 139f
 tsunami sediments and erosion,
 138f
Beach, 397

Bed C, 294
Bed D
 characteristics, 286–291, 289f,
 293t
 semiquantitative modeling of
 depositional process,
 291–294
 shear velocities *vs.* wave
 heights, 293t
Bed I, 290f, 291
Bed load, 25–26
Bedforms reflecting tsunami
 waveform, 55
Beloc, K/T-boundary tsunami
 deposits, 345
Bioclasts, 187–188
 characteristics, 286–291
 sedimentary structures, 294
 semiquantitative modeling,
 291–294
 graded skeletal sheet, 286
 grain-size distribution, 308
 microspherules, 288, 298
 shear velocity, 293t
 shocked quartz grains, 284f, 288
 turbidity currents, 293
Bioturbation, 165, 266
Black Sea Basin, 193
Bolide impacts, 410
Boso Peninsula, storm waves
 around, 67
Bottom of bays, 397–398
Boulder deposits, 355–356,
 360–362
 deposited by paleotsunami
 events, 363–372
 bioerosive rock pools on
 limestone boulders, 365f
 field of platy boulders, 368f
 giant boulders near Glass
 Window Bridge, 372f

 imbrication of very large
 platy boulders, 366f
 platy boulders, 370f
 ridge-like boulder
 accumulation, 369f
 imbrication of, 356f
Boulders, 10, 353, 354f,
 355–356
 at Laem Pakarang, south-west
 Thailand, 37f
 long-time large stable boulders,
 364f
 tongue of, 359f
Bouma sequence, 227, 229, 233,
 347, 391
Brazos River section, 343
Bronze Age Homogenites, 194
Bronze-Age destructive eruption,
 207

C

Cacarajícara Formation,
 317–318
 deposited in Yucatán platform,
 307
 grain and mineral compositions,
 323
 sedimentary logs, 317f, 342f
 thickness of, 317–318
 in western Cuba, 307, 316
Calcareous nannofossils, 230
Camebax SX50 EPMA, 272–273
Cape, 16
Cape Pakarang boulder field,
 360–362, 361f, 361t
Casares, site survey for tsunami,
 91
Cascadia Subduction Zone
 (CSZ), 110, 194
Catastrophic events, 1
Chaotic changes, 409–410

Index

Chelegenic Cycle, 412
Chicxulub crater, 305, 307,
    339–347, 341f
  genesis of tsunamis, 305–307,
    324–327
  K/T-boundary tsunami deposits
    around, 341f, 347–348
    on edge of Yucatán platform,
    340–343
    Gulf of Mexico region,
    343–344
    K/T-boundary impact event,
    339–340
    proto-Caribbean sea region,
    344–346
    sedimentary logs, 342f
Chicxulub event, 353–355
Chilean earthquake tsunami
    (1960), 9, 20–21, 31, 36,
    40, 383
CHIRP. *See* Compressed high-
    intensity radar pulse
Cidra
  calcirudites in, 312
  distribution of Peñalver
    formation, 308–309
Clay-sized particles, 29, 36
Coarse-grained dunes, 57
Coastal beach role, in tsunami
    records, 397
Coastal boulder, 358
Coastal lacustrine basin, 396
Coastal plain, 395–396
"Cobblestone" scientific program,
    208–209
Complex craters, 332–333
Compressed high-intensity radar
    pulse (CHIRP), 214
Contusotruncana contusa, 277
Convergent boundaries, 11
Corinto, site survey for
    tsunami, 90
Costal boulders, 357–359
Crater-generated tsunamis,
    326–327, 334–336.
    See also Oceanic impact
    craters
Cretaceous Cuban arc,
    307, 345
Cretaceous Miyako Group, 417

Cretaceous/Tertiary boundary (K/
    T boundary), 14–15, 263,
    305–306, 353–355
Cacarajícara Formation,
    317–318
characteristic sedimentary
    successions, 14–15
comparison of deep-sea tsunami
    deposits, 316–324
compositional variations deep-
    sea tsunami deposits, 324
deep-sea tsunamiites, 6
DSDP sites
  comparison of thickness and
    sedimentary structures,
    323–324
  lithology, 320–321
  sedimentary logs, 321f
  sedimentary mechanism, 322
  stratigraphical setting, 320
impact event, 339–340
implications for genesis and
    number of tsunami
    currents, 324–327
mass extinction, 305
meteorite impact-induced
    tsunamiites, 2
Moncada Formation, 318–319
paleogeography of proto-
    Caribbean sea and
    geological setting,
    307–308
Peñalver Formation
  paleogeotectonic setting of,
    306f
  sedimentary processes of,
    308–316
sandstone complex, 343–344
tsunami deposits, 332, 347–348
  around Chicxulub crater, 341f
  on edge of Yucatán platform,
    340–343
  Gulf of Mexico region,
    343–344
  proto-Caribbean sea region,
    344–346
  sedimentary logs, 342f
tsunamiite, 2
variations in thickness and
    lithology, 313t

Cretaceous–Paleogene boundary
    (K/Pg boundary),
    263–264
Cricotriporites almadaensis, 277
Critical condition, 23–24
Cross-stratification, 28–29
CSZ. *See* Cascadia Subduction
    Zone
Cuba, K/T-boundary tsunami
    deposits, 344–346
Cuban volcanic arc, 345
Current ripples, 346–347
Current velocity (U_{max}), 16, 22

D

DART, 156
Debris avalanches, 14
Debris-flow deposits, 6
Deep Sea Drilling Project
    (DSDP), 307–308, 342f,
    344–346
  comparison of thickness and
    sedimentary structures,
    323–324
  lithology, 320–321
  sedimentary logs, 321f
  sedimentary mechanism, 322
  stratigraphical setting, 320
Deep-sea
  bottom erosion, 230–231
  environments, 399
  tsunami deposits, 2
Deep-sea homogenites, 185,
    208–215
  absence of tephra Z-2, 216
  data, 216
  type A homogenites, 209–213,
    220
  type B homogenites, 213–215,
    220
Deep-sea tsunamiite, 225–226
  comparison with other deep-sea
    tsunamiites, 234–235
  K/Pq-boundary, 235
Deep-towed echo sounder, 212
Deep-water highs, 226
Deformation, shallow-water,
    17–18
Density-flow materials, 232
Deposit grading, 105

Index

Depositional model in shallow water, 60–61
Depositional sorting, 103
Deposits, 5
"Diary of Moriai Family", 198–199, 198f
Diastrophic occurrence, 383
Diatom assemblages, 159–160, 160f
Dinogymnium, 277
Distal ejecta, 337–338
Distal turbidites, 214
Divergent boundaries, 11
Documented Jogan tsunami of (July 13, 869 CE), 201
Dosinella penicillata, 67
DSDP. *See* Deep Sea Drilling Project

E

Earthquake-induced tsunamiite occurrences, 411–416
and tectonics, 407–408
Earthquake-induced tsunamis, 11–12
Earthquakes, 1, 12–13, 411
changes in frequency, 413f
Eastern Mediterranean
Bronze-Age destructive eruption, 207, 216
deep-sea record, 209
Island of Santorini, 208f
map showing physiographic features, 210f
stratigraphic correlation of eight long cores, 219f
Echitriporites trianguliformis, 277
Edge waves, 79
EDS. *See* Energy Dispersive Spectrometer/Spectrometry
El Transito, site survey for tsunami, 90
Electron probe microanalysis (EPMA), 272–273, 283t
Eltanin impact, 333
Encrustation, salt, 137f, 138
End-of-a-geological-period meteorite-impact tsunamiites, 410–411

Energy Dispersive Spectrometer/Spectrometry (EDS), 272–273
Environmental change, 405
Eoglobigerina edita, 277
Eoglobigerina eobulloides, 277
Ephemeral characteristics, 385
EPMA. *See* Electron probe microanalysis
Epsilon cross-stratification, 394
Erosion
of deep-sea bottom, 230–231
surfaces, 397
Erosional sorting, 103
Eruption, volcanic, 13, 239–240, 389–390
characteristics, 243
in Minoan age, 2
pumiceous sand layers, 249
pyroclastic flows, 244
tsunami wave heights, 242
Explosive eruptions, 13–14

F

False dating of Paleotsunami events due to tsunami-induced erosion, 165–167
Farallon-Izanagi ridge, 408
"Fire of rice sheaves", 202
First-order high stand sea-level, tsunamiites occurrence, 413
Flores (Indonesia) earthquake tsunami (1992), 31
Flow velocity, of storm waves, 334–336
Foraminifers, 277
Force, tractive, 22–24
Free falling, 25
Fukushima-Daiichi nuclear power plant, 163
Fulvia mutica, 67

G

Gaia, 405
Garanduwa, Sri Lanka, 135–137
Gas hydrates, 12–13
Geochemical inferences of offshore tsunami deposits, 187–188

Geological Society of America (GSA), 263
"Ghost forest", 195–196
Gigantic tsunami energy, 225–226
Global Positioning System, 146–148
Grading
in layer and fining upward through stacked layer, 392
of tsunami deposits, 96–105
landward grading, 96–101
vertical grading, 101–105
Grain-size distribution, 146–149
of 1992 Nicaragua tsunami deposits, 97f
homogenite, 227–229
at Phra Thong Island and Khao Lak, 146–149
relationship with tsunami waveform, 75–78, 80f
of tsunami deposits, 71–74
sampling and methodology, 71–72
T3 tsunami deposit, 72, 73f, 74
tsunami deposits with saw-toothed, 79–81
Grasses, 137
GSA. *See* Geological Society of America
Guembelitria cretacea, 277
Gulf of Mexico region
Brazos River section, 343
North-Eastern Mexico, 343–344

H

Haiti, K/T-boundary tsunami deposits, 344–346
Hakuho Nankai Earthquake (684), 200
Halimeda, 213–214
Harbour resonance, 18
Harbour wave, 11
Havana–Matanzas anticlinorium, 308–309
HCS. *See* Hummocky cross-stratification
HCS mimics. *See* Hummock-mimics

Index

Hemipelagic surficial sediments liquefaction, 220
Herodotus abyssal plain
absence of homogenites in, 216–219
high-resolution seismic reflection profiles, 218f
Herringbone structures, in tidal sediments, 388, 393
High tsunami-induced suspension cloud, 231–233
Hirota Bay, 162
Historic tsunamis, 409
Hjulström diagram, 23–24, 24f, 36
Hokkaido earthquake tsunami, south-west (1993), 31, 36, 38, 43
Hokkaido-Komagatake eruption and tsunami (1640 CE), 240–241
Hokkaido-Nansei-Oki earthquake tsunami (1993), 110
damage caused by tsunami, 113f
field survey, 113–114
general characteristics, 115–116
general setting, 110–112
ideal model of 1993 tsunami sedimentation, 122–123, 123f
location maps, 111f
occurrences of tsunami deposits, 115f
sedimentary characteristics and facies, 120–122
sedimentary description, 114–115
sedimentary structures, 116, 117f–118f
sedimentary units and facies, 116–120, 121f
Holocene coastal sand dunes, 30
Holocene deposits, 66–67
paleo-Tomoe Bay and, 66–67, 68f
Holocene subaqueous tsunami deposits in east Japan, 58f
Holocene tsunamiites, 407–408
Homogenite-like deep sea beds, 234–235

Homogenites, 208–209, 225, 235–236
absence in Herodotus abyssal plain, 216–219
accumulation rate of suspended load, 233
constituents, 230
distribution, 226
erosion of deep-sea bottom, 230–231
high tsunami-induced suspension cloud, 231–233
records of shuttle movement and backwash current, 233–234
sedimentary properties
grain-size distribution, 227–229
sedimentary petrographic characteristics, 228f
structures, 227–229
setting, 226
type A, 220, 226, 229
carbonate content, 211f
characteristics, 209–212
correlation, 211f
depositional model, 212f
grain-size distribution, 211f
interpretation, 212–213
type B, 220, 226, 229–230
carbonate content, 213f
characteristics, 213–214
depositional model, 215f
genesis of sandy division, 231
grain-size distribution, 213f
interpretation, 214–215
types, 226
Huehuete, site survey for tsunami, 91
Hummock-mimics (HCS mimics), 394
Hummocky cross-stratification (HCS), 57, 67–70, 78
sand unit, 76f
Hummocky structures of tsunamiites, 394
Hurricanes, 359
Hydrates, gas, 12–13
Hydraulic effect, 13

I

IGCP. *See* International Geological Correlation Programme
Imbrication of boulder deposits, 356f
Impact-generated tsunami deposits
deposits formed by water flowing into oceanic impact crater, 336–337
outside oceanic impact crater, 337–339
Impact-generated tsunamis, 331–332
Impact-induced tsunamiites studies, 406–407
Indian Ocean tsunami (2004), 1, 3, 12, 38–39, 127–128, 128f, 145, 155–156, 188, 207
discussion, 141–142
distribution and significance of tsunami deposits
Bang Sak beach, Thailand, 135–139
Garanduwa, Sri Lanka, 135–137
Pakarang Cape, Thailand, 132–134
localities and methods of study, 129–132
measured tsunami heights, 129t
Tsunami boulders at Laem Pakarang, south-west Thailand, 36–38, 37f
Instrumental neutronic activation analysis, 272
Internal architecture of offshore tsunami deposits, 186
International Geological Correlation Programme (IGCP), 266
International Union of Geological Sciences/International Commission on Stratigraphy (IUGS/ICS), 263

Index

Intraclast
  characteristics, 286–291
    sedimentary structures, 294
    semiquantitative modeling, 291–294
  graded skeletal sheet, 286
  grain-size distribution, 308
  microspherules, 288, 298
  shear velocity, 293t
  shocked quartz grains, 284f, 288
  turbidity currents, 293
Iridium (Ir) anomaly, 265, 272, 277–280, 278f, 295–296
Iridium (Ir)-bearing dust, 331–332
IUGS/ICS. *See* International Union of Geological Sciences/International Commission on Stratigraphy

J

Japan Sea earthquake tsunami (1983), 9–10, 31
Jiquilillo, site survey for tsunami, 89
Joban "coal-seam-bearing" deposits, 417
Jogan earthquake (869 CE), 158–159

K

K/Pg boundary. *See* Cretaceous –Paleogene boundary
K/Pq boundary
  deep-sea tsunamiite, 235
  tsunami sediment layers, 407
K/T boundary. *See* Cretaceous/ Tertiary boundary
Kamchatka earthquake tsunami (1923), 31
Kanto tsunami (1923 CE), 74, 75f, 77
Kanto tsunami deposit (1703 CE), 59
Karymskoye tsunami deposits (1996 CE), 255–256
Khao Lak, 146
  graded tsunami deposits, 150f
  study areas, 146

thickness and grain-size distribution, 146–149
tsunami impact, 146, 149f
tsunami waves, 150
Kinematic viscosity of fluid, 24–25
Komagatake tsunami deposits (1640 CE), 250–251, 250f
Krakatau (Sunda Sea, 1883), volcanic eruptions in, 13
Krakatau eruption and tsunami (1883 CE), 242–243, 251–252
Krakatoa volcanic explosion tsunami (1883), 353–355
Kura-Pacific ridge, 408
Kyoto University, 72

L

Lacustrine tsunami deposits, 29–30
Landslide-generated tsunamis, 336. *See also* Oceanic impact craters
Landward grading, 96–101
Las Salinas, site survey for tsunami, 91–92, 104–105
  stratigraphy and interpretation, 104f
Lichen-covered boulders or boulders, 363–365
Linear long-wave approximation, 17–18
Lisbon earthquake tsunami (1755), 29–30
Long wave, 15
  velocity profile of, 22f
Los Alamos National Laboratory, 272
Lowe's approach, 233

M

Maastrichtian limestone bed, 346–347
Makaha, 194–195
Managua tsunami deposits (3000 –6000 BP), 246–247
Manning's relative roughness, 17–18

Marine sediment
  dilution, 161
  source, 187
Marine-target impact events, 331
Masachapa, site survey for tsunami, 91
Mateare Sand, 246–247
Maximum orbital velocity (U_{max}), 338–339
Mediterranean abyssal plains
  location map of long cores, 217f
  Stratigraphic correlation of eight long cores, 217f
Mediterranean homogenites, 229–230, 233–235
  calcareous material, 235
  genetic model, 232f
  reflections on terminology, 235–236
  sedimentary properties at basal contact, 229
  submarine topography, 227f
Megalithic cultures, 353
Megatsunami
  formed in deep-water accretionary prisms, 220
  submarine earthquakes as cause of, 207
  triggered by collapse of suprasubduction volcanic edifice, 221
Megaturbidite, 12–13, 235–236
Mesozoic-type animals, extinction of, 406–407
Meteorite-impact frequency and tsunamiites, 410–411, 410f
Methane hydrates, 12–13
Microspherules, 281f, 288, 298
Middle Miocene upper bathyal tsunami-induced current conglomeratic deposit, 6
Minoan volcanic event, 20–21
Miocene Tanabe Group, 417
Miocene Tsubutegaura boulder, 384–385
Miyagi-oki earthquake, 156
Miyakojima typhoon in northern Japan (1959), 42

Index

Moncada Formation, 318–319, 344–345
  paleocurrents direction, 318–319, 325
  by repeated Tsunami currents, 318–319
  sedimentary log, 319f
  between Yucatán platform and Cretaceous Cuban arc, 307
Muddy layers, 28, 57
Multiple-bed deposits, 56–60
  fining and thinning upwards series of sand sheets, 59–60
  repeated reversal of current directions, 57–58
  succession of sand sheets capped by mud drapes, 57
Multiple-bed tsunami deposits, 57
Multiple-graded units, 146, 150

N

National Geophysical Data Center, 353–355
Navier–Stokes equation, 17–18
Nearshore tsunamiites, 397
Ni-rich spinel, 331–332
Nicaragua tsunami (1992), 85
  field and laboratory protocols, 105–106
  location map, 86f
  tide-gauge record, 87f
  tsunami deposits
    grading of, 96–105
    grain-size distributions, 97f
    along Nicaragua coast, 87–92
    near Playa de Popoyo, 92–96
  tsunami elevation and distance from shore, 88f
Normal grading, 288
North-Eastern Mexico, 343–344
Northernmost Costa Rica, site survey for tsunami, 92
NOWPHAS, 156

O

Observed tsunamis, 20
Occurrence frequency, 414

Ocean Drilling Program (ODP), 342f, 346
Ocean spreading, 408
Ocean waves, 357
Oceanic impact craters, 331, 332f
  Atlantic Ocean, 346–347
  generation of tsunamis by oceanic impacts, 333–336
    crater-generated tsunamis, 334–336
    landslide-generated tsunamis, 336
  impact-generated tsunami deposits
    deposits formed by water flowing, 336–337
    impact-generated deposits, 337–339
  K/T-boundary tsunami deposits, 347–348
    around Chicxulub crater, 341f
    on edge of Yucatán platform, 340–343
    Gulf of Mexico region, 343–344
    K/T-boundary impact event, 339–340
    proto-Caribbean sea region, 344–346
    sedimentary logs, 342f
  morphology, 332–333
  Pacific Ocean, 347
  resurge process of ocean water, 334f
ODP. *See* Ocean Drilling Program
Offshore fluvial flood deposits, 189
Offshore Portugal biomarkers, 188
Offshore storm *vs.* tsunami deposits, 189
Offshore tsunami deposits, 183–184
  current knowledge, 184–185
  features, 185–189
  geochemical inferences, 187–188
  internal architecture, 186

larger extent and lower preservation potential of, 162–165
  palaeontological features, 188
  textural and compositional aspects, 187
Onshore tsunami, 42
Onshore tsunami deposits, 31–32
  Hokkaido-Nansei-Oki earthquake tsunami, 110
  field survey, 113–114
  general characteristics, 115–116
  general setting, 110–112
  sedimentary characteristics and facies, 120–122
  sedimentary description, 114–115
  sedimentary structures, 116, 117f–118f
  sedimentary units and facies, 116–120, 121f
Onshore tsunami sedimentation, 32–40
  sedimentological features associated with repetition and waning of waves, 40
  associated with tsunami backwash, 38–40
  associated with tsunami run-up, 32–35
  source and particle composition of tsunami deposits, 35–38
"Orphan tsunami", 194, 197, 200, 202–203
Oshima-Ohshima eruption and tsunami (1741 CE), 241–242
Outsized clasts, 59

P

Pacific Ocean, 347
Pacific Rim, 11
Pakarang Cape, Thailand, 132–134
  tsunami-induced sedimentary features, 134f
  tsunami-transported reef blocks, 130f

Index

Palaeontological features of offshore tsunami deposits, 188
Palaeotsunami events, 41
Paleo-Tomoe Bay and Holocene deposits, 66–67, 68f
PALEOFLUX EU-funded MAST project, 216–217
Paleogeography of proto-Caribbean sea and geological setting, 307–308
Paleotsunami, 363–365
  deposits, 370
  research, 353–356
  in Tohoku, 173
Palms, 277
Palynomorphs, 277
Parallel lamination, 28–29
Particle
  Hjulström diagram, 23–24, 24f, 36
  motion, mode of, 25–26
  properties and geological configuration, 23–24
  pumice, 22
  sedimentary, 25
    quantity of, 26
    suspension, 29
  size of tsunami layers, 29
Parvularugoglobigerina eugubina, 277
Patinopecten yessoensis, 118
PDR. *See* Precision depth recorder
Peñalver Formation, 306, 306f, 309f–310f, 345
  paleogeotectonic setting of, 306f
  sedimentary logs, 342f
  sedimentary processes, 308–316
    lateral and vertical variations in lithology, composition and grain size, 312–314
    lithology and petrography at type locality near Havana, 310–312
    origin and sedimentary mechanism, 314–316
    stratigraphic setting and studied localities, 308–309

Periodicity, 383, 414
Pernambuco Basin, 264–266, 272, 282, 288, 297f
Philips XL30 SEM, 272–273
Phra Thong Island, 146
  graded tsunami deposits, 150f
  study areas, 146, 147f
  thickness and grain-size distribution, 146–149
  tsunami impact, 146, 149f
  tsunami waves, 150
Plane beds, 57
Planorbulina mediterranea, 213–214
Plate boundary, 11
Playa de Popoyo
  large clasts, 95t
  overview map, 93f
  profiles, 94–96, 94f
  site survey for tsunami, 92
  vertical grading, 101–104
    stratigraphy and interpretation, 103f
Playa Hermosa, site survey for tsunami, 90
Plume tectonics, 408
Pochomil, site survey for tsunami, 91
Poneloya, site survey for tsunami, 90
Post-tsunami field surveys, 155
Poty quarry K/Pg boundary, 295–297
  characteristics of, 274–282
  controversy about the position of K/Pg boundary, 282–286
  field data and petrographic analyses, 275t–277t
  late Cretaceous paleogeographic map, 288f
  litho-and chemostratigraphy, 279f
Pre-Quaternary impact-induced tsunamiites, 406–407
Precision depth recorder (PDR), 212
Propagation speed, 16
Proto-Caribbean sea, 316–324

comparison of K/T-boundary deep-sea tsunami deposits, 316–324
compositional variations of K/T-boundary deep-sea tsunami deposits, 324
Paleogeography of, 307–308
Protothaca (*Neochione*) *jedoensis*, 67
Proxapertites cursus, 277
Proximal ejecta, 337–338
Pseudoguembelina costulata, 277
Pumiceous sand, 249–250
Punta Teonoste, site survey for tsunami, 91
Pyroclastic flows, 14
Pyroclastic sediment, 6

Q

Quartz, 187–188

R

Rabaul eruption and tsunamis (1994 CE), 244–245
Rabaul tsunami deposits (1994 CE), 252–255, 253f
Rain-water dissolution, 363–365
Resonance periods, 18
Resurge deposits, 336–337
Reynolds number, 24–25
Ridge subduction, 408
Rim wave, 333. *See also* Oceanic impact craters
Ritter tsunami (1888 CE), 243–244
Rugoglobigerina ex gr. *rugosa*, 277
Run-up, 28–29
  sedimentological feature associated with, 32–35

S

Salt encrustation, 137f, 138
Sand layers in intertidal lakes and lagoons, 29–30
Sand sheets, 109
Sand-sized particles, 36
Sandy division, 231
Sandy mud, 38–39

Index

Sanriku Coast, 158
Santa Isabel areas
  calcirudites in, 312
  distribution of Peñalver
    formation, 308—309
  Subunit B and Subunit A, 314
Santorini Bronze-Age
  catastrophic eruption, 209
Santorini Island, 207—208, 208f
Santorini tsunami deposits (3500
  BP), 225—226, 247—249
Sapropel S-1, 209
Scanning Electron Microscope
  (SEM), 272—273
Schizeoisporites eocenicus, 277
Sea-level change, 416
  earthquake-induced tsunamiite
    occurrences and,
    411—416
Sea-level history, 357—360
Seawater incursion, 158—159
Sediment gravity flow, 235—236,
  392
Sediment transport
  mechanics of, 22—26
  tsunami sediment transport
    modeling, 161
Sedimentary characteristics of
  Hokkaido-Nansei-Oki
  earthquake tsunami,
  120—122
Sedimentary facies
  of Hokkaido-Nansei-Oki
    earthquake tsunami,
    116—122, 121f
  of tsunami deposits, 67—71
    Unit Tna, 67—70, 69f
    Unit Tnb, 70—71
    Unit Tnc, 71
    Unit Tnd, 71
Sedimentary particles, 25
  quantity of, 26
  suspension, 29
Sedimentary records
  earthquake-induced tsunamiite
    occurrences, 411—416
  earthquake-induced tsunamiites
    and tectonics, 407—408
  impact-induced tsunamiites
    studies, 406—407

meteorite-impact frequency and
  tsunamiites, 410—411, 410f
slump-induced tsunamiites, 416
volcanic-eruption-induced
  tsunamiites, 408—409
Sedimentary relic, lack of,
  202—203
Sedimentary structures, 65, 69f,
  249—250
  Hokkaido-Nansei-Oki
    earthquake tsunami, 116,
    117f—118f
  length in tsunamiites, 385—386
  reflecting tsunami waveform, 55
  in tsunamiite beds, 390—394
    current structures, 393—394
    grading in layer and fining
      upward through stacked
      layer, 392
    in sedimentary set, 391—392
Sedimentary units of Hokkaido-
  Nansei-Oki earthquake
  tsunami, 116—120, 121f
Sedimentology, 6
  of tsunamiites
    patterns of tsunamiite
      occurrence in time,
      409—416
    preservation potential,
      416—418
    submarine slides and
      tsunamiites, 409
    tsunamiites as records of
      ancient events, 406—409
Sediments, 145
  associated, 390
  sequences in tsunami deposits,
    355f
  storm, 42
  tidal, 388, 393
  tsunami, 5
Seismic sea wave, 11, 338
"Seismic-Driving" processes,
  173—174
Seismites, 5—6
SEM. *See* Scanning Electron
  Microscope
Sendai Bay, 162—163
  spatial distribution and
    stratigraphy, 164f

Sendai Plain, 158—160, 160f,
  162—163, 168—170
Shallow sea
  including continental shelf,
    398—399
  tsunami deposits, 2
Shallow-focus earthquakes, 12
Shallow-water deformation,
  17—18
Shallow-water long-wave theory,
  17—18, 24—25. *See also*
  Tsunami—propagation
Shallow-water wave, 338—339,
  338f
Shear stress, 22—24. *See also*
  Sediment transport
Shear velocity, 293t
Shock waves, 9
Shocked quartz, 284f, 331—332.
  See also Impact-generated
  tsunami deposits
Shuttle movement
  gravity flows and tsunamiite
    variety, *vs.*, 388—389
  records, 233—234
  of tsunamis, 387—388
Silt-sized particles, 29, 36
Single-bed deposits, 55—56
Slump-induced tsunamiites, 416
Sorites, 213—214
Sorting
  depositional, 103
  erosional, 103
Spherules, 331—332
Sri Lanka
  natural beaches, 131—132
  stratigraphy and sedimentology
    of deposits, 128
  study area around Garanduwa,
    128f
Storegga slides, 13
Storegga tsunami deposits, 396
Storm deposits, 6, 65
  discriminating tsunami deposits
    from, 78—79
Storm sediments, 42
Storm waves, 54, 338—339
  around Southern Boso
    Peninsula, 67
  storm-induced waves, 54—55

Index

Stratigraphy, 128
Stromboli volcano, 239
Subduction slab, 6
Submarine earthquakes, 11, 207, 410
Submarine impact events, 331
Submarine slides and tsunamiites, 409
Submarine tsunami deposits, 27–29, 165
Sumatra–Andaman Islands earthquake (December, 2004), 127
Sunda Strait, 353–355, 409
Suspension load, 25–26, 170
    accumulation rate of, 233
    high tsunami-induced, 231–233
Swaley-like structures of tsunamiites, 394
Swash marks, 388–389

T

TCs. *See* Tropical cyclones
Tectonic activity, 413–414
Tempestites, 5–6
Tephra Z-2 absence, 216
Terminal falling velocity, 24–25
Thailand
    stratigraphy and sedimentology of deposits, 128
    tsunami-transported reef blocks, 130f
Theora fragilis, 67
Thera (Santorini), volcanic eruptions in, 13, 20–21
Thick breccias, 347–348
Thickness, 209
    of Phra Thong Island and Khao Lak, 146–149
    spatial variability of deposit thickness, 170–171
Tidal deposit, 6
Tide-gauge, 85–87, 87f
Tides around Southern Boso Peninsula, 67
Tohoku-oki tsunami (2011), 156, 362
    learned lessons

challenges to estimating earthquake size and extent, 172–175
conceptual model of tsunami-induced erosion and deposition, 167f
false dating of Paleotsunami events due to tsunami-induced erosion, 165–167
larger extent and lower preservation potential, 162–165
limitation of marine materials as evidence for tsunami inundation, 159–161
spatial variability of deposit thickness, 170–171
uncertainties in tsunami inundation distance, 168–170
and precursors, 156–159
rupture area, 157f
Tokachi-Oki earthquake, M8.0 (September, 2003), 27–28
Total organic carbon (TOC) anomaly, 265–266, 277–280, 280f
Trace height, 19
Traction carpet, 76
Tractive force, 22–24. *See also* Sediment transport
Transgression, 413–414
Trees, 137
Tropical cyclones (TCs), 359
Tsubutegaura tsunamiite, 408, 413–414
    succession, 386
Tsunami, 1–2, 9, 145, 193, 230, 239, 292f, 338–339, 353–356
    boulders, 357–359
    distinction between tsunami and storm deposits, 41–43
    earthquake, 85
    generation, 11–15
        by oceanic impacts, 333–336
    geological investigations, 9
    geology, 53
    hydrodynamic modeling, 172
    imprints of, 183

incursion, 32
induced by
    asteroid impact, 14–15
    submarine sliding, 12–13
inflow, 110, 111f, 113f, 116, 118–120, 122
inundation
    limitation of marine materials as evidence for, 159–161
    uncertainties in tsunami inundation distance, 168–170
layers, 29
occurrences, 40–46
outflow, 110, 111f, 113f, 117, 120
process, 294–295
propagation, 15–18, 16f
    in deep sea, 15–17
    numerical modeling of, 161
    in shallow seas, 17–18
quantification, 18–20
    evaluation of ancient tsunamis, 43–46
    run-up height, 19
    tsunami height, 19
    tsunami intensity, 20
    tsunami magnitude, 19–20
    tsunami period, 19
quantitative evaluation of ancient tsunamis, 43–46
research, 356
sediment, 5
    transport modeling, 161
sedimentation, 26, 27f
setup of experimental channel, 44f
wave period of, 74
waveform, 74
    relationship between grain-size distribution and, 75–78, 80f
waves, 150
Tsunami boulder deposition, 359–372
boulders deposited by paleotsunami events, 363–372

461

Index

Tsunami boulder deposition
(*Continued*)
documented case studies of recent tsunami events, 360–362
Tsunami deposits, 5–6, 10, 53, 183, 245. *See also* Boulders
bedforms and sedimentary structures reflecting tsunami waveform, 55
characteristics, 26–32
depositional model in shallow water, 60–61, 60f
differences of waveforms between tsunami-and storm-induced waves, 54–55
discrimination from storm deposits, 78–79
grain-size distribution of, 71–74
Lacustrine tsunami deposits, 29–30
landward fining of, 35f
multiple-bed deposits, 56–60
fining and thinning upwards series of sand sheets, 59–60
repeated reversal of current directions, 57–58
succession of sand sheets capped by mud drapes, 57
along Nicaragua coast, 87–92
onshore tsunami deposits, 31–32
reference map, 21f
regional setting, 66f
paleo-Tomoe Bay and Holocene deposits, 66–67, 68f
storm waves and tides around Southern Boso Peninsula, 67
with saw-toothed grain-size distribution, 79–81
sediment sequences in, 355f
sedimentary facies of, 67–71
single-bed deposits, 55–56
source and particle composition, 35–38

submarine tsunami deposits, 27–29
Tsunami sedimentology, 9–10, 20–46
characteristics of Tsunami deposits, 26–32
mechanics of sediment transport, 22–26
critical condition, 23–24
mode of particle motion, 25–26
quantity of sedimentary particles, 26
shear stress, 22–23
terminal falling velocity, 24–25
tractive force, 22–23
turbulence, 24–25
onshore tsunami sedimentation, 32–40
Tsunami traces of 1700 Cascadia earthquake, 194–200
date identification, 197–199
discussions, 200
estimation of magnitude, 200
estimation of occurrence time, 199
geological tsunami evidences, 195f
modern traces of sudden subsidence of Cascadia coast, 195–196
sedimentary traces of tsunami flow, 196–197
tsunami told in stories down through tradition by native tribes of Canada, 194–195
Tsunami-induced current, quantity of sedimentary particles in, 26
Tsunami-induced waves, 54–55
Tsunami-related turbidites, 164–165
Tsunami-reworked deposits, 5
Tsunami-reworked sediment, 5
Tsunami-worked deposit, 5
Tsunami-worked sediment, 5
Tsunamigenic transport, 371–372

Tsunamiites, 2, 5–6, 235–236, 245, 264, 392, 399
approach toward identification, 266–272
Bed D
characteristics, 286–291, 289f, 293t
semiquantitative modeling of the depositional process, 291–294
shear velocities *vs.* wave heights, 293t
characteristics of, 268t–269t
constituents of, 394–395
features of, 383, 395–399
beach, 397
bottom of bays, 397–398
coastal lacustrine basin, 396
coastal plain, 395–396
deep-sea environments, 399
nearshore, 397
shallow sea including continental shelf, 398–399
geological setting of study area, 273–274
impact-derived, 268t–269t
as marker horizons, 389–390
methods and data analyses, 272–273
occurrences, 40–46
patterns of tsunamiite occurrence in time, 409–416
earthquake-induced tsunamiite occurrences, 411–416
meteorite-impact frequency and tsunamiites, 410–411, 410f
sea-level change, 411–416
slump-induced tsunamiites, 416
Poty quarry K/Pg boundary, 295–297
characteristics of, 274–282
controversy about the position of K/Pg boundary, 282–286
field data and petrographic analyses, 275t–277t

462

Index

late Cretaceous
paleogeographic map, 288f
litho-and chemostratigraphy,
279f
preservation potential, 416–418
as records of ancient events,
406–409
earthquake-induced
tsunamiites and tectonics,
407–408
impact-induced tsunamiites
studies, 406–407
volcanic-eruption-induced
tsunamiites, 408–409
sedimentary structures in
tsunamiite beds, 390–394
sedimentation, 1
sedimentological profile, 387f
sedimentology, 3
submarine slides and, 409
theoretical considerations on,
268t–269t
tsunami process, 294–295
updated information on area,
295–297
Tsunamis and tsunami deposition
characteristics, 384–390
associated sediments, 390
diastrophic nature, 384–385
length of tsunami waves and
sedimentary structures,
385–386
shuttle movement
gravity flows and tsunamiite
variety, *vs.*, 388–389
of tsunamis, 387–388
tsunamiites as marker horizons,
389–390
Turbidites, 5–6, 231
Turbidity currents, 5
Turbulence, 24–25. *See also*
Onshore tsunami
sedimentation; Sediment
transport
Turbulent flow, 24–25
Typhoons, 359

U

Unit Tna, 61, 67–70, 69f, 72, 75
Unit Tnb, 70–72, 76

Unit Tnc, 71–72, 77
Unit Tnd, 61, 71–72, 77
Unobserved tsunamis, 21
Unzen eruption and tsunami
(1792 CE), 242
Usubetsu River, 112–114, 123

V

VEI. *See* Volcanic Explosivity
Index
Velocity profile of long wave, 22f
Vertical grading, 101–105
grain-size distributions, 102f
Las Salinas, 104–105
Playa de Popoyo, 101–104
Very shallow-water wave,
338–339, 338f
Vía Blanca Formation, 308,
310–311, 314–316.
See also Peñalver
formation
Violent volcanism, 410
Viscosity, kinematic, 24–25
Volcanic debris avalanches, 14
Volcanic eruption, 13, 239–240,
389–390
characteristics, 243
in Minoan age, 2
pumiceous sand layers, 249
pyroclastic flows, 244
tsunami wave heights, 242
Volcanic Explosivity Index
(VEI), 242–243
Volcanic temblors, 12
Volcanic-eruption-induced
tsunamiites, 408–409
Volcanism-induced tsunamiites,
244–245
Aniakchak tsunami deposits
(3500 BP), 249–250
Karymskoye tsunami deposits
(1996 CE), 255–256
Komagatake tsunami deposits
(1640 CE), 250–251, 250f
Krakatau tsunami deposits (1883
CE), 251–252
Managua tsunami deposits (3000
–6000 BP), 246–247
Rabaul tsunami deposits (1994
CE), 252–255, 253f

Santorini tsunami deposits (3500
BP), 247–249
Volcanism-induced tsunamis,
13–14, 239–245
Hokkaido-Komagatake eruption
and tsunami (1640 CE),
240–241
Krakatau eruption and tsunami
(1883 CE), 242–243
locations, 240f
Oshima-Ohshima eruption and
tsunami (1741 CE),
241–242
Rabaul eruption and tsunamis
(1994 CE), 244–245
Ritter tsunami (1888 CE),
243–244
Unzen eruption and tsunami
(1792 CE), 242
Volcanogenic tsunami deposits,
240f, 249, 256–257

W

Water elevation, 15–18
Water flux, 44–46
Wave, 353
amplification of,
17–18
heights, 293t
storm, 54–55, 67
tsunami, 242
motion, transient,
10–11
period of tsunamis, 74
Waveforms, 54–55
bedforms and sedimentary
structures reflecting, 55
of storm surges, 55
between tsunami and wind
wave, 54f
of tsunamis, 60f
Wavelength, 10–12, 15, 17
Wavelength Dispersive
Spectrometer/
Spectrometry (WDS),
272–273
WDS. *See* Wavelength Dispersive
Spectrometer/
Spectrometry
Welded tuff, 6

Index

X

X-radiographs, 186
X-ray fluorescence scanner (XRF scanner), 187—188

Y

YAXCOPOIL-1 (YAX-1), 325, 326f
Yucatan Peninsula, 14—15, 285
Yucatan platform, 14—15, 307, 318, 324—325

Printed in the United States
By Bookmasters